READ AT A GLANCE

All Learning is Self-Teaching

How do you or does everyone 12-90 years old in the family relate to the following?

1. (a) Do you sometimes write checks in words and numerals for your bank account?.
   (b) Do you know how to accomplish (a)?
   If No, you need to study arithmetic.

2. (a) Do you sometimes change measurements in yards to meters and vice versa?
   (b) Do you know how to accomplish (a)?
   If No, you need to study arithmetic.

3. (a) Do you sometimes have to determine for example, the elapsed time from 1:30 PM to 10:15 PM?
   (b) Do you know how to accomplish (a)?
   If No, you need to study arithmetic.

4. (a) Do you sometimes want to work at a mortgage office?
   (b) Do you have the arithmetic preparation for this office?
   If No, you need to study arithmetic.

5. (a) Do you sometimes have to determine the time for traveling a certain distance at a certain speed?
   (b) Do you know how to accomplish (a)?
   If No, you need to study arithmetic.

6. (a) Do you sometimes compare a price per a certain quantity to a price per a different quantity?
   (b) Do you know how to accomplish (a)?
   If No, you need to study arithmetic.

7. (a) Do you sometimes have to determine the quantities of ingredients (e.g., salt, sugar, tomatoes) for food recipes for different parties?.
   (b) Do you know how to accomplish (a)?
   If No, you need to study arithmetic.

8. (a) Do you sometimes as a nurse have to calculate the correct dosage of a drug for a patient?
   (b) Do you have the arithmetic preparation to accomplish (a)?
   If No, you need to study arithmetic.

9. (a) Do you sometimes have to do calculations in chemistry?
   (b) Do you have the arithmetic background for (a)?
   If No, you need to study arithmetic.

10. (a) Do you sometimes want to work at the post office?
    (b) Do you have the arithmetic preparation for (a)?
    If No, you need to study arithmetic.

11. (a) Do you sometimes calculate employment and unemployment benefits?
    (b) Do you have the arithmetic preparation to accomplish (a)?
    If No, you need to study arithmetic.

12. (a) Do you sometimes work as a builder or a contractor?
    (b) Do you have the arithmetic preparation to accomplish (a)?
    If No, you need to study arithmetic.

13. (a) Do you sometimes compare interest rates and accompanying conditions?
    (b) Do you know how to accomplish (a)?
    If No, you need to study arithmetic.

14. (a) Do you sometimes have to divide profits among investors according to investment amounts?
    (b) Do you know how to accomplish (a)?
    If No, you need to study arithmetic

15. (a) Do you sometimes have to compare apartment or house prices?
    (b) Do you know how to accomplish (a)?
    If No you need to study arithmetic

16. (a) Do you sometimes have to deal with percents such as 20% of 60?
    (b) Do you know how to accomplish (a)?
    If No, you need to study arithmetic.

17. (a) Do you sometimes arrange items labeled with fractions or decimals in ascending or descending order?
    (b) Do you know how to accomplish (a)?
    If No, you need to study arithmetic

18. (a) Do you sometimes have to do estimations such as the area of the living room floor of your house?
    (b) Do you know how to accomplish (a)?
    If No, you need to study arithmetic.

19. (a) Do you sometimes have to work with fractions or decimals on the job?
    (b) Do you know how to accomplish (a)?
    If No, you need to study arithmetic.

20. (a) Do you sometimes have to prepare a budget?
    (b) Do you know how to accomplish (a)?
    If No, you need to study arithmetic.

21. (a) Do you sometimes want to work in finance?
    (b) Do you have the arithmetic preparation for (a)?
    If No, you need to study arithmetic.

22. (a) Do you sometimes want to go to Wall Street and do some trading?
    (b) Do you have the arithmetic preparation for (a)?
    If No, you need to study arithmetic.

23 (a) Do you sometimes work as a carpenter?
    (b) Do you have the arithmetic preparation for (a)?
    If No, you need to study arithmetic.

24. (a) Do you sometimes work as a bricklayer?
    (b) Do you have the arithmetic preparation for (a)?
    If No, you need to study arithmetic.

25. Do you sometimes take a civil service exam with a mathematics component?
    (b) Do you know how to accomplish (a)?
    If No, you need to study arithmetic.

26 (a) . Do you sometimes want to work as a bank teller?
    (b) Do you have the arithmetic background for (a)?
    If No, you need to study arithmetic.

27. (a) Do you sometimes have to work on personal finances?
    (b) Do you know how to accomplish (a)?
    If No, you need to study arithmetic.

28. (a) Do you sometimes have to compare college tuitions?
    (b) Do you know how to accomplish (a)?
    If No, you need to study arithmetic.

29. (a) Do you sometimes buy and sell goods for profit?
    (b) Do you have the arithmetic preparation for (a)?
    If No, you need to study arithmetic.

Read More

If you are very competent in **arithmetic** and you do not even have a formal college education, and you get a job as a banker, as a nurse, as an auto mechanic, as a construction worker, or a job in a finance company, or a mortgage company, you will be able to train quickly, and move up in rank in any company. If you are not competent in arithmetic, you will always guess, make mistakes, and make wrong decisions and you may be cheated, for example, if you cannot do the following problems correctly.

1. Suppose you and a friend invested $2,000 and $3,000 respectively, in a company, and you are to divide a profit of $1,500, how much is your share of this profit based on the amount you invested?

Answer: $600

2. Suppose you and a friend worked 8 hours and 12 hours respectively, as security guards at a trade show. If the total payment for the 20 hours is $260, what is your pay for the hours you worked?

Answer: $104

3. According to the decision criteria of a mortgage company, an applicant for a loan will qualify for the loan if the applicant's monthly income-expense ratio is greater than 1. If your monthly income is $5,000; and your monthly expenses are $4,900, do you qualify for the loan?   Answer: Yes

Some people look down upon the subject "arithmetic" as being so elementary and may hesitate to buy a book with such a title. However, most everyday life and business decisions are based on arithmetic. There are various levels of arithmetic from grade school, through high school, to college, and at the work place. For example, the following is an arithmetic problem.

If you deposited $30,000 in a savings account at an interest rate of 5%, and the interest is compounded annually for 2 years, what is the amount in your account after 2 years?

Answer: $33,075

We use arithmetic in making financial decisions, in accounting practices, in business management, in income statement preparation, and in determining the cost of goods sold, operating expenses,, net profit, and in managing accounts receivable, accounts payable, and cash flow.

1	2	3	4	5	6
$1 \times 1 = 1$	$2 \times 1 = 2$	$3 \times 1 = 3$	$4 \times 1 = 4$	$5 \times 1 = 5$	$6 \times 1 = 6$
$1 \times 2 = 2$	$2 \times 2 = 4$	$3 \times 2 = 6$	$4 \times 2 = 8$	$5 \times 2 = 10$	$6 \times 2 = 12$
$1 \times 3 = 3$	$2 \times 3 = 6$	$3 \times 3 = 9$	$4 \times 3 = 12$	$5 \times 3 = 15$	$6 \times 3 = 18$
$1 \times 4 = 4$	$2 \times 4 = 8$	$3 \times 4 = 12$	$4 \times 4 = 16$	$5 \times 4 = 20$	$6 \times 4 = 24$
$1 \times 5 = 5$	$2 \times 5 = 10$	$3 \times 5 = 15$	$4 \times 5 = 20$	$5 \times 5 = 25$	$6 \times 5 = 30$
$1 \times 6 = 6$	$2 \times 6 = 12$	$3 \times 6 = 18$	$4 \times 6 = 24$	$5 \times 6 = 30$	$6 \times 6 = 36$
$1 \times 7 = 7$	$2 \times 7 = 14$	$3 \times 7 = 21$	$4 \times 7 = 28$	$5 \times 7 = 35$	$6 \times 7 = 42$
$1 \times 8 = 8$	$2 \times 8 = 16$	$3 \times 8 = 24$	$4 \times 8 = 32$	$5 \times 8 = 40$	$6 \times 8 = 48$
$1 \times 9 = 9$	$2 \times 9 = 18$	$3 \times 9 = 27$	$4 \times 9 = 36$	$5 \times 9 = 45$	$6 \times 9 = 54$
$1 \times 10 = 10$	$2 \times 10 = 20$	$3 \times 10 = 30$	$4 \times 10 = 40$	$5 \times 10 = 50$	$6 \times 10 = 60$
$1 \times 11 = 11$	$2 \times 11 = 22$	$3 \times 11 = 33$	$4 \times 11 = 44$	$5 \times 11 = 55$	$6 \times 11 = 66$
$1 \times 12 = 12$	$2 \times 12 = 24$	$3 \times 12 = 36$	$4 \times 12 = 48$	$5 \times 12 = 60$	$6 \times 12 = 72$
$1 \times 13 = 13$	$2 \times 13 = 26$	$3 \times 13 = 39$	$4 \times 13 = 52$	$5 \times 13 = 65$	$6 \times 13 = 78$
$1 \times 14 = 14$	$2 \times 14 = 28$	$3 \times 14 = 42$	$4 \times 14 = 56$	$5 \times 14 = 70$	$6 \times 14 = 84$
$1 \times 15 = 15$	$2 \times 15 = 30$	$3 \times 15 = 45$	$4 \times 15 = 60$	$5 \times 15 = 75$	$6 \times 15 = 90$
$1 \times 16 = 16$	$2 \times 16 = 32$	$3 \times 16 = 48$	$4 \times 16 = 64$	$5 \times 16 = 80$	$6 \times 16 = 96$
$1 \times 17 = 17$	$2 \times 17 = 34$	$3 \times 17 = 51$	$4 \times 17 = 68$	$5 \times 17 = 85$	$6 \times 17 = 102$
$1 \times 18 = 18$	$2 \times 18 = 36$	$3 \times 18 = 54$	$4 \times 18 = 72$	$5 \times 18 = 90$	$6 \times 18 = 108$
$1 \times 19 = 19$	$2 \times 19 = 38$	$3 \times 19 = 57$	$4 \times 19 = 76$	$5 \times 19 = 95$	$6 \times 19 = 114$
$1 \times 20 = 20$	$2 \times 20 = 40$	$3 \times 20 = 60$	$4 \times 20 = 80$	$5 \times 20 = 100$	$6 \times 20 = 120$

7	8	9	10	11	12
$7 \times 1 = 7$	$8 \times 1 = 8$	$9 \times 1 = 9$	$10 \times 1 = 10$	$11 \times 1 = 11$	$12 \times 1 = 12$
$7 \times 2 = 14$	$8 \times 2 = 16$	$9 \times 2 = 18$	$10 \times 2 = 20$	$11 \times 2 = 22$	$12 \times 2 = 24$
$7 \times 3 = 21$	$8 \times 3 = 24$	$9 \times 3 = 27$	$10 \times 3 = 30$	$11 \times 3 = 33$	$12 \times 3 = 36$
$7 \times 4 = 28$	$8 \times 4 = 32$	$9 \times 4 = 36$	$10 \times 4 = 40$	$11 \times 4 = 44$	$12 \times 4 = 48$
$7 \times 5 = 35$	$8 \times 5 = 40$	$9 \times 5 = 45$	$10 \times 5 = 50$	$11 \times 5 = 55$	$12 \times 5 = 60$
$7 \times 6 = 42$	$8 \times 6 = 48$	$9 \times 6 = 54$	$10 \times 6 = 60$	$11 \times 6 = 66$	$12 \times 6 = 72$
$7 \times 7 = 49$	$8 \times 7 = 56$	$9 \times 7 = 63$	$10 \times 7 = 70$	$11 \times 7 = 77$	$12 \times 7 = 84$
$7 \times 8 = 56$	$8 \times 8 = 64$	$9 \times 8 = 72$	$10 \times 8 = 80$	$11 \times 8 = 88$	$12 \times 8 = 96$
$7 \times 9 = 63$	$8 \times 9 = 72$	$9 \times 9 = 81$	$10 \times 9 = 90$	$11 \times 9 = 99$	$12 \times 9 = 108$
$7 \times 10 = 70$	$8 \times 10 = 80$	$9 \times 10 = 90$	$10 \times 10 = 100$	$11 \times 10 = 110$	$12 \times 10 = 120$
$7 \times 11 = 77$	$8 \times 11 = 88$	$9 \times 11 = 99$	$10 \times 11 = 110$	$11 \times 11 = 121$	$12 \times 11 = 132$
$7 \times 12 = 84$	$8 \times 12 = 96$	$9 \times 12 = 108$	$10 \times 12 = 120$	$11 \times 12 = 132$	$12 \times 12 = 144$
$7 \times 13 = 91$	$8 \times 13 = 104$	$9 \times 13 = 117$	$10 \times 13 = 130$	$11 \times 13 = 143$	$12 \times 13 = 156$
$7 \times 14 = 98$	$8 \times 14 = 112$	$9 \times 14 = 126$	$10 \times 14 = 140$	$11 \times 14 = 154$	$12 \times 14 = 168$
$7 \times 15 = 105$	$8 \times 15 = 120$	$9 \times 15 = 135$	$10 \times 15 = 150$	$11 \times 15 = 165$	$12 \times 15 = 180$
$7 \times 16 = 112$	$8 \times 16 = 128$	$9 \times 16 = 144$	$10 \times 16 = 160$	$11 \times 16 = 176$	$12 \times 16 = 192$
$7 \times 17 = 119$	$8 \times 17 = 136$	$9 \times 17 = 153$	$10 \times 17 = 170$	$11 \times 17 = 187$	$12 \times 17 = 204$
$7 \times 18 = 126$	$8 \times 18 = 144$	$9 \times 18 = 162$	$10 \times 18 = 180$	$11 \times 18 = 198$	$12 \times 18 = 216$
$7 \times 19 = 133$	$8 \times 19 = 152$	$9 \times 19 = 171$	$10 \times 19 = 190$	$11 \times 19 = 209$	$12 \times 19 = 228$
$7 \times 20 = 140$	$8 \times 20 = 160$	$9 \times 20 = 180$	$10 \times 20 = 200$	$11 \times 20 = 220$	$12 \times 20 = 240$

13	14	15	16	17	18
$13 \times 1 = 13$	$14 \times 1 = 14$	$15 \times 1 = 15$	$16 \times 1 = 16$	$17 \times 1 = 17$	$18 \times 1 = 18$
$13 \times 2 = 26$	$14 \times 2 = 28$	$15 \times 2 = 30$	$16 \times 2 = 32$	$17 \times 2 = 34$	$18 \times 2 = 36$
$13 \times 3 = 39$	$14 \times 3 = 42$	$15 \times 3 = 45$	$16 \times 3 = 48$	$17 \times 3 = 51$	$18 \times 3 = 54$
$13 \times 4 = 52$	$14 \times 4 = 56$	$15 \times 4 = 60$	$16 \times 4 = 64$	$17 \times 4 = 68$	$18 \times 4 = 72$
$13 \times 5 = 65$	$14 \times 5 = 70$	$15 \times 5 = 75$	$16 \times 5 = 80$	$17 \times 5 = 85$	$18 \times 5 = 90$
$13 \times 6 = 78$	$14 \times 6 = 84$	$15 \times 6 = 90$	$16 \times 6 = 96$	$17 \times 6 = 102$	$18 \times 6 = 108$
$13 \times 7 = 91$	$14 \times 7 = 98$	$15 \times 7 = 105$	$16 \times 7 = 112$	$17 \times 7 = 119$	$18 \times 7 = 126$
$13 \times 8 = 104$	$14 \times 8 = 112$	$15 \times 8 = 120$	$16 \times 8 = 128$	$17 \times 8 = 136$	$18 \times 8 = 144$
$13 \times 9 = 117$	$14 \times 9 = 126$	$15 \times 9 = 135$	$16 \times 9 = 144$	$17 \times 9 = 153$	$18 \times 9 = 162$
$13 \times 10 = 130$	$14 \times 10 = 140$	$15 \times 10 = 150$	$16 \times 10 = 160$	$17 \times 10 = 170$	$18 \times 10 = 180$
$13 \times 11 = 143$	$14 \times 11 = 154$	$15 \times 11 = 165$	$16 \times 11 = 176$	$17 \times 11 = 187$	$18 \times 11 = 198$
$13 \times 12 = 156$	$14 \times 12 = 168$	$15 \times 12 = 180$	$16 \times 12 = 192$	$17 \times 12 = 204$	$18 \times 12 = 216$
$13 \times 13 = 169$	$14 \times 13 = 182$	$15 \times 13 = 195$	$16 \times 13 = 208$	$17 \times 13 = 221$	$18 \times 13 = 234$
$13 \times 14 = 182$	$14 \times 14 = 196$	$15 \times 14 = 210$	$16 \times 14 = 224$	$17 \times 14 = 238$	$18 \times 14 = 252$
$13 \times 15 = 195$	$14 \times 15 = 210$	$15 \times 15 = 225$	$16 \times 15 = 240$	$17 \times 15 = 255$	$18 \times 15 = 270$
$13 \times 16 = 208$	$14 \times 16 = 224$	$15 \times 16 = 240$	$16 \times 16 = 256$	$17 \times 16 = 272$	$18 \times 16 = 288$
$13 \times 17 = 221$	$14 \times 17 = 238$	$15 \times 17 = 255$	$16 \times 17 = 272$	$17 \times 17 = 289$	$18 \times 17 = 306$
$13 \times 18 = 234$	$14 \times 18 = 252$	$15 \times 18 = 270$	$16 \times 18 = 288$	$17 \times 18 = 306$	$18 \times 18 = 324$
$13 \times 19 = 247$	$14 \times 19 = 266$	$15 \times 19 = 285$	$16 \times 19 = 304$	$17 \times 19 = 323$	$18 \times 19 = 342$
$13 \times 20 = 260$	$14 \times 20 = 280$	$15 \times 20 = 300$	$16 \times 20 = 320$	$17 \times 20 = 340$	$18 \times 20 = 360$

19	20	21	22	23	24
$19 \times 1 = 19$	$20 \times 1 = 20$	$21 \times 1 = 21$	$22 \times 1 = 22$	$23 \times 1 = 23$	$24 \times 1 = 24$
$19 \times 2 = 38$	$20 \times 2 = 40$	$21 \times 2 = 42$	$22 \times 2 = 44$	$23 \times 2 = 46$	$24 \times 2 = 48$
$19 \times 3 = 57$	$20 \times 3 = 60$	$21 \times 3 = 63$	$22 \times 3 = 66$	$23 \times 3 = 69$	$24 \times 3 = 72$
$19 \times 4 = 76$	$20 \times 4 = 80$	$21 \times 4 = 84$	$22 \times 4 = 88$	$23 \times 4 = 92$	$24 \times 4 = 96$
$19 \times 5 = 95$	$20 \times 5 = 100$	$21 \times 5 = 105$	$22 \times 5 = 110$	$23 \times 5 = 115$	$24 \times 5 = 120$
$19 \times 6 = 114$	$20 \times 6 = 120$	$21 \times 6 = 126$	$22 \times 6 = 132$	$23 \times 6 = 138$	$24 \times 6 = 144$
$19 \times 7 = 133$	$20 \times 7 = 140$	$21 \times 7 = 147$	$22 \times 7 = 154$	$23 \times 7 = 161$	$24 \times 7 = 168$
$19 \times 8 = 152$	$20 \times 8 = 160$	$21 \times 8 = 168$	$22 \times 8 = 176$	$23 \times 8 = 184$	$24 \times 8 = 192$
$19 \times 9 = 171$	$20 \times 9 = 180$	$21 \times 9 = 189$	$22 \times 9 = 198$	$23 \times 9 = 207$	$24 \times 9 = 216$
$19 \times 10 = 190$	$20 \times 10 = 200$	$21 \times 10 = 210$	$22 \times 10 = 220$	$23 \times 10 = 230$	$24 \times 10 = 240$
$19 \times 11 = 209$	$20 \times 11 = 220$	$21 \times 11 = 231$	$22 \times 11 = 242$	$23 \times 11 = 253$	$24 \times 11 = 264$
$19 \times 12 = 228$	$20 \times 12 = 240$	$21 \times 12 = 252$	$22 \times 12 = 264$	$23 \times 12 = 276$	$24 \times 12 = 288$
$19 \times 13 = 247$	$20 \times 13 = 260$	$21 \times 13 = 273$	$22 \times 13 = 286$	$23 \times 13 = 299$	$24 \times 13 = 312$
$19 \times 14 = 266$	$20 \times 14 = 280$	$21 \times 14 = 294$	$22 \times 14 = 308$	$23 \times 14 = 322$	$24 \times 14 = 336$
$19 \times 15 = 285$	$20 \times 15 = 300$	$21 \times 15 = 315$	$22 \times 15 = 330$	$23 \times 15 = 345$	$24 \times 15 = 360$
$19 \times 16 = 304$	$20 \times 16 = 320$	$21 \times 16 = 336$	$22 \times 16 = 352$	$23 \times 16 = 368$	$24 \times 16 = 384$
$19 \times 17 = 323$	$20 \times 17 = 340$	$21 \times 17 = 357$	$22 \times 17 = 374$	$23 \times 17 = 391$	$24 \times 17 = 408$
$19 \times 18 = 342$	$20 \times 18 = 360$	$21 \times 18 = 378$	$22 \times 18 = 396$	$23 \times 18 = 414$	$24 \times 18 = 432$
$19 \times 19 = 361$	$20 \times 19 = 380$	$21 \times 19 = 399$	$22 \times 19 = 418$	$23 \times 19 = 437$	$24 \times 19 = 456$
$19 \times 20 = 380$	$20 \times 20 = 400$	$21 \times 20 = 420$	$22 \times 20 = 440$	$23 \times 20 = 460$	$24 \times 20 = 480$

If you can remember a needed information, you make decisions faster, you work faster, and you are more productive.

Yes, you can memorize them.
Squares of Natural Numbers

$1 \times 1 = 1$	$26 \times 26 = 676$
$2 \times 2 = 4$	$27 \times 27 = 729$
$3 \times 3 = 9$	$28 \times 28 = 784$
$4 \times 4 = 16$	$29 \times 29 = 841$
$5 \times 5 = 25$	$30 \times 30 = 900$
$6 \times 6 = 36$	$31 \times 31 = 961$
$7 \times 7 = 49$	$32 \times 32 = 1024$
$8 \times 8 = 64$	$33 \times 33 = 1089$
$9 \times 9 = 81$	$34 \times 34 = 1156$
$10 \times 10 = 100$	$35 \times 35 = 1225$
$11 \times 11 = 121$	$36 \times 36 = 1296$
$12 \times 12 = 144$	$37 \times 37 = 1369$
$13 \times 13 = 169$	$38 \times 38 = 1444$
$14 \times 14 = 196$	$39 \times 39 = 1521$
$15 \times 15 = 225$	$40 \times 40 = 1600$
$16 \times 16 = 256$	$41 \times 41 = 1681$
$17 \times 17 = 289$	$42 \times 42 = 1764$
$18 \times 18 = 324$	$43 \times 43 = 1849$
$19 \times 19 = 361$	$44 \times 44 = 1936$
$20 \times 20 = 400$	$45 \times 45 = 2025$
$21 \times 21 = 441$	$46 \times 46 = 2116$
$22 \times 22 = 484$	$47 \times 47 = 2209$
$23 \times 23 = 529$	$48 \times 48 = 2304$
$24 \times 24 = 576$	$49 \times 49 = 2401$
$25 \times 25 = 625$	$50 \times 50 = 2500$

Natural Numbers = $\{1, 2, 3, 4, 5, 6, 7, 8, 9, 10, 11, 12, 13, ...\}$
The natural numbers are also known as the counting numbers or the positive integers.

Whole numbers = $\{0, 1, 2, 3, 4, 5, 6, 7, 8, 9, 10, 11, 12, 13, ...\}$

Integers = $\{..., -7, -6, -5. -4, -3, -2, -1, 0, 1, 2, 3, 4, 5, 6, 7, ...\}$

Rational number : A number which can be written as the ratio of two integers.

Signed number operations: **1**. -7 - 9 = -16 (add); **2**. -7 - (-9) = -7 + 9 = 2 (subtract); **3**. -7(-9) = +63 (multiply).

Slope, m, of the line passing through the points (x_1, y_1) and (x_2, y_2) is given by $m = \dfrac{y_2 - y_1}{x_2 - x_1}$

Slope-intercept form of the **equation** of a line with slope m and y-intercept b, is **$y = mx + b$.**

An **equation** of the line with y-intercept b, and x-intercept a, is given by **$y = -\dfrac{b}{a} x + b$.**

Factoring: $a^2 - b^2 = (a + b)(a - b)$; **2**. $3a^2 - 12 = 3(a^2 - 4) = 3(a + 2)(a - 2)$.

Quadratic equations: **1**. If $(x - 2)(x + 3) = 0$, then $x = 2$ or $x = -3$.

Radicals: **1**. $\sqrt{12} = \sqrt{4}\sqrt{3} = 2\sqrt{3}$; **2**. $\sqrt{18} = \sqrt{9}\sqrt{2} = 3\sqrt{2}.$

Scientific notation: **1**. $436.7 = 4.367 \times 10^2$; **2**. $.00721 = 7.21 \times 10^{-3}$

Ratio: The ratio $a{:}b$ is the fraction $\dfrac{a}{b}$

Proportion: If a is to b as c is to d, then $\dfrac{a}{b} = \dfrac{c}{d}$

Exponents: **1**. $x^2 \cdot x^3 = x^5$; **2**. $(x^2)^3 = x^6$; **3**. $\dfrac{x^6}{x^2} = x^4$; **4**. $(xy)^3 = x^3 y^3$.

Factoring: 3. $a^3 + b^3 = (a + b)(a^2 - ab + b^2)$. **4**. $a^3 - b^3 = (a - b)(a^2 + ab + b^2)$.

Quadratic equations: **1**. If $(x - 2)(x + 3) = 0$, then $x = 2$ or $x = -3$; **2.** If $x^2 = 4$, then $x = \pm 2$.

Quadratic formula: $x = \dfrac{-b \pm \sqrt{b^2 - 4ac}}{2a}$

Complex numbers: **1**. $i = \sqrt{-1}$; $i^2 = -1$; **2.** If $x^2 = -4$, then $x = \pm 2i$

Logarithms: $\log_2 16 = 4$, because $2^4 = 16$.

$\log_b x = y$ if and only if $b^y = x$

Absolute value: $|x| = \begin{cases} x \text{ if } x \geq 0 \\ -x \text{ if } x < 0 \end{cases}$

Absolute value equations: If $|x - 2| = 6$, then $x - 2 = 6$ or $-(x - 2) = 6$; and from which $x = 8$ or $x = -4$.

Absolute value inequalities

Case **1**: $|ax + b| < c$. where c is positive. Example: $|x - 2| < 1$. (There is no solution if c is negative)

Case **2**: $|ax + b| > c$, where c may be positive or negative. Example: $|x - 2| > 1$.

Equation of the **circle** with center at (h,k) and radius r is given by $(x - h)^2 + (y - k)^2 = r^2$

Area & Perimeter of a Circle: Area, A, is given by $A = \pi r^2$; Perimeter, P, is given by $P = 2\pi r$

Heron's Formula : For any triangle of sides a, b, c, the area A is given by

$$A = \sqrt{s(s-a)(s-b)(s-c)} \quad \text{where} \quad s = \tfrac{1}{2}(a+b+c)$$

Distance Formula: $d = \sqrt{(x_2 - x_1)^2 + (y_2 - y_1)^2}$

The **mid-point** of the line, P_1P_2, connecting the points $P_1(x_1,y_1)$ and $P_2(x_2,y_2)$ has the coordinates given by the following formulas:

x-coordinate, x_m, of the midpoint is given by $x_m = \dfrac{x_1 + x_2}{2}$

y-coordinate, y_m of the midpoint is given by $y_m = \dfrac{y_1 + y_2}{2}$

Trigonometric Identities

Reciprocal Identities

1. $\sec\theta = \dfrac{1}{\cos\theta}$

2. $\csc\theta = \dfrac{1}{\sin\theta}$

3. $\cot\theta = \dfrac{1}{\tan\theta}$

Ratio Identities

1. $\tan\theta = \dfrac{\sin\theta}{\cos\theta}$

2. $\cot\theta = \dfrac{\cos\theta}{\sin\theta}$

Pythagorean Identities

1. $\sin^2\theta + \cos^2\theta = 1$

2. $1 + \tan^2\theta = \sec^2\theta$

3. $1 + \cot^2\theta = \csc^2\theta$

Addition formulas

1. $\cos(A - B) = \cos A \cos B + \sin A \sin B$

2. $\cos(A + B) = \cos A \cos B - \sin A \sin B$

3. $\sin(A + B) = \sin A \cos B + \cos A \sin B$

4. $\sin(A - B) = \sin A \cos B - \cos A \sin B$

5. $\tan(A - B) = \dfrac{\tan A - \tan B}{1 + \tan A \tan B}$

6. $\tan(A + B) = \dfrac{\tan A + \tan B}{1 - \tan A \tan B}$

Double angle identities

1. $\sin 2\theta = 2\sin\theta \cos\theta$;

2. $\cos 2\theta = \cos^2\theta - \sin^2\theta$;

3. $\cos 2\theta = 2\cos^2\theta - 1$

4. $\cos 2\theta = 1 - 2\sin^2\theta$;

5. $\tan 2\theta = \dfrac{2\tan\theta}{1 - \tan^2\theta}$

FREMPONG'S STEP-BY-STEP SERIES IN MATHEMATICS

Elementary Mathematics
(Arithmetic, Algebra)
&
Intermediate Mathematics (US)
(Algebra, Geometry & Trigonometry)

Excellent for Exam Preparation

SEVENTH EDITION

A.A. FREMPONG

Elementary Mathematics and Intermediate Mathematics

ISBN 978-1-946485-35-9

Seventh Edition

Printed in the United States of America

Faculty/Student Suggestions for Future Editions

Send Suggestions to the Publisher

Copy and complete the following. This survey will help improve future editions of this book.

1. Which Topics would you like to be added to future Editions? You may include sample problems (and solutions).

2. Which Topics in this book do you think need more or better coverage?.

3. Which coverage in this book impressed you most?

.

How useful was this book in preparing for and taking the Final Exam?

Check one: A (Excellent) ; B (Good) ; C (Average) ;   D (Fair)

In Memory of My Parents

Mom:
She was a devoted mother, sharing, kind, kinder to strangers and generous to a fault. She never cursed, she never hated; she never cheated, and she never envied. She never lied, and she never got angry. Once, she nursed an almost dying stranger renting a room in her house back to good health to the extent that the relatives of this renter later travelled one hundred miles just to thank mom. She was always peace-loving and forever forgiving.
An angel once lived on this earth to serve others.

Dad:
A great dad, kind, generous and forgiving. He emphasized and was an example of both formal education and self-education. A veterinarian, a bacteriologist, an Associate of the Institute of Medical Laboratory Technology (UK), a Fellow of the Royal Society of Health (UK); an incorruptible civil servant; his book on ticks has always inspired me to write whenever the need arises.

NOTE TO THE STUDENT

This book was written with you in mind at all tines. You may use this book as the course textbook or as a review book since the book gets to the point quickly on all relevant topics and yet covers these topics in detail.

Begin to master the definitions and the solutions of the sample problems thoroughly. (You have mastered a sample problem if you can solve the sample problem and similar problems without any reference to this book or any other source). For some problems, two or more methods are presented. Read the various methods and decide which methods you would like to remember; but always be aware of the existence of the other methods, in case the need arises. After having mastered the sample problems, try the exercise problems. The answers to these problems are presented immediately after the problems. You may cover the answers with paper before you attempt these problems, if the answers are too obvious. You may refer back and forth to the solved problems when you do not remember how to proceed.

You may also attempt some of the sample problems first, before reading the solution methods, and in this approach, the sample problems become more practice problems for you.

As a reminder, in any book, do not dwell on the few inadvertent errors you may find, but rather concentrate on what is useful to you.

For this book to be useful both as the course textbook, as well as review for exams, it is **important** to **Understand, Remember, Apply**, and **Remember** the material covered.

Wishing you Good Luck on all the exams
A.A.Frempong

Books in the series by the author: Arithmetic; Elementary Algebra; Intermediate Algebra, Elementary Mathematics; Intermediate Mathematics, Combined **Elementary Mathematics & Intermediate Mathematics,** College Algebra; College Trigonometry; College Algebra & Trigonometry and Calculus 1 & 2.

PREFACE TO THE SEVENTH EDITION

This edition improves on the previous edition and combines both the **Elementary Mathematics** and the **Intermediate Mathematics** of the fifth editions into a single volume because of the recent trends in overlapping topics at these two levels. The topics have been expanded and covered quite extensively, especially those topics students find difficult to understand or sort out. For such topics, I have endeavored to organize the coverage in such a way as to minimize student confusion. One such topic whose coverage is well-organized is "finding an equation of a straight line" (page 257-265).

In writing this edition, I was encouraged by comments from students who had used the previous editions. One student remarked: "Your book gets to the point. It does not beat about the bush."

In addition to the regular programs, this book can also be used for short term programs such as mini-sessions and immersion programs, and distance learning programs, with the instructor, perhaps, supplementing with hand-out homework problems, when necessary.

I gratefully acknowledge the help and encouragement of students, colleagues, and friends.

A. A. Frempong
New York, May 2012

CONTENTS
Arithmetic

Elementary Algebra

CHAPTER 22
Areas and Perimeters

CHAPTER 23
ALGEBRAIC FRACTIONS

CHAPTER 24
Inequalities

Intermediate Mathematics

CHAPTER 25 (Use Chapters 16A &16B)
Factoring Polynomials

CHAPTER 26 (Use Chapters 17A & 17B)

CHAPTER 1

Lesson 1: **Place-Value Relationships; Basic Definitions; Terminology; and Types of Numbers**
Lesson 2: **Introduction to Fractions, Decimals, Ratio and Proportion Percent Symbol and Some Interpretations**

Lesson 1

Symbols for Writing Numbers

The symbols for writing numbers are called the **digits** or **numerals**. The **ten basic digits** for writing numbers, in the decimal system, are 0, 1, 2, 3, 4, 5, 6, 7, 8, and 9. Using these digits we can write all the numbers that we cover in arithmetic.

Place-value relationships

When a number is written using digits, each digit has a value according to its position in the number. For example, in 45, the 4 represents four tens (or 4×10) and we say that the 4 is in the tens place. However, in the number 415, the 4 represents four hundreds (or 4×100) and we say that the 4 is in the hundreds place. Below, we present a number showing the place-value-relationships.

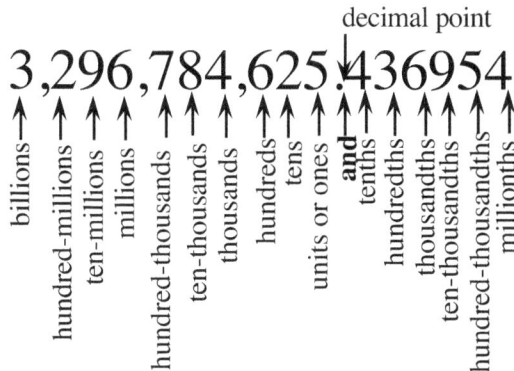

The above number is read " three billion, two hundred ninety-six million, seven hundred eighty-four thousand, six hundred twenty-five **and** four hundred thirty-six thousand, nine hundred fifty-four millionths".

We shall briefly look at the different kinds of numbers that we deal with in mathematics, aided by a number flow chart below, and then we will branch-off to the numbers that we deal with in arithmetic.

From the Entire Picture of Numbers to Arithmetic

The basic elements we deal with in our study of mathematics are numbers. It is important that we obtain a good understanding of the types of numbers we deal with in mathematics. It is also important that we are able to distinguish between the different kinds of numbers and their associated terminology. A very good grasp of the terminology will help us read and understand subsequent material much more quickly than otherwise.

As shown in the number flow chart below, all the numbers we deal with in mathematics can be divided into two main sets, namely the set of **real numbers** and the set of **non-real numbers.**(We must note that the terms "real", "non-real", and similar terms are only names we use in mathematics to distinguish between numbers and we must note that the real numbers are **not** more real than the non-real numbers (literally). These terms are only names for convenient distinction between some numbers.) The real numbers are also divided into two main sets, namely the **rational numbers** and the **irrational numbers.**

The rational numbers are further subdivided into **integers** (strictly integers) and **fractions** (strictly fractions).   The integers consist of the **negative integers, zero,** and the **positive integers** (natural numbers); and the positive integers consists of the **prime numbers** and the **composite numbers**.

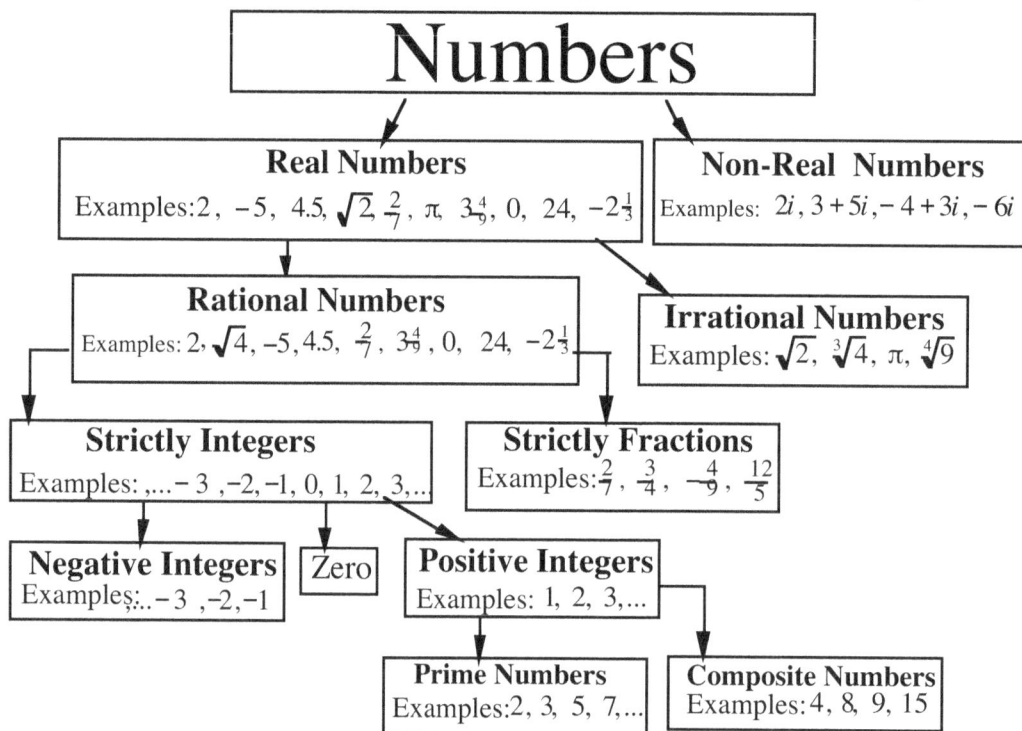

Numbers

Real Numbers

Examples:2, -5, 4.5, $\sqrt{2}$, $\frac{2}{7}$, π, $3\frac{4}{9}$, 0, 24, $-2\frac{1}{3}$

Non-Real Numbers

Examples: $2i$, $3+5i$, $-4+3i$, $-6i$

Rational Numbers

Examples: 2, $\sqrt{4}$, -5, 4.5, $\frac{2}{7}$, $3\frac{4}{9}$, 0, 24, $-2\frac{1}{3}$

Irrational Numbers

Examples: $\sqrt{2}$, $\sqrt[3]{4}$, π, $\sqrt[4]{9}$

Strictly Integers

Examples: $,...-3$, -2, -1, 0, 1, 2, 3,...

Strictly Fractions

Examples:$\frac{2}{7}$, $\frac{3}{4}$, $-\frac{4}{9}$, $\frac{12}{5}$

Negative Integers

Examples: $...-3$, -2, -1

Zero

Positive Integers

Examples: 1, 2, 3,...

Prime Numbers

Examples:2, 3, 5, 7,...

Composite Numbers

Examples: 4, 8, 9, 15

Number Flow Chart

In **arithmetic**, we will deal mainly with **non-negative real numbers** (i.e., zero, and positive real numbers which include the positive integers, positive fractions, mixed numbers and decimals).

We shall now define some terms that we use very often to communicate in arithmetic.

Lesson 1: Place-Value Relationships; Basic Definitions; Terminology; Types of Numbers

We define a **set of numbers** as a well-defined collection of numbers.

The set of the **natural numbers** consists of the numbers $1, 2, 3, 4, 5, 6, 7, 8, 9, 10, 11, 12, 13,...,$
If we know a natural number, to obtain the next natural number we add 1. The smallest natural
number is 1, but we do not know the largest natural number, since given even a very large natural
number, we can always obtain the next natural number by adding 1. The natural numbers are also
known as the **counting numbers** or the **positive integers**.

The set of **whole numbers** consists of the numbers $0, 1, 2, 3, 4, 5, 6, 7, 8, 9, 10, 11, 12, 13,...,$ If we
 know a whole number, to obtain the next whole number we add 1. The smallest whole number
is 0, but we do not know the largest whole number, since given even a very large whole number,
we can always obtain the next whole number by adding 1.

Even though we do not cover a set of numbers called **negative integers** in arithmetic, we briefly
 mention them here for completeness. If we take the opposites of the set of natural numbers also
 called positive integers, we obtain a set of numbers called **negative integers** (such as -1, -2, and
-12). If we combine the whole numbers with the negative integers we obtain a set of numbers
called the **integers**. The set of integers therefore consists of the set of numbers $...,-6, -5. -4, -3,$
$-2, -1, 0, 1, 2, 3, 4, 5, 6,...$
(The three dots preceding the -6 on the left indicates that the numbers continue to decrease to the left
 and the three dots after the 6 on the right indicates that the numbers continue to increase to the right)

Ratio: The ratio a is b is the fraction $\frac{a}{b}$. **Example:** The ratio 3 is to 4 is the fraction $\frac{3}{4}$.

Rational number: A rational number (a fraction) is a number which **can** be written as
the ratio of two integers. The word **rational** pertains to the word **ratio.**

 Examples are (a) $\frac{2}{3}$;   (b) $\frac{1}{5}$;   (c)   4   (since $4 = \frac{4}{1}$)

(d)   0   (since $0 = \frac{0}{7} = \frac{0}{3}$ ... or $0 = \frac{0}{b}$, where b is an integer and b $\neq$ 0)

(e) $\sqrt{4}$ (because $\sqrt{4} = 2 = \frac{2}{1}$)

A rational number can also be written either as a terminating decimal or as a repeating decimal.

Examples of terminating decimals: $\frac{1}{4} = .25$; $\frac{13}{2} = 6.5$; $\frac{37}{8} = 4.625$.

Examples of repeating decimals: $\frac{1}{3} = .333...$ or $.\overline{3}$ and $\frac{1}{6} = .1666...$ or $.1\overline{6}$; $\frac{2}{3} = .66...$ or $.\overline{6}$.

Note the bar (vinculum) placed over the repeating digit or block of digits.
We may also regard a terminating decimal as non-terminating if we attach zeros to the right of the
decimal. Examples are $.25 = .250000...$, $.5 = .5000...$

Other types of numbers we cover in arithmetic are fractions, mixed numbers, and decimals.
There is another set of numbers, namely the irrational numbers. We define an **irrational number** as a
number which **cannot** be written as the ratio of two integers. However, we can approximate irrational
numbers as closely as we wish by rational numbers or decimals.

 Examples of irrational numbers are $\sqrt{2}$, $\sqrt[3]{4}$, and π (pi). We can for example, approximate
$\sqrt{2}$ by 1.414, and π by $\frac{22}{7}$ or 3.142.

When written in decimal form, an irrational number is non-repeating and non-terminating.

If we combine the natural numbers, the fractions, decimals, irrational numbers, the negatives of these
numbers, and zero , we obtain the set of **real numbers**. Simply, the real numbers consists of the rational
numbers and the irrational numbers.

We can represent numbers by points on a horizontal line called the **real number line** (Fig.1)   4

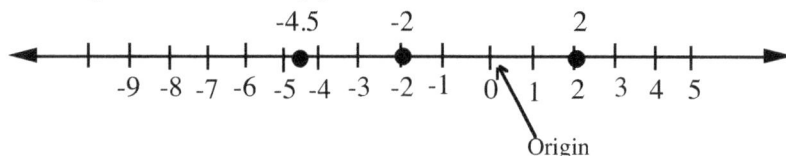

Figure 1

The **real number line** is a horizontal straight line with equally spaced intervals as in Figure 1 above. We label a point called the origin, 0 (zero). Points to the right of the origin are labeled positive and points to the left of the origin are labeled negative. The numbers increase as one moves from the left to the right on the real number line. Roughly speaking, a real number is a number that can be represented by a point on the real number line. The real numbers consists of the integers, fractions, mixed numbers, decimals, and radicals. In Figure 1, if the real numbers, -4.5, -2, and 2 are of interest, we can represent them by the dots shown. Every point on this line is associated with a real number; and every real number is associated with a point on this line. We can also say that the set of real numbers consists of the signed numbers and zero.

The arithmetic number line

Since in arithmetic, we deal with the numbers from zero to the right on the real number line, we can consider the **arithmetic number line** as the part of the real number line beginning from zero and extending to the right.

Summary for Some Number Terminology

Positive integers, or natural numbers or counting numbers $= \{1, 2, 3, 4, ...\}$

Negative integers $= \{..., -4, -3, -2, -1\}$

Non-negative integers or **whole numbers** $= \{0, 1, 2, 3, 4, ...\}$

Non-positive integers $= \{..., -4, -3, -2, -1, 0\}$

Non-negative real numbers: Examples are $0, \ 4, 7.5, \ 3\frac{1}{2}, \ \frac{1}{4}, \ 6,$ and $\sqrt{2}$

Non-positive real numbers: Examples are $-\sqrt{11}, -4, -3\frac{1}{2}, -2, -1, -\frac{1}{4}, .126,$ and 0.

We must understand thoroughly the above terms because they will be used over and over in the future. Anytime we meet any of these terms, we should try to form a quick mental picture of representative examples.

Lesson 1 Exercises

1. Define the following:

   Natural numbers, counting numbers, positive integers, whole numbers , negative integers, integers, rational number

2. From memory, draw the number flow chart in Lesson 1.

Lesson 2

Introduction to Fractions, Decimals, Ratio and Proportion; Percent Symbol and Some Interpretations

Fractions (In **chapter 3**, we cover the various operations on fractions and mixed numbers)

Fraction: A fraction is a number representing one or more of the equal parts into which a whole (thing) has been divided.

Unit fraction One part of a whole which has been divided (or broken) into equal parts.

The numerator of a unit fraction is 1. Examples of unit fractions are $\frac{1}{2}, \frac{1}{8}$ and $\frac{1}{100}$.

Terms and Meaning of a Fraction

Example In the fraction, $\frac{2}{3}$, the 2 and 3 are called the terms of the fraction; the 2 is called the numerator and the 3 is called the denominator.

In a fraction, the numerator indicates how many of the unit fraction the fraction represents; the denominator indicates into how many equal parts a whole has been divided; the denominator also indicates the size of the unit fraction: the smaller the denominator, the larger the size, and the larger the denominator, the smaller the size of the unit fraction.

Note that in addition to representing parts of a whole, a **fraction** is also used to express **division**, and also to **compare** numbers.

Types of Fractions

Proper Fraction: A fraction in which the numerator is less than the denominator.

Examples are $\frac{4}{5}, \frac{2}{7}$ and $\frac{34}{45}$. (A proper fraction is always less than 1)

Improper Fraction: A fraction in which the numerator is greater than or equal to the denominator.

Examples are $\frac{8}{5}, \frac{9}{3}$ and $\frac{7}{7}$. (An improper fraction is always greater than or equal to 1)

Mixed Number: A mixed number consists of a whole number part and a proper fraction part. We obtain a mixed number from an improper fraction when the denominator does not divide the numerator exactly (i.e. without a remainder).

Converting an Improper Fraction to a Mixed Number

Example Convert $\frac{91}{17}$ to a mixed number.

Solution $\frac{91}{17}$ = 5 remainder 6 (using long division)

$$\begin{array}{r} 5 \\ 17\overline{)91} \\ -85 \\ \hline 6 \end{array}$$

Therefore, $\frac{91}{17} = 5\frac{6}{17}$

Converting a Mixed Number to an Improper fraction

Example Convert $6\frac{3}{4}$ to an improper fraction.

Solution Step 1: $6 \times 4 = 24$

Step 2: $24 + 3 = 27$

Step 3: $6\frac{3}{4} = \frac{27}{4}$ (Normally, we do the conversion mentally and show only Step 3.

Decimal Fractions and Decimals

6

(In **Chapter 4**, we cover the various operations on decimals)

Powers of 10

Examples: $10^1 = 10$

$10^2 = 100$

$10^3 = 1,000$

$10^4 = 10,000$

A **decimal fraction** is a fraction whose denominator is a power of 10.

Examples are $\frac{3}{10}$, $\frac{5}{100}$, and $\frac{23}{1000}$ ..

A **decimal** is the symbol for an equivalent decimal fraction.

Examples: $\frac{3}{10} = .3$ ($\frac{3}{10}$ is the decimal fraction; .3 is the decimal)

$\frac{5}{100} = .05$

$\frac{23}{10000} = .0023$

Ratio and Proportion

(In **chapters 5 & 6**, we cover ratio and proportion problems)

The **ratio** of two quantities a and b (written $a:b$) is the fraction $\frac{a}{b}$ (i.e., is the first quantity (mentioned) divided by the second quantity (mentioned)). The quantities a and b are called the terms of the ratio.

Proportion : A proportion is a statement that two ratios are equal.

The proportion "a is to b as c is to d " (or $a:b::c:d$) is the equality

$$\frac{a}{b} = \frac{c}{d} \tag{1}$$

In the above proportion, a and d are called the extremes and b and c are called the means of the proportion.

We may also define a proportion as a mathematical statement indicating the equality between two **equivalent fractions.**   Example $\frac{2}{3} = \frac{4}{6}$ is a proportion

Percent symbol "%" and some interpretations

(In **chapter 7**, we cover problems involving percent (%)

1. Over hundred: For example, 20% means $\frac{20}{100}$.   (Twenty over hundred or 20 divided by 100)

2. For each hundred: For example, a savings account with a 5% interest rate pays the depositor $5 for each $100 in the account. (Five for each hundred)

3. Hundredths: For example, 20% means $\frac{20}{100}$ or .20   (Twenty hundredths)

4. As a number out of 100: For example, a grade of 80% on a test means a student got 80 points out of 100 points.

Converting a decimal (or any fraction) **to percent**   7

Procedure: Multiply by 100%. (i.e., multiply by 100 and attach the percent symbol "%".

To change a decimal to percent, move the decimal point two places to the right and attach the percent symbol. (Note that moving the decimal point two places to the right is equivalent to multiplying by 100)

Example 1 Convert .74 to percent.

Solution $.74 = 74\%$.

Example 2 Convert .008 to percent.

Solution $.008 = .8\%$

Example 3 Convert 1 to percent.

 Solution $1 = 100\%$

Example 4 Convert 12 to percent.

Solution $12. = 1200\%$

Example 5   Change $\frac{1}{4}$ to percent

Method 1 $\frac{1}{4} \times \frac{100\%}{1} = 25\%$

Method 2 $\frac{1}{4} = .25 = .25 \times 100\%$ (Changing to decimal first, and then changing to percent)

Converting a percent to a decimal

Procedure: Divide by 100%. (i.e., drop the percent symbol "%" and divide by 100) or apply the meaning of the percent symbol to change the percent to a decimal fraction and then easily to a decimal.

To change a percent to a decimal, drop the percent symbol and move the decimal point two places to the left .   (**Note** that moving the decimal point two places to the left is equivalent to dividing by 100)

Example 1 $74\% = .74$ (or $74\% = \frac{74}{100} = .74$)

Example 2 $.8\% = .008$ (or $.8\% = \frac{.8}{100} = .008$)

Example 3 $145\% = 1.45$

Example 4 $14.5\% = .145$

Example 5 $.145\% = .00145$

Note above: Attaching the % symbol is equivalent to dividing by 100, and dropping the % symbol is equivalent to multiplying by 100.

Lesson 2 Exercises

A. **1** Convert to mixed numbers: (a). $\frac{22}{7}$;   (b). $\frac{17}{5}$;   (c) $\frac{79}{18}$.

2 Verify the answers in **A** by converting your answers back to improper fractions.

3. Change to improper fractions. (a) $8\frac{5}{6}$;   (b). $9\frac{3}{8}$

Answers: **1.** **(a).** $3\frac{1}{7}$;   **(b)**. $3\frac{2}{5}$; (c) $4\frac{7}{18}$;   **3**: **(a)** $\frac{53}{6}$;   **(b).** $\frac{75}{8}$.

B. Write as decimals: (a) $\frac{36}{1000}$;   (b) $\frac{40}{100}$;   (c) $\frac{9}{100}$

Answers (a) .036 ;   (b).40 ;   (c) .09

C Change each of the following to a fraction in its lowest terms:
  (a) .64 ;   (b) 6.4 ;   (c) 0.036 ;   (d) 1.36 ;   (e) .63;   (f) 0.0825

Answers (a) $\frac{16}{25}$;   (b) $\frac{32}{5}$;   (c) $\frac{9}{250}$);   (d) $\frac{34}{25}$;   (e) $\frac{63}{100}$;   (f) $\frac{33}{400}$

CHAPTER 2

Lesson 3

Writing Whole Numbers Using Numerals and Words

Writing Whole Numbers Using Numerals

Example 1   Write, using numerals: Forty-five million, seven hundred sixty-two thousand, three hundred twenty-five.

 Note   1 — One billion has nine zeros: 1,000,000,000
     2 — One million has six zeros: 1,000,000
     3 — One thousand has three zeros: 1,000
     4 — One hundred has two zeros: 100
     5 — One ten has one zero: 10

By applying these notes, let us write the sample number given above.

Step 1:   Forty-five million   = 45,000,000   (1 million has six zeros)
Step 2:   Seven hundred sixty-two thousand=   762,000   (1 thousand has three zeros)
Step 3:   Three hundred   =   300   (1 hundred has two zeros)
Step 4:   Twenty-five   =   25
Step 5:   Add the digits in the columns   _____
               45,762,325

     Answer is: 45,762,325

Example 2   Write, using numerals: Fifty-four million, six hundred forty-five.

Step 1:   Fifty-four million: 54,000,000   (1 million has six zeros)
Step 2:   Six hundred: 600   (1 hundred has two zeros)
Step 3:   Forty-five: 45
Step 4:   Add the digits in the columns_____
          54,000,645

     Answer is: 54,000,645

Example 3   Write, using numerals: Two billion, eleven million, five thousand, seven.

     2,000,000,000   (1 billion has nine zeros)
      11,000,000   (1 million has six zeros)
        5,000   (1 thousand has three zeros)
          7
     ‾‾‾‾‾‾‾‾‾‾‾‾‾‾
     2,011,005,007

     Answer is: 2,011,005,007

Writing Whole Numbers in Words

In writing whole numbers in words, we will be guided by the group names:

$$\overbrace{3,}^{\text{billions}}\overbrace{296,}^{\text{millions}}\overbrace{784,}^{\text{thousands}}\overbrace{625.}^{\text{ones}}436954$$

Example 1 Write in words: 534

Solution   We proceed as follows:

534 = 500 + 34 or 500 + 30 + 4

Five hundred thirty-four   (Hyphenate a number between 21 and 99. e.g., thirty-four)

Example 2 Write 12,379,534 in words

We proceed as follows (considering the digits in groups of three from right to left):
12,379,534

= 12,000,000 + 379,000 + 500 + 34

= Twelve million, three hundred seventy-nine thousand, five hundred thirty-four

Example 3 Write in words: 8,604,009

We proceed as follows:

8,604,009

= 8,000,000 + 604,000 + 9

= Eight million, six hundred four thousand, nine.

Lesson 3 Exercises

A. Write, using numerals: **1.** Five million, six hundred eight thousand, nine hundred seven.

2. Seven hundred twenty-nine thousand, four hundred.

3.. Thirty-six million, seventy thousand, two hundred thirty-four

Answers   **1.** 5,608,907 ;   **2.**   729, 400 ;   **3.** 36,070,234

B. Write the following numbers using words:

1. 54,660,007;   **2.** 735,038,429;   **3.** 892,607

Answers   **1..** Fifty-four million, six hundred sixty thousand, seven.
2. Seven hundred thirty-five million, thirty-eight thousand, four hundred twenty-nine.
3. Eight hundred ninety-two thousand, six hundred seven.

C. Write using numerals: **1.** Six billion, three hundred thousand, nine.

2. Eight hundred thirty-five thousand, nine hundred fifty-seven.

3. Two million, forty-seven thousand;   **4.** One billion, ninety-five.

5. Four hundred thousand, twenty.

D. Write the following in words:

6. 834,689 ;   **7.** 2,608,007;   **8.** 629,353,086 ;   **9.** 857.23 ;   **10.** $487\frac{1}{2}$

Answers: **1.** 6,000,300,009;   **2.** 835,957;   **3.** 2,047,000;   **4.** 1,000,000,095;   **5.** 400,020
6. Eight hundred thirty-four thousand, six hundred eighty-nine; **7.** Two million, six hundred eight thousand, seven.
8. Six hundred twenty-nine million, three hundred fifty-three thousand, eighty-six;
9. Eight hundred fifty-seven and twenty-three hundredths; **10.** Four hundred eighty-seven and one half.

Lesson 4

Basic Arithmetic Operations and Properties

(The material in this lesson will guide us to justify the basic operations we perform in mathematics)

The four fundamental operations of arithmetic are addition (+), subtraction (-), multiplication (×) and division (÷). Each operation is an operation on two numbers. Two other operations are power finding (for example, $2^3 = 8$) and root extraction (for example, $\sqrt[3]{8} = 2$).

Addition

Example: $4 + 5 = 9$. The 4 and the 5 are called the **addends**, the 9 is called the **sum**.

Commutative Property: Addition is commutative. By this property, changing the order in which two numbers are added yields the same sum.
Example: $4 + 5 = 5 + 4$
$9 = 9$

Associative Property: Addition is associative. By this property, changing the grouping of numbers in addition yields the same sum.
Example: $(2 + 3) + 4 = 2 + (3 + 4)$.
$5 + 4 = 2 + 7$
$9 = 9$

Additive identity element The additive identity element is 0. For example, $5 + 0 = 5$

Identity property of 0. When zero is added to a given number, the sum is the given number.
Example: $4 + 0 = 4$.

Subtraction

Example: $7 - 2 = 5$. The 7 is called the **minuend**, the 2 is called the **subtrahend**, and the 5 is called the **difference**. Subtraction is the inverse of addition.
Subtraction is **not** commutative, That is, order is important in subtraction
For example, $6 - 2 \neq 2 - 6$ (That is, 6 - 2 is not equal to 2 - 6.)
$4 \neq -4$ (That is, 4 is not equal to -4.)
Subtraction is also **not a**ssociative.
Example: $(8 - 5) - 2 \neq 8 - (5 - 2)$
$3 - 2 \neq 8 - 3$

Subtraction with Borrowing

Example Subtract: $506 - 39$

Procedure: We use a vertical arrangement and indicate the decimal point explicitly in all the whole numbers.
(Note that the explicit indication of the decimal point is **not** necessary)

Then, lining up the decimal points vertically, we obtain

$$
\begin{array}{r}
506. \\
- \ 39. \\
\hline
467.
\end{array}
$$
(Check by adding 39 and 467 to obtain 506)

The answer is 467.

Note that, for a whole number, the decimal point is behind the units digit (last digit). The use of the decimal point, above, was to make sure that the "9" was under the "6" and the "3" was under the "0". (i.e., the units are under the units, and the tens are under the tens and so on.) Therefore, your may ignore the decimal points.

Multiplication

Implication of Multiplication   Example: $2 \times 3 = 2(3) = (2)3 = (2)(3) = (2) \bullet (3) = 2 \bullet 3 = 6$
The 2 and the 3 are called the **factors,** and the 6 is called the **product.**

Commutative Property: Multiplication is commutative. By this property, changing the order   in which two numbers are multiplied yields the same product.

Example: $2 \times 3 = 3 \times 2$
$6 = 6$

Associative Property: Multiplication is associative. By this property, changing the grouping   of numbers in multiplication yields the same product.

Example: $(2 \times 3) \times 4 = 2 \times (3 \times 4)$.
$6 \times 4 = 2 \times 12$
$24 = 24$

Multiplicative identity: The multiplicative identity element is 1. For example, $6 \times 1 = 6$

Multiplicative inverse: The multiplicative inverse of a given number is that number which if multiplied by the given number yields a product equal to 1.

Example: The multiplicative inverse of 4 is $\frac{1}{4}$. (Note: $4 \times \frac{1}{4} = 1$.)
The multiplicative inverse of a number is also known as the **reciprocal** of that number.

Multiplicative property of zero   Zero multiplied by any number equals zero.
Symbolically, for any whole number a, $a \times 0 = 0$.

Multiplication as repeated addition The consideration of multiplication as **repeated addition** can be a useful guide in solving problems such as the following: How many centimeters are there in 1000 meters? To solve this problem, we could add 100 centimeters a thousand times as:
$\underbrace{100 + 100 + 100 + 100 + ... + 100}_{\textbf{1000 addends}}$ (since 100 centimeters = 1 meter). However, instead of adding 100 a thousand times (a tedious computational process) we could simply multiply 100 by 1000 and obtain:
$100 \times 1000 = 100,000$ centimeters .

Division

Implication of division   Example: $8 \div 2 = 4$ or $\frac{8}{2}$ or $2\overline{)8}^{\,4}$. The 8 is called the **dividend**, the 2 is called the **divisor**, and the 4 is called the **quotient**. Division is the inverse of multiplication. Note that $\frac{8}{2} = 4$ because $2 \times 4 = 8$.

Division is **not** commutative (That is, order matters in division.)
For example, $8 \div 2 \neq 2 \div 8$ (That is, $8 \div 2$ is **not** equal to $2 \div 8$)
$4 \neq \frac{1}{4}$ (That is, 4 is **not** equal to $\frac{1}{4}$.

Division is also **not** associative. For example,
$(18 \div 9) \div 3 \neq 18 \div (9 \div 3)$

$\frac{2}{3} \neq 6$

Division by 1: A given number divided by 1 equals the given number. For example, $\frac{4}{1} = 4$

Division involving zero

Examples   1. $\frac{0}{5} = 0$ (because $5 \times 0 = 0$)   **2**. $\frac{0}{-6} = 0$

But   3. $\frac{5}{0}$ is undefined.(There is **no** number such that $0 \times$ "that number" = 5. Do not divide by 0)

4. $\frac{0}{0}$ is indeterminate. (Any number will do since $0 \times$ "any number" = 0)

Division as repeated subtraction The consideration of division as **repeated subtraction** can be a useful guide in solving problems such as the following:
 How many 60-gallon water containers can be filled from a tank containing 2,400 gallons of water? To solve this problem, we could subtract 60 gallons repeatedly from 2,400 gallons and add up the number of subtractions required for the tank to be either completely empty or to contain less than 60 gallons of water?   However, instead of subtracting and recording the number of subtractions
 (40 subtractions), it is faster to divide 2,400 by 60 in one step to obtain 40 containers..

Powers of Whole Numbers

Examples 1. 2^5 $= (2)(2)(2)(2)(2) = 32$

2^5 is read as 2 to the fifth power. (or 2 raised to the fifth power)
The exponent "5" indicates how many times the base "2" is being used as a factor.

2. $4^3 = 4 \times 4 \times 4$
$\qquad = 64$
3. $10^4 = 10 \times 10 \times 10 \times 10$
$\qquad = 10000$
4. $0^2 = 0 \times 0$
$\qquad = 0$
5. $0^3 = 0 \times 0 \times 0$
$\qquad = 0$

Zero as an Exponent

Any (nonzero) number raised the to power zero is 1.
Examples

1. $5^0 = 1$
2. $4^0 = 1$
3. $10^0 = 1$

Note that 0^0 is indeterminate. Example: $0^0 = \frac{0^2}{0^2} = \frac{0}{0}$ which is indeterminate (see also division involving zero)

Roots of Whole Numbers 14

Root finding and **Power finding** are inverse operations (in much the same way as multiplication and division are inverse operations).

Square Root: Symbol " $\sqrt{}$ " or " $\sqrt[2]{}$ "

Finding the square roots of perfect squares by inspection or by factoring

1. In arithmetic, we may cheaply define the square root of a number as one of the two equal factors of that number. (In algebra, we shall modify this definition)

2. The square root of zero is zero (i.e. $\sqrt{0} = 0$).

Examples

1. The square root of 9 is symbolized $\sqrt{9}$, and

$$\sqrt{9} = 3$$
$$\text{because } 3^2 = 9$$

Note that , according to the above definition,
$$\sqrt{9} = \sqrt{(3)(3)} = 3$$

2. The square root of 64 is written $\sqrt{64}$, and

$$\sqrt{64} = 8$$
because $8^2 = 64$.

Also, $\sqrt{64} = \sqrt{(8)(8)} = 8$

3. Note: $\sqrt{0} = 0$
because $0^2 = 0$

If the given number is not a perfect square, we will use tables or a calculator to find the square root.

The **Cube Root** : Symbol " $\sqrt[3]{}$ "

From the above square root definition, we can similarly define the cube root of a number as one of the three equal factors of that number. (In algebra, we shall modify this definition.)

Examples

1. $\sqrt[3]{8} = 2$ Note that $\sqrt[3]{8} = \sqrt[3]{(2)(2)(2)} = 2$

because $2^3 = 8$

2. $\sqrt[3]{64} = 4$ Note that $= \sqrt[3]{4 \times 4 \times 4} = 4$

because $4^3 = 64$

More formal definition of square root

1. $\sqrt{A} = r$ if $r^2 = A$, where A, and r are both real and positive.

2. The square root of zero is zero (i.e. $\sqrt{0} = 0$).

Lesson 4 Exercises

A. Subtract: **1.** $703 - 86$; **4.** $8,765,210 - 65,432$

  2. $420 - 109$; **5.** $76,521 - 65,432$

  3. $6035 - 247$; **6.** $894,352 - 743,956$

Answers **1.** 617; **2.** 311 ; **3.** 5,788; **4.** 8,699,778; **5.** 11,089; **6.** 150,396.

B. Evaluate the following:

  1. 3^3 ; **2.** 2^4 ; **3.** 8^2 ; **4.** 4^0 ; **5.** 0^0 ; **6.** 0^3 ; **7.** 4^3 ; **8.** 2^4 ; **9.** 5^3

Answers: 1. 27; **2.** 16 ; **3.** 64 ; **4.** 1 ; **5.** Indeterminate ; **6.** 0 ; **7.** 64 ; **8.** 16; **9.** 125.

C. Simplify: **1.** $\sqrt{25}$; **2.** $\sqrt{81}$; **3.** $\sqrt{36}$; **4.** $\sqrt{256}$ **5.** $\sqrt{1024}$

  6. $\sqrt[3]{64}$ **7.** $\sqrt[4]{81}$; **8.** $\sqrt{121}$; **9.** $\sqrt{\dfrac{25}{49}}$; **10.** $\sqrt[3]{8}$

Answers: **1.** 5 ; **2.** 9 ; **3.** 6 ; **4.** 16 ; **5.** 32 ; **6.** 4 ; **7.** 3 ; **8.** 11 ; **9.** $\dfrac{5}{7}$; **10.** 2.

Word Problems

D

1. What must be added to 3 to make 7.?
2. What must be added to 86 to make 703.?
3. James wants to buy a computer costing $703. At present, he has only $86. How much more money does he need to have in order to buy this computer?
4. Mary had $806 in her bank account this morning. In the afternoon she withdrew $86 from this account to buy a printer, How much money is left in her bank account?

Answers: **1.** 4; **2..** 617; **3..** $ 617; **4.** $ 617

Lesson 5

Order of Operations and Evaluation of Arithmetic Expressions

Order of Operations in Performing Arithmetic Operations

Perform the operations according to the following order:

1. Grouping Symbols: If there are grouping symbols, evaluate within the grouping symbols first;
2. Then, evaluate the powers and the roots in any order ; followed by

3. Division and Multiplication from left to right (or invert the divisors and multiply in any order); and then

4. Addition and Subtraction from left to right.

Note: If grouping symbols occur within other grouping symbols, evaluate within the innermost grouping symbols first, followed by the next innermost symbols and so on.

Example 1

$$5 + 2(4 + 7) - 9$$
$$= 5 + 2(11) - 9 \qquad \text{Evaluating within the parentheses first}$$
$$= 5 + 22 - 9 \qquad\quad \text{Multiplying next}$$
$$= 27 - 9 \qquad\qquad \text{Adding from left to right}$$
$$= 18$$

Example 2

$$2^3 + 9(4 + 3) + 8 \qquad \text{Evaluating within the parentheses first}$$
$$= 2^3 + 9(7) + 8$$
$$= 8 + 63 + 8 \qquad \textbf{Scrapwork: } 2^3 = (2)(2)(2) = 8$$
$$= 71 + 8$$
$$= 79$$

Example 3

$$18 \div 6 \times 3 \qquad \text{From left to right, we encounter division first and}$$
$$= 3 \times 3 \qquad\qquad \text{therefore, we divide first, according to}$$
$$= 9 \qquad\qquad\quad \text{the order of operations.}$$

Note in Example 3 above that, if we had multiplied first, we would **not** obtain the correct answer of 9.
In multiplying first, $18 \div 6 \times 3 = 18 \div 18 = 1$ <-----**This is the wrong solution**.

Example 4

$$6 \times 3 \div 18$$
$$= 18 \div 18 \qquad \text{(From left to right, we encounter multiplication}$$
$$= 1 \qquad\qquad \text{first, therefore, we multiply first and then divide)}$$

Note however that, if we divide first, we will still get the correct answer.

$$\text{Thus, } 6 \times 3 \div 18 = \frac{6}{1} \times \frac{3}{18} = \frac{6}{1} \times \frac{1}{6} = 1$$

We conclude, from Examples 3 and 4, that if we always divide first (whether or not division comes first) , we will obtain the correct answer; but if multiplication does not come first in going from left to right, then we should **not** multiply first. Thus if we always perform division first, we will obtain the correct answer. Note that Example 3 and Example 4 are different problems.

EXTRA. Simplify: $20 \div \frac{1}{2}$ of 10 (Hint: Perform the "of" operation first before dividing.)

Solution $\qquad 20 \div \frac{1}{2}$ of 10

$$= 20 \div (\frac{1}{2} \times 10) \qquad \text{(The notion of changing "of" to multiplication and performing the operations from}$$
$$= 20 \div 5 = 4 \qquad \text{left to right will lead to the \textbf{wrong result} in this problem: "of" precedes} \times \text{and} \div \text{.)}$$

Example 5

$$10 + 7[7 + 3(5 - 1)]$$
$$= 10 + 7[7 + 3(4)]$$
$$= 10 + 7[7 + 12]$$
$$= 10 + 7[\ 19\ ]$$
$$= 10 + [\ 133]$$
$$= 143$$

Example 6 $\dfrac{18 + 6}{9 + 3} = \dfrac{24}{12}$

$$= 2$$

(Add the terms in the numerator, then add the terms in the denominator before dividing)

Example 7   Evaluate $3^2 + 7(3)$

Procedure: Apply the order of operations.

$$3^2 + 7(3)$$
$$= 9 + 21$$
$$= 30$$

Scrapwork: $3^2 = 9$

Example 8

$$3\sqrt{16} + 2(6) \div 3$$
$$= 3(4) + 12 \div 3$$
$$= 12 + 4$$
$$= 16$$

Example 9   Evaluate: 2×3^4

Solution : $= 2 \times 3 \times 3 \times 3 \times 3$
$$= 2 \times 81$$
$$= 162$$

Example 10   Evaluate: $2 + 4 \times (3 + 7)^2$

Solution: Apply the order of operations (see p.16)

$$2 + 4 \times (3 + 7)^2$$
$$= 2 + 4 \times (10)^2$$
$$= 2 + 4 \times (100)$$
$$= 2 + 400$$
$$= 402$$

(Scrapwork $3 + 7 = 10$)

(**note** that the addition of the "2" was done last)

Lesson 5 Exercises

A. 1. $4 + 5 \times (6 + 5)$;   **2.** $8 \times (9 - 3) + 7^2$;   **3.** $38 + 7 \times (4 + 2)^2$

Answers   **1.** 59; **2.** 97; **3.** 290

B. Evaluate the following: **1.** $7 + 5 \times (3 + 4)$;   **2.** $8 + 3(5 + 1)^2$;   **3.** $6\sqrt{81} + 5(4)(2)$

   4. $3[9 + (5 - 1) - 8]$;   **5.** $15 \div 3 \times 5$;   **6.** $12 \times 8 \div 4$;   **7.** $\dfrac{4 + 7}{18 + 4}$;

8. $5 \times 2^3 + 7 \times 4$;   **9.** $2 \times 4 \times 6 \times 0 \times 9$;   **10.** $\dfrac{4 \times (3 - 1) - 8}{6}$; **11.** $\dfrac{5 + 2}{5 \times (3 - 2) - 5}$

Answers: **1.** 42; **2.** 116; **3.** 94; **4.** 15; **5.** 25; **6.** 24; **7.** $\dfrac{1}{2}$; **8.** 68; **9.** 0; **10.** 0; **11.** Undefined

---

C. Simplify the following:

   1. $6 + 2(3 + 8)$;   **2.** $12 \div 6 \times 2$;   **3.** $6 \times 2 \div 12$;   **4.** $\sqrt{4^2 + 3^2}$;

   5. $8^2 + 3(0)$;   **6.** $\dfrac{8 + 2}{2}$;   **7.** $\dfrac{6 + 3}{2 + 3}$;   **8.** $7 \times 6 + 2 \times 4$;

   9. $10^2 + 2 \times 4$;   **10.** $6\sqrt{9} + 2 \times 4$;   **11.** $3[4 + 2(3 + 1)] - 5$

Answers: **1.** 28 ; **2.** 4 ; **3.** 1 ; **4.** 5 ; **5.** 64 ; **6.** 5 ; **7.** $1\frac{4}{5}$; **8.** 50 ; **9.** 108 ; **10** 26 ; **11.** 31

Word Problems

D   Last week, Paul had two part-time jobs. One job paid \$5 per hour, while the other job paid \$6 per hour.. He worked 5 hours last week at each job. In addition he was paid a \$4 bonus at one of the jobs. What is the total amount of money Paul was paid, last week?

Answers: \$59

Lesson 6

Rounding off Whole Numbers and Estimation

Rounding off Whole numbers

Procedure:

Step 1: Locate the digit in the round-off place.
(The round-off place is the place to which we want to round-off the number)

Step 2: Drop all digits to the right of the round-off place, and if the digit immediately to the right of the round-off place is 5 or more, add 1 to the round-off place digit (i.e. we round-up); but if the digit immediately to the right of the round-off place is less than 5, the round-off place digit remains unchanged (i.e. we round-down). Also, replace each digit dropped by a zero.

Example 1 Round-off 85,376 to the following places:
(a) To the nearest ten.
(b) To the nearest hundred.
(c) To the nearest thousand.
(d) To the nearest ten-thousand.

Solution
(a) To the nearest ten

Step 1: The round-off place digit is 7 (since 7 is in the tens place)

Step 2: We drop the 6, add 1 to the 7 (since the 6 is more than 5) and replace the 6 by a zero.
Then 85,376
$\approx$ 85,380   (Note the importance of replacing the 6 by zero. Without the zero, the answer
would be 8,538, which is far off the correct approximation of **85,380**.

Therefore 85,376 $\approx$ **85,380** (To the nearest ten)

(The symbol " $\approx$ " means " is approximately equal to".

(b)   85,376 to the nearest hundred

Step 1: The round-off place digit is 3. (since 3 is in the hundreds place)

Step 2: We drop the 7 and the 6, add 1 to the 3 (since 7 is more than 5) and replace the 7 and 6 by zeros.

Then 85,376
$\approx$ 85,400

(c) 85,376 to the nearest thousand

Step 1: The round-off place digit is 5. (since 5 is in the thousands place)

Step 2: We drop the 3, the 7 and the 6; replace each of them by a zero, and retain the 5 unchanged (since the 3 dropped is less than 5. The 3 is the digit immediately after the round-off place digit)

Then 85,376
$\approx$ 85,000   (To the nearest thousand)

(d)   85,376 to the nearest ten-thousand

Step 1: The round-off place digit is 8.

Step 2: We drop the 5, the 3, the 7, and the 6; replace each of them by a zero and add 1 one to the 8.

Then 85,376 $\approx$ 90,000 (to the nearest ten-thousand)

Example 2   Round-off 73,964 to the nearest hundred.

Step 1:   The round-off place digit is 9.(since 9 is in the hundreds place)

Step 2:   We drop the 6, the 4; replace them by zeros, add 1 to the 9.
(which becomes 10, and we write a zero to replace the 9 and add 1 to the 3 as usual)

Then 73,964

$\approx$ 74,000 (To the nearest hundred)

Rounding off Decimals

Even though decimals are not be covered in this chapter (see Chapter 4), for some completeness, and comparison with rounding off whole numbers, we round-off some decimals, below.

We use the same rule for rounding off whole numbers, except that after the decimal point, we do **not** replace any digits dropped by zeros.

Example   We round off **85376.7463** to the following places:

1. 85376.7463 to the nearest **thousandth** becomes **85376.746** (We do **not** replace the 3 dropped by a zero)

2. 85376.7463 to nearest **hundredth** becomes **85376.75** (We added 1 to the digit in the round-off place)

3. 85376.7463 to the nearest **tenth** becomes **85376.7** (The 7 is unchanged since the 4 dropped is less than 5)

4. 85376.7463 the nearest **unit** becomes **85377**.   (Adding 1 to the 6)

5. 85376.7463 to the nearest **ten** becomes **85380**. (Replacing the 6 dropped by a zero)

6. 85376.7463 to the nearest **hundred** becomes **85400.** (Replacing the digits (6 and 7) dropped by zeros)

7. 85376.7463 to the nearest **whole number** becomes **85377**. (same as to the nearest unit)

Estimation

In estimation, we **round-off** the numbers **before** carrying out the operation (which may be addition, subtraction, multiplication, division etc). We will round-off each number to the specified place; but if no round-off place is specified, we will round-off to the place of the first non-zero digit.
In the following example, we will round-off each number to the place of the first digit and carry out the indicated operation.

Example Estimate the product of 8,765 and 382.

Solution $9,000 \times 400$ (8,765 was rounded off to 9,000; and 382 to 400)

$= 3,600,000$

Lesson 6 Exercises

A. Round-off **369,528** to the following places:
1. To the nearest ten.
2. To the nearest hundred.
3. To the nearest thousand.
4. To the nearest ten-thousand.
5. To the nearest hundred-thousand.

Answers: 1. 369,530; **2**. 369,500; **3**. 370,000 ; **4**. 370,000 ; **5**. 400,000

B. Round-off **975.387** to the following places:

1. To the nearest hundredth.

2. To the nearest tenth.

3. To the nearest unit.

4. To the nearest ten.

5. To the nearest hundred.

6. To the nearest whole number.

Answers: **1**. 975.39 ; **2**. 975.4 ; **3**. 975; **4**. 980; **5**. 1,000 ; **6**. 975

C. Estimate the following: **1.**. $685,000 \times 364$; **2.** $6,457 + 346,238 + 7,398$; **3.**. $8,672 \div 32$.

Answers: 1. 280,000,000; **2**. 313,000 ; **3**. 300

Lesson 7

Factors, Product, Prime Numbers, Divisibility Rules, Prime Factorization

Factors and Product

Example   In $4 \times 9 = 36$, 4 and 9 are called the factors , and 36 is called the product.
(That is, the numbers being multiplied are called the factors, and the result of multiplying the numbers is called the product. Also, 4 and 9 are called divisors of 36. Therefore, saying that 4 is a factor of 36 implies that 4 divides 36 exactly (i.e., without a remainder or with zero remainder).

The process of finding the factors (divisors) of a given number is called **factoring.**

Prime Number: A natural number (greater than 1) whose natural number divisors are only itself and 1. (or a natural number (greater than 1) whose natural number factors are only itself and 1.)   Examples are $2, 3, 5, 7, 11, 13, 17, 19, 23,...$

Composite Number: A natural number that has more natural number factors other than itself and 1. Examples are $4, 6$, and 10.
$$((4 = 2 \times 2 \times 1; \qquad 10 = 5 \times 2 \times 1))$$

Divisibility Rules for 2, 3, and 5 (A guide for finding the factors 2, 3 and 5)

For 2: If the last (units) digit of a number is divisible by 2, then 2 is a divisor (or factor) of that number.
  Examples: 2 is a divisor of the following:
    5378; 6796; 310; 5552.

For 3: If the sum of the digits of a number is divisible by 3, then 3 is a divisor of that number.
  Example: 3 is a divisor of **471**
    Test: $4 + 7 + 1 = 12$. The sum 12 is divisible by 3.
    The following are also divisible by 3 : **582; 7881; and 43263.**

      The following are **not** divisible by 3: 673; 5431.

For 5: If the last digit of a number is 0 or 5, then 5 is a divisor (or factor) of that number.

  Examples: 5 is a divisor of the following: **33420; 7775; 985; 111220.**

(See also next page for divisibility rules for 4, 6, 8 and 10.)

Prime Factor: A prime factor of a given number is a prime number which divides the given
number exactly.

Example 3 is a prime factor of 12 (because 3 is a prime number and it also divides 12 exactly).

However, 4 is **not** a prime factor of 12. (Although 4 is a factor of 12, it is not a prime number.)
Also, 5 is not a prime factor of 12. (Although 5 is a prime number, it is not factor of 12.)

Prime Factorization of a Number (Indicated product of all the prime factors of the number)

We will write a given number as the product of all the prime numbers which when multiplied produce the given number.

Example: Find the prime factorization of 84.

Solution: We will divide 84 successively by prime numbers which can divide it without remainders.

Step 1: We use 2 as divisor (since the last digit, 4, of 84 is exactly divisible by 2), followed by division by 3 and 7 (applying our knowledge of the multiplication tables), until the quotient is 1; and then, the division process is complete.

$$
\begin{array}{r|r}
2 & 84 \\
2 & 42 \\
3 & 21 \\
7 & 7 \\
& 1
\end{array}
$$

Step 2: The prime factorization of 84 is the product of the prime divisors used.

Answer: $84 = 2 \times 2 \times 3 \times 7$ or $2^2 \times 3 \times 7$

Another Method

Step 1: $84 = 7 \times 12$
Step 2: We factor the 12 into primes.
$84 = 7 \times 2 \times 2 \times 3$
Answer: $84 = 2^2 \times 3 \times 7$.

Divisibility Rules for $4, 6, 8, 9,$ and 10 (See also previous page for the rules for $2, 3,$ and 5)

The divisibility rules for $4, 6, 8, 9$ and 10 are similar to the divisibility rules for $2, 3,$ and 5.

For 4: If the number named by the last **two digits** (from left to right) of a number is divisible by 4, then 4 is a divisor of that number. Example: 4 is a divisor of 1512 since 12 is divisible by 4.
(You may remember this rule as using the divisibility rule for 2 twice: Note that $2 \times 2 = 4$)

For 8: If the number named by the last **three digits** (from left to right) of a number is divisible by 8, then 8 is a divisor of that number. Example: 8 is a divisor of 2,1168 since 168 is divisible by 8.
(You may remember this rule as using the divisibility rule for 2 three times. Note that $2 \times 2 \times 2 = 8$)

For 6: If the sum of the digits of a number is divisible by 3 **and** the last digit is an even number, then 6 is a divisor of that number (This rule is a combination of the divisibility rules for 2 and 3).
Example: 6 is a divisor of 10584. (The sum of the digits, 18, is divisible by 3, and also the last digit is even)

For 9: A number is divisible by 9 if the sum of the digits is divisible by 9.
(You may remember this rule as using the divisibility rule for 3 twice. Note that $3 \times 3 = 9$)
Example: 12096 is divisible by 9 (because the sum $1 + 2 + 0 + 9 + 6 = 18$ is divisible by 9).

For 10. If the last digit of a number is 0, then 10 is a divisor of that number.
Example: 10 is a divisor of 3790.

Multiples and Common Multiples

Example   36 is a **multiple** of 4 because 36 can be obtained by multiplying 4 by a whole number, which in this case is 9. Similarly, 36 is a multiple of 9 because 36 can be obtained by multiplying 9 by a whole number, which in this case is 4. Also, 36 is divisible by 4 and by 9. Furthermore, 36 is a multiple of 12, because 36 can be obtained by multiplying 12 by a whole number (which in this case is 3). Also, 36 is divisible by 12.

Zero is a multiple of every whole number.

A **common multiple** is a number that is a multiple of two or more given whole numbers.

Also, a number (say, 36) is a common multiple of two or more given numbers (say, 9 and 12) if the number is divisible by each of the two or more given numbers (say, 9 and 12).
Thus, 36 is a common multiple of 9 and of 12, because 36 is divisible by 9 and by 12.

Definition: The **Least Common Multiple (LCM)** of two or more given (natural) numbers is the smallest (natural) number which is a multiple of the given numbers.
OR
The **Least Common Multiple (LCM)** of two or more given (natural) numbers is the smallest (natural) number which is divisible by each of the given numbers.

Finding the LCM (Least Common Multiple) of Two or More Numbers

Example Find the LCM of 16, 18, and 27.

Solution: We will cover two methods

Method 1

We will use a direct method which is the application of the prime factorization technique discussed previously. We will use **prime numbers** successively as divisors. At each step of the division process, each prime number used **must divide at least** one of the dividends (i.e., it may divide one, two or more, or even all the dividends), and the other dividends which are not divisible, are brought down for the subsequent steps. The division process ends when we have 1's in all the columns. The LCM is the product of the divisors used.

```
divisors
    │   dividends
    ↓   ↓  ↓  ↓
    2│16│18│27
    2│ 8│ 9│27
    2│ 4│ 9│27
    2│ 2│ 9│27
    3│ 1│ 9│27
    3│ 1│ 3│ 9
    3│ 1│ 1│ 3
        1  1  1
```

Answer: The LCM of 16, 18 and 27 $= 2 \times 2 \times 2 \times 2 \times 3 \times 3 \times 3$
$$= 432$$

Method 2 Finding the LCM by factoring

Step 1:   Factor each number into prime factors and write the repeated factors in power form.

$16 = 2 \times 2 \times 2 \times 2 = 2^4$
$18 = 2 \times 3 \times 3 = 2 \times 3^2$
$27 = 3 \times 3 \times 3 = 3^3$

Step 2: Consider every prime factor and choose the highest power of each prime factor.
The LCM is the product of the highest powers.

The LCM of 16, 18, and 27 $= 2^4 \times 3^3 = 432$

Lesson 7 Exercises

A. Find the prime factorization of the following:

1. 72;   **2.** 105;   **3.** 180 ;   **4.** 1024;   **5.** 432; **6.** 168 ; **7.** 420

Answers:

1. $2^3 \times 3^2$; 2. $3 \times 5 \times 7$; 3. $2^2 \times 3^2 \times 5$; 4. 2^{10}; 5. $2^4 \times 3^3$; 6. $2^3 \times 3 \times 7$; 7. $2^2 \times 3 \times 5 \times 7$

B. Find the LCM of the following: **1.** 9, 12 and 16;   **2.** 3, 7 and 8; **3.** 14, 15, and 20.

4. 12, 18 and 32 ; **5.** 30 and 36; **6.** 15 and 24.

Answers **1**. 144 ; **2.** 168; **3.** 420; 4. 288 ; **5.** 180; **6.** 120

CHAPTER 3

Fractions and Mixed Numbers

Lesson 8

Definitions; Writing Fractions and Mixed Numbers in Words; Reducing Fractions to lowest terms; Equivalent Fractions

Fraction: A fraction is a number representing one or more of the equal parts into which a whole (thing) has been divided.

Unit fraction One part of a whole which has been divided (or broken) into equal parts.

The numerator of a unit fraction is 1. Examples of unit fractions are $\frac{1}{2}, \frac{1}{8}$ and $\frac{1}{100}$.

Terms and Meaning of a Fraction

Example In the fraction, $\frac{2}{3}$, the 2 and 3 are called the terms of the fraction; the 2 is called the numerator and the 3 is called the denominator.

In a fraction, the numerator indicates how many of the unit fraction the fraction represents; the denominator indicates into how many equal parts a whole has been divided; the denominator also indicates the size of the unit fraction: the smaller the denominator, the larger the size, and the larger the denominator, the smaller the size of the unit fraction.

Note that in addition to representing parts of a whole, a **fraction** is also used to express **division**, and also to **compare** numbers.

Types of Fractions

Proper Fraction: A fraction in which the numerator is smaller than the denominator.

A proper fraction is always less than 1.   Examples are $\frac{1}{2}, \frac{2}{7}$ and $\frac{34}{45}$

Improper Fraction: A fraction in which the numerator is greater than or equal to the denominator.

An improper fraction is greater than or equal to 1. Examples are $\frac{8}{5}, \frac{9}{3}$ and $\frac{7}{7}$.

Mixed Number: A mixed number consists of a whole number part and a proper fraction part. We obtain a mixed number from an improper fraction when the denominator does not divide the numerator exactly (i.e. without a remainder).

Converting an Improper Fraction to a Mixed Number

Example Convert $\frac{91}{17}$ to a mixed number.

Solution $\frac{91}{17}$ = 5 remainder 6 (using long division)

$$\begin{array}{r} 5 \\ 17\overline{)91} \\ -85 \\ \hline 6 \end{array}$$

Therefore, $\frac{91}{17} = 5\frac{6}{17}$

Converting a Mixed Number to an Improper fraction   2 7

Example Convert $6\frac{3}{4}$ to an improper fraction.

Step 1: $6 \times 4 = 24$
Step 2: $24 + 3 = 27$
Step 3: $6\frac{3}{4} = \frac{27}{4}$

(Normally, we do the conversion mentally and show only Step 3 (skipping Steps 1 & 2)).

Writing Fractions and Mixed Numbers in Words

Example 1   Write in words: $\frac{4}{5}$
Solution: Four fifths.

Example 2 Write in words: $7\frac{3}{10}$
Solution :   Seven **and** three tenths.

Example 3 Write in words: $568\frac{9}{100}$
Solution: Five hundred sixty-eight **and** nine hundredths.

Example 4 Write in words: $43\frac{2}{7}$
Solution   Forty-three **and** two sevenths.

Example 5: Write in words: $64\frac{102}{200}$
Solution: Sixty-four **and** one hundred two two-hundredths

A note about hyphenation

The author suggests that one should hyphenate only for clarity and to avoid confusion.. Unnecessary hyphenation sometimes leads to confusion in communication. The objective must be to communicate unambiguously.

Reduction of Fractions to Lowest terms

A fraction is in its lowest terms if the numerator and the denominator do not have any common factors (divisors) other than 1.

Examples of fractions in lowest terms: $\frac{2}{3}$, $\frac{5}{4}$ and $\frac{1}{6}$

Example   Reduce $\frac{14}{21}$ to lowest terms.

Solution $\frac{14}{21}$

$= \frac{2 \times 7}{3 \times 7}$ <-------You may skip this step.

$= \frac{2}{3}$ (Dividing out or canceling the common factor, 7.)

Equivalent Fractions

Equivalent fractions are fractions that have the same value but differ in the numerators and the denominators.

Examples: $\frac{1}{2}$, $\frac{2}{4}$, and $\frac{5}{10}$ are equivalent fractions. On reducing equivalent fractions to their lowest terms, they become identical. In the above examples, $\frac{1}{2}$ is in the lowest terms; $\frac{2}{4}$ and $\frac{5}{10}$ are higher terms of $\frac{1}{2}$

Forming Equivalent Fractions

Given a fraction **in** its lowest terms, to form an equivalent fraction, multiply both the numerator and the denominator by the same nonzero number.

Example 1 $\quad \frac{1}{2} = \frac{1 \times 3}{2 \times 3} = \frac{3}{6}$

Given a fraction **not in** its lowest terms, to form an equivalent fraction, multiply (to higher terms) or divide (to lower terms) both the numerator and the denominator by the same nonzero number.

Example 2 $\quad \frac{8}{12} = \frac{8 \times 2}{12 \times 2} = \frac{16}{24}$ <-----------Going to higher terms.

Example 3 $\quad \frac{8}{12} = \frac{8 \div 2}{12 \div 2} = \frac{4}{6} = \frac{2}{3}$ <-------Going to lower terms.

Example 4 Find the missing number: $\frac{8}{12} = \frac{?}{9}$

Solution

Step 1: Reduce to lowest terms.

$$\frac{8}{12} = \frac{2}{3} = \frac{?}{9}$$

Step 2: Form the required equivalent fraction.

$$\frac{2}{3} = \frac{2 \times 3}{3 \times 3} = \frac{6}{9}$$

The missing number is 6.

Note above (as in Example 4) that in some cases, if the given fraction is not in its lowest terms, first reduce to lowest terms before forming the equivalent fraction.

Example 5 Find x: $\quad \frac{8}{12} = \frac{x}{9}$

Use exactly the same procedure as in Example 4. In the future, we will learn how to find x using algebraic methods.

Answer: $x = 6$.

Some applications of equivalent fraction formation

Equivalent fraction formation has a number of applications:

1. In adding and subtracting unlike fractions. (see p.30-31)

2. In forming proportions and solving proportion problems.(see p.59)

3. In rationalizing denominators of fractions involving radicals and complex numbers.

Lesson 8 Exercises

A. (a) Convert the following to mixed numbers: 1. $\frac{22}{7}$; 2. $\frac{17}{5}$; 3. $\frac{79}{18}$.

  (b) Verify the answers in **(a)** by converting your answers back to improper fractions.

  (c) Change the following to improper fractions: .1. $8\frac{5}{6}$ 2. $9\frac{3}{8}$

Answers: (a). **1.** $3\frac{1}{7}$; **2.** $3\frac{2}{5}$; **3.** $4\frac{7}{18}$; (c): **1.** $\frac{53}{6}$; **2.** $\frac{75}{8}$

B Write in words: **1.** $\frac{5}{8}$; **2.** $63\frac{4}{100}$; **3.** $2\frac{5}{17}$ **4.** $\frac{24}{25}$; **5.** $9\frac{3}{5}$; **6.** 63,400.

Answers: 1. Five eighths;   **2.** Sixty-three and four hundredths;   **3.** Two and five seventeenths;
  4. Twenty-four twenty-fifths;   **5.** Nine and three fifths;   **6.** Sixty-three thousand, four hundred.

C. Write using numerals:: **1.** Eight and nine hundredths. **2.** Eight hundred nine.
  3. Eight hundred nine and nine tenths.

Answers: : **1.** $8\frac{9}{100}$; **2.** 809; **3.** $809\frac{9}{10}$

D. Reduce the following to lowest terms:

1. $\frac{21}{28}$ **2.** $\frac{24}{30}$ **3.** $\frac{20}{16}$ **4.** $\frac{512}{1024}$ **5.** $\frac{210}{1050}$

Answers: **1.** $\frac{3}{4}$; **2.** $\frac{4}{5}$; **3.** $\frac{5}{4}$); **4.** $\frac{1}{2}$; **5.** $\frac{1}{5}$

E 1. Convert $\frac{3}{4}$ to 28ths (That is, change $\frac{3}{4}$) to an equivalent fraction whose denominator is 28.)

2. Convert $\frac{3}{2}$ to 16ths.   **6.** Convert $\frac{11}{12}$ to 24ths.

3. Change $\frac{5}{6}$ to 36ths   **7.** Change $\frac{6}{12}$ to 20ths

4. Find the missing number: $\frac{15}{16} = \frac{?}{32}$ **8.** Find the missing number: $\frac{?}{56} = \frac{5}{8}$

5. Find x: $\frac{15}{16} = \frac{x}{32}$ **9.** Find x: $\frac{x}{56} = \frac{5}{8}$

Answers: Answers: **1.** $\frac{21}{28}$; **2.** $\frac{24}{16}$; **3.** $\frac{30}{36}$; **4.** 30; **5.** 30; **6.** $\frac{22}{24}$); **7.** $\frac{10}{20}$; **8.** 35; **9.** 35.

Lesson 9

Addition and Subtraction of Fractions; Addition of Mixed Numbers

Addition of Fractions

Like Fractions (Fractions with the same denominator).

Example: Add $\frac{5}{8} + \frac{4}{8}$

Solution   1. Add the numerators and keep the common denominator.
            2. Reduce the resulting fraction to lowest terms (if reducible) and change
               any improper fractions to mixed numbers.

$$\frac{5}{8} + \frac{4}{8} = \frac{9}{8}$$
$$= 1\frac{1}{8}$$

Unlike fractions (Fractions with different denominators).

Example 1 Add: $\frac{2}{7} + \frac{5}{9}$

In this case, we change the unlike fractions to like fractions, and then add the numerators

---

Format 1 (Sideways Format)

Step 1:   Find the LCD of the fractions.
     (The LCD of the fractions is the LCM of
     the denominators. See p.24)
     The LCD of the fractions is 63.

Step 2: Change the fractions to equivalent
     fractions with LCD of 63.

$$\frac{2}{7} = \frac{2 \times 9}{7 \times 9} = \frac{18}{63}$$

$$\frac{5}{9} = \frac{5 \times 7}{9 \times 7} = \frac{35}{63}$$

Step 3: Add the numerators and keep
     the LCD, 63.

$$\frac{18}{63} + \frac{35}{63} = \frac{53}{63}$$

Therefore, $\frac{2}{7} + \frac{5}{9} = \frac{53}{63}$.

Format 2 (Vertical Format)

$$\frac{2}{7} + \frac{5}{9}$$

Step 1: Find the LCD of the fractions
     The LCD = 63

Step 2: Form equivalent fractions using
     the same LCD of 63 to obtain

$$\frac{2 \times 9}{7 \times 9} + \frac{5 \times 7}{9 \times 7}$$
$$= \frac{18}{63} + \frac{35}{63}$$

Step 3: Add the numerators and keep
     the LCD, 63. to obtain

$$\frac{53}{63}$$

Therefore, $\frac{2}{7} + \frac{5}{9} = \frac{53}{63}$.

Example 2 Add: $\frac{5}{6} + \frac{1}{2} + \frac{3}{4}$

Solution

Method 1

Example 2 Add: $\frac{5}{6} + \frac{1}{2} + \frac{3}{4}$

Format 1 (Sideways Format)

Step 1: Find the LCD of the fractions
The LCD = 12

Step 2: Form equivalent fractions using
the same LCD of 12.

Then $\frac{5 \times 2}{6 \times 2} = \frac{10}{12}$
$\frac{1 \times 6}{2 \times 6} = \frac{6}{12}$
$\frac{3 \times 3}{4 \times 3} = \frac{9}{12}$

Step 3: Add the numerators and keep
the common denominator to obtain
$$\frac{10 + 6 + 9}{12} = \frac{25}{12} = 2\frac{1}{12}$$

Format 2 (Vertical Format)

$\frac{5}{6} + \frac{1}{2} + \frac{3}{4}$

Step 1: Find the LCD of the fractions
The LCD = 12

Step 2: Form equivalent fractions using
the same LCD of 12 to obtain

$= \frac{5 \times 2}{6 \times 2} + \frac{1 \times 6}{2 \times 6} + \frac{3 \times 3}{4 \times 3}$

$= \frac{10}{12} \quad + \quad \frac{6}{12} \quad + \quad \frac{9}{12}$

$= \frac{25}{12} = 2\frac{1}{12}$

Therefore $\frac{5}{6} + \frac{1}{2} + \frac{3}{4} = 2\frac{1}{12}$

**Method 2 Using prime factorization of the denominators followed by
making the denominators of the fractions the same**

Add: $\frac{5}{6} + \frac{1}{2} + \frac{3}{4}$

Step 1: Factor the denominators into primes.

$$\frac{5}{2 \times 3} + \frac{1}{2} + \frac{3}{2 \times 2}$$

Step 2: Make the denominators have the same factors by looking for missing factors and
multiplying accordingly; but note that whenever, you multiply the denominator by any factor,
you must also multiply the numerator by the same factor (in order to keep the value of
the original fraction unchanged).

$\frac{5 \times 2}{2 \times 3 \times 2} + \frac{1 \times 3 \times 2}{2 \times 3 \times 2} + \frac{3 \times 3}{2 \times 2 \times 3}$ <---- The denominator of the first fraction needs a "2."
The denominator of the second fraction needs
a "2" and a "3".

$= \frac{10}{12} \quad + \quad \frac{6}{12} \quad + \quad \frac{9}{12}$ The denominator of the third fraction needs a "3".

$= \frac{10 + 6 + 9}{12}$

$= \frac{25}{12}$

$= 2\frac{1}{12}$

Subtraction of Fractions

The process in the subtraction of fractions is exactly like that of the addition of fractions, except that, in
this case, we **subtract** the numerators (instead of adding) of the like fractions.

Addition of Mixed Numbers 3 2

Procedure: 1. Add the whole number parts.
2. Add the fractional parts and simplify.

Example Add: $2\frac{3}{4} + 3\frac{7}{8}$

Format 1 (Sideways Format)	**Format 2 (Vertical Format)**
$2\frac{3}{4} + 3\frac{7}{8}$	$2\frac{3}{4} + 3\frac{7}{8}$
$2\frac{3}{4} = 2\frac{6}{8}$	$= 2\frac{3\times 2}{4\times 2} + 3\frac{7}{8}$
$3\frac{7}{8} = 3\frac{7}{8}$	$= 2\frac{6}{8} + 3\frac{7}{8}$
Scrapwork	$= 5\frac{6+7}{8}$
$5\frac{13}{8} = 5 + 1\frac{5}{8}$ $6 + 7 = 13$	$= 5\frac{13}{8}$
$= 5 + 1\frac{5}{8}$	$= 5 + 1\frac{5}{8}$
$= 6\frac{5}{8}$	$= 6\frac{5}{8}$

Lesson 9 Exercises

A Add the following: **1.** $\frac{5}{7} + \frac{4}{7}$; **4.** $\frac{3}{10} + \frac{4}{15}$; **7.** $\frac{2}{45} + \frac{5}{18} + \frac{7}{30}$; **2.** $\frac{5}{12} + \frac{7}{9}$;

5. $\frac{3}{16} + \frac{5}{14} + \frac{7}{8}$; **3.** $\frac{1}{6} + \frac{3}{5}$; **6.** $\frac{8}{21} + \frac{1}{3} + \frac{9}{14}$

Answers 1. $1\frac{2}{7}$; **2.** $1\frac{7}{36}$; **3.** $\frac{23}{30}$; **4.** $\frac{17}{30}$; **5.** $1\frac{47}{112}$; **6.** $1\frac{5}{14}$; **7.** $\frac{5}{9}$

B Subtract: **1.** $\frac{9}{11} - \frac{5}{33}$; **2.** $\frac{3}{7} - \frac{2}{7}$; **3.** Subtract $\frac{1}{16}$ from $\frac{11}{48}$; **4.** $\frac{7}{8} - \frac{5}{12}$;

5. $\frac{13}{18} - \frac{5}{12}$; **6.** $\frac{5}{18} - \frac{2}{9}$; **7.** $\frac{17}{48} - \frac{5}{16}$

Answers **1.** $\frac{2}{3}$; **2.** $\frac{1}{7}$; **3.** $\frac{1}{6}$; **4.** $\frac{11}{24}$; **5.** $\frac{11}{36}$; **6.** $\frac{1}{18}$); **7.** $\frac{1}{24}$

C Add: **1.** $5\frac{2}{3} + 7\frac{3}{5}$; **3.** $6\frac{4}{5} + 7\frac{1}{5}$ **5.** $4\frac{5}{8} + 7\frac{3}{4}$; **7.** $4\frac{8}{11} + 2\frac{1}{5}$

2. $4\frac{5}{12} + 9\frac{5}{8}$; **4.** $3\frac{2}{5} + 6\frac{7}{15}$; **6.** $16\frac{3}{4} + 5\frac{1}{6}$; **8.** $2\frac{3}{4} + 3\frac{7}{8}$

Answers **1.** $13\frac{4}{15}$; **2.** $14\frac{1}{24}$; **3.** 14; **4.** $9\frac{13}{15}$ **5.** $12\frac{3}{8}$; **6.** $21\frac{11}{12}$; **7.** $6\frac{51}{55}$; **8.** $6\frac{5}{8}$.

Lesson 10   33
Comparison of Fractions; Subtraction of Mixed Numbers

Comparison of Fractions

Example 1   Which fraction is the larger:

$$\frac{2}{5} \text{ or } \frac{4}{7} \text{ ?}$$

There are a number of methods for comparing fractions. We will use a method which we can call "cross-multiplication of the numerators and denominators". We will multiply the numerator of one fraction by the denominator of the other fraction. The fraction whose numerator (in the multiplication) results in the larger product is the larger of the two fractions; and the fraction whose numerator results in the smaller product is the smaller of the two fractions.

$$7 \times 2 = 14 \qquad 5 \times 4 = 20$$

$$\frac{2}{5} \diagdown\!\!\!\!\diagup \frac{4}{7}$$

(**Note** that the **directions** of the **arrows** are from the denominators to the numerators.)

Since 20 is larger than 14, $\frac{4}{7}$ is the larger of the two fractions (but of course, $\frac{2}{5}$ is the smaller of the two fractions). **Note** that $\frac{2}{5} = \frac{14}{35}$

and $\frac{4}{7} = \frac{20}{35}$

Example 2 Which is the larger:

$$\frac{7}{16} \text{ or } \frac{3}{5} \text{ ?}$$

$$5 \times 7 = 35 \qquad 16 \times 3 = 48$$

$$\frac{7}{16} \diagdown\!\!\!\!\diagup \frac{3}{5}$$

Since 48 is larger than 35, $\frac{3}{5}$ is the larger fraction (It is the fraction whose **numerator** when multiplied by the denominator of the other fraction resulted in the larger 48).

By using the above method , we do not have to find the least common denominator. We however, produce equivalent fractions, in which the common denominator may or may not be the least common denominator.

The objective of the above method was to find a fast method of comparing fractions.

Note: In some cases, if the LCD of the fractions can easily be found, we can form equivalent fractions using the same LCD, and then compare the numerators.

Determining the Smallest or Largest Fraction   3 4

Example 3 Determine the smallest fraction.

$$\frac{2}{3}, \ \frac{3}{5}, \ \frac{5}{7}, \ \frac{10}{13}$$

We will compare two fractions at a time, and after any comparison, we will keep the smaller fraction (since we want to determine the smallest fraction) and compare it with the other fractions.

Step 1:

$$5 \times 2 = 10 \qquad 3 \times 3 = 9$$

$$\frac{2}{3} \diagup\!\!\!\!\diagdown \boxed{\frac{3}{5}}$$

(We "box "the fraction to keep for the next step)

$\frac{3}{5}$ is the smaller (since 9 is smaller than 10) and we keep $\frac{3}{5}$ for the next step.

Step 2:

$$7 \times 3 = 21 \qquad 5 \times 5 = 25$$

$$\boxed{\frac{3}{5}} \diagup\!\!\!\!\diagdown \frac{5}{7}$$

(bringing down the next fraction, $\frac{5}{7}$, for comparison)

$\frac{3}{5}$ is the smaller (since 21 is smaller than 25)

Step 3:

$$13 \times 3 = 39 \qquad 5 \times 10 = 50$$

$$\boxed{\frac{3}{5}} \diagup\!\!\!\!\diagdown \frac{10}{13}$$

$\frac{3}{5}$ is the smaller (since 39 is smaller than 50)

The smallest fraction is $\frac{3}{5}$.

Note, above, that after every comparison, we 'box" the fraction to keep for the next comparison.

Lesson 10: Comparison of Fractions and Subtraction of Mixed Numbers

Example 4   For the last example, we will determine the largest fraction

$$\frac{2}{3}, \ \frac{3}{5}, \ \frac{5}{7}, \ \frac{10}{13}.$$

Bear in mind that in this example, we will always keep the larger fraction after every comparison (and compare it with the other fractions) since we want to determine the largest fraction.

Step 1: $5 \times 2 = 10$ $3 \times 3 = 9$

(We "box" the fraction to keep for the next step)

$\frac{2}{3}$ is the larger

Step 2: $7 \times 2 = 14$ $3 \times 5 = 15$

$\frac{5}{7}$ is the larger

Step 3: $13 \times 5 = 65$ $7 \times 10 = 70$

$\frac{10}{13}$ is the larger

The largest fraction is $\frac{10}{13}$

Arranging Fractions in Descending (decreasing) or Ascending (increasing) Order

To arrange in descending order: Determine the largest fraction first, remove it from the list and write it down; then determine the largest of the remaining fractions, and similarly remove it from the list and write it after the largest fraction. Repeat the process until only one fraction remains, and write this fraction last. (It is the smallest.)

To arrange in ascending order: The process here is similar to the descending order case, but in this case, we determine the smallest fraction first, followed by the next smallest and so on. The last remaining fraction becomes the largest, and it is written last.

Subtraction of Mixed Numbers

Case 1: We do **not** have to borrow

    Example 1   Subtract $9\frac{5}{8} - 2\frac{1}{4}$

In this problem, because the fractional part being subtracted is smaller than the fractional part from which we are subtracting, we do **not** have to borrow from the whole number part; (Of course, if you borrow, you can still arrive at the correct answer, at the expense of extra simplification work).

$$9\frac{5}{8} = 9\frac{5}{8}$$
$$-2\frac{1}{4} = -2\frac{2}{8}$$
$$\overline{\qquad 7\frac{3}{8}}$$

Case 2: We have to borrow

    Example 2   **Subtract:**

$$14\frac{3}{7} - 4\frac{5}{8}$$

A quick check (see p.33) indicates that $\frac{5}{8}$ is larger than $\frac{3}{7}$ $(8 \times 3 = 24; \ 7 \times 5 = 35)$

Therefore we borrow 1 from the 14. (**The fractional part of the subtracted mixed number is larger than the fractional part of the mixed number from which we are subtracting.**)

    Then, $14\frac{3}{7} = 13\frac{7}{7} + \frac{3}{7} = 13\frac{10}{7}$

    Now, we subtract: $13\frac{10}{7} - 4\frac{5}{8}$

LCD = 56
$$13\frac{10}{7} = 13\frac{80}{56}$$
$$-4\frac{5}{8} = -4\frac{35}{56}$$
$$\overline{\qquad 9\frac{80-35}{56} = 9\frac{45}{56}}$$

Case 3: We have to borrow

    Example 3 $12 - 4\frac{2}{5}$

$$\downarrow \qquad \downarrow$$
$$11\frac{5}{5} - 4\frac{2}{5} \quad (\text{we borrow 1 from the 12 and change it to } \tfrac{5}{5})$$

$$11\frac{5}{5}$$
$$-4\frac{2}{5}$$
$$\overline{\quad 7\frac{3}{5}}$$

(To check, add $4\frac{2}{5}$ and $7\frac{3}{5}$ to obtain 12)

Case 4: We do **not** have to borrow.   **Example 4** Subtract $25\frac{4}{7} - 10$

$$25\frac{4}{7}$$
$$-10$$
$$\overline{\quad 15\frac{4}{7}}$$

In this case we do not have to borrow, since the number to be subtracted is a whole number (i.e., does not have a fractional part)

Lesson 10 Exercises

A. Determine the larger fraction:

1. $\frac{2}{7}$ or $\frac{4}{15}$;   2. $\frac{5}{8}$ or $\frac{7}{11}$;   3. $\frac{7}{8}$ or $\frac{8}{11}$;   4. $\frac{3}{7}$ or $\frac{4}{9}$;   5. $\frac{5}{6}$ or $\frac{9}{11}$

Answers 1. $\frac{2}{7}$; 2. $\frac{7}{11}$; 3. $\frac{7}{8}$; 4. $\frac{4}{9}$; 5. $\frac{5}{6}$

B Determine the smallest fraction: 1. $\frac{3}{5}$, $\frac{4}{7}$, $\frac{9}{13}$, $\frac{8}{11}$;   2. $\frac{3}{4}$, $\frac{2}{3}$, $\frac{11}{15}$, $\frac{19}{30}$.

Answers 1. $\frac{4}{7}$; 2. $\frac{19}{30}$

C Determine the largest fraction:

1. $\frac{3}{5}$, $\frac{4}{7}$, $\frac{9}{13}$, $\frac{8}{11}$; 2. $\frac{3}{4}$, $\frac{2}{3}$, $\frac{11}{15}$, $\frac{19}{30}$; 3. $\frac{3}{8}$, $\frac{5}{16}$, $\frac{1}{4}$

Answers   1. $\frac{8}{11}$; 2. $\frac{3}{4}$; 3. $\frac{3}{8}$

D   Subtract the following:

1. $6\frac{5}{12} - 2\frac{1}{4}$;   5. $3\frac{1}{4} - 1\frac{5}{8}$
2. $15\frac{2}{9} - 6\frac{5}{18}$;   6. $51\frac{4}{9} - 7\frac{1}{6}$
3. $48 - 7\frac{3}{8}$ 7. $89 - 5\frac{3}{7}$
4. $76\frac{3}{4} - 23$ 8. $43\frac{2}{5} - 15$

Answers   1. $4\frac{1}{6}$; 2. $8\frac{17}{18}$;   3. $40\frac{5}{8}$; 4. $53\frac{3}{4}$; 5. $1\frac{5}{8}$;   6. $44\frac{5}{18}$;   7. $83\frac{4}{7}$;   8. $28\frac{2}{5}$

E
Arrange the fractions in the above, B, and C in descending order; and in ascending order.

Word Problems

F

1. What must be added to 2 to make 5.?

2. What must be added to $2\frac{1}{4}$ to make $6\frac{5}{12}$.?

3. Maria needs $6\frac{5}{12}$ pounds of sugar to make some cakes, At present, she has only $2\frac{1}{4}$ pounds of sugar. How many more pounds of sugar does she need to have in order to make the cakes?

4. Mary had $6\frac{5}{12}$ pints of milk in her refrigerator this morning. In the afternoon she used $2\frac{1}{4}$ pints of this milk. How much milk is left in the refrigerator?

Answers: **1.** 3; **2.** $4\frac{1}{6}$ **3.** $4\frac{1}{6}$ pounds   **4.** $4\frac{1}{6}$ pints

Lesson 11   3 8

Multiplication and Division: of Fractions and Mixed Numbers

Multiplication of Fractions

Procedure:

Step 1: Divide out (cancel) **any** common factors in the numerators and the denominators.

Step 2: Multiply the remaining numerators, and multiply the remaining denominators.

Example 1   Multiply $\frac{3}{5} \times \frac{5}{16} \times \frac{2}{9}$

Solution:

Step 1:   Canceling the common factors, we obtain

$$\overset{1}{\cancel{3}} \times \overset{1}{\cancel{5}} \times \overset{1}{\cancel{2}}$$
$$\frac{}{\underset{1}{\cancel{5}}} \times \frac{}{\underset{8}{\cancel{16}}} \times \frac{}{\underset{3}{\cancel{9}}}$$

Step 2: Multiply the remaining factors.

$$\frac{1 \times 1 \times 1}{1 \times 8 \times 3}$$
$$= \frac{1}{24}$$

Example 2   Multiply: $\frac{2}{3} \times \frac{2}{5}$

Solution: There are no common factors to cancel.

Therefore, we multiply the numerators, and then multiply the denominators.

$$\frac{2}{3} \times \frac{2}{5}$$
$$= \frac{2 \times 2}{3 \times 5}$$
$$= \frac{4}{15}$$

Division of Fractions

Reciprocals: The reciprocal of a real number A is $\frac{1}{A}$

Examples: The reciprocal of $\frac{2}{3}$ is $\frac{3}{2}$. The reciprocal of 4 is $\frac{1}{4}$. The reciprocal of $\frac{1}{4}$ is 4. Thus, to find the reciprocal of a number, invert the number (or interchange the numerator and the denominator)
 The reciprocal of number is also known as the multiplicative inverse of that number. The product of a number and its reciprocal is 1. Example $\frac{1}{4} \times \frac{4}{1} = 1$

Example 1 $\frac{4}{9} \div \frac{2}{15}$ ($\frac{4}{9}$ is the dividend; $\frac{2}{15}$ is the divisor)

Procedure: Invert the divisor and multiply. (or multiply the dividend by the reciprocal of the divisor)

$$\text{Then } \frac{4}{9} \div \frac{2}{15}$$
$$= \frac{4}{9} \times \frac{15}{2}$$

$$= \frac{\overset{2}{\cancel{4}}}{\underset{3}{\cancel{6}}} \times \frac{\overset{5}{\cancel{15}}}{\underset{1}{\cancel{2}}}$$

$$= \frac{2 \times 5}{3 \times 1}$$

$$= \frac{10}{3}$$

$$= 3\frac{1}{3}$$

(Always, cancel any common factors in the numerators and the denominators before multiplying)

Division of Complex Fractions

Example   Simplify: $\dfrac{\frac{3}{4}}{\frac{16}{27}}$

Solution: $\dfrac{\frac{3}{4}}{\frac{16}{27}} = \frac{3}{4} \div \frac{16}{27}$

$$= \frac{3}{4} \times \frac{27}{16} \qquad \text{(inverting the divisor and multiplying)}$$

$$= \frac{81}{64}$$

$$= 1\frac{17}{64}$$

Multiplication of Mixed Numbers

Procedure: Step 1:   Change each mixed number to an improper fraction.
Step 2:   Multiply the resulting fractions.

Example $2\frac{4}{5} \times 2\frac{1}{7}$

Step 1: $= \frac{14}{5} \times \frac{15}{7}$

Step 2: $= \frac{\overset{2}{\cancel{14}}}{\underset{1}{\cancel{5}}} \times \frac{\overset{3}{\cancel{15}}}{\underset{1}{\cancel{7}}}$

$$= 6$$

Division of Mixed Numbers

Example $5\frac{2}{3} \div 4\frac{1}{2}$ ($4\frac{1}{2}$ is the divisor)

Procedure: Change each mixed number to an improper fraction and divide.

$$5\frac{2}{3} \div 4\frac{1}{2}$$

Step 1: $= \frac{17}{3} \div \frac{9}{2}$

Step 2: $= \frac{17}{3} \times \frac{2}{9}$ (inverting the divisor and multiplying)

$$= \frac{34}{27} = 1\frac{7}{27}$$

Lesson 11 Exercises

A. Multiply:: **1.** $\frac{21}{32} \times \frac{8}{15}$; **2.** $\frac{3}{7} \times \frac{35}{36}$; **3.** $\frac{4}{5} \times \frac{6}{7}$;; **4.** $12 \times \frac{2}{3}$; **5.** $\frac{9}{8} \times \frac{4}{45}$

6. $\frac{3}{5} \times \frac{5}{16} \times \frac{2}{9}$

Answers **1.** $\frac{7}{20}$; **2.** $\frac{5}{12}$; **3.** $\frac{24}{35}$; **4.** 8; **5.** $\frac{1}{10}$; **6.** $\frac{1}{24}$

B. Divide: **1.** $\frac{4}{5} \div \frac{3}{7}$; **2.** $\frac{3}{8} \div \frac{5}{9}$; **3.** $\frac{5}{12} \div \frac{5}{4}$; **4.** $\frac{3}{4} \div \frac{4}{3}$; **5.** $\frac{7}{15} \div \frac{21}{20}$; **6.** $\frac{7}{8} \div \frac{3}{16}$

Answers: **1.** $1\frac{13}{15}$; **2.** $\frac{27}{40}$; **3.** $\frac{1}{3}$; **4.** $\frac{9}{16}$; **5.** $\frac{4}{9}$; **6.** $4\frac{2}{3}$

C. Multiply: (a) $3\frac{3}{4} \times 6\frac{2}{3}$; (b) $1\frac{2}{5} \times 1\frac{1}{2}$; (c) $1\frac{7}{9} \times \frac{3}{4}$

(d) $2\frac{1}{3} \times 5\frac{1}{4}$; (e) $3 \times 4\frac{1}{2}$; (f) $\frac{4}{5} \times 2\frac{1}{7}$; (g) $3\frac{3}{7} \times \frac{3}{4} \times 14$

Answers: (a) 25; (b) $2\frac{1}{10}$; (c) $1\frac{1}{3}$; (d) $12\frac{1}{4}$; (e) $13\frac{1}{2}$; (f) $1\frac{5}{7}$; (g) 36

D. Divide: (a) $5\frac{1}{2} \div 3\frac{1}{7}$; (d) $2\frac{2}{3} \div 5\frac{1}{3}$; (g) $7 \div 4\frac{2}{3}$

(b) $12\frac{3}{4} \div 2\frac{1}{8}$; (e) $1\frac{7}{8} \div 2\frac{3}{16}$; (h) $3\frac{1}{2} \div 1\frac{4}{5}$

(c) $14\frac{1}{4} \div 3$; (f) $2\frac{1}{8} \div 12\frac{3}{4}$

Answers (a) $1\frac{3}{4}$; (b) 6; (c) $4\frac{3}{4}$; (d) $\frac{1}{2}$; (e) $\frac{6}{7}$; (f) $\frac{1}{6}$; (g) $1\frac{1}{2}$; (h) $1\frac{17}{18}$

Test # 1   Student's Self-Test (**Always**, **Test yourself before you are tested**)   4 1
Attempt all questions on clean sheets of paper, **Do not write in the book** Show all necessary work.

---

1. Write, using numerals:

 Seven million, five hundred eight thousand, nine hundred eight.

2. Subtract: 8,765,210 – 64,434

---

3. Round-off **965.467** to the following places:

(a). To the nearest hundredth.

(b)   To the nearest tenth.

4. Write the following in words:

(a). 72,935,307;   (b) $586\frac{2}{3}$.

---

5. Estimate the following:

(a) 674,000 × 524;

(b) 7,457 + 356,238 + 7,393.

6. Find the prime factorization of the following:

 (a)   160 ;   (b)   540.

---

7. Find the LCM of the following:
 (a) 18, 12 and 16;   (b) 2, 7 and 9.
(c) Find the smallest whole number that is divisible by each of 14, 42, and 56.

8. Convert the following to mixed numbers:

(a). $\frac{23}{7}$;   (b). $\frac{18}{5}$;   (c) $\frac{89}{18}$.

---

9. Change the following to improper fractions.

(a) $7\frac{5}{6}$;   (b) $8\frac{3}{8}$

Verify the answers in by converting your answers back to mixed numbers.

10. Write in words:

(a) $83\frac{4}{400}$; (b) $\frac{34}{35}$; (c) 63,400.

---

11. Write using numerals:

(a) Seven and three hundredths.

(b). Four hundred four and four tenths.

12. Reduce the following to lowest terms:

(a) $\frac{18}{30}$;   (b) $\frac{256}{512}$;   (c) $\frac{340}{1700}$

---

13.   Convert $\frac{3}{4}$ to 76ths

14. Convert $\frac{3}{12}$ to 28ths

Lesson 11: Multiplication and Division: of Fractions and Mixed Numbers

15. Find x: $\frac{x}{24} = \frac{5}{8}$

16. Add: $\frac{7}{10} + \frac{3}{15}$

17. Add: $\frac{2}{15} + \frac{5}{18} + \frac{7}{30}$

18. Add: $6\frac{5}{12} + 9\frac{5}{8}$

19. Determine the smallest fraction:

$$\frac{3}{7}, \quad \frac{4}{9}, \quad \frac{9}{13}, \quad \frac{8}{11}$$

20. Subtract: $6\frac{5}{12} - 2\frac{1}{4}$

21. Subtract $98 - 6\frac{2}{9}$

22. Multiply: $\frac{7}{32} \times \frac{28}{15}$

23. Divide: $\frac{8}{15} \div \frac{21}{20}$

24. $4\frac{4}{7} \times \frac{3}{4} \times 14$

25. Divide: $14\frac{3}{4} \div 1\frac{1}{3}$

Bonus: Subtract: $86\frac{4}{7} - 38$

Answers: **1.** 7,508,908; **2.** 8,700,776; **3.** (a). 965.47, (b) 965.5;
4. (a) Seventy-two million, nine hundred thirty-five thousand, three hundred seven;
(b) Five hundred eighty-six and two thirds; **5.** (a) 350,000,000; (b) 414,000;
6. (a) $2^5 \times 5$; (b) $2^2 \times 3^3 \times 5$; **7.** (a) 144; (b) 126; (c) 168; **8.** (a) $3\frac{2}{7}$; (b) $3\frac{3}{5}$; (c) $4\frac{17}{18}$;
9. (a) $\frac{47}{6}$; (b) $\frac{67}{8}$; **10.** (a) Eighty-three and four four-hundredths; (b) Thirty-four thirty-fifths;
(c) Sixty-three thousand, four hundred; **11.** (a) $7\frac{3}{100}$ (b) $404\frac{4}{10}$;
12 (a) $\frac{3}{5}$; (b) $\frac{1}{2}$; (c) $\frac{1}{5}$; **13.** $\frac{57}{76}$; **14.** $\frac{7}{28}$; **15.** 15; **16.** $\frac{9}{10}$; **17.** $\frac{29}{45}$; **18.** $16\frac{1}{24}$;
19. $\frac{3}{7}$; **20.** $4\frac{1}{6}$; **21.** $91\frac{7}{9}$; **22.** $\frac{49}{120}$; **23.** $\frac{32}{63}$; **24.** 48; **25.** $11\frac{1}{16}$; **Bonus:** $48\frac{4}{7}$

CHAPTER 4
DECIMALS

Lesson 12: **Definitions, Addition and Subtraction; Multiplication and Division by Powers of 10, Multiplication of Decimals; Comparison of Decimals ; converting decimals to fractions**

Lesson 13: **Complex Decimals; Dividing Decimals; Converting Fractions to Decimals**

Lesson 12

Definitions, Addition and Subtraction of Decimals; Multiplication and Division by Powers of 10; Multiplication of Decimals; Comparison of Decimals ; Converting Decimals to Fractions

Definitions
Powers of 10

Examples:
$$10^1 = 10$$
$$10^2 = 100$$
$$10^3 = 1,000$$
$$10^4 = 10,000$$

A **decimal fraction** is a fraction whose denominator is a power of 10.

Examples are $\frac{3}{10}$, $\frac{5}{100}$, and $\frac{23}{1000}$ . .

A **decimal** is the symbol for an equivalent decimal fraction.

Examples: $\frac{3}{10} = .3$ ($\frac{3}{10}$ is the decimal fraction; .3 is the decimal)

$\frac{5}{100} = .05$

$\frac{23}{10000} = .0023$

Some Common Fractions, Decimal Fractions, Decimals and Names

Common Fraction	Decimal Fraction	Decimal	Name
$\frac{1}{2}$	$\frac{5}{10}$	.5	Five tenths
$\frac{1}{4}$	$\frac{25}{100}$	.25	Twenty-five hundredths
$\frac{61}{100}$	$\frac{61}{100}$	.61	Sixty-one hundredths
$\frac{490}{1000}$	$\frac{490}{1000}$	.490	Four hundred ninety thousandths

Number of decimal places or decimal digits: The number of decimal places is the number of digits to the right of the decimal point.

Examples: (a) 0.62 has two decimal places;   (b) 46.3 has one decimal place;
(c) 53.005 has three decimal places.

Also note for example that $0.62 = .62$ (The zero preceding the decimal point on the left-hand side is used to remind the reader that a decimal point follows.)

Equivalent decimals   Equivalent decimals are decimals that have the same value but differ in the number of decimal places.

(see also p. 28 for a similar definition of equivalent fractions)

Example: .5, .50, and .500 are equivalent decimals, but note that .5 is the simplest form

(in much the same way as $\frac{1}{2}$ is the lowest terms of the equivalent fractions $\frac{2}{4}$ and $\frac{5}{10}$).

Similarly, 4.2 and 4.20 and 4.2000 are equivalent decimals

Addition of Decimals

Example   Add: $13.23 + 7.1 + 9$

Procedure: Indicate the decimal point explicitly in all the whole numbers.
Arrange the numbers vertically, while lining up the decimal points.
Note: $9 = 9$.

Then we obtain:

```
                    ——— decimal points in line
                 ↓
         13 .23    (Add as is done in whole
          7 .10      number addition)
          9.00
         29.33
```

Subtraction of Decimals

Example: Subtract: $15 - 2.43$

Indicating the decimal point in the 15 and arranging the numbers vertically, while lining up the decimal points, we obtain:

```
    15.00<-------------------------(attach zeros)
    −2.43
    12.57
```

Multiplying decimals or whole numbers by a power of 10

Procedure: Move the decimal point to the right as many places as there are zeros in the power of 10.   Examples of powers of 10

Example 1 $3.45 \times 10 = 34.5$ $10^1 = 10$

Example 2 $3.45 \times 10^2 = 345.$ $10^2 = 100$

Example 3 $3.45 \times 10^3 = 3450.$ $10^3 = 1000$

Example 4 $23 \times 10000 = 230000.$ or 230000
(In Examples 3 and 4, we wrote zeros to hold places)

Dividing decimals or whole numbers by a power of 10   4 5

Procedure: Move the decimal to the left as many places as there are zeros in the power of 10.

Example 1 $53.4 \div 10 = 5.34$

Example 2 $53.4 \div 100 = .534$

Example 3 $53.4 \div 1000 = .0534$

Example 4 $534 \div 10 = 53.4$

Example 5 $\dfrac{672.89}{100} = 6.7289$

Example 6 $\dfrac{89675}{100} = 896.75$

Multiplication of a decimal by a decimal or by a whole number

Step 1: Ignore the decimal points and multiply the whole numbers.

Step 2: Count the total number of decimal places (decimal digits) in the decimals being multiplied.

Step 3: Insert the decimal point in the product from Step 1, so that the number of decimal places equals the total number of decimal places counted in Step 2. (We may have to write zeros to hold places in some cases.)

Example Multiply 4.21 by 3.5   **Note**: $4.21 \times 3.5 = \dfrac{421}{100} \times \dfrac{35}{10} = \dfrac{14735}{1000} = 14.735$

```
    4.2 1
  × 3.5 __        (The total number of decimal places is 3)
    2105
    1263
   ─────
   14.735
```

Comparison of Decimals

Example   Determine which decimal is the smallest: .234, .098, .0725

There are a number of methods for comparing decimals.
First, we will use a method which we can call " comparison of place-value digits", and then we will discuss another method.

Method 1

Step 1:   Arrange the numbers vertically while lining up the decimal points **as if** one were to add the numbers.

.2340   (we can attach zeros)
..0980
.0725

Step 2:   We will compare the digits from left to right (tenths, hundredths, and thousandths places etc).
In the tenths' column, 2 is the largest, and therefore .2340 cannot be the smallest.
We cross out .2340 or ignore it.

Step 3: Compare   .0980
.0725
Since the tenths' places are the same, we go to the next column
(the hundredths' place) and there, 7 is smaller than 9 and therefore .
.0725 is smaller than .098 . Hence .0725 is the smallest decimal.
(of course, .2340 is the largest decimal)

Method 2   46

Multiply each decimal by a power of 10, so that all the decimals become whole numbers.
In multiplying by a power of 10, move the decimal point to the right as many places as there are zeros in the power of 10. The power of 10 to use is based on the decimal with the most number of decimal places.

In the above example, .0725 has the most number of decimal places, four decimal places.

Proceeding,

$$.2340 \times 10000 = 2340$$
$$..0980 \times 10000 = 980$$
$$.0725 \times 10000 = 725$$

Since .0725 gave the smallest whole number 725, .0725 is the smallest decimal.

Arranging Decimals in Descending (decreasing) or Ascending (increasing) Order

To arrange in descending order: Determine the largest decimal first, remove it from the list and write it down; then determine the largest of the remaining decimals, and similarly remove it from the list and write it after the largest decimal. Repeat the process until only one decimal remains, and write this decimal last. (It is the smallest decimal.)

To arrange in ascending order: The process here is similar to the descending order case, above, but in this case, we determine the smallest decimal first, write it down, followed by the next smallest, and so on. The last remaining decimal becomes the largest, and it is written last.

Converting a Decimal to a Fraction

Example   Change .96 to a fraction in its lowest terms.

Step 1: Write .96 as a decimal fraction .

$$.96 = \frac{96}{100}$$

(The number of zeros in the denominator equals the number of decimal places in the decimal)

Step 2:   Reduce the fraction to its lowest terms.

Then, $\dfrac{\overset{\overset{24}{\cancel{48}}}{\cancel{96}}}{\underset{\underset{25}{\cancel{50}}}{\cancel{100}}} = \dfrac{24}{25}$

$$\therefore \quad .96 = \frac{24}{25}$$

Lesson 12 Exercises

A. Write as decimals: (a) $\frac{36}{1000}$; (b) $\frac{40}{100}$; (c) $\frac{9}{100}$

Answers (a) .036 ; (b) .40 ; (c) .0 9

B. Add the following:: (a) 13.456 + 4.03 + 9 + .463;   (d) 19 + .018 + 298

(b) 43.050 + 73.4 + 7 + .0304 + 29 ;   (e) .5 + .54 + .6407

(c) 7.8 + 47 + 5.605 + .006; (f) What must be added to 3.48 to make 9?.

Answers (a). 26.949; (b) 152.4804 ; (c) 60.411 ; (d) 317.018 ; (e) 1.6807 ; (f) 5.52

C. Subtract: **(a)** 60.54 - 3.679;   **(b)** 28 - 16.603; **(c)** 46.07 - 9; **(d)** 8.23 - 5.025

(e) Subtract 37.5 from 76.4; **(f)** From 8, subtract 5.96; **(g)** Subtract .65 from 1.1.

Answers **(a)** 56.861; **(b)** 11.397 ; **(c)** 37.07; (d) 3.205; (e) 38.9 ; (f) 2.04 ; (g) .45

D. Multiply: (a) 65.8 × 100; (b) 4.306×10^2; (c) .006 × 10000; (d) 26 × 1000

Answers (a) 6580. or 6580; (b) 430.6 ; (c) 60. or 60; (d) 26,000.

E. Divide:(a) 74.68 ÷ 100; (b) $34.07 \div 10^4$; (c) 68953 ÷ 100 ; (d) $\frac{245.6}{10000}$

Answers (a) .7468 ; (b) .003407 ; (c) 689.53 (d) .02456

F. Which decimal is the larger? (a) .0764 or .04685 (b) 3.245 or .986753

(c) 4 or .8954

Answers (a) .0764 ; (b) 3.245 ; (c) 4

G. Determine the smallest decimal: (a) .067, .0096, .12402; (b) 3.008, .094, .4509

(c) 1.007, .753, .087

Answers (a) .0096; (b) .094; (c) .087

H. Use the inequality symbol " > " to compare the following:

(a) .0764 or .04685; (b) 3.245 or .986753; (c) 4 or .8954

I. Order the following using the symbol " < ": (a) .067, .0096, .12402;

b) 3.008, .094, .4509; (c) 1.007, .753, .087

J. 1. Arrange the decimals in **G** and **H** (above) in descending order..

2. Arrange the decimals in **G** and **H** (above) in ascending order..

K. Change each of the following to a fraction in its lowest terms:

(a) .64 ; (b) 6.4 ; (c) 0.036 ; (d) 1.36 ; (e) .63; (f) 0.0825

Answers (a) $\frac{16}{25}$) ; (b) $\frac{32}{5}$; (c) $\frac{9}{250}$; (d) $\frac{34}{25}$; (e) $\frac{63}{100}$; (f) $\frac{33}{400}$

Lesson 13

Complex Decimals; Dividing Decimals; Converting Fractions to Decimals

Complex Decimals

A complex decimal consists of a decimal part and a common fraction part, Examples are
$.11\frac{5}{7}$ and $.117\frac{1}{7}$.

Using complex decimals allows one to write the exact quotient of a division problem in which no round-off place for the quotient is specified. Example: $.82$ divided by $7 = .11\frac{5}{7}$ or $.117\frac{1}{7}$

In a complex decimal, the **number of decimal places** equals the number of digits to the right of the decimal point, **ignoring** the fractional part.

Examples **1.** $.8\frac{2}{3}$ has **one** decimal place. **2.** $.14\frac{1}{3}$ has **two** decimal places.
3. $.3\frac{1}{3}$ has **one** decimal place. **4.** $.33\frac{1}{3}$ has **two** decimal places.
5. $.166\frac{2}{3}$ has **three** decimal places.

Converting a Complex Decimal to a Fraction

Example 1 Convert $.8\frac{2}{3}$ to a fraction in its lowest terms.

$$.8\frac{2}{3} = \frac{8\frac{2}{3}}{10} \qquad (.8\frac{2}{3} \text{ has one decimal place; and note, for example, that } .8 = \frac{8}{10})$$

$$= \frac{26}{30} \qquad \left(\frac{26}{3} \div \frac{10}{1} = \frac{26}{3} \times \frac{1}{10} = \frac{26}{30}\right)$$

$$= \frac{13}{15}$$

Example 2 Convert $.14\frac{1}{3}$ to a fraction in its lowest terms.

$$.14\frac{1}{3} = \frac{14\frac{1}{3}}{100} \qquad (.14\frac{1}{3} \text{ has two decimal places})$$

$$= \frac{43}{300}$$

How to divide a Decimal by a Whole number   49

Procedure Align the decimal point vertically in the quotient and divide as in whole number division. You may attach zeros (one at a time) if needed to continue the division process.

Case 1: Round-off place for the quotient is specified

Example 1 Divide .82 by 7 and round-off quotient to the nearest hundredth.

$$\begin{array}{r} .117 \approx \mathbf{.12} \qquad \text{(We divided to the thousandths place before rounding-off)} \\ 7\overline{).82} \\ \underline{-7} \\ 12 \\ \underline{-7} \\ 50 \\ \underline{-49} \\ 1 \end{array}$$

Conclusion: To the nearest hundredth, $\dfrac{.82}{7} \approx \mathbf{.12}$

Case 2: No Round-off place for the quotient is specified

In this case, we will leave the answer as a complex decimal, since this will be equivalent to the exact quotient. Note that when we round-off, the quotient obtained is an approximation.

Example 2 Divide .82 by 7.   (Round-off-place is **not** specified)

$$\begin{array}{r} .11\tfrac{5}{7} \quad \text{(dividing to \textbf{two} decimal places)} \\ 7\overline{).82} \\ \underline{-7} \\ 12 \\ \underline{-7} \\ 5 \end{array} \qquad \text{or} \qquad \begin{array}{r} .117\tfrac{1}{7} \quad \text{((dividing to \textbf{three} decimal places)} \\ 7\overline{).820} \\ \underline{-7} \\ 12 \\ \underline{-7} \\ 50 \\ -\underline{49} \\ 1 \end{array}$$

To check: Let us convert $.11\tfrac{5}{7}$ and $.117\tfrac{1}{7}$ back to the original division problem.

$$.11\tfrac{5}{7} = \frac{82}{700} \qquad\qquad .117\tfrac{1}{7} = \frac{820}{7000} = \frac{82}{700}$$
$$= \frac{.82}{7} \qquad\qquad = \frac{.82}{7}$$

Therefore, $.11\tfrac{5}{7}$ and $.117\tfrac{1}{7}$ are equivalent.

We will make the following agreement: In computations, unless otherwise instructed, we will leave the decimal as a complex decimal of two decimal places. This will produce sufficient decimal places for most applications and yet provide additional information for either obtaining more decimal places or for rounding-off.

Lesson 13: Complex Decimals; Dividing Decimals; Converting Fractions to Decimals

To **round-off** a complex decimal by **dropping only** the **fractional part,** add 1 to the
digit immediately to the left of the fractional part if the numerator of the fractional part
is greater than or equal to half of the denominator; but if the numerator is less than half
of the denominator, the digit to the left of the fractional part remains unchanged.

Examples (a): $.11\frac{5}{7}$ rounded-off to the nearest hundredth becomes **.12** $\left(5 > \frac{7}{2}\right)$

(b) $.117\frac{1}{7}$ rounded-off to the nearest thousandth becomes **.117** $\left(1 < \frac{7}{2}\right)$

Note however that $.117\frac{1}{7}$ rounded-off to the nearest hundredth becomes **.12**

Dividing a Number by a Decimal
Procedure

Step 1: Make the divisor a whole number by moving the decimal point to the right as many places
as the number of decimal places in the divisor, and at the same time, move the decimal point
in the dividend to the right as many places as the decimal point was moved in the divisor.
Note however, that for a divisor such as .20 or .200, we need to move the point only one
place to the right to make it a whole number, since 2 is equivalent to 2.0 or 2.00.

Step 2: Rewrite the dividend and the divisor and divide as by a whole number.

Example 1 Divide 1.23 by .8 and round-off quotient to the nearest hundredth.

Step 1: $.8\overline{)1.23} = .8\overline{)1.23}$ (moving the decimal point 1 place to the
right in the divisor and dividend)

Step 2: $8\overline{)12.3}$

Step 3:
$$\begin{array}{r} 1.537 \approx \mathbf{1.54} \\ 8\overline{)12.300} \\ \underline{-8} \\ 43 \\ \underline{-40} \\ 30 \\ \underline{-24} \\ 60 \\ \underline{-56} \\ 4 \end{array}$$

Conclusion: To the nearest hundredth, $\frac{1.23}{.8} \approx 1.54$

Example 2   Divide .86 by .20

Step 1: $.20\overline{).86} = .20\overline{).86}$ (We need to move the decimal point only one place to make
it a whole number)

Step 2: $2\overline{)8.6}$ quotient 4.3

Writing Fractions as Decimals

When a fraction (a rational number) is converted to a decimal, the decimal is either terminating or repeating

Examples of terminating decimals: **1.** $\frac{1}{4} = .25$; **2.** $\frac{1}{8} = .125$

Examples of repeating decimals: **1.** $\frac{1}{3} = .\overline{3}$ or .33...; **2.** $\frac{3}{11} = .\overline{27}$ or .2727...

Converting a Fraction to a Decimal
Case 1: Round-off place of the decimal is specified

Example   Change $\frac{3}{7}$ to a decimal, rounding off the quotient (answer) to the nearest
hundredth (two decimal places).

Procedure: Using long division, divide the numerator by the denominator, attaching zeros to the dividend during the
division process. We will carry the division to three decimal places (that is, one extra place) before
rounding off the quotient. Do **not** just stop at the hundredth place.

$$
\begin{array}{r}
0.428 \\
7\overline{)3.000} \\
-\underline{2\,8} \\
20 \\
-\underline{1\,4} \\
60 \\
\underline{-56} \\
4
\end{array}
$$

Then, 0.428 to the nearest hundredth becomes 0.43.

Note: To round off quotient to the nearest tenth, we carry out the division to two decimal places before rounding off;
to the nearest thousandth, the division is to four decimal places.

Case 2: Round-off place of the decimal is **not** specified

In this case, we will leave the decimal as a complex decimal of two decimal places, since this will produce sufficient
decimal places for most applications and yet provide additional information for either obtaining more decimal places or for
rounding-off.

Example 2   Change $\frac{3}{7}$ to a decimal.   (Round-off place is **not** specified)

Procedure: Using long division, divide the numerator by the denominator, attaching zeros to the dividend during the
division process. No round-off place is specified and therefore we will carry the division to two decimal
places, and write the remainder divided by the divisor as a fraction.

$$
\begin{array}{r}
0.42\frac{6}{7} \quad \longleftarrow \text{(complex decimal with two decimal places)} \\
7\overline{)3.000} \\
-\underline{2\,8} \\
20 \\
-\underline{1\,4} \\
6
\end{array}
$$

Thus $\frac{3}{7} = 0.42\frac{6}{7}$ (a complex decimal with two decimal places)

Example 3: Change $\frac{2}{3}$ to a decimal rounded-off to two decimal places.

$$
\begin{array}{r}
0.666 \approx .67 \\
3\,\overline{)\,2.000} \\
-\underline{18} \\
20 \\
-\underline{18} \\
20 \\
-\underline{18} \\
2
\end{array}
$$

Example 4: Change $\frac{2}{3}$ to a decimal.

$$
\begin{array}{r}
0.66\frac{2}{3} \quad \longleftarrow \text{ (complex decimal with two decimal places)} \\
3\,\overline{)\,2.00} \\
-\underline{18} \\
20 \\
-\underline{18} \\
2
\end{array}
$$

Note: In Example 4, we could write the quotient as $0.\overline{6}$ or $0.6...$ instead of $0.66\frac{2}{3}$. However, in computations, it is much easier to convert $0.66\frac{2}{3}$ to a fraction than to convert $0.\overline{6}$ or $0.6...$ to a fraction. The general method for converting $0.\overline{6}$ or $0.6...$ to a fraction is algebraic.

Lesson 13 Exercises

A. Determine the number of decimal places

1. $.2\frac{2}{3}$; 2. $.02\frac{2}{3}$; 3. $1.2\frac{1}{3}$; 4. $.4\frac{3}{11}$; 5. $43.4\frac{2}{7}$

Answers: **1**. 1 ; **2**. 2; **3**. 1 ; **4**. 1 ; **5**. 1 .

B. Convert to a fraction in its lowest terms.

1. $.2\frac{2}{3}$; 2. $.02\frac{2}{3}$; 3. $1.2\frac{1}{3}$; 4. $.4\frac{3}{11}$; 5. $43.4\frac{2}{7}$

Answers: **1**. $\frac{4}{15}$; **2**. $\frac{2}{75}$; **3**. $\frac{37}{30}$; **4**. $\frac{47}{110}$; **5**. $\frac{304}{7}$

C. Divide: **1.** $0.72 \div 5$; **2.** $4.32 \div 12$; **3.** $72.1 \div 0.20$; **4.** $0.738 \div 0.6$; **5.** $30.03 \div 2.1$

Divide and round-off quotient to the nearest hundredth.

6. $0.463 \div 0.8$; **7.** $62.82 \div 14$; **8.** $56.6536 \div 23$; 9. $46.95 \div 1.1$; **10.** $38.93 \div 0.08$

Answers: **1**.. 0.144; **2**. 0.36; **3**. 360.5; 4, 1.23; **5**.14.3; **6**. 0.58; **7**. 4.49; **8**. 2.46; **9**. 42.68; **10**. 486.63

D. 1. Change $\frac{6}{7}$ to a decimal, rounding off the quotient to the nearest hundredth.

2. Change $\frac{5}{9}$ to a decimal, rounding off quotient to the nearest tenth.

Answers **1**. $0.857 \approx 0.86$ **2**. $0.55 \approx 0.6$

CHAPTER 5
Ratio

Lesson 14: **Definition; Reduction of Ratios to Lowest Terms**
Lesson 15: **Using Ratios to Compare Quantities**
Lesson 16: **Using Ratios to Divide a Quantity into Parts**

Note: This chapter and the next chapter are very important for everyday life, whether you have your own business or you work for someone. If you master these chapters very well, you would have a good mathematical preparation for a lot of endeavors both at home or at the workplace., We have used both arithmetic and algebraic methods in covering these chapters. If you have no algebraic background, you may skip the algebraic parts and cover only the arithmetic parts. However, you can study the algebraic material in chapters 9-13 before covering these chapters.

Lesson 14

Definition and Reduction of Ratios to Lowest Terms

Whenever you go shopping to buy apples, you always want to know the selling price, and the first question you usually ask the seller is "How much are these apples?" A typical answer may be "4 apples for one dollar". As soon as you have agreed to this price, you and the seller have established a ratio between the number of apples the seller is willing to give you and the number of dollars you are willing to pay. Simply, the ratio you have agreed to is 4 apples to 1 dollar, and we can symbolize this as 4**:**1. The seller might also have replied "1 dollar for 4 apples", but in this case, the ratio of dollars to apples is 1 to 4 or 1:4. Note the order in which the terms 1 and 4 of the ratio follow "dollar and apples".

If your grandmother left $1000, in her will, to be divided between you and your brother so that you receive $600 and your brother receives $400, then the ratio of your share to your brother's share is $600 to $400 and in simplest terms, this ratio is 3 **to** 2, which means that whenever you receive $3, your brother receives $2.

If the simple interest rate for your savings account at a bank is 5% (5 percent or 5 per hundred), then after one year, the ratio of the interest to the deposit you had in your account is 5 to 100 (That is, your interest is $5 for every $100 in your account.)

We can observe from the above examples that, in everyday life, we are always dealing with ratios, consciously or unconsciously. Uses of ratios include comparison of quantities, division of a quantity into parts, in business decision making and in forming proportions.

Definition

The **ratio** of two quantities a and b or of a to b (written $a:b$) is a comparison of the two quantities and written as a fraction is $\frac{a}{b}$ (i.e., is the first quantity mentioned divided by the second quantity mentioned), where a and b may or may not be of the same units. The quantities a and b are called the terms of the ratio. However, the ratio of b to a is $\frac{b}{a}$.

Examples

1. The ratio, 3**:**4 is $\frac{3}{4}$ but the ratio 4**:**3 is $\frac{4}{3}$. **Note** that order is important.

2. The ratio, 4 inches to 12 inches is the fraction $\frac{4 \text{ inches}}{12 \text{ inches}} = \frac{1}{3}$
   (Terms have the same units and cancel out)

3. The ratio, 4 inches to 1 foot is the fraction $\dfrac{4 \text{ inches}}{1 \text{ foot}} = \dfrac{4 \text{ inches}}{12 \text{ inches}} = \dfrac{1}{3}$ (1 ft = 12 in.)

  (One unit can be converted to the other)

4. The ratio, 4 apples to 1 dollar is the fraction $\dfrac{4 \text{ apples}}{1 \text{ dollar}}$

 (Terms do not have the same units and one unit cannot be converted to the other)

5. The ratio, 50 miles to 2 hours is the fraction $\dfrac{50 \text{ miles}}{2 \text{ hours}} = \dfrac{25 \text{ miles}}{1 \text{ hour}}$ (25 mph) <-different units

Note that since ratios can be written as fractions, ratios can be reduced to lowest terms in the same way that we reduce fractions to lowest terms.

Since ratios are used for comparing quantities, a ratio may have three or more terms.

Example The number of pencils, the number of pens, and the number of notebooks are
     in the ratio of 2:3:4. (i.e., 2 pencils to 3 pens to 4 notebooks)

Rate

 If a and b are **not** of the same units **and** one unit **cannot** be converted to the other unit, some authors call the ratio a **rate.**

1. The rate, 50 miles to 1 hour is the fraction $\dfrac{50 \text{ miles}}{1 \text{ hour}}$ (50 mph)

Even though **rate** is a common terminology for quantities with different units, in the following example, note that it is actually a ratio, since the terms have the same units or no units.

2. An interest rate of 5% for a savings account (at a bank in the US.) means that the depositor receives

$5 for every $100 in the account. Note also that 5% is also the fraction $\dfrac{5}{100}$ or the ratio, 5:100, which

is a ratio. Perhaps we should call this interest rate interest ratio.

Example: Reduce the ratio 12:8 to lowest terms

Solution: $12:8 = \dfrac{12}{8} = \dfrac{3}{2}$

Answer: In lowest terms 12:8 = 3:2.

Lesson 14 Exercises

Reducing Ratios to Lowest Terms

Reduce to lowest terms (Two terms)
(a). The ratio 2:4
(b). The ratio 6:3
(c). The ratio 8:12
(d) The ratio 12:8
(e). The ratio 5 inches to 10 inches
(f) The ratio 6 apples to 3 dollars

Answers: (a) 1:2; (b) 2:1; (c) 2:3; (d) 3:2; (e) 1:2; (f) 2 apples: 1 dollar;

Reduce to lowest terms (Three Terms)
(a) .The ratio 2:4:6   (b) The ratio 6:3:18
(c) The ratio 8:12:6
(d) The ratio 12:8:6
(e) . The ratio 5 inches to 10 inches to 30 inches
(f) The ratio 6 apples to 10 apples to 24 apples
Answers: (a) 1:2:3; (b) 2:1:6; (c) 4:6:3; (d) 6:4:3; (e) 1:2:6; (f) 3:5:12;

Lesson 15

Using Ratios to Compare Quantities

Example 4

At a certain college, there are 2,000 male students and 6,000 female students.

(a) What is the ratio of the number of male students to the number of female students at this college?

(b) What is the ratio of number of female students to the number of male students at this college?

(c) What is the ratio of the number of female students to the total number of students at this college?

Solution

(a) The ratio of males to females is $2000:6000 = \frac{2000}{6000} = \frac{1}{3}$ or $1:3$

(b) The ratio of females to males is $6000:2000 = \frac{6000}{2000} = \frac{3}{1}$ or $3:1$

(c) Total number of students $= 2000 + 6000 = 8000$
The ratio of the number of females to the total number of students is $6000:8000$
$= \frac{6000}{8000} = \frac{3}{4}$ or $3:4$

Example 5

Now let us verify the results of the above problem by doing it backwards.

There are altogether 8,0000 students at a certain college. The ratio of the number of males to the number of females at this college is 1:3.

(a) Find the number of males.   **(b)** Find the number of females at this college.

Solution

(a) Fraction of students who are males $\frac{1}{1+3} = \frac{1}{4}$ (Fraction corresponds to first term, 1, of ratio)

The number of males $= \frac{1}{4} \times \frac{8000}{1} = 2,000$

(b) Fraction of students who are females $\frac{3}{1+3} = \frac{3}{4}$ (Fraction corresponds to second term, 3)

The number of females $= \frac{3}{4} \times \frac{8000}{1} = 6,000$

(c) Let us recalculate the number of females given that the ratio of number of females to the total number of students is $3:4$.

Solution: Fraction of students who are females $= \frac{3}{4}$.

The number of females $= \frac{3}{4} \times \frac{8000}{1} = 6,000$ <-----This is the same solution as is (b)

Note the difference between how the denominator, 4, of the fraction in (c) was obtained by using only the second term of the ratio, and how the denominator of the fraction in (b) was obtained by adding 1 and 3, the terms of ratio. Note also that in (b) and (a), we added the terms of the ratio to obtain the denominators of the fractions, since the terms, 1 and 3 correspond to the individual genders involved in the problem; while in (c), the second term, 4, of the ratio corresponds to the total number of students. Note particularly that we did **not** add the 3 and the 4 to obtain the denominator of the fraction in (c), but we rather used only the 4 as the denominator. See also p.73.

The problem in (c) could have been posed as " 3 out of 4 students at a certain college are females. If there are 8,000 students at this college, how many of the students are females. The solution method is the same as in (b). We could also algebraically, use proportion to obtain the same solution: $\frac{3}{4} = \frac{x}{8000}$, where x is the number of females.

$4x = 3 \times 8000$ and $x = \frac{3}{4} \times \frac{8000}{1} = 6,000$.

As we shall soon observe, Example 5 actually belongs to the next lesson.

Lesson 15 Exercises

Comparing two quantities

1. At a certain college, there are 18,000 male students and 32,000 female students.

(a) What is the ratio of the number of male students to the number of female students at this college?

(b) What is the ratio of number of female students to the number of male students at this college?

c) What is the ratio of the number of female students to the total number of students at this college?   **Answer** (a) 9:16;   (b) 16:9 ; (c) 16:25

2. In a math class, class, there are 15 females and 8 males.

(a) What is the ratio of the number of males to the number of females in this class?

(b) What is the ratio of the number of females to the number of males?

(c) What is the ratio of the number of males to the total number of students?

Answer: (a) 8:15; (b) 15:8; (c) 8:23

3. Rameltha and Albert are to divide $45,000

If Rameltha receives $18,000, what is the ratio of Albert's share to Rameltha's share?

Answer: 3:2

4. A piece of black tape is 4 inches long, and a similar piece of red tape is 2 feet long..

What is the ratio of the length of the black tape to the length of the red tape?

Answer: 1:6

5. Out of 20 games that a team played, the team won 12 games and lost the rest.

(a) What is the ratio of the number of games won to the number of games played?

(b) What is the ratio of the number of games won to the number of games lost?

Answer: (a) 3:5; (b) 3:2

Comparing three quantities

6. In a bag, there are 16 red marbles, 20 black marbles and 12 green marbles.

(a) What is the ratio of red marbles to black marbles to green marbles?

(b) What is the ratio of black marbles to red marbles to green marbles?

(c) What is the ratio of black marbles to the sum of the red and green marbles?

(d) What is the ratio of green marbles to the total number of marbles in the bag?

Answer: (a) red:black:green = 4:5:3 ; (b) black:red:green = 5:4:3; (c) 5:7; (d) 1:4

7. Rameltha, Albert and James are to divide $45,000. If Rameltha receives $18,000, and Albert receives $17,000, what is the ratio of Rameltha's share to James' share to Albert's share?   Answer: 18:10:17

8. A piece of black tape is 4 inches long, and similar pieces of red and green tapes are 2 feet , and 3 feet long respectively.

(a) What is the ratio of the length of the black tape to green tape to red tape?

(b) What is the ratio of the length of green tape to red tape to black tape?

Answer: (a) 1:9:6   (b) 9:6:1

9. Out of 20 games that a team played, the team won 12 games, lost 4 and drew the rest

(a) What is the ratio of games won to games lost to games drawn?

(b) What is the ratio of games lost to games won to games played?

(c) What is the ratio of games drawn to games won?

Answer: (a) 3:1:1 ; (b) 1:3:5 ; (c) 1:3

Lesson 16

Using Ratios to Divide a Quantity into Parts

Example 1   Divide 28 in the ratio 3: 4

Method 1:   **Using Arithmetic**

Step 1:   Fraction of the number corresponding to the first term: $\dfrac{3}{3+4} = \dfrac{3}{7}$

Fraction of the number corresponding to the second term: $\dfrac{4}{3+4} = \dfrac{4}{7}$

Step 2: The number corresponding to the first term (of the ratio) $= \dfrac{3}{7} \times \dfrac{28}{1} = 12$

The number corresponding to the second term (of the ratio) $= \dfrac{4}{7} \times \dfrac{28}{1} = 16$

The numbers are 12 and 16.

Method 2: Using algebra

Let the number corresponding to the first term (of the ratio) $= 3k$ *(k a constant)*
Let the number corresponding to the second term (of the ratio) $= 4k$
"The sum of the two numbers is 28" translates to

$3k + 4k = 28$
$\quad 7k = 28$
$\quad\quad k = 4$
One of the numbers $= 3k = 3(4) = 12$, and
the other number $= 4k = 4(4) = 16$

The numbers are 12 and 16.

Method 3: Using Algebra

See Method 3 of Example 2,below.

Example 2 A bag contains red and white marbles in the ratio 3:4. There are altogether 28 marbles in this bag. How many of these marbles are red?

Solution Method 1:   Using arithmetic

The number of red marbles $= \dfrac{3}{7} \times \dfrac{28}{1} = 12$

Method 2: Using algebra

Let the number of red marbles $= 3k$ *(k a constant)*
Let the number of white marbles $= 4k$

$3k + 4k = 28$
$\quad\quad 7k = 28$ The number of red marbles $= 3k = 3(4) = 12$
$\quad\quad\quad k = 4$

Method 3: Using Algebra

Let the number of red marbles $= x$
Then the number of white marbles $= (28 - x)$
The proportion "3:4 as $x : (28 - x)$ translates to: (ratios of red marbles to white marbles)

$\dfrac{3}{4} = \dfrac{x}{(28 - x)}$

Solve for x:

$3(28 - x) = 4x$
$84 - 3x = 4x$
$\quad\quad 84 = 7x$
$\quad\quad 12 = x$

The number of red marbles $= 12$

Lesson 16 Exercises

Dividing a quantity into two parts

1. The ratio of male students to female students at a certain college is 2:5. if there are 2,800 students at this college, how many of the students are females? Answers: **1.** 2,000;

2. If the ratio (by weight) of hydrogen to oxygen in water is 1:8, find the weight of hydrogen in 27 grams of water. Answer: 3 grams

3. The sum of two numbers is 40. The numbers are in the ratio 2:3. Find these numbers.

Answers: **1.** 16 and 24

4. James and Mary are to divide $72 in the ratio 5:7. How much does each receive?

Answer: James: $30, Mary: $42

5. A grandmother left $45,000 in her will to be divided between two granddaughters, Maria and Diana in the ratio 2 to 3. How much does each receive?

Answer Maria: $18,000; Diana: $27,000

6. John invested $2,000 in a company while James invested $3,000 in the same company. If after 3 years, a profit of $2,000 is to be shared between them in the ratio of their investments. How much does each receive? **Answer** John: $800; James: $1200

7. Mary wants to prepare a solution consisting of alcohol and water in the ratio 1**:**2 by volume. If the total volume of the solution is 12 liters, find the volumes of alcohol and water to be used

Answer alcohol: 4 liters; water: 8 liters

9. A farmer wants to prepare 24 liters of gasohol which consists of alcohol and gasoline in the ratio 1**:**9. How many liters of alcohol and gasoline are to be mixed?

Answer alcohol: 2.4 liters; gasoline: 21.6 liters

9. A bag contains p fruits, made up of oranges and apples only in the ratio q**:**r. Obtain an expression for the number of oranges in terms of p, q, and r.Answer: $\frac{pq}{q+r}$

10. In a math class, the ratio of males to females is 1**:**3.: If there are 28 students in this class, how many males are in this class? Answer: 7 males

Dividing a quantity into three parts

11. A grandmother left $45,000 in her will to be divided between three granddaughters, Ana, Elizabeth, and Diana in the ratio 1 to 2 to 3. How much does each receive?

Answer: Ana: $7,500, Elizabeth: $15,000, Diana: $22,500

12. John invested $3,000 in a company while James and Alexander invested $4,000 and $5,000 respectively in the same company. If after 4 years, a profit of $4,800 is to be shared between them in the ratio of their investments. How much does each receive? **Answer** : John: $1,200, James: 1,600, Alexander: $2,000

13. Mary wants to prepare a solution consisting of alcohol, water and orange juice in the ratio 1**:**2**:3** by volume. If the total volume of the solution is 36 liters, find the volumes of alcohol, water and juice to be used. **Answe**r Alcohol: 6 liters, water: 12 liters, orange juice: 18 liters

14. A farmer wants to prepare 44 liters of fuel blend which consists of fuel A, fuel B and fuel C in the ratio 1**:**1**:**9. How many liters of each type of fuel are to be mixed?

Answer : fuel A: 4 liters, : fuel B: 4 liters, fuel C: 36 liters

15. A bag contains p fruits, made up of oranges, grapefruits, and apples in the ratio q**:**r**:**s.

Obtain an expression for the number of oranges in terms of p, q, r, and s. Answer: $\frac{pq}{q+r+s}$

16. On a math test, the number of A's, B's, and C's are in the ratio of 1**:**4**:**2 If there are 28 students in this class, determine the number of (a) A's, (b) B's, and (c) C's. ?

Answer: A's: 4, B's: 16, C''s: 8:

CHAPTER 6

PROPORTION (VARIATION)

Lesson 17
Introduction to Direct and Inverse Proportion

(Definitions, Terminology, Types of Proportion and Principles)

Basically, there are two main **types** of **proportion** (or **variation**):
1. **Direct proportion** or simply, **proportion** (or direct variation).
2. **Inverse proportion** (or indirect variation).
   There are also problems which involve both direct and inverse proportion simultaneously.

Direct proportion

Discussion: In the introduction of the lesson on ratios, we discussed an example involving buying apples. There, the seller quoted the price of apples as "4 apples for 1 dollar", and we established the ratio of the number of apples to the number of dollars as 4 to 1. Now, if the buyer wants to buy exactly 4 apples, then the buyer pays exactly 1 dollar. However, if the number of apples is different from 4, then the number of dollars to pay would be different from 1, and the seller has to determine how much the buyer pays for the new number of apples. For example, if the buyer wants 8 apples (instead of 4 apples), then the seller will determine how many dollars the buyer pays (in this case, $2), or if the buyer wants 2 dollars worth of apples, the seller has to determine how many apples to give the buyer for $2 (in this case, 8 apples). We note above that if the buyer wants to buy exactly 4 apples, then only two quantities are involved in the problem, namely 4 apples and 1 dollar.

However, if the buyer wants 8 apples, then four quantities would be involved in the problem, namely 4 apples, 1 dollar, 8 apples, and the price for 8 apples. We will determine the fourth quantity, namely, the price for 8 apples using these three known quantities and a relationship called a **direct proportion**. In this relationship, as the number of apples increases, the number of dollars also increases; or as the number of apples decreases, the number of dollars also decreases. When the number of apples is zero, the number of dollars is also zero. Uniform or constant changes in the number of apples result in uniform changes in the number of dollars. For example, as the number of apples is doubled, the number of dollars is also doubled; but if he number of apples is halved, the number of dollars is also halved.

   Another application of proportion is as follows: Suppose the roof of your house has a triangular shape, and you want to build a much bigger or smaller house such that the corresponding roof has the same shape as your present house, then **direct proportion** can be used to determine the lengths of the sides of the new roof.

Direct proportion occurs very often in the arithmetic of everyday life. It may not be the number of apples and number of dollars, but it may be the number of miles traveled and number of gallons of gasoline used, or the number of miles traveled and number of hours, or the number of grams of hydrogen and the number of grams of oxygen. Direct proportion is usually termed **proportion** in most textbooks, and it is this type that you meet very often.

We should note above that problems stated in terms of proportion can also be stated in terms of variation and conversely. In the examples covered in this chapter, we cover both cases.

Simple Direct Proportion (Direct Variation) 60

Definition: A **proportion** is a statement that two ratios are equal.

The direct proportion "a is to b as c is to d " (or $a{:}b{::}c{:}d$) is the equality

$$\frac{a}{b} = \frac{c}{d} \quad \textbf{(quotient = quotient)}$$

In the above proportion, a and d are called the extremes and b and c are called the means of the proportion. We may also define a proportion as a mathematical statement indicating the equality between two **equivalent fractions.**   Example: $\frac{2}{3} = \frac{4}{6}$ is a proportion

Proportion Principles

1. We can invert both ratios of a proportion: From (1) above, we obtain after the inversion,

$$\frac{b}{a} = \frac{d}{c}$$

2. We can interchange the extremes of a proportion: From (1) above we would obtain:

$$\frac{d}{b} = \frac{c}{a}$$

3. We can also interchange the means to obtain:

$$\frac{a}{c} = \frac{b}{d}$$

Also $ad = bc$ (That is the product of the extremes = the product of the means) <-- also called cross-multiplication

The proportion principles can be used to obtain desired ratios. For example, if we have $\frac{3}{2}$ as the ratio of one side of a proportion, and we desire the ratio $\frac{2}{3}$ then proportion principle #1 above (inversion of both sides of a proportion) can be used to obtain this ratio, $\frac{2}{3}$, from $\frac{3}{2}$.

More Principles (From $\frac{a}{b} = \frac{c}{d}$, we obtain the following:

4. $\dfrac{a+b}{b} = \dfrac{c+d}{d}$ (by addition). Note: $\dfrac{a+b}{b} = \dfrac{a}{b} + \dfrac{b}{b} = \dfrac{a}{b} + 1$ **and** $\dfrac{c+d}{d} = \dfrac{c}{d} + \dfrac{d}{d} = \dfrac{c}{d} + 1$

5. $\dfrac{a-b}{b} = \dfrac{c-d}{d}$ (by subtraction)

6. $\dfrac{a-b}{a+b} = \dfrac{c-d}{c+d}$ (by subtraction and addition)

7. $\dfrac{a+b}{a-b} = \dfrac{c+d}{c-d}$ (by addition and subtraction)

8. If $\dfrac{a}{b} = \dfrac{c}{d} = \dfrac{e}{f}$, then $\dfrac{a+c+e}{b+d+f} = \dfrac{a}{b} = \dfrac{c}{d} = \dfrac{e}{f}$

e.g.,($\dfrac{2}{10} = \dfrac{3}{15} = \dfrac{4}{20} = \dfrac{2+3+4}{10+15+20} = \dfrac{9}{45} = \dfrac{1}{5}$)

To show that **8** is true, see Appendix, p. 412

Note:: For 6 and 7, divide LHS and RHS of 4 and 5,

Problems Involving Proportion

Example 1   Find x from the proportion x is to 4 as 3 is to 2.

Solution $\dfrac{x}{4} = \dfrac{3}{2}$

$2x = 12$ (cross multiplication)

$x = 6$

(You could also solve the above equation by multiplying both sides of the equation by 4)

Example 2   Find x from the proportion: 6 is to x as 4 is to 12.

Solution $\dfrac{6}{x} = \dfrac{4}{12}$

$4x = 72$; and $x = 18$

Simple Inverse Proportion (Inverse Variation)   61

Discussion: Another type of a relationship called **inverse proportion** (or indirect proportion) also occurs in everyday life. For example, suppose that when you drive your car from City A to City B at a speed of 25 mph, it takes you 4 hours. Now, if you always drive at a speed of 25 mph from City A to City B, and it takes you 4 hours, then only two quantities are involved, namely 25 mph and 4 hours. However, if you want to travel the same distance at a speed of say, 50 mph, then you would like to know the time needed to drive from City A to City B at 50 mph, and in this case, four quantities would be involved in the problem, namely, 25 mph, 4 hours, 50 mph, and the time, t, for 50 mph. Since for the same distance, as the speed is increased, the time taken decreases, or if the speed is decreased, the time taken increases, speed is said to be inversely proportional to the time.

Recall the main difference between direct proportion and inverse proportion: for direct proportion, both quantities increase or decrease at the same time; whereas for inverse proportion, as one quantity increases, the other quantity decreases and vice versa.
 For example, as one quantity is doubled, the other quantity is halved; but if one quantity is halved, the other quantity is doubled.

Example 1 The time taken by a number of people to do a piece of work is **inversely proportional** to the number of people (assuming that each person works at the same rate as everyone else). Thus, as the number of people increases, the time taken to do the work decreases but as the number of people decreases, the time taken increases (i.e., more people, less time; and less people, more time)

Example 2 At constant temperature, the volume of a given mass of a gas is inversely proportional to the pressure on the gas. (This relationship is known as Boyle's law.)

Definition

If a is inversely proportional to b. (same as a is proportional to the **reciprocal** of b)

Then a_1 corresponds to b_1, and a_2 corresponds to b_2. and

$$a_1 : \frac{1}{b_1} = a_2 : \frac{1}{b_2}$$

$$a_1 \div \frac{1}{b_1} = a_2 \div \frac{1}{b_2}$$

$$a_1 b_1 = a_2 b_2 \qquad \textbf{(product = product)}$$

Or if we use the familiar terms a, b, c, d as the terms of the inverse proportion

Then $a : \frac{1}{b} = c : \frac{1}{d}$

$$a \div \frac{1}{b} = c \div \frac{1}{d}$$

$ab = cd$ **(product = product)** **<------** inverse proportion

We will use the product = product form since this form is easy to remember

Compare:

Direct proportion $\frac{a}{b} = \frac{c}{d}$ **<--------** **(quotient = quotient)**

Indirect proportion $ab = cd$ **<------(product = product)**

Determining the type of Proportion (Variation) from the Wording   6 2

Explicit Specification of direct proportion (variation)

There are a number of ways of **explicitly** specifying direct proportion (variation)
If y is a function of x, then the following are equivalent to one another:

(a) y is directly proportional to x, or simply
(b) y is proportional to x. (The word "directly" is omitted.)
(c) y varies directly as x, or simply,
(d) y varies as x, (The word "directly" is omitted.)

The methods we will cover for solving direct proportion problems can be used to solve any of the problems worded as in (a), (b), (c) and (d) above.

Explicit Specification of inverse proportion (variation)

If y is a function of x, then the following are equivalent to one another:

(a) y is inversely or indirectly proportional to x.
(b) y varies inversely as x,

Any of the methods (we will cover) for solving indirect proportion problems can be used to solve any of the problems stated as in (a) and (b) above.

Implicit Specification of proportion (variation)

In most problems, the terms " directly proportional to" , "inversely proportional to", " varies directly as" or "varies inversely as" and similar terms are not used in wording the problem; but rather, we have to deduce from the problem whether it is a direct or an indirect proportion from experience. For example, if 15 oranges cost \$3, what is the cost of 10 oranges?. From life experience, we know that at the same price per an orange, as the **number** of oranges **increases**, the **cost** of oranges **increases** accordingly. Therefore, the relationship in this problem is that of direct **proportion** (variation).

Lesson 17 Exercises A

A. Solve for x: **1.** $\dfrac{x}{8} = \dfrac{3}{2}$;   **2.** $\dfrac{5}{8} = \dfrac{20}{x}$;   **3.** $\dfrac{9}{x} = \dfrac{3}{5}$;   **4.** $\dfrac{6}{8} = \dfrac{x}{2}$;   **5.** $\dfrac{x}{9} = \dfrac{3}{2}$;

6. $\dfrac{10}{x} = \dfrac{3}{5}$;   **7.** $\dfrac{x}{8} = \dfrac{4\frac{1}{2}}{2}$; **8.** $\dfrac{x}{8\frac{2}{3}} = \dfrac{3\frac{2}{3}}{2\frac{3}{4}}$;   **9.** $\dfrac{x}{1.8} = \dfrac{3}{2.1}$;   **10.** $\dfrac{x}{8} = \dfrac{12\%}{96}$

Answers:**1.** $x = 12$; **2.** $x = 32$; **3.** $x = 15$; **4.** $x = \dfrac{3}{2}$; **5.** $13\frac{1}{2}$; **6.** $16\frac{2}{3}$; **7.** $x = 18$; **8.** $11\frac{5}{9}$;

9. $2\frac{4}{7}$; **10.** 1% or $\dfrac{1}{100}$

Lesson 17 Exercises B

Determine which of the following are direct proportion or inverse proportion:

1. If 10 gallons of gasoline cost 15 dollars, find the cost of 24 gallons of gasoline.

2. If on 2 gallons of gasoline, a car can travel 24 miles, how many miles can a car travel on 20 gallons of gasoline?

3. If 8 oranges sell for 2 dollars, find the selling price for 24 oranges.

4. If 2 people can paint a house in 6 days, how many people would paint the house in 3 days?

5. At a speed of 60 miles per hour, a car takes 2 hours to travel from City A to City B. At a speed of 40 miles per hour, how long will it take the car to ravel from City A to City B?

6. If 6 mail men can sort a given quantity of mail in 4 hours, how many mail men would be needed to sort the same quantity of mail in 2 hours?

Answers , 1, 2, 3 are direct proportion; 4, 5 and 6 are inverse proportion

Lesson 18

Methods for Solving Direct Proportion Problems
(Direct Variation Problems)

There are a number of methods or approaches for solving **direct proportion** problems.
In simple proportion problems, four quantities are involved, and we are (usually) given three quantities and we are required to find the fourth quantity.
We cover **five** methods with examples, and then discuss the relative merits of the methods.

Method 1: Quotient = Quotient or Ratio = Ratio method (Algebraic)
In this method, we word or reword the problem as discussed below.

Consider a direct proportion problem involving four quantities such that a_1 corresponds to b_1, and a_2 corresponds to b_2. We word the proportion as follows:

The proportion a_1 is to b_1 as a_2 is to b_2 translates into the equation

$$\frac{a_1}{b_1} = \frac{a_2}{b_2} \quad \textbf{(quotient = quotient or } \text{ratio = ratio)} \textbf{ or}$$

Using the familiar notation "a is to b as c is to d", we obtain

$$\frac{a}{b} = \frac{c}{d} \quad \textbf{(quotient = quotient)}$$

Example 1
If 2 dollars can buy 8 apples, how many dollars are needed to buy 24 apples?

Method 1: **Quotient = Quotient Method** (Algebraic)
We are required to find the number of dollars needed to buy 24 apples.
Step 1: Let the number of dollars needed to buy 24 apples = x.
(we represent the unknown by a variable)
2 dollars correspond to 8 apples
x dollars correspond to 24 apples.
(Note the order in which the units in the second correspondence is mentioned.
That is, dollars before apples, and not apples before dollars)
Step 2: 2 dollars are to 8 apples as x dollars are to 24 apples

Step 3: Translating into an equation, $\dfrac{2 \text{ dollars}}{8 \text{ apples}} = \dfrac{x \text{ dollars}}{24 \text{ apples}}$

Step 4: Cross-multiplying, $8x = 2(24)$ $\qquad$ ($8x$(dollars)(appless) = 2(24)(dollars)(apples))

Step 5 Solve for x: $\qquad x = \dfrac{2(24)}{8}$ $\qquad$ (Note above that the units cancel out)
$$= 6$$

Conclusion: 6 dollars are needed to buy 24 apples.
Note in the above example that the order in which we mention the units, dollars and apples is important.
Note also, however, that in the above example, we could word the proportion as follows: The proportion,

8 apples are to 2 dollars as 24 apples are to x dollars" translates to $\dfrac{8}{2} = \dfrac{24}{x}$

$$8x = 2(24)$$
$$x = \frac{2(24)}{8}$$
$$= 6 \qquad \text{(Again, we obtain the same solution as before.)}$$

Method 2 Units Label Method or Dimensions Method   (Arithmetic method)
 This method is highly recommended in applications such as converting from one unit to another
 unit in measurements as well as calculations in chemistry. In Chemistry, some authors call this
 method " The factor label method". In this method, we use the units of the given quantities to
 guide us in obtaining an expression which on simplification yields the desired result.

We repeat the question: If 2 dollars can buy 8 apples, how many dollars are needed to buy
24 apples?

Solution: We can consider this problem as converting 24 apples to dollars. The justification
 of this method is in parentheses.

Step 1: $\dfrac{24 \text{ apples}}{1} \times \dfrac{?}{?}$

Step 2: Multiply $\dfrac{24 \text{ apples}}{1} \times \dfrac{?}{?}$ by a fraction formed by using the quantities 2 dollars and 8 apples as

the terms of the fraction such that the denominator has the same units as the 24 apples (the quantity in the
numerator), and under such conditions, we can divide out (cancel) the common units in the numerator and
the denominator, leaving us the units, dollars.

Then, $\dfrac{24 \text{ apples}}{1} \times \dfrac{2 \text{ dollars}}{8 \text{ apples}}$

Step 3: $\dfrac{24 \ \cancel{\text{apples}}}{1} \times \dfrac{2 \text{ dollars}}{8 \ \cancel{\text{apples}}}$ (= Number of apples $\times$ cost per each apple)

$= 6$ dollars

Conclusion: 6 dollars are needed to buy 24 apples.

Let us do the above problem backwards, with the justification of this method in parentheses
If 2 dollars can buy 8 apples, how many apples can one buy for 6 dollars?
Solution   Here, we are converting 6 dollars to apples.

$\dfrac{6 \text{ dollars}}{1} \times \dfrac{?}{?}$

$= \dfrac{6 \text{ dollars}}{1} \times \dfrac{8 \text{ apples}}{2 \text{ dollars}}$ $\left(= \text{total cost} \div \text{cost per 1 apple} = \text{total cost} \times \dfrac{1}{\text{cost per 1 apple}}\right)$

$= 24$ apples, which is what was given in the original problem.

Method 3 Unitary Method (Arithmetic)
We repeat the question: If 2 dollars can buy 8 apples, how many dollars are needed to buy 24 apples?

Solution: Cost of 8 apples = 2 dollars

Cost of 1 apple = $\dfrac{2 \text{ dollars}}{8}$ (Divide to obtain the cost for 1 apple))

Cost of 24 apples = $\dfrac{2}{8} \times \dfrac{24}{1}$ dollars (Multiply to obtain the cost for other than 1)

$= 6$ dollars

Conclusion: 6 dollars are needed to buy 24 apples.

Method 4 "More or Less Method" (Arithmetic method. This method the shorter form of Method 3.
We repeat the question: If 2 dollars can buy 8 apples, how many dollars are needed to buy 24 apple?

If cost of 8 apples = 2 dollars

Then cost of 24 apples = $\dfrac{2 \text{ dollars}}{1} \times \dfrac{24 \text{ \sout{apples}}}{8 \text{ \sout{apples}}}$

= 6 dollars

Conclusion: 6 dollars are needed to buy 24 apples.

In method 4, we used a very useful principle which states that " If more, the smaller divides, and if less, the larger divides". That is, since we expect 24 apples to cost more than 2 dollars, in forming the fraction involving 8 apples and 24 apples, the smaller of 8 apples and 24 apples is the divisor (smaller divides), and hence we used the fraction,

$\dfrac{24 \text{ apples}}{8 \text{ apples}}$ <----- " smaller divides" (We multiplied by an **improper fraction**)

However, if we had expected the number (answer) to be less than 2 dollars, we would have used the fraction $\dfrac{8 \text{ apples}}{24 \text{ apples}}$

<----- "larger divides" (i.e., we would have multiplied by a **proper fraction**)

The More or Less Method above is very powerful once you have mastered it, because it can be applied to inverse proportion problems as well as to problems involving both inverse and direct proportion and any number of quantities (variables), especially in science.

Method 5 Variation Method (Algebraic)
We reword the problem: The number of apples a person buys varies directly as the number of dollars paid. If when the number of apples is 8, the number of dollars paid is 2, find the number of dollars paid when the number of apples is 24.

Step 1: Let the number of apples = A
Let the number of dollars paid = D
Then $A = kD$, where k is called the proportionality constant.
Step 2: Substituting $A = 8, D = 2 \text{ in } A = kD$,
$8 = 2k$
$4 = k$
Substituting $k = 4 \text{ in } A = kD$, we obtain the formula,
$A = 4D$.
Step 3: When $A = 24$
$24 = 4D$ (substituting 24 for A in the formula, $A = 4D$)
$6 = D$ (solving for D)
$D = 6$
Therefore, 6 dollars would be paid for 24 apples.

Example 2 On 3 gallons of gasoline, one can drive 60 miles. How far can one drive on 8 gallons of gasoline?

Solution
Method 1   Quotient = Quotient Method (Algebraic)
Let the number of miles for 8 gallons of gasoline = x
Proportion statement: 3 gallons are to 60 miles as 8 gallons are to x miles.

Translating, $\dfrac{3 \text{ gallons}}{60 \text{ miles}} = \dfrac{8 \text{ gallons}}{x \text{ miles}}$

$\dfrac{3}{60} = \dfrac{8}{x}$

$3x = 60(8)$

$$x = \frac{60(8)}{3}$$

$$x = 160$$

Therefore, on 8 gallons of gasoline, one can drive 160 miles.

Note above that the ratio, gallons to miles = the ratio, gallons to miles; or
the ratio miles to gallons = the ratio miles to gallons
Incorrect: the ratio miles to gallons = the ratio gallons to miles. (violation of the order of the units)

All other proportion problems involving four quantities can be reworded and translated into equations by imitating the examples covered.. Only, the units involved might be different. We may have apples and pounds as the units involved, or hours and dollars, or years and people.

Method 2 Units Label Method. We repeat the problem: On 3 gallons of gasoline, one can drive 60 miles. How far can one drive on 8 gallons of gasoline?

Solution: We can consider this problem as converting 8 gallons to miles.

Step 1: $\dfrac{8 \text{ gallons}}{1} \times \dfrac{?}{?}$

Step 2: Multiply $\dfrac{8 \text{ gallons}}{1} \times \dfrac{?}{?}$ by a fraction formed by using the quantities 3 gallons and 60 miles

$$\frac{8 \text{ gallons}}{1} \times \frac{60 \text{ miles}}{3 \text{ gallons}}$$

Step 3: $\dfrac{8 \, \cancel{\text{gallons}}}{1} \times \dfrac{60 \text{ miles}}{3 \, \cancel{\text{gallons}}}$

$= 160 \text{ miles}$

Therefore, on 8 gallons of gasoline, one can drive 160 miles.

Method 3 Unitary Method (Arithmetic)

We repeat the problem: On 3 gallons of gasoline, one can drive 60 miles. How far can one drive on 8 gallons of gasoline?

Solution

Number of miles driven on 3 gallons of gasoline = 60 miles

Number of miles driven on 1 gallon of gasoline = $\dfrac{60}{3}$ miles

(Divide to obtain the miles for 1 gallon)

Number of miles driven on 8 gallon of gasoline = $\dfrac{60}{3} \times 8$ miles (Multiply for other than 1 gallon)

$= 160 \text{ miles}$

Therefore, on 8 gallons of gasoline, one can drive 160 miles.

Method 4 More or Less Method We repeat the problem:

On 3 gallons of gasoline, one can drive 60 miles. How far can one drive on 8 gallons of gasoline?

Solution Number of miles driven on 3 gallons gasoline = 60 miles

Number of miles driven on 8 gallons gasoline = $\dfrac{60}{1} \times \dfrac{8}{3}$ miles

$= 160 \text{ miles}$

Therefore, on 8 gallons of gasoline, one can drive 160 miles.

Method 5 Variation Method (Algebraic)   We reword the problem:
    The number of miles one drives is directly proportional to the number
    of gallons of gasoline used. If when the number of gallons of gasoline
    is 3, the number of miles is 60, find the number of miles when the number of gallons
    of gasoline is 8.

Solution

Step 1: Let the number of gallons of gasoline $= n$
    Let the number of miles driven $= d$
    Then since d is directly proportional to n,
    $d = kn$, where k is the proportionality constant

Step 2: Substitute for $d = 60$, $n = 3$, in $d = kn$ and solve for k.
    Then $60 = k(3)$ and from which $k = 20$

Step 3: Substitute for $k = 20$ in $d = kn$ to obtain a formula relating d and n.
    Then $d = 20n$

Step 4: Substitute for $n = 8$ in $d = 20n$ and determine d
    $d = (20)(8)$
    $= 160$
    Therefore, on 8 gallons of gasoline, one can drive 160 miles.

Example 3   The ratio of an astronaut's weight on the earth to weight on the moon is 6:1. If
        an astronaut weighs 180 lb (on earth), what is the astronaut's weight on the moon?

Solution

Method 1 Quotient = Quotient Method (Algebraic)
        Let the weight on the moon be x lb, (i.e., x lb corresponds to 180 lb on the earth).

Step 1:   The ratio of the weight on earth to weight on the moon is 6:1 or $\frac{6}{1}$.

Step 2:   Also, the ratio of the weight on earth to the weight on the moon is 180 lb to x lb
        or $\frac{180}{x}$.

Step 3 :   Equate the two ratios and solve for x.
        $$\frac{6}{1} = \frac{180}{x}$$
        $$6x = 180(1)$$
        $$x = \frac{180(1)}{6}$$
        $$x = 30$$

Therefore, the astronaut 's weight on the moon is 30 lb.

Method 2   Reword the original problem as a proportion:
        6 is to 1 as 180 is to x.

Solution

        Translating, $\frac{6}{1} = \frac{180}{x}$
        Solving, $x = 30$, and we obtain the same solution as by Methods 1 and 2.

Method 3 Unitary Method (Arithmetic)

Whenever the astronaut weighs 6 lb on the earth, the astronaut weighs 1 lb on the moon.

Whenever the astronaut weighs 1 lb on the earth, the astronaut weighs $\frac{1}{6}$ lb on the moon; and

Whenever the astronaut weighs 180 lb on the earth, the astronaut weighs $\frac{1}{6}$ × 180 lb = 30 lb on the moon.

Method 4: More or Less Method We reword the problem
The weight of an astronaut on earth is directly proportional to the weight on the moon. When the weight on earth is 6 lb, the weight on the moon is 1 lb, find the weight on the moon when the weight on earth is 180 lb..

Solution Weight on the moon when the weight on earth is 6 lb = 1 lb

Weight on moon when weight on earth is 180 lb = 1 ℓb × $\frac{180\ \ell b}{60\ \ell b}$ = 30 lb

Therefore, the astronaut 's weight on the moon is 30 lb.

Method 5   Variation Method (Algebraic)
We reword the problem   The weight of an astronaut on earth is directly proportional to the weight on the moon. When the weight on earth is 6 lb, the weight on the moon is 1 lb, find the weight on the moon when the weight on earth is 180 lb..

Solution
Step 1: Let the weight on earth = h lb
Let the weight on moon = n lb
Then since h is directly proportional to n,
$h = kn$
Step 2: Determine the value of k by substituting $h = 6, n = 1$ in $h = kn$
Then $6 = k(1)$ and from which $k = 6$
Step 3: Substitute $k = 6$ in $h = kn$ to obtain a formula relating h and n.
Then we obtain $h = 6n$
Step 4: Substitute $h = 180$ in the formula, $h = 6n$ and solve for n.
Then $180 = 6n$
$30 = n$
Therefore, the astronaut 's weight on the moon is 30 lb.

More Examples on the Variation Method
(Note that these examples can also be solved using Methods 1, 2, 3 and 4)

Example A: y varies directly as x. When $y = 48, x = 3$. Find x when $y = 96$

Solution

Step 1: $y = kx$ (1)
Step 2: Substitute $y = 40, x = 2$ in equation (1) and solve for k.
$48 = k(3)$
$48 = 3k$
$16 = k$
Substitute $k = 16$ in equation (1) above to obtain
the formula $y = 16x$ (2)
Step 3: Now, to find x when $y = 96$, replace y by 96 in (2)
Then $96 = 16x$
$6 = x$
Therefore when $y = 96, x = 6$.

Example B The distance y an object falls from rest is directly proportional to the square of the time x. When $y = 128, x = 4$. (*a*) Find x when $y = 200'$ (*b*) Find y when $x = 6$.

Solution

Step 1: $y = kx^2$ ....(1)

Step 2: Substitute $y = 128, x = 4$ in equation (1) and solve for k.

Then $128 = k(4)^2$

$128 = 16k$

$k = 8$

Substitute $k = 8$ in equation (1) above to obtain

the formula $y = 8x^2$ .... (2)

Step 3: (a) To find x when $y = 200$. Replace y by 200 in equation (2) and solve for x.

$200 = 8x^2$

$\dfrac{200}{8} = x^2$

$25 = x^2$

$5 = x$

$x = 5$ (Since the time taken must be positive in this problem, we reject -5 as a solution)

(b) To find y when $x = 6$, replace x by 6 in equation (2) and evaluate.

$y = 8(6)^2$

$y = 8(36)$

$y = 288$.

Therefore when $x = 6, y = 288$.

EXTRA

Example 4 Assume volume V is directly proportion to temperature T.

Then V_1 corresponds to T_1 as V_2 corresponds to T_2

Since this is direct proportion, we apply **quotient = quotient**

Then $\dfrac{V_1}{T_1} = \dfrac{V_2}{T_2}$ (quotient = quotient)

The above equation is a familiar formula for gases in physics.

Relative Merits of the Methods for Solving Proportion Problems

Method 1: Quotient = quotient Method

The quotient = quotient method such as $\frac{a}{b} = \frac{c}{d}$ is very popular and is efficient for a one-step problem involving only four quantities, but is not efficient in a problem such as in Method 2 below, since the method has to be repeated in steps with the introduction of a new variable in each step.

Method 2: Units Label Method (Arithmetic)

The Units Label Method, an arithmetic method, can readily do what Method 1 can do, if the quantities involved have units, and is more efficient than Method 1 in doing repeated calculations. This method is particularly very efficient in calculations involving measurements and in chemistry as in the following example.

Example How many yards are there in 96 inches?

Solution We reword this problem as: Convert 96 in. to yd.

Plan: This plan is based on what is available in the conversion chart or table.

(Using a conversion chart is analogous to using a map to go from one location to another)

The sequence is inches--> feet---> yards.

We use these relationships: 12 in. = 1 ft; 3 ft = 1 yd (We could use 36 in = 1 yd)

  (Note: inch = in.)

Step 1: $\frac{96 \text{ in.}}{1} \times \frac{?}{?}$

Step 2: Multiply the $\frac{96 \text{ in.}}{1}$ by a fraction formed by using the quantities 12 in. and 1 ft as   the terms of the fraction such that the denominator has the same units as the 96 in.;

followed by similar multiplication by a fraction using 3 ft and 1 yd as the terms of the fraction. Then, we obtain

$$\frac{96 \text{ in.}}{1} \times \frac{1 \text{ ft}}{12 \text{ in.}} \times \frac{1 \text{ yd}}{3 \text{ ft}} \qquad \left(\frac{96 \text{ in.}}{1} \times \frac{1 \text{ ft}}{12 \text{ in.}} = \frac{96 \text{ in.}}{1} \times \frac{1 \text{ ft}}{12 \text{ in.}} \times \frac{1 \text{ yd}}{3 \text{ ft}} = \frac{96 \times 1 \times 1 \text{ yd}}{1 \times 12 \times 3} \right)$$

$$= 2\tfrac{2}{3} \text{ yd}$$

$\therefore$ 96 in. = $2\tfrac{2}{3}$ yd

Justification of the Units Label Method in the above problem:

From 12 in. = 1 ft, If we divide both sides of this equality by 12 in., we obtain

$$\frac{12 \text{ in.}}{12 \text{ in.}} = \frac{1 \text{ ft}}{12 \text{ in.}}$$

$$1 = \frac{1 \text{ ft.}}{12 \text{ in.}}$$

Similarly, from 3 ft = 1 yd, If we divide both sides by 3 ft., we obtain

$$\frac{3 \text{ ft}}{3 \text{ ft}} = \frac{1 \text{ yd}}{3 \text{ ft}}$$

$$1 = \frac{1 \text{ yd}}{3 \text{ ft}}$$

In step 2, multiplying by $\frac{1 \text{ ft.}}{12 \text{ in.}}$ or by $\frac{1 \text{ yd}}{3 \text{ ft}}$ is equivalent to multiplying by 1, and therefore, the given value remains equivalent.

Note: In the above example, if we were to use the quotient = quotient method, we would have to apply this method twice, with the introduction of a variable in each step. Note above also that it is easy to generalize the justification of the "Units Label Method" when we have equality such as 12 in. = 1 ft, than when we have equivalence such as 8 apples are equivalent to 2 dollars.

Method 3: Unitary Method (Arithmetic)
This is an arithmetic method and does not require any knowledge of algebra and can be taught as the first arithmetic method for solving proportion problems.

Method 4: More or Less Method (Arithmetic)
This method is a shorter form of Method 3, as one step in Method 3 is skipped. When mastered well, the "More or Less Method" can be extended to solve inverse proportion as well as compound proportion problems, and very efficient in stoichiometric calculations in chemistry as well as conversions in measurements.

Method 4: Variation Method (Algebraic)
This has applications at all levels of mathematics. Some textbooks ignore this method at the elementary level and cover it at the intermediate level.

EXTRA: Two Cases of the Statement of Ratio and the Corresponding Proportion Problems

In solving problems involving ratios and proportion, we must distinguish between the cases in which the terms of a given ratio refer to **disjoint sets** and the case in which the ratios refer to sets which are **not** disjoint.

Case 1

The terms of the ratios refer to disjoint sets (example: the set of females and the set of males in a school)
Example: The ratio of the number of females to the number of males is 3 to 2.
Proportion: The ratio 3 **females** to 2 **males** = the ratio 6 **females** to 4 **males**
    or ratio 2 **males** to 3 **females** = ratio 4 **males** to 6 **females**

(Two sets are disjoint if the sets do not have any members common to both sets)
Problems such as " If 8 apples cost 2 dollars, what is the cost of 24 apples" are of case 1

Case 2

The terms do **not** refer to disjoint sets (example: the set of females and the set of the total number of students at a school)

Example

The ratio of the number of **females** to the **total number** of students is 3 to 5.
Proportion: The ratio 3 **females** to 5 **students** = the ratio 6 **females** to 10 **students.**
Case 2 is sometimes worded as **a number** out of a **total number**
For example, 3 out of 5 students are females or out of every 5 students, 3 are females, which is the same as $\frac{3}{5}$ of the students are females.

Problems such as " if 3 out of 5 students in a school are females, and there are 600 females in this school, how many students are in this school?" can also be set up using

$$\frac{\text{what number out of}}{\text{what number}} = \frac{\text{what number out of}}{\text{what number}} \text{ as well as other methods.}$$

Example for Case 2
Example 1   3 out of 5 students at a certain college are females.

(a) If 600 students at this college are females,
    how many students are there at this college?

(b) If one were to collect a sample of 200 students at this
    college, how many of these students would be females?

Solution (a):
Method 1 Using ratio = ratio method
        Let the number of students at this college = x
Then the proportion 3 females are to 5 students as 600 females are to x students translates to: $\frac{3 \text{ females}}{5 \text{ students}} = \frac{600 \text{ females}}{x \text{ students}}$

$$\frac{3}{5} = \frac{600}{x}$$

$$3x = 5(600)$$

$$x = \frac{5(600)}{3}; \; x = 1000$$

Therefore, there are 1000 students at this college.

Method 2: Using $\dfrac{\text{what number out of}}{\text{what number}} = \dfrac{\text{what number out of}}{\text{what number}}$

Let the number of students at this college $= x$

$$\frac{3 \text{ out of}}{5} = \frac{600 \text{ out of}}{x}$$

$$\frac{3}{5} = \frac{600}{x}$$

$$3x = 5(600)$$

$$x = \frac{5(600)}{3}$$

$$x = 1000$$

Therefore, there are 1000 students at this college

Method 3   Since 3 out of 5 means $\dfrac{3}{5}$, we can reword this part of the question as: if $\dfrac{3}{5}$ of a number is 600, what is the number?

Let the number be x

Then $\dfrac{3x}{5} = 600$

$$3x = 3000$$

$$x = 1000$$

There are 1000 students at this college.

Method 4   Divide 600 by $\dfrac{3}{5}$ (If $\dfrac{3}{5}$ of a number is 600, what is the number?)

$$600 \div \frac{3}{5} = \frac{600}{1} \times \frac{5}{3}$$

$$= 1000$$

There are 1000 students at this college.

Method 5: The Units Label Method

Note: 3 out 5 students are females means that for every 5 students, there are 3 females.

Step 1: $\dfrac{600 \text{ females}}{1} \times \dfrac{?}{?}$

Step 2: $= \dfrac{600 \text{ females}}{1} \times \dfrac{5 \text{ students}}{3 \text{ females}}$

$= 1000 \text{ students}$

Part (b)

Method 1:   We can reword this part of the question as: Find $\dfrac{3}{5}$ of 200.

$$\frac{3}{5} \text{ of } 200 = \frac{3}{5} \times \frac{200}{1} = 120$$

That is, of a sample of 200 students, 120 students would be females.

Method 2:: $\dfrac{\text{what number out of}}{\text{what number}} = \dfrac{\text{what number out of}}{\text{what number}}$

Let the number of females $= x$

$$\frac{3 \text{ out of}}{5} = \frac{x \text{ out of}}{200}$$

$$\frac{3}{5} = \frac{x}{200}$$

$$5x = 3(200)$$

$$x = \frac{3(200)}{5}$$

$$x = 120$$

Therefore, there are 120 females.

Method 3: The Units Label Method

Step 1: $\dfrac{200 \text{ students}}{1} \times \dfrac{?}{?}$

Step 2: $\dfrac{200 \text{ students}}{1} \times \dfrac{3 \text{ females}}{5 \text{ students}}$

$= 120 \text{ females}$

Therefore, there are 120 females.

Method 4

Let the number of females. $= x$

Then the proportion 3 females are to 5 students as x females are to 200 students translates to:

$$\frac{3 \text{ females}}{5 \text{ students}} = \frac{x \text{ females}}{200 \text{ students}}$$

$$\frac{3}{5} = \frac{x}{200}$$

$$5x = 3(200)$$

$$x = \frac{3(200)}{5} = 120$$

Therefore, there are 120 females..

Extra Method

We repeat the question in **Example 1**   3 out of 5 students at a certain college are females.

(a) If 600 students at this college are females,
how many students are there at this college?

(b) If one were to collect a sample of 200 students at this
college, how many of these students would be females?

This method follows the algebraic method for dividing a quantity into parts, given the ratio.
Since the ratio of females to the total number of students is 3**:**5,
let the number of females $= 3x$, and
let the total number of students $= 5x$.
(a) Then $3x = 600$ (there are 600 females); and solving, $x = 200$
Therefore, total number of students $=5x = 5(200) =$**1000**
Therefore, there are 1000 students at this college.

(**b**) $5x = 200$ (there are 200 students) , and

$x =40$

Therefore, the number of females $= 3x = 3(40) =120$.

Lesson 18 Exercises A

Problems on direct proportion in which the terms refer to distinct sets

1. If 1 gallon of gasoline costs 2 dollars, how many gallons of gasoline can one buy for 15 dollars?   Answer: $7\frac{1}{2}$ gallons

2. If 10 gallons of gasoline cost 15 dollars, find the cost of 24 gallons of gasoline.
Answer: 36 dollars

3. If 10 gallons of gasoline cost 15 dollars, how many gallons of gasoline can one buy for 45 dollars?   Answer: 30 gallons

4. If x gallons of gasoline cost y dollars, what is the cost of p gallons of gasoline
(Express your answer in terms of x y and p)   Answer: $\frac{py}{x}$ dollars

5. If on a gallon of gasoline, a car can travel 24 miles, how many miles can the car travel on 10 gallons of gasoline?   Answer: 240 miles

6. If on 2 gallons of gasoline, a car can travel 24 miles, how many miles can a car travel on 20 gallons of gasoline?   Answer: 240 miles

7. If on 3 gallons of gasoline, a bus can travel 36 miles, how many gallons of gasoline are needed to travel 60 miles?   Answer: 5 gallons

8. If on p gallons of gasoline, a bus can travel r miles, how many gallons of gasoline are needed to travel m miles? (Express your answer in terms of the variables.)   Answer: $\frac{mp}{r}$ gallons

9. If 4 oranges cost 1 dollar, how many oranges can one buy for 20 dollars?
Answer: 80 oranges

10. If 8 oranges sell for 2 dollars, find the selling price for 24 oranges.   Answer: 6 dollars

11. If 12 oranges cost 3 dollars, what is the cost of 60 oranges?   Answer: 15 dollars

12. If r oranges cost p dollars, what is the cost of m oranges?   Answer: $\frac{mp}{r}$ dollars
(Express your answer in terms of the variables.)

13. If a household uses one pound of sugar in 3 weeks, how many pounds of sugar would be used in 6 weeks?   Answer: 2 pounds

14. If a household uses 2 pounds of sugar in 3 weeks, how many pounds of sugar would be used in 51 weeks?   Answer: 34 pounds

15. If a household uses 2 pounds of sugar in 3 weeks, 12 pounds of sugar would be used in how many weeks?   Answer: 18 weeks

16. If a household uses x pounds of sugar in r weeks, how many pounds of sugar would be used in y weeks? (Express your answer in terms of the variables.)   Answer $\frac{xy}{r}$ pounds

17. If one teacher is needed for a class of 22 students, how many teachers would be needed to teach 88 students, assuming that each class has 22 students?   Answer: 4 teachers

18. If 6 teachers are needed to teach 102 students, how many students would be taught by 3 teachers?   Answer: 51 students

19. If t teachers are needed to teach s students, how many students would be taught by b teachers? (Express your answer in terms of the variables.)   Answer: $\frac{bs}{t}$ students

20. If 4 grams of hydrogen are required to produce 36 grams of water, how many grams of hydrogen would be required to produce 108 grams of water?   Answer:: 12 grams

21, If from 4 grams of hydrogen, 36 grams of water are produced, how many grams of hydrogen are needed to produce 216 grams of water?   Answer: 24 grams

22. If from t grams of hydrogen, s grams of water are produced, how many grams of hydrogen are needed to produce b grams of water (Express your answer in terms of the variables   Answer: $\frac{bt}{s}$ grams

23. If p gallons of water are needed to produce x pounds of a concrete mixture, how many gallons of water are needed to produce y pounds of this mixture?   Answer: $\frac{py}{x}$ gallons

24. 1f each room in a building has 3 windows, how many windows are there in 6 rooms,?   Answer: 18 weeks

25. If 3 rooms in a building have 12 windows, how many windows are in 36 rooms, assuming that each room has the same number of windows?
Answer: 144 windows

26. If p rooms in a building have q windows, how many windows are in r rooms, assuming that each room has the same number of windows? (Express answer in terms of the variables.)
Answer: $\frac{qr}{p}$

27. A buyer purchased 20 items for 60 dollars. How many items can be purchased for 30 dollars, assuming that the cost is the same for each of the items.   Answer: 10 items

28. A buyer purchased r items for s dollars. How many items can be purchased for t dollars, assuming that the cost is the same for each of the items. (Answer in terms of the variables.)
Answer: $\frac{rt}{s}$

29. In 2 hours, a car can travel 120 miles. Find the distance the car can travel in 3 hours, assuming the same speed   Answer:: 180 miles

30. In x hours, a car can travel y miles. Find the distance the car can travel in k hours, assuming the same speed.   Answer: $\frac{ky}{x}$

31. In 2 hours, a car can travel 120 miles, At the same speed, how long will it take the car to travel 180 miles?.   Answer: 3 hours

32. In x hours, a car can travel y miles. At the same speed, how long will take the car to travel t miles? (Express your answer in terms of the variables)   Answer: $\frac{xt}{y}$

33. If 40 liters of air are needed to ventilate a patient for 2 minutes, how many liters of air are needed to ventilate this patient for 24 minutes? Answer: 480 liters

34. If a liters of air are needed to ventilate a patient for b minutes, how many liters of air are needed to ventilate this patient for c minutes? Answer $\frac{ac}{b}$

35. If $4\frac{1}{2}$ gallons of gasoline cost 6 dollars, how many gallons of gasoline can one buy for 16 dollars?   Answer: 12 gallons

36. If there are 12 inches in 1 foot, how many inches are there in 4 feet?   Answer: 48 inches

37. If there are 12 inches in I foot, how many inches are there in x feet?
(answer in terms of x)   Answer: $12x$ inches

38. If a person's heart beats 4 times every 3 seconds, how many times will it beat in 21 seconds.
Answer: 28 times

39. To prepare an optimum mixture of gasohol which consists of alcohol and gasoline, Alexander used 20 liters of alcohol and 180 liters of gasoline. How many liters of alcohol are needed if 270 liters of gasoline are used and the composition ratio of the mixture remains unchanged?   Answer: 30 liters

40. Charles' Law states that at constant pressure, the volume of a gas is directly proportional the temperature of the gas. If when the volume of a given mass of a gas is 4.2 L, the temperature is $309°K$, find the volume when the temperature is $297°K$ Answer : 4.0 L

41. Gay-Lussac's Law states that at constant volume, the pressure on a gas is directly proportional to the temperature of the gas. If when the pressure on a given mass of a gas is 720 mm Hg, the temperature is $303°K$, find the pressure when the temperature is $309°K$?   Answer: 734 mm Hg

4 2 The ratio of a man's weight on the earth to his weight on the moon is 6:1. If the man weighs (on earth) 240 lb, what is his weight on the moon?   Answer: 40 l

Lesson 18 Exercises B
Problems on direct proportion in which some terms refer to non distinct sets

1. If 2 out of 5 students at a college are males and there are 2,800 students at this college, how many of the students are males?   Answers: 1,120

2. If 2 out of 5 students at a college are males and there are 2,800 students at this college, how many of the students are females?   Answers: 1,680

3. If 2 out of 5 students at a college are males and there are 1,120 males at this college, how many students are at this college?   Answers: 2,800

4.. If 40% of students at a college are males and there are 1,120 males at this college, how many students are at this college?   Answers: 2,800

5. If 40% of the students at a college are males and there are 2,800 students at this college, how many of the students are males? Answers: 1,120

6. Is $\frac{2}{5}$ of a number is 16, what is the number? Answer: 40

7. James and Mary are to divide some amount of money and James is to receive $\frac{5}{12}$ of this amount. If James receives 30 dollars, what is the amount of money to be divided?
Answer: 72 dollars

8. If 2 out 3 dollars in a bank account are to be given to Diana, If Diana was given $30,000. how much money is in this bank account? Answer: $45,000

9. If 1 out of 3 liters of an alcohol-water solution is pure alcohol, and this solution contains 12 liters of pure alcohol, what is the total volume of this alcohol-water solution?
Answer 36 liters

10. If q out of p fruits are oranges, and there are r oranges, how many fruits are there. Obtain an expression for the number of fruits in terms of p, q, and r. Answer: $\frac{pr}{q}$

11. In a math class, 2 out of 4 students passed the final exam. If 32 students passed the exam, how many students are in this class? Answer: 64

Lesson 18 Exercises C

1. y varies as x. When $y = 12$, $x = 4$. Find y when $x = 10$.

2. The distance S an object falls from rest is directly proportional to the square of the time t. If when $S = 64$, $t = 2$, find S when $t = 4$.

3. W is proportional to V. When $W = 120$, $V = 3$, find W when $V = 8$.

4.. S is directly proportional to t.. When $S = 64$, $t = 2$. Find t when $S = 288$.

5. Express by means of an equation: $(a - b)$ is proportional to $(c - d)$.

6. The distance S an object falls from rest varies directly as the square of the time t. If when $S = 32$, $t = 1$, find S when $t = 5$.

Answers: 1. 30; **2.** 256; **3.** 320; **4.** 9; **5.** $a - b = k(c - d)$, where k is a constant; **6.** 800.

Lesson 18 Exercises D

1. At constant pressure, the volume V of a perfect gas varies as the absolute temperature T.
If when the temperature is 2730 Absolute, the volume of one gram of a gas is 1.7 cm^3, find the volume when the absolute temperature is 4830°.

2. According to Hooke's law, the force required to stretch a spring is directly proportional to the elongation of the spring. If a 20 lb force stretches a spring 6 in., what force will be required to stretch it 8 in.?

3. The cost C of gasoline is proportional to the number of gallons N of gasoline. If 10 gallons of gasoline cost $12, what is the cost of 25 gallons of gasoline? How many gallons of gasoline can $18 purchase?

Answers: 1. 3.0 cm^3; **2.** 27 lb; **3.** $30; 15 gallons

Lesson 19
8 0

Methods for Solving Inverse Proportion Problems

It is very interesting but not surprising that the operations in the steps for solving inverse proportion problems are the opposite operations (inverse operations) in the steps for solving direct proportion problems. By operations, we mean multiplication and division are opposite (inverse) operations. Thus if we know the operations in the steps for solving a direct

proportion problem, we can deduce the operations in the steps for solving an inverse proportion problem (by using the opposite operations). For every method for solving direct proportion problems, there is a corresponding method for solving inverse proportion problems, with the steps in the inverse method obtained from the direct proportion method by using opposite operations.

Method 1: **Product = Product Method** (Algebraic)

      Assume a is inversely proportional to b.

    (same as a is proportional to the **reciprocal** of b)

      Then a_1 corresponds to to b_1, and a_2 corresponds to b_2. and

$$a_1 : \frac{1}{b_1} \text{ as } a_2 : \frac{1}{b_2}$$

$$a_1 \div \frac{1}{b_1} = a_2 \div \frac{1}{b_2}$$

$$a_1 b_1 = a_2 b_2 \qquad (\textbf{product = product})$$

      Or if we use the familiar terms a, b, c, d as the terms of the inverse proportion

      Then $a : \frac{1}{b}$ as $c : \frac{1}{d}$

$$a \div \frac{1}{b} \text{ as } c \div \frac{1}{d}$$

      $ab = cd$ (**product = product**) **<------** inverse proportion

We will use the product = product form since this form is easy to remember.

Compare: **Direct proportion** $\frac{a}{b} = \frac{c}{d}$ **<------**(**quotient = quotient**)

           Inverse proportion $ab = cd$ **<------**(**product = product**)

Example 1   2 people can paint a house in 6 days. How many people would be required to paint the house in 3 days? Derive a formula relating the number of people and the number of days; and use this formula to find the number of people required to paint the house in 3 days.

Method 1 Product = Product Method

It is known that as the number of people increases, the number of days required decreases, and vice versa.

      Let the number of people $= P$,

      Let the number of days D.

      Then P is indirectly proportional to D,

      P_1 corresponds indirectly to D_1 as P_2 corresponds indirectly to D_2.

      Since this is indirect proportion, $P_1 D_1 = P_2 D_2$.

      From the problem, $P_1 = 2$, $D_1 = 6$, $P_2 = ?$. $D_2 = 3$.

      Substituting these values in the equation

$$2(6) = P_2(3)$$

$$\frac{2(6)}{3} = P_2$$

$$4 = P_2$$

Therefore, 4 people are required to paint the house in 3 days.

Method 2: Units Label Method for Inverse Proportion
By reversing the steps in the corresponding direct proportion method (Method 2), we obtain the approach covered below.

We repeat the question: 2 people can paint a house in 6 days. How many people would be required to paint the house in 3 days?

Step 1: Find the product, 2 people •(6 days)

Step 2: Divide : $\dfrac{2 \text{ people} \cdot (6 \text{ days})}{3 \text{ days}}$

$$\frac{2 \text{ people} \cdot (6 \overset{2}{\cancel{\text{ days}}})}{3 \cancel{\text{ days}}}$$

$= 4$

Therefore, 4 people are required to paint the house in 3 days.

Method 3: Unitary Method (Arithmetic)

We repeat the question: 2 people can paint a house in 6 days. How many people would be required to paint the house in 3 days?

Solution

The number of people required for 6 days = 2

The number of people required for 1 day = 2×6 (Multiply. More people are needed)

The number of people required for 3 days = $\dfrac{2 \times 6}{3}$ (We divide. Less people are needed)

$= 4$

Therefore, 4 people are required to paint the house in 3 days.

Method 4: More or Less Method

We repeat the question: 2 people can paint a house in 6 days. How many people would be required to paint the house in 3 days?

Solution

Step 1: The number of people needed to paint the house in 6 days = 2

Step 2: The number of people needed to paint the house in 3 days = $2 \times \dfrac{6}{3}$

$= 4$ (more people needed)

Therefore, 4 people are required to paint the house in 3 days.

(Note that in Step 2, in forming a fraction using 3 and 6, we used an improper faction, $\dfrac{6}{3}$ (fraction greater than or equal to 1), since we expect the number of people to be greater than 2. For the painting to be done faster, more people are needed. If we had expected the number of people to be less than 2, we would have used the proper fraction $\dfrac{3}{6}$). Master this arithmetic method.

Method 5: Variation Method (Algebraic) **We reword the question:**

The number of people needed to paint a house varies inversely as the number of days taken to paint the house. If when the number of people is 2, the number of days is 6, find the number of people when the number of days is 3.

Step 1: Let the number of people = n
Let the number of days = d
Then since n varies inversely as d, (or n is proportional to the **reciprocal** of d

$$n = \frac{k}{d}$$

Step 2: Determine the value of k by substituting $n = 2, d = 6$ in $n = \dfrac{k}{d}$

Then $2 = \dfrac{k}{6}$ and from which $k = 12$.

Step 3:   Substitute $k = 12$ in $n = \dfrac{k}{d}$ to obtain a formula relating n and d.

Then we obtain $n = \dfrac{12}{d}$

Step 4: Substitute for $d = 3$ in the formula . $n = \dfrac{12}{d}$ and solve for n

Then $n = \dfrac{12}{3} = 4$

Thus when $d = 3, n = 4$

Therefore, 4 people are required to paint the house in 3 days.

One more example on Method 5

Example 2 The number of people required to sort a quantity of letters varies inversely as the time taken to sort the letters. If 8 people can sort a quantity of letters in 3 hours, how many people would be required to sort the same quantity of letters in 2 hours?

Let the number of people required to sort the letters in x hours be y

Step 1: Then $y = \dfrac{k}{x}$ (1)

Step 2: When $x = 3$, $y = 8$. Substitute these values in equation (1) and solve for k.

$8 = \dfrac{k}{3}.$

Solving, $k = 24$.

Substitute for $k = 24$ in equation (1) above to obtain

the formula $y = \dfrac{24}{x}$ ....(2)

Step 3. when $x = 2, y = \dfrac{24}{2}$ (Replacing x by 2 in equation (2))

$y = 12$

Therefore, to sort the letters in 2 hours would require 12 people.

Example 4 Assume pressure P is indirectly proportional to volume V 8 3

Then P_1 corresponds to V_1 as P_2 corresponds to V_2
Since P is inversely proportional to V,

$P_1:\dfrac{1}{V_1}$ as $P_2:\dfrac{1}{V_2}$ and from which

Then $P_1V_1 = P_2V_2$ (product = product)
The above equation is a familiar formula for gases in physics.

Lesson 19 Exercises A (Inverse Proportion)

1. If 2 people can paint a house in 6 days, how many people would paint the house in 3 days?
                                                                    Answer: 4 people

2. At a speed of 60 miles per hour, a car takes 2 hours to travel from City A to City B. At a speed of 40 miles per hour, how long will it take the car to ravel from City A to City B?
                                                                    Answer: 3 hours

3. If 6 mail men can sort a given quantity of mail in 4 hours, how many mail men would be needed to sort the same quantity of mail in 2 hours?
                                                                    Answer: 12 men

4. According to Boyle's law, at constant temperature, the volume of a given mass of a gas is inversely proportional to pressure on the gas. When the volume of a gas is 50 cu. in, the pressure is 10 lb per sq. in. What is the pressure (a) when the volume is 80 cu. in.? ; (b) when the volume is 15 cu. in.?; (c) What is the volume when the pressure is 40 lb per sq. in.?   Answer: (a) $6\frac{1}{4}$ lb per sq. in.; (b) $33\frac{1}{3}$ lb per sq. in; (c) $12\frac{1}{2}$ u. in;

5. The speed of rotation of meshed gears is inversely proportional to the number of teeth. A gear with 12 teeth rotates at 840 rpm. What is the rotation speed of a 16-tooth gear?

                                                                    Answer: 630 rpm
6. The electric current in a circuit is inversely proportional to the resistance of the circuit. If the current is 22 amps when the resistance is 5 ohms, find the current when the resistance is 10 ohms.   Answer: 11 amps

Lesson 19 Exercises B

1. At constant pressure, the volume V of a perfect gas varies as the absolute temperature T.

   If when the temperature is 2730° Absolute, the volume of one gram of a gas is 1.7 cm^3, find the volume when the Absolute temperature is 4830°.

2. According to Hooke's law, the force required to stretch a spring is directly proportional to the elongation of the spring. If a 20 lb force stretches a spring 6 in., what force will be required to stretch it 8 in.?

3. The cost C of gasoline is proportional to the number of gallons N of gasoline.
   If 10 gallons of gasoline cost $12, (a) what is the cost of 25 gallons of gasoline? and (b) how many gallons of gasoline can $18 buy?

Answers: **1.** 3.0 cm^3; **2.** $26\frac{2}{3}$ lb; **3.** (a) $30, (b) 15 gallons

Lesson 19 Exercises C

1. y varies inversely as x. If when $y = 6, x = 2$, determine y when $x = 8$.

2. F is indirectly proportional to the square of D. When $F = 30, D = 3$. Find F when $D = 2$.

3. W is indirectly proportional to V. When $W = 15, V = 45$ find W when $V = 60$.

Answers: **1.** $\frac{3}{2}$; **2.** $67\frac{1}{2}$; **3.** $11\frac{1}{4}$;

Lesson 19 Exercises D

1 According to Boyle's law, at constant temperature, the volume of a given mass of a gas varies inversely as the pressure on the gas. When the volume of a gas is 50 in.^{3}, the pressure is 10 lb per $in.^2$. What is the pressure (a) when the volume is 80 in.^{3}? ; (b) when the volume is15 in.^{3} ?; (c) Find the volume when the pressure is 40 lb per in.^{2}.?

2. If 8 people can complete a piece of work in 40 days, how many people will complete the same piece of work in I0 days? Assume that all the people work at the same rate as one another.

3. At constant temperature, the volume of a given mass of a gas is inversely proportional to the pressure on the gas. When the volume is 150 cu. in,. the pressure is 30 lb per $in.^2$. Find the pressure (a) when the volume is 60 in.^{3} ?; (b) when the volume is 300 in.^{3}.

Ans:**1.** (a) $6\frac{1}{4}$ lb; (b) $33\frac{1}{3}$ lb per in.^{2}; (c) $12\frac{1}{2}$ in.^{3} **2.** 32 people; **3.** (a) 75 lb per in.^{2}.; (b) 15 lb per in.^{2}.

Lesson 20

Compound Proportion (Variation) Problems

Deriving formulas for any proportion problem, with any number of variables and involving both direct and inverse relationships

Case 1: Joint Proportion (Joint Variation)

If a quantity varies **directly** as **two** or more other quantities (i.e. as the product of two or more other quantities), we call such a proportion **joint proportion or joint variation**.

Example 1: If z varies directly as x and y, then

$$z = kxy \text{ or } \frac{z}{xy} = k \text{ (joint variation)}$$

or using subscripted variables, $\frac{z_1}{x_1 y_1} = \frac{z_2}{x_2 y_2}$ <-- Faster form since you do not have to find k first.

Case 2: Joint Inverse Proportion (Joint Inverse Variation)

The following type of proportion may not be found in current textbooks. However, the author believes that for completeness, this proportion should be included. See the two application examples below.

If a quantity varies **indirectly** as **two** or more other quantities (i.e. as the inverse of the product of two or more other quantities), we will call such a proportion inverse **joint proportion or** inverse **joint variation**.

Example   If z varies **indirectly** as x and y, then

$$z = \frac{k}{xy} \text{ or } xyz = k \text{ (inverse joint variation)}$$

or using subscripted variables, $z_1 x_1 y_1 = z_2 x_2 y_2$ <-- Faster form

Examples

1. A weight watcher exercising 3 hours per day, 5 times a week lost 10 pounds in 4 months. How many months will it take the weight watcher to lose 10 pounds exercising 2 hours per day, 6 times a week?   **Answer:** 5 months

2. A car traveling at 48 mph and for 6 hours per day covered a certain distance in 10 weeks. How many weeks will it take the car traveling at 60 mph and for 4 hours per day, to cover the same distance?   **Answer:** 12 weeks

Case 3 Combined Proportion (Combined Variation)

If a quantity varies **directly** as one quantity (or as two or more quantities) and **inversely** as another quantity (or other quantities), we call such a variation **combined variation**.

Example 2: If z varies directly as x and inversely as y, then

$$z = k\frac{x}{y} \text{ or } \frac{zy}{x} = k \text{ (combined variation)}$$

or using subscripted variables, $\frac{z_1 y_1}{x_1} = \frac{z_2 y_2}{x_2}$ <-- Faster form since you do not have to find k first.

Lesson 20: Compound Proportion (Compound Variation) Problems

Example 3: Assume that pressure P is directly proportional to the temperature T, and inversely proportional to the volume V.

Step 1: Then P_1 corresponds to T_1 directly, and to V_1 inversely as P_2 corresponds to T_2 directly, and to V_2 inversely.

Step 2: Write down P_1 (the first quantity.)

Step 3: For each of the other quantifies, if it is directly proportional to P_1, then it divides P_1 (i.e., write it as a divisor), but if it is inversely proportional to P_1, it multiplies P_1 (that is write it as a factor in the numerator). Similarly, repeat the process beginning with P_2.

$$\text{Then } \frac{P_1 V_1}{T_1} = \frac{P_2 V_2}{T_2}, \text{ a familiar gas law in physics.}$$

Example 4: Assume that A is directly proportional to B and directly proportional to C.

Step 1: Then A_1 corresponds to B_1 directly, and to C_1 directly as A_2 corresponds to B_2 directly, and to C_2 directly.

Step 2: Write down A_1 (the first quantity).

Step 3: For each of the other quantities if it is directly proportional to A_1, then it divides A_1 (That is write it as a divisor) but if it is inversely proportional to A_1, it multiplies A_1 (i.e., write it as a factor in the numerator). Similarly, on the right side of the equality symbol, repeat the process beginning with A_2.

$$\text{Then } \frac{A_1}{B_1 C_1} = \frac{A_2}{B_2 C_2}$$

Example 5 Assume that the temperature T is directly proportional to the pressure, P, and directly proportional to the volume, V.

Step 1: Then T_1 corresponds to P_1 directly, and to V_1 directly as T_2 corresponds to P_2 directly, and to V_2 directly.

Step 2: Write down T_1 (the first quantity).

Step 3: For each of the other quantifies if it is directly proportional to T_1, then it divides T_1 (i.e., write it as a divisor) but if it is inversely proportional to T_1, it multiplies T_1 (i.e., is write it in the numerator as a factor) Similarly, repeat the process beginning with T_2.

$$\text{Then } \frac{T_1}{P_1 V_1} = \frac{T_2}{P_2 V_2}$$

By inverting both sides of the above equation, we obtain

$$\frac{P_1 V_1}{T_1} = \frac{P_2 V_2}{T_2}, \text{ the familiar gas law.}$$

Question: Can we obtain the same equation in above example, given the following information?:
Assume volume V is directly proportional to temperature T. and inversely proportional to the pressure P. Then V_1 corresponds to T_1 directly, to P_1 inversely as V_2 corresponds to T_2 directly, and to P_2 inversely.

$$\text{Translating into an equation } \frac{V_1 P_1}{T_1} = \frac{V_2 P_2}{T_2} \text{ or equivalently }, \frac{P_1 V_1}{T_1} = \frac{P_2 V_2}{T_2}. \text{ Answer: Yes.}$$

Example 6: Assume that A is inversely proportional to B and inversely proportional to C.

Then A_1 corresponds to B_1 inversely, and to C_1 inversely as A_2 corresponds to B_2 inversely, and to C_2 inversely.

Then, following the agreement made in the previous examples, we obtain

$A_1 B_1 C_1 = A_2 B_2 C_2$ (**product = product**)

Solving Compound Proportion Problems Using Arithmetic   8 7

In the previous examples, we covered how to solve compound proportion problems using the variation method. However, we can also solve compound proportion problems using **arithmetic.**
We can apply **Method 4** of Lesson 18 (p.66) and Lesson 19 (p.81), (The "More or Less Method")

Example 7. If 8 men can make 20 tables in 6 days, how many men would be required to make 60 tables in 9 days?

Solution:

The number of men required to make 20 tables in 6 days = 8 men

The number of men required to make 60 tables in 9 days = 8 men $\times \dfrac{60 \text{ tables}}{20 \text{ tables}} \times \dfrac{6 \text{ days}}{9 \text{ days}}$

(See p.66 and p.81 for setting up the fractions) $= 8 \text{ men } \times \dfrac{60}{20} \times \dfrac{6}{9} = 16 \text{ men}$

Lesson 20 Exercises A
(Compound Proportion)

1. If 5 people working 6 hours per day can complete a job in 8 days, how many people working 2 hours per day can complete the same job in 10 days?   **Answer:** 12 people

2   If 12 people working 4 hours per day, 3 days per week can complete a job in 4 years, how many people working 9 hours per day, 4 days per week can complete the same job in 2 years?   **Answer:** 8 people

3. A weight watcher exercising 3 hours per day, 5 times a week lost 10 pounds in 4 months. How many months will it take the weight watcher to lose 10 pounds exercising 2 hours per day, 6 times a week?   **Answer:** 5 months

4. A car traveling at 48 mph and for 6 hours per day covered a certain distance in 10 weeks. How many weeks will it take the car traveling at 60 mph and for 4 hours per day, to cover the same distance?   **Answer:** 12 weeks

5   If 10 people working 6 hours per day, 7 days per week earned a certain amount of income each week, how many people working 4 hours per day, 5 days per week would earn the same income in one week?   **Answer:** 21 people

6   Six men working 2 days a week can build a fence in 8 weeks. How many men working 4 days a week will build the same fence in 12 weeks.   **Answer:** 2 men

7. Six men working 2 days a week can build a fence in 8 weeks. How many weeks will it take 3 men working 4 days a week to build the same fence.   **Answer:** 8 weeks

8. If 8 men can make 20 tables in 6 days, how many men would be required to make 60 tables in 9 days?   **Answer:**   16 men

Lesson 20 Exercises B

1. Z varies jointly as x and y. If when $x = 2$, $y = 3$, $Z = 16$, what is Z when $x = 5$ and $y = 4$?

2. V is indirectly proportional to P and directly proportional to T. When $V = 30$, $T = 25$, $P = 15$. Find V when $T = 40$, and $P = 3$.

3. F is directly proportional to G and varies inversely as L. When $F = 256$, $G = 36$, and $L = 30$. What is G, when $F = 128$ and $L = 15$?

4. The kinetic energy E of a moving object varies as the mass M. of the object and the square of the velocity V. If when $E = 50$, $M = 4$, $V = 5$, find E when $M = 16$ and $V = 3$.

5. If 12 people working 2 days a week can complete a piece of work in 4 weeks, how many people working 3 days per week can complete the same piece of work in 2 weeks? (Assume that all the people involved work at the same rate as one another)

Answers: **1.** $53\frac{1}{3}$; **2.** 240; **3.** 9; **4.** 72; **5.** 16 people;

EXTRA

Determining the type of relationship, given an equation

Determine how the first variable is related to each of the other variables in each equation.

(a) $\dfrac{P_1 V_1}{T_1} = \dfrac{P_2 V_2}{T_2}$;   **(b)** $\dfrac{A_1 B_1}{C_1} = \dfrac{A_2 B_2}{C_2}$;   **(c)** $\dfrac{B_1}{D_1} = \dfrac{B_2}{D_2}$;   **(d)** $\dfrac{A_1 B_1 C_1}{D_1 E_1} = \dfrac{A_2 B_2 C_2}{D_2 E_2}$

Solution
(a) P is inversely proportional to V and directly proportional to T
(b) A is inversely proportional to B and directly proportional to C
(c) B is directly proportional to D.
(d) A is inversely proportional to B, inversely proportional C, directly proportional to D, and directly proportional E.

Comparison of Sample Formulations of Proportion Problems

Proportion Formulation (subscript notation)	Variation Formulation	Statement of relationship
1. $\dfrac{V_1}{T_1} = \dfrac{V_2}{T_2}$	$\dfrac{V}{T} = k$ or $V = kT$	V is directly proportional to T
2. $P_1 V_1 = P_2 V_2$	$PV = k$ or $P = \dfrac{k}{V}$	P is inversely proportional to V
3. $\dfrac{P_1 V_1}{T_1} = \dfrac{P_2 V_2}{T_2}$	$\dfrac{PV}{T} = k$ or $PV = kT$	P is inversely proportional to V and directly proportional to T.
4. $I_1 R_1 = I_2 R_2$	$IR = k$ or $I = \dfrac{k}{R}$	I is inversely proportional to R
5. $E_1 I_1 = E_2 I_2$	$EI = k$ or $E = \dfrac{k}{I}$	E is inversely proportional to I
6. $\dfrac{E_1}{N_1} = \dfrac{E_2}{N_2}$	$\dfrac{E}{N} = k$ or $E = kN$	E is directly proportional to N
7. $I_1 N_1 = I_2 N_2$	$IN = k$ or $I = \dfrac{k}{N}$	I is inversely proportional to N
8.		
9.		

CHAPTER 7

Lesson 21: **Percent (%) and Inter-conversions**
Lesson 22: **Calculations Involving Percent**
Lesson 23: **More Applications Involving Percent**

Lesson 21

Percent (%) and Inter-conversions

Some interpretations of the percent symbol "%":

1. Over hundred: For example, 20% means $\frac{20}{100}$ (Twenty over hundred or 20 divided by 100)

2. For each hundred: For example, a savings account with a 5% interest rate pays the depositor $5 for each $100 in the account. (Five for each hundred)

3. Hundredths: For example, 20% means $\frac{20}{100}$ or .20   (Twenty hundredths)

4. As a number out of 100: For example, a grade of 80% on a test means a student got 80 points out of 100 points.

Changing a decimal (or any fraction) to percent

Procedure: Multiply by 100%. (i.e., multiply by 100 and attach the percent symbol "%".

To change a decimal to percent, move the decimal point two places to the right and attach the percent symbol. (Note that moving the decimal point two places to the right is equivalent to multiplying by 100)

Example 1 Convert .74 to percent.
Solution $.74 = 74\%$.

Example 2 Convert .008 to percent.
Solution $.008 = .8\%$

Example 3 Convert 1 to percent.
 Solution $1 = 100\%$

Example 4 Convert 12 to percent.
Solution $12. = 1200\%$

Example 5   Change $\frac{1}{4}$ to percent

Method 1 $\frac{1}{4} \times \frac{100\%}{1} = 25\%$

Method 2 $\frac{1}{4} = .25 = .25 \times 100\%$ (Changing to decimal first, and then changing to percent)

Changing a percent to a decimal

Procedure: Divide by 100%. (i.e., drop the percent symbol "%" and divide by 100) or apply the meaning of the percent symbol to change the percent to a decimal fraction and then easily to a decimal.

To change a percent to a decimal, drop the percent symbol and move the decimal point two places to the left . (**Note** that moving the decimal point two places to the left is equivalent to dividing by 100)

Case 1: Example 1 $\quad 74\% = .74 \quad (74\% = \frac{74}{100} = .74)$

      Example 2 $\quad .8\% = .008 \quad$ (or $.8\% = \frac{.8}{100} = .008$)

      Example 3 $\quad 145\% = 1.45$

      Example 4 $\quad 14.5\% = .145$

      Example 5 $\quad .145\% = .00145$

Case 2: Example 6 Convert $84\frac{2}{3}\%$ to a decimal

$$84\frac{2}{3}\% = \frac{84\frac{2}{3}}{100}$$

$$= .84\frac{2}{3} \text{ (moving the decimal point two places to the left)}$$

Example 7 Convert $84\frac{2}{3}\%$ to a decimal rounded-off to the nearest **hundredth.**

$$84\frac{2}{3}\% = 0.846... \approx 0.85 \quad \left(\frac{254}{3}\% = 84.6...\% = .846... \approx .85\right)$$

Example 8 Convert $8\frac{1}{4}\%$ to a decimal.

$$8\frac{1}{4}\% = 0.08\frac{1}{4} \quad \longleftarrow \text{-------- Complex decimal.}$$

However, since $\frac{1}{4} = 0.25$, a terminating decimal,

$$8\frac{1}{4}\% = 0.08\frac{1}{4} = 0.0825 \quad \text{(Also, } 8\frac{1}{4}\% = \frac{33}{4}\% = 8.25\% = \frac{8.25}{100} = .0825.\text{)}$$

Example 9 Convert $4\frac{2}{3}\%$ to a decimal

$$4\frac{2}{3}\% = \frac{4\frac{2}{3}}{100}$$

$$= .04\frac{2}{3} \quad \text{(moving the decimal point two places to the left and writing a zero to hold place)}$$

To check: Let us convert $.04\frac{2}{3}$ to a percent.

$$.04\frac{2}{3} = .04\frac{2}{3} \times 100\%$$

$$= 4\frac{2}{3}\% \quad \text{(moving the decimal point two places to the right and attaching the percent symbol)}$$

Changing a Percent to a Fraction (in its lowest terms)

Example 1. Convert 25% to a fraction in its lowest terms.

Solution: $25\% = \frac{25}{100} = \frac{1}{4}$.

Example 2. Convert 23% to a fraction in its lowest terms.

Solution: $23\% = \frac{23}{100}$.

Example 3. Convert 74% to a fraction in its lowest terms

Solution: $74\% = \frac{74}{100} = \frac{37}{50}$.

Example 4 Convert $4\frac{2}{3}\%$ to a fraction

$$4\frac{2}{3}\% = \frac{4\frac{2}{3}}{100}$$

$$= \frac{14}{300}$$

$$= \frac{7}{150}$$

Example 5 Convert $.4\frac{2}{3}\%$ to a fraction

$$.4\frac{2}{3}\% = \frac{.4\frac{2}{3}}{100}$$

$$= .4\frac{2}{3} \div 100$$

$$= \frac{14}{30} \div 100$$

$$= \frac{14}{3000}$$

$$= \frac{7}{1500}$$

Note: Attaching the % symbol is equivalent to dividing by 100; and dropping the % symbol is equivalent to multiplying by 100.

Lesson 21 Exercises

A. Convert to a decimal: **1.** 23%; **2.** 8.25%; **3.** 0.8%; **4.** 10%; **5.** 10.5%; **6.** 8%

Answers: **1.** 0.23 ; **2.** 0.0825 **3.** 0.008; **4.** 0.10 or 0.1; **5.** 0.105; **6.** 0.08

B. Convert to a decimal: **1.** $2\frac{2}{3}\%$; **2.** $25\frac{2}{7}\%$; **3.** $4.3\frac{5}{11}\%$; **4.** $0.16\frac{2}{3}\%$; **5.** 8.2%; **6.** $64\frac{3}{11}\%$

Answers: **1.** $0.02\frac{2}{3}$; **2.** $0.25\frac{2}{7}$; **3.** $0.043\frac{5}{11}$; **4.** $0.0016\frac{2}{3}$. **5.** 0.082; **6.** $0.64\frac{3}{11}$

C. Convert to a fraction in its lowest terms

1. 24% ; **2.** 63% ; **3.** 96%; **4.** 8.2%

Answers: **1.** $\frac{6}{25}$; **2.** $\frac{63}{100}$; **3.** $\frac{24}{25}$; **4.** $\frac{41}{500}$.

D. Convert to a fraction in its lowest terms

1. $2\frac{2}{3}\%$; **2.** $25\frac{2}{7}\%$; **3.** $4.3\frac{5}{11}\%$; **4.** $0.16\frac{2}{3}\%$.

Answers: **1.** $\frac{2}{75}$; **2.** $\frac{177}{700}$; **3.** $\frac{239}{5500}$; **4.** $\frac{1}{600}$.

Lesson 22
Calculations Involving Percent (%)

In calculations involving percent, three main quantities are involved, namely the percentage, the base, and the rate percent. Some authors call the percentage the amount. n these problems, you are usually given two of these quantities and you are asked to find the third quantity.

*Percentage: This is what is obtained when a percent is taken of a number.

Base: This is the number **of** which a percent is taken.

Rate: This the **percent** that is taken of a number.

There are formulas relating these three quantities:
  1. **percentage = base × rate%**
  2. **base = percentage ÷ rate %**
  3. **rate% = $\frac{\text{percentage}}{\text{base}}$ × 100%**   (i.e. rate% = (the ratio of percentage to base) × 100%)

You do not need to memorize the first formula, provided you note that "**of**" implies multiply. **Memorize** the second and the third formulas (even though some of the methods discussed below do not need the recall of these formulas). *Some authors call the percentage the **amount** and suggest the proportion

$$\frac{r}{100} = \frac{\text{Percentage}}{\text{base}} = \frac{r}{100} = \frac{A}{B}$$, where r = rate, A = Amount and B = Base.

Finding the Percentage

Example 1   Find 20% of 72   (i.e. **Finding the Percentage**)

(Note: "%" means over 100. Example: 20% = $\frac{20}{100}$

20% of 72   ("of" means multiply)

Step 1: $= \frac{20}{100} \times \frac{72}{1}$ <---------Simplify this by any of the methods
                         discussed below.

Step 2:

Method 1: $\frac{20}{100} \times \frac{72}{1}$ = .20 × 72 = 14.40 = 14.4   (Using decimals)

Method 2: $\frac{20}{100} \times \frac{72}{1} = \frac{20 \times 72}{100} = \frac{1440}{100}$ = 14.40 or 14.4 (Multiplying numerators
                                          and dividing by 100)

Method 3: $\frac{\overset{1}{\cancel{20}}}{\underset{5}{\cancel{100}}} \times \frac{72}{1} = \frac{72}{5} = 14\frac{2}{5}$ or 14.4   (using cancellation)

The most convenient method will depend on the type of numbers involved, and whether we want the answer as a decimal, as a fraction, or as a mixed number.
For example, if there are common factors in the numerator and the denominator, cancellation, (Method 3) may be more convenient; but if there are no common factors , use Method 1 or Method 2.
  In any case, it is a good practice to set up the problem as in Step 1. The next example will show the usefulness of setting up the problem before proceeding to simplify.

Example 2 Find $4\frac{2}{7}\%$ of 28000 .

Step 1 : Translating: $\qquad 4\frac{2}{7}\%$ of 28000

$$= \frac{4\frac{2}{7}}{100} \times \frac{28000}{1}$$

Step 2: $\qquad = \frac{30}{700} \times \frac{28000}{1}$ <-----------Simplify this by any method.

The easiest method is by cancellation, since there are common factors
in both the numerator and the denominator.

$$\frac{30}{\cancel{700}_1} \times \frac{\cancel{28000}^{40}}{1} = 30 \times 40 = 1200$$

Note that steps 1 and 2 are important. Do not round off $4\frac{2}{7}\%$ as .04%
because you will not get the exact answer. If the rate %
were say, 70%, then you could immediately write 70% = .70 and then
similarly write $70\frac{1}{4}\% = .7025$; but better, $70\frac{1}{4}\% = \frac{281}{400}$.

Finding the Base (Finding the Original Number)

Note that the next problem is different from the last two examples both in the wording of the problem and how we solve it. In the last two problems, we multiplied. In the next problem; we will divide, but, we must note which number is the divisor.

Example If 20% of a number is 64, what is the number?

There are a number of methods for solving this problem. We will discuss five methods, which include algebraic methods. You may skip the algebraic methods if you do not have the algebraic background to allow you to follow the procedure. Later, in Chapter 7 (percent problems) we will repeat the algebraic method.

Method 1:   **Using Formula**

Base = Percentage ÷ Rate %

In the above problem, 64 is the percentage. (The percentage is sometimes called the " is number". It is the number that (usually) immediately follows or precedes the word "is " in the word problem.

20% is the rate%

base = percentage ÷ rate %

$$= 64 \div 20\%$$

$$= \frac{64}{1} \div \frac{20}{100} \quad \text{<-----you can simplify this by approach 1 or 2 below.}$$

Approach 1. $\frac{64}{1} \times \frac{\overset{5}{\cancel{100}}}{\cancel{20}} = 64 \times 5 = 320$

Approach 2. $64 \div .20$ (by long division) $20\overline{)6400}^{\,320}$

The number is 320.

Method 2   **Using Algebra**

Let the number be x.

Then, "20% of the number is 64" translates to $\frac{20x}{100} = 64$

i.e. $\frac{20x}{100} = 64$

Solve for x: $\frac{\overset{1}{(\cancel{100})}}{\underset{1}{(\cancel{20})}} \frac{\overset{1}{\cancel{20}x}}{\cancel{100}} = 64\frac{\overset{5}{(\cancel{100})}}{\underset{1}{(\cancel{20})}}$

(or $20x = 100 \times 64$

$x = 320$ (or $x = \frac{100 \times 64}{20} = 320$)

Again, the number is 320 .

Note also that since 20% of the number is 64, the number must be greater than 64.

Method 3   If 20% of the number = 64

then 1% of the number $= \dfrac{64}{20}$

and 100% of the number $= \dfrac{64}{20} \times \dfrac{100}{1}$

$= 320.$

Method 4   "**Ratio Method**" (**Arithmetic**)
(This method follows from method 3.)

If 20% of a number = 64

then, 100% of the number $= \dfrac{64}{1} \times \dfrac{\cancel{100\%}^{5}}{\cancel{20\%}}$

$= 320$

In method 4, we used a very useful principle which states that " If more, the smaller divides, and if less, the larger divides". That is, since we expect 100% of the number to be greater than 64, in forming the fraction involving 20% and 100%, the smaller of 20% and 100% is the divisor (smaller divides), and hence we used the fraction,

$\dfrac{100\%}{20\%}$ <----- " smaller divides" (We multiplied by an improper fraction)

However, if we had expected the number (answer) to be less than 64, we would have used the fraction

$\dfrac{20\%}{100\%}$ <----- "larger divides" (i.e., we would have multiplied by a proper fraction)

Method 5   **Using Proportion**

20% is to 64 as 100% is to x, where 100% of the
original number (the base) is x.
Translating the proportion,

$$\dfrac{20\%}{64} = \dfrac{100\%}{x} \qquad \left(\text{ or } \dfrac{.20}{64} = \dfrac{1}{x}\right)$$

Solve for x: 20% $x = 100\%$ (64)   (or $.20x = 64$)

$$x = \dfrac{100\%}{20\%}(64) \qquad \left(\text{or } x = \dfrac{64}{.20}\right)$$

$x = 320.$

Note: Methods 3, 4 and 5 are basically the same.

Note that 20% as a fraction is $\frac{20}{100} = \frac{1}{5}$.

Therefore, in the above problem, instead of asking "if 20% of a number is 64, what is the number?", we could have asked "if $\frac{1}{5}$ of a number is 64, what is the number? "

Example   If $\frac{1}{5}$ of a number is 64, what is the number? We can solve this by any of the methods discussed above. Let us use the algebraic method.

Solving Algebraically:

Let the number be x.

Then $\frac{x}{5} = \frac{64}{1}$ (Note that $\frac{1}{5}x = \frac{x}{5}$)

$(5)\frac{x}{5} = \frac{64}{1}(5)$ or $x \times 1 = 5 \times 64$ (by cross-multiplication)

$x = 320$ or $x = 320$.

The number is 320.

(See also Method 1 of the preceding example)

Go over the last two problems, and note the differences between how the questions are worded and how they are solved. You **must rememb**er how each problem is worded and how to proceed to solve it.

Note also that the previous example was a direct proportion problem. Direct proportion involves a relationship between two quantities whereby as one quantity **increases** the other quantity also **increases.** See p.64 for more examples on direct proportion and also examples on inverse (indirect) proportion whereby as one quantity **increases** the other quantity **decreases** and vice versa.

Example   Mary spends 20% of her weekly income on food. If she spends $64 on food every week, what is her weekly income?

Solution : We could reword this problem in the familiar form as " If 20% of a number is 64, what is the number?"; and then use exactly the same method as in the example, above.

Answer: $320 (numerically, the same answer as for the last example).

Finding the Rate %

Example 1 What rate % of 24 is 15?
Method 1

$$\text{By formula: } \text{Rate}\% = \frac{\text{Percentage}}{\text{Base}} \times 100\%$$

In this problem, the percentage is 15. The percentage is sometimes referred to as the " is number".
It is the number that follows or precedes the word "is" if the problem is worded in the above form.
The base is 24. The base is sometimes referred to as the "of number". It is the number that
(usually) follows the word "of" if the problem is worded in the above form.

$$\begin{aligned} \text{Then, rate } \% &= \frac{15}{24} \times \frac{100\%}{1} \\ &= \frac{5\,\cancel{15}}{\cancel{24}\,8} \times \frac{100\%}{1} \\ &= .625 \times 100\% \\ &= 62.5\% \text{ or } 62\tfrac{1}{2}\% \end{aligned}$$

Method 2

Step 1: Form a fraction using the mnemonic device " $\dfrac{\text{is number}}{\text{of number}}$ "

$$\text{Then, we obtain } \frac{15}{24}$$

Step 2: Change $\frac{15}{24}$ to percent by multiplying by 100 and
attaching the '"%" symbol

$$\frac{5\,\cancel{15}}{\cancel{24}\,8} \times \frac{100\%}{1} = \frac{500\%}{8}$$

$$= 62\tfrac{1}{2}\% \text{ or } 62.5\%$$

Method 3 Using algebra <--You may skip this method if you do not have the algebraic background to allow
you to follow the procedure. Later, in Chapter 7, we will repeat this method.

Let $x\%$ be the rate %. We will write an equation in terms of x, and solve for x.
From the wording, "$x\%$ of 24 is 15" translates to:

$$\frac{x}{100} \times \frac{24}{1} = 15$$

$$\frac{24x}{100} = \frac{15}{1}$$

$24x = 100 \times 15$ (by cross-multiplication or by multiplying both sides of the equation by 100)

$x = \dfrac{100 \times 15}{24}$ (by dividing both sides of the equation by 24)

$x = 62\tfrac{1}{2}$

$\therefore$ the required rate $= 62\tfrac{1}{2}\%$.

Example 2   What rate % of 18 is 44?

Method 1   base = 18 (the "of" number)
                        percentage = 44 (the "is" number)

$$\text{rate \%} = \frac{\text{percentage}}{\text{base}} \times 100\%$$

$$= \frac{22\,\cancel{44}}{\cancel{18}\,9} \times \frac{100\ \%}{1}$$

$\therefore$ the required rate is $244\frac{4}{9}\%$

Note in the above problem that, the larger number does NOT have to be in the denominator.
 The base (the "of" number) must always be in the denominator. Therefore, in forming the fraction, ignore the relative sizes of the numbers.

Method 2 (Using algebra)
Let the required rate be $x\%$.
Then, "$x\%$ of 18 is 44" translates to:

$$\frac{x}{100} \times \frac{18}{1} = 44$$

$$\frac{18x}{100} = \frac{44}{1}$$

$18x = 100 \times 44$ (by cross-multiplication or by multiplying both sides of the equation by 100)

$x = \frac{100 \times 44}{18}$ (by dividing both sides of the equation by 18)

$x = 244\frac{4}{9}$

$\therefore$ the required rate is $244\frac{4}{9}\%$

Example 3 A family's annual income last year was \$20,000. This year, the income
              is \$33 ,000. What is the percent increase in the annual income?

Solution

        Step 1: The increase in income = \$33,000 - \$20,000
                              = \$13,000
        Step 2: The percent increase in income $= \frac{13000}{20000} \times 100\%$ (Finding the rate percent)
                              = 65%

Note: In Step 2 , the question could have been posed as: What percent of 20,000 is 13.000 ?

Example 4   On a class test , out of 20 questions, a student answered
              16 questions correctly. What was the student's grade in percent?

Step 1:   Fraction of questions answered correctly $= \frac{16}{20}$

Step 2: Change $\frac{16}{20}$ to percent by multiplying by 100 and attaching the % symbol.

$$\frac{16}{20} \times 100\% = 80\%$$

The student's grade was 80%.

Lesson 22 Exercises

A. (*a*) Find 80% of 25;   (*b*) Find 23% of 60;   (*c*) Find $5\frac{2}{9}$% of 1800

Answers   (*a*)   20; (*b*)   13.8 or $13\frac{4}{5}$; (*c*) 94

B. 1. If 20% of a number is 72, what is the number?

2. 53 is 25% of what number?

3. 16% of what number is 82?

4. If 25% of a number is 140, what is the number?

5. If 120% of a number is 103.2, what is the number?

Answers 1. 360 ; **2.** 212; **3.** 512.5 or $512\frac{1}{2}$; **4.** 560 ;   **5.** 86.

C. 1. The monthly rent for Maria's apartment is $800. If Maria spends 25% of her monthly income on this rent, what is her monthly income?

2. A homeowner borrowed money from a bank at the interest rate of 12% per year. If the homeowner pays the bank an interest of $6,000 per year, how much money did the homeowner borrow?

3.   If $3\frac{2}{7}$ of a number is 46, What is the number?

Answer: 1. $3,200;   **2.** $50,000;   **3.**   14.

D. 1.   What rate percent of 32 is 12?

2.   What rate% of 20 is 80?

3.   On a math test, there were 40 questions. James answered 32 questions correctly. What was his grade in percent?

Answers   **1.** 37.5% or $37\frac{1}{2}$%; **2.** 400% ; **3.** 80%

Lesson 23
More Applications Involving Percent:
Discount , Salary Change and Sales Tax Problems

Applications of Base Finding and Percentage Finding

In these applications, the questions are **not** worded in forms such as "find 20% of a number; "if 30% of a number is 45, what is the number?". A good approach is to reword the problem in any of these familiar forms and then proceed accordingly.

Note 1. The fraction involved in the problem may very likely be the **rate** (but it may **not** be the rate).

Note for example that $\frac{3}{5} = 60\%$

2. The quantity following the word "of" may very likely be the **base**.
If you know any two of the three quantities, then third quantity is easily deduced.

Example 1   3 out of 5 students at a certain college study Biology.

    (a) If 600 students at this college study Biology, how many students are there at this college?

    (b) If one were to collect a sample of 200 students at this college, how many of these students would study Biology?

Solution:

There are a number of approaches for solving this problem. We will cover two methods.

Note that 3 out of 5 means $\frac{3}{5}$

Part (a): We can reword this part of the question as: if $\frac{3}{5}$ of a number is 600, what is the number?

Method 1   Let the number be x

Then $\frac{3x}{5} = 600$

$3x = 3000$

$x = 1000$

There are 1000 students at this college.

Method 2 (see also p. 94)   Divide 600 by $\frac{3}{5}$

$600 \div \frac{3}{5} = \frac{600}{1} \times \frac{5}{3}$

$= 1000$

Part (b)   We can reword this part of the question as: Find $\frac{3}{5}$ of 200.

$\frac{3}{5}$ of $200 = \frac{3}{5} \times \frac{200}{1} = 120$

That is, of a sample of 200 students, 120 students would study Biology.

Example 2

Note that since $\frac{3}{5} = 60\%$, the above problem could have been posed as 60% of students at a certain college study Biology.

(a) If 600 students study Biology, how many students are there?

(b) If one were to collect a sample of 200 students, how many would study Biology?
   Solution: Proceed exactly as in Example 1 above.

Example 3   At a certain college, 53% of students registered for chemistry, and 24% registered for Biology. If there are 500 students at this college, how many students did not register for Chemistry or Biology?

Solution: We will use two methods to solve this problem.

Method 1

Step 1:   Percent of students registering for Chemistry or Biology is
$(53\% + 24\%) = 77\%$

Step 2:   Percent of students **not** registering for Chemistry or Biology is
$100\% - 77\% = 23\%$

Step 3: Number of students who did not register for Chemistry or Biology is

$$23\% \text{ of } 500$$
$$= \frac{23}{100} \times 500$$
$$= 115$$

Method 2

Step 1:   Percent of students registering for Chemistry or Biology is
$(53\% + 24\%) = 77\%$

Step 2 :   Find 77% of 500 and subtract the result from 500.

$$\text{i.e. } \frac{77}{100} \times 500 = 385$$

Then, the number of students not registering for Chemistry or Biology is

Step 3: $500 - 385 = 115$

By either Method 1 or Method 2, the number of students not registering for Chemistry or Biology is 115.

Discount Problem

Example   A bag originally sold for $85.00

The selling price was reduced by 30%
What is the new selling price?

Method 1

Step 1: Find 30% of $85.00

i.e. $\frac{30}{100} \times \frac{85}{1} = 25.50$ or $\begin{array}{r} 85 \\ \times .30 \\ \hline 25.50 \end{array}$

Step 2: New Price = $85.00 - $25.50
                    = $59.50

Method 2   Assuming a 100% rate % for the original price,

Step 1:   Subtract 30% from 100%
                 i.e. 100% - 30% = 70%

Step 2 :   Find 70% of $85.00

i.e. $\frac{70}{100} \times \frac{85}{1} = \$59.50$ or $\begin{array}{r} 85 \\ \times .70 \\ \hline 59.50 \end{array}$

Salary Change Problem

Example

A year ago, Mary's annual salary was $45,000. This year she received a 15% raise. What is her new salary?

The approach in solving the problem is similar to that of the above discount problem, except that in this case, we add any change (in salary).

Method 1   Step 1:   Find 15% of 45,000
$\frac{15}{100} \times \frac{45000}{1} = 6,750$
The increase (raise) in salary is $6,750

Step 2: New Salary = Original Salary + Increase in salary
                     = $45,000 + $6,750
                     = $51,750

Method 2

Step 1:   Add 15% to 100%.
            15% + 100 % = 115%

Step 2:   Find 115% of 45,00
            $\frac{115}{100} \times \frac{45000}{1} = \$51,750$

Sales Tax Problem

Example

A book sells for $50.00
and the sales tax is 8%.
How much does the purchaser pay?

Method 1

Step 1: Find 8% of $50

$$\frac{8}{100} \times \frac{50}{1} = 4 \qquad \text{or} \qquad \begin{array}{r} 50 \\ \times .08 \\ \hline 4.00 \end{array}$$

the sales tax is $4.00

Step 2:   Total cost = $50.00 + $4.00
=$54.00
∴ The purchaser pays $54.00

Method 2   Step 1: Add 8% to 100%
i.e.   8% + 100% = 108%

Step 2: Find 108% of $50.00
$$\frac{108}{100} \times \frac{50}{1} = \$54.00$$

Lesson 23 Exercises

1. A new math textbook sells for $32.00. However, a used edition of this book is sold at a 6% discount. What is the selling price of the used edition?

Answer $30.08

2, Last year , the president of a corporation was earning $160,000 per year. This year, because of financial problems, the annual salary is to be reduced to $140,000.
What is the percent decrease in salary?

Answer 12.5% or $12\frac{1}{2}\%$

3. A book sells for $65.00 and the sales tax is 8.25%.
What is the total price?

Answer $70.36

CHAPTER 8
Miscellaneous Topics

Lesson 24
Averages, Duration, Profit and Loss,

Finding Averages

Example 1   Find the average of 6, 8, 11, and 15.

We must **always remember** that in finding an average, we add all the quantities and divide by the number of quantities.

$$\begin{array}{r} 6 \\ 8 \\ 11 \\ 15 \\ \hline 40 \end{array}$$

The number of quantities is 4.
We divide 40 by 4

Then, $\dfrac{40}{4}$ The average is 10

Example 2   During the semester, James scored 85, 70, 75 on the first three tests. He wants his class average for the term to be 80. What is the score he needs on the next and last test ?

Solution   We cover three methods.

Method 1: Arithmetic approach.

Step 1:   For an average of 80 on 4 tests, total score must be $80 \times 4 = 320$

Step 2:   Total score on the first three tests is $75 + 70 + 75 = 230$

Step 3:   Score needed on the 4th test to make total score 320 is
$320 - 230 = 90$

∴   James needs a score of 90 on the next test.

Method 2: Layperson's approach
   Step 1:

1st test	2nd est	3rd test	4th test
85	**70**	**75**	**?**
Here, there are an extra 5 points to spare	Here, we need 10 points	Here, we need 5 points	Here, we need 80 points

Step 2: If we add the 5 extra points from the first test to the 3rd test, we
         would have

1st test	2nd test	3rd test	4th test
80	**70**	**80**	**?**
This is O.K	Here, we need 10 more points	This is O.K	Here, we need 80 points

Since there is only one more test to be taken, James must make up the 10 points needed for the
 2nd test, and obtain 80 points needed for the 4th test, all on the last test.

Hence, James needs a score of 10 + 80 = 90 on the last test.

Method 3: **Using algebra.**

   Let x be the next and last score.
   Then, the total score for the four tests will be $85 + 70 + 75 + x$

   Average score $= \dfrac{\text{Total score}}{\text{Number of tests}} = \dfrac{85 + 70 + 75 + x}{4}$

   Now we want this average score to be 80

$$\dfrac{85 + 70 + 75 + x}{4} = 80 \qquad \text{(Solve for } x\text{)}$$

$$\dfrac{230 + x}{4} = 80 \qquad\qquad (85+70+75 = 230)$$

$$\dfrac{4(230 + x)}{4} = 80(4)$$

$$230 + x = 320$$
$$x = 90$$

   ∴ James needs a score of 90 on the next and last test.

Travel time, and Duration Problems

   Subtract :   4 hrs.   10 mins
                 2 hrs.   40 mins

Solution: Note that 1 hr.= 60 mins)..Borrow 1 hr from 4 hr, change it to 60 minutes and add it to 10 mins

      3   (60 + 10) = 70 − − − −− >   3hrs   70 mins
   ~~4hrs~~   ~~10 mins~~
   - 2hrs   40 mins − − − − − − − >   −2hrs   40 mins
                                         1hr   30 mins

   As a word problem, the above problem could have been posed as:
      (a)   A student left school at 2:40 pm and arrived home at
              4:10 pm. How long did it take the student to get home?   or
      (b)   A movie began at 2: 40 pm and ended at 4:10 pm., how
              long was the movie?   For (a) or (b), the answer is 1 hr. 30 mins

Profit or Loss Problems

Example

Tickets to a concert cost $0.45 each .
Answer the following questions.

(a)   How many tickets must be sold in order to collect $47.25

(b)   How much money is collected if 135 tickets are sold?

(c) If expenses for organizing the concert is $120.00   and a 1000 tickets are sold,
how much profit does the organizer of the concert make?

Solution: (a) Divide the amount collected by the cost of each ticket . $\dfrac{\$47.25}{\$.45}$.

$$.45\overline{)47.25}$$

$$\downarrow$$

$$\begin{array}{r} 105 \\ 45\overline{)4725} \end{array}$$

105 tickets must be sold.

(To check multiply 105 by .45 to obtain $47.25)

(b)   Multiply the cost of each ticket by the number of tickets sold.

Then, $\$.45 \times 135 = 60.75$

or

$$\begin{array}{r} 135 \\ \times .45 \\ \hline 675 \\ 540 \\ \hline 60.75 \end{array}$$

$\therefore$ $60.75 are collected.

(c) Profit = Amount collected for tickets -   Expenses
        $= (1000 \times \$.45) - \$120$
        $=$ $450 - $120
        $=$ $330
The organizer makes a profit of $330.

*A Problem Worth Knowing (Time and Work Problem)

Working alone, Maria can paint an apartment in 4 hours, while James can paint the same apartment in 6 hours, working alone. How long it will take Maria and James to paint the same apartment if they both paint together?

Solution

In 1 hour, Maria will paint $\frac{1}{4}$ of the apartment

In 1 hour, James will paint $\frac{1}{6}$ of the apartment

In 1 hour, Maria and James (working together) will paint $\frac{1}{4} + \frac{1}{6} = \frac{5}{12}$ of the apartment.

Now, If $\frac{5}{12}$ of the apartment is painted in 1 hour,

then, the whole apartment will be painted in 1 hour $\div \frac{5}{12}$

$$= 1 \text{ hour} \times \frac{12}{5} = 2.4 \text{ hours}$$

*Note: In elementary algebra, this problem will be solved using algebra.

Lesson 24 Exercises

1. A student's grades for class tests were 73, 64, 0, 85, and 78. What was the student's average class grade ?

Answer 60

2. Betty has $850, $700, $750 in three different bank accounts. This morning, she has decided to open a fourth bank account, today. What is the total amount of money Betty needs to deposit so that at the end of the day, she would have exactly $800 in each of the four bank accounts?

Answer: $900

3. Today, Hilda worked from 7:45 a.m. to 11:15 a.m. For how long did she work today?

Answer: 3 hrs. 30 mins

4. A trader bought 45 bags at $4.20 each and sold each bag for $5.80.
  (a) What was the profit made ?   b) What was the profit percent ?

Answer (a) $72; (b) 38.1%

Lesson 25
Areas and Perimeters; Bar, Line and Circle (Pie) Graphs;

Areas and Perimeters of Rectangles

Example 1 Find the area of the rectangle shown below.

Solution: Area of a rectangle = length × altitude (or width)
Area = (15)(6)
= 90 square units

(Area, $A = L \times W$, where L is the length and W is the altitude or the width)

Example 2 In Example 1 above, find the perimeter of the figure.

Solution: The perimeter is the distance around the figure.
The perimeter is (15 + 6 + 15 + 6) = 42 units.
(Perimeter, $P = L + W + L + W = 2L + 2W$

Example 3 How many yards of fence are needed to enclose a rectangular plot
of land of dimensions 15 yards by 6 yards?
Answer: 42 yards. (see Example 2 above)
Note that yards of fence is this problem implies the perimeter of the rectangular plot.

Example 4 The dimensions of the floor of a room are 15 yards by 12 yards.
If carpet costs $4.00 per sq. yard, what is the cost of carpet needed
to cover the floor of this room?

Solution: Area of the floor = (15×12) sq. yards

Cost of carpet needed = $\$((15 \times 12) \times 4)$
= $720.00

Note that area is a measure of the **size** of a two-dimensional **surface,** but perimeter is a
measure of the total distance (**length**) **around** the figure.

Example 5 Find the area of the figure below.

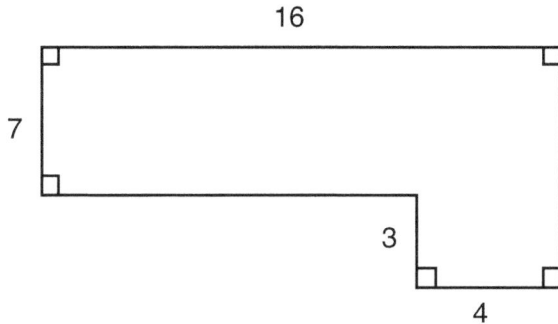

Solution: We will break the figure into two rectangles ; find the area of each rectangle (using the
Method 1   area formula for a rectangle) and then add the individual areas to obtain the total area
of the figure.

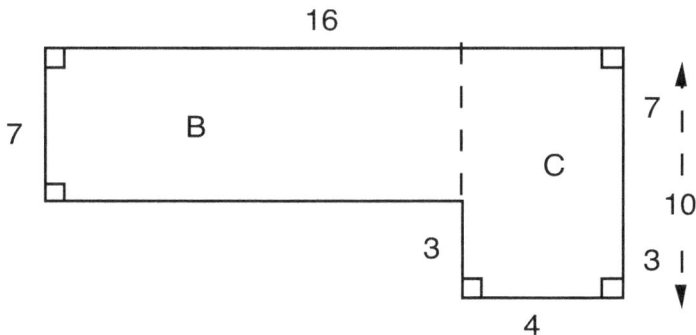

Area of rectangle B = 7(12)   Scrapwork: (16 - 4) =12
= 84 square units
Area of rectangle C   = 4(10)
= 40 square units
Total area = area of rectangle B + area of rectangle C
= 84 sq. units + 40 sq. units
= 124 square units
Method 2

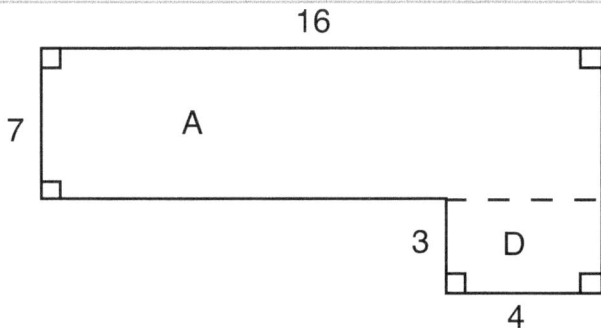

Area of rectangle A = 7(16)
= 112 square units
Area of rectangle D   = 3(4)
= 12 square units

Total area =
area of rectangle A + area of rectangle D
= 112 sq. units + 12 sq. units
= 124 square units

Bar Graphs
Example

The bar graph below shows the number of students studying chemistry at certain school during the four years 1980 - 1983.

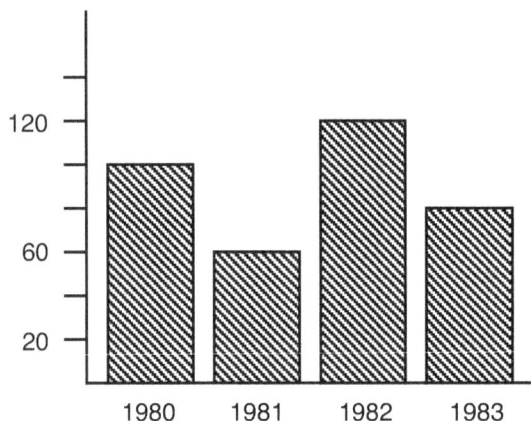

(a) How many students (approximately) studied chemistry in the two years 1980 and 1982?

(b) How many students (approximately) studied chemistry in the years 1980 - 1983?

Solution:

To determine the number of students for any particular year, pick the year (labeled horizontally) and move up vertically to the top of the box for that year. The corresponding reading (labeled) vertically on the left is the number of students for that year.

then, (a)   For the year 1980 we read 100 students (approximately)
for the year 1982, we read 120 students (approximately)

The total number of students for the years 1980 and 1982 is $(100 + 120) = 220$.
Therefore, 220 students studied chemistry in the two years 1980 and 1982.

(b) Similarly, for   1980, we read 100 students
1981, we read   60 students
1982, we read 120 students
1983, we read <u>80 students</u>

Total number of students who studied
chemistry during 1980 - 1983 $=$ 360 students

Line Graphs

We will reformulate the bar graph example (last example) in the form of a line graph but with respect to the population of a small village.

Example   The line graph below shows the population of a small village during the years 1980- 1983.
What is the difference between the populations in 1983 and in 1981?

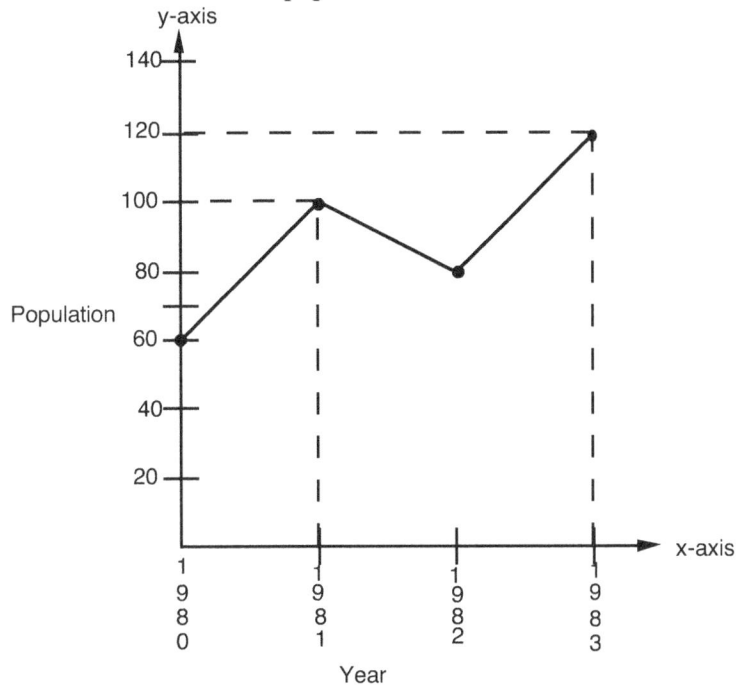

Solution : Here, we consider each point on the line as an ordered pair, where the x-coordinate is th year and y-coordinate the population.
To determine the population for any year, locate on the x-axis, the point corresponding to that year; then, from this point, move up **vertically** until the line graph is met (intersected) . From this intersection point, move horizontally to the left to the y-axis (the vertical axis) to read the corresponding population for that year.

For the year 1981, we read approximately 100 people.
For the year 1983, we read approximately 120 people.

The difference between the 1983 population and that of 1981 = 120 - 100
i.e. 20 people

Note above that for the year 1980 ,we do not have to move horizontally, since the point of intersection happens to be on the y-axis. For 1980, we read 60.

Circle or Pie Graphs
Example
The pie graph below shows the percent distribution of grades for a math test.

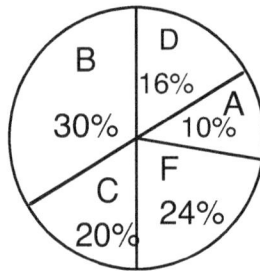

(*a*)   What percent of students received grades of C or better?

Solution: Receiving a grade of C or better means the student received either
    a C, a B or an A.
    From the graph, % of C grade =   20 %
                  % of B grade =   30 %
                  % of A grade =   10 %
(a) % of students receiving grades C or better   =   60 %

(*b*)   If there are 90 students in this class, how may students received grades of C or better?

Solution: Percent receiving grades of C or better is 60 % (from part (a))

    .: Number of students receiving grades of C or better is 60 % of 90

    i.e. $\dfrac{60}{100} \times \dfrac{90}{1} = 54$

Lesson 25 Exercises

A. 1. Find the area of the figure below.   **2.** Find the perimeter of the figure.

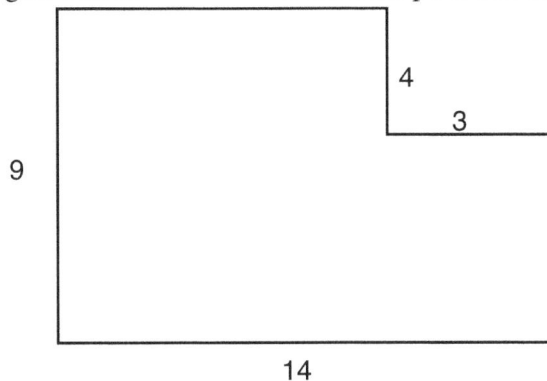

Answers: **1.** The area = 114 sq. units;   **2.** The perimeter = 46 units

B. 1. . Find the area of the figure below.   **2.** Find the perimeter of the figure below.

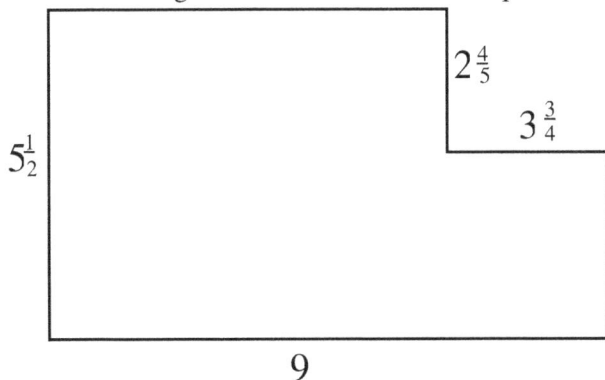

Answers: **1.** The area = 39 sq. units;   **2.** The perimeter = 29 units

C. In the bar graph below, find the difference between the number of students who studied chemistry in 1982 and in 1983.

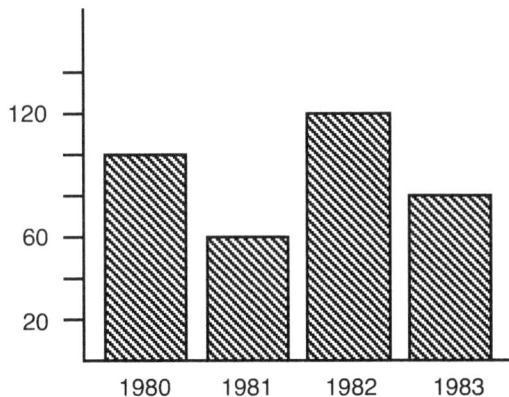

Answer : 40 students

D. The line graph below shows the population of a small village during the years 1980- 1983

1. Find the ratio of the 1982 population to the 1981 population.

2. Find the product of the 1983 population and the 1980 population

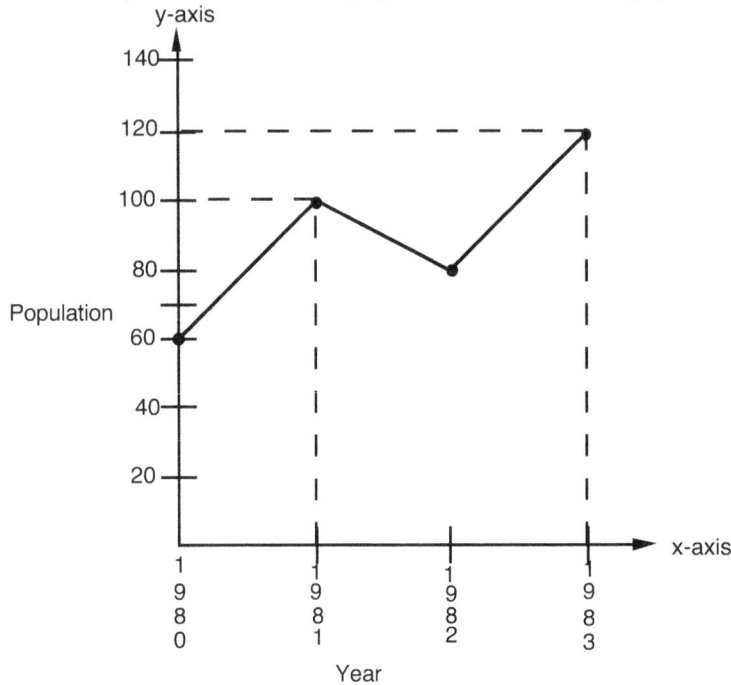

y-axis

140

120

100

80

Population

60

40

20

x-axis

1
9
8
0

1
9
8
1

1
9
8
2

1
9
8
3

Year

Answers: 1. 0.8 or $\frac{4}{5}$; **2.** 7200

E The pie graph below shows the percent distribution of grades for a math test.

1. Determine the percent of students with grades of C and D.

2. If there are 25 students in the math class, how many students received grades of C and D?

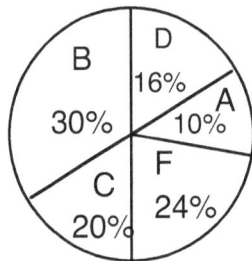

B
30%

D
16%

A
10%

C
20%

F
24%

Answers: **1.** 36% **2.** 9

Lesson 26
Scientific Notation

Preliminaries:

Powers of 10

Examples:
$$10^1 = 10$$
$$10^2 = 100$$
$$10^3 = 1000$$
$$10^4 = 10,000$$
} positive integral powers of 10.

Also,

$$10^{-1} = \tfrac{1}{10}$$
$$10^{-2} = \tfrac{1}{100}$$
$$10^{-3} = \tfrac{1}{1000}$$
$$10^{-4} = \tfrac{1}{10000}$$
} negative integral powers of 10.

The inverse of multiplying by 1000 is dividing by 1000 (or multiplying by $\tfrac{1}{1000}$)

The inverse of multiplying by 10^3 is multiplying by 10^{-3} (Since $10^3 = 1000$ and $10^{-3} = \tfrac{1}{1000}$)

Definition: A number is in scientific notation if it is written as the product of a number between 1 and 10 and an integral (integer) power of 10.

Examples The following are in scientific notation:

(a) 2.45×10^4;   (b) 5.6×10^{-3} ;   (c) 6.023×10^{23}

Procedure for writing a given number in scientific notation:

To write a number in scientific notation, place the decimal point immediately after the first non-zero digit, and multiply by an appropriate power of 10 so that the value of the original number remains unchanged.

Example   Write the following in scientific notation:
(a)   6434 ;   (b) .27898   (c) 830,000 ;   (d) .004039

Solution

(a) $6434. = \boxed{6.434 \times 10^3}$ (Move the decimal point three places to the left (i.e., $\times 10^{-3}$) and on the right-hand side of the equation, do the opposite of $\times 10^{-3}$, which is $\times 10^3$.)

$\times 10^{-3}$

Check: $6.434 \times 10^3 = 6.434 \times 1000 = 6434$ (It is good practice to always check the result.)

(b) $.27898 = \boxed{2.7898 \times 10^{-1}}$ (Move the decimal point one place to the right (i.e., $\times 10^1$) and on the right-hand side of the equation, do the opposite of $\times 10^1$, which is $\times 10^{-1}$.)

$\times 10^1$

(c) $830000. = 8.3 \times 10^5$ (Move the decimal point five places to the left (i.e., $\times 10^{-5}$) and on the right-hand   side of the equation, do the opposite of $\times 10^{-5}$, which is $\times 10^5$.)

$\times 10^{-5}$

(d) $.004039 = 4.039 \times 10^{-3}$ (Move the decimal point three places to the right (i.e., $\times 10^{3}$), and on the
                                    right-hand side of the equation, do the opposite of $\times 10^{3}$, which is $\times 10^{-3}$.)

$\times 10^{3}$

Note above, for example, that: in (d) .004039

$$= .004039 \times 10^{3} \times 10^{-3} \qquad \text{(Note: } 10^{3} \times 10^{-3} = 10^{0} = 1)$$
$$= (.004039 \times 10^{3}) \times 10^{-3}$$
$$= 4.039 \times 10^{-3}$$

Lesson 26 Exercises

A. Write in scientific notation: (*a*) 7856.3; (*b*) 890400; (*c*) .000562; (*d*) 45.35; (*e*) 0.0046

Answers: (*a*) 7.8563×10^{3}; (*b*) 8.904×10^{5}; (*c*) 5.62×10^{-4}; (*d*) 4.535×10; (*e*) 4.6×10^{-3}

B. Simplify and leave answer in scientific notation:

1. $[9.3 \times 10^{5}] \times [4.0 \times 10^{-8}]$ **2.** $\dfrac{[8.1 \times 10^{4}] \times [5.1 \times 10^{-8}]}{[1.7 \times 10^{-7}] \times [9.0 \times 10^{2}]}$

Answers: 1. 3.72×10^{-2};   **2.** 2.7×10

Lesson 27
Measurements
(see also Appendix B, p. 401)
American and Metric Systems
Measurement of Length, Mass and Volume

When one goes shopping to buy fabric, the length of the fabric has to be determined. When one goes to buy meat, the weight (mass) of the meat has to be determined, and when one goes to buy gasoline or orange juice, the volume of the gasoline or orange juice has to be determined. For the fabric, we measure its length, for the meat we measure its mass (weight), and for the gasoline or orange juice, we measure its volume. In everyday life as well as in science, we therefore measure length, mass (weight) and volume.

There are at present two main systems of measurement, namely the American system (or the British system) and the metric system. At present, most countries in the world use the metric system of measurement. However, in the United States, most measurements are made using the American system, except in science.

In measurements, we may have to convert from one unit to another unit within the American system or metric system; or we may have to convert from a unit in the American system to a unit in the metric system. We therefore need relationships between units **within** each system, which we shall call **intraconversion factors** and relationships between units **not in** the same system, which we shall call **interconversion factors**. On pages 119-120, we present two charts for some of the basic conversion factors for the two systems. Below, two conversion examples are presented.

Conversion from one unit to another unit

Converting from one unit to another unit is a direct proportion problem, and therefore all the techniques for solving proportion problems we have learned so far are applicable. In particular, the Units Label method (page 65) is highly recommended, especially, for conversions in which we use two or more conversion factors. Below, we apply the Units Label method (Review page 65).

Case 1

Example 1   Convert 96 in. to yd. (Intraconversion: conversion within the American system)

Solution   Plan: This plan is based on what is available in the conversion chart or table.
(Using a conversion chart is analogous to using a map to go from one location to another)
The sequence is inches--> feet---> yards.
We use these relationships: 12 in. = 1 ft;   3 ft = 1 yd (We could use 36 in = 1 yd) (Note: inch = in.)

Step 1: $\dfrac{96 \text{ in.}}{1} \times \dfrac{?}{?}$

Step 2: Multiply the $\dfrac{96 \text{ in.}}{1}$ by a fraction formed by using the quantities 12 in. and 1 ft as the terms of he fraction such that the denominator has the same units as the 96 in.; followed by similar multiplication by a fraction using 3 ft and 1 yd as the terms of the fraction. Then, we obtain

$$\dfrac{96 \text{ in}}{1} \times \dfrac{1 \text{ ft}}{12 \text{ in}} \times \dfrac{1 \text{ yd}}{3 \text{ ft}} \qquad \left(\dfrac{96 \text{ in}}{1} \times \dfrac{1 \text{ ft}}{12 \text{ in}} = \dfrac{96 \text{ in}}{1} \times \dfrac{1 \text{ ft}}{12 \text{ in}} \times \dfrac{1 \text{ yd}}{3 \text{ ft}} = \dfrac{96 \times 1 \times 1 \text{ yd}}{1 \times 12 \times 3} \right)$$

$$= 2\tfrac{2}{3} \text{ yd}$$

$\therefore$ 96 in. = $2\tfrac{2}{3}$ yd

Case 2

Example 2: Convert 96 inches to meters (Interconversion: from the American system to the metric system))

Solution Plan: The sequence is inches ----> centimeters --->meters

(The expressions within the parentheses are the steps for the expression on the left.)

$$\frac{96 \text{ in}}{1} \times \frac{2.54 \text{ cm}}{1 \text{ in}} \times \frac{1 \text{ m}}{100 \text{ cm}} \left(\frac{96 \text{ in}}{1} \times \frac{2.54 \text{ cm}}{1 \text{ in}} = \frac{96 \text{ in}}{1} \times \frac{2.54 \text{ cm}}{1 \text{ in}} \times \frac{1 \text{ m}}{100 \text{ cm}} = \frac{96 \times 2.54 \text{ m}}{100} \right)$$

$$= 2.44 \text{ m}$$

$\therefore$ 96 in. = 2.44 m (to the nearest hundredth)

Conversion Factors for Measurements

American System (British System)   Interconversion (Factors)   Metric System

Some "**bridge**s" for converting from one system to the other

1 kilometer (km)	$= 10^3$ m $= 1000$ m
1 hectometer (hm)	$= 10^2$ m $= 100$ m
1 dekameter (dam)	$= 10^1$m $= 10$ m
1 meter (m)	$= 10^0$ m $= 1$m
1 decimeter (dm)	$= 10^{-1}$ m $= 0.1$m
1 centimeter (cm)	$= 10^{-2}$ m $= 0.01$m
1 millimeter (mm)	$= 10^{-3}$ m $= 0.001$m

Length

12 inches (in) = 1 foot (ft.)
3 feet (ft.) = 1 yard (yd)
5280 feet = 1 mile (mi)
1760 yards = 1 mile

1 in. = 2.54 cm
1 yd = 0.9144 m
1 mi = 1.61 km
1 km = 0.62 mi

Some "**bridge**s" for converting from one system to the other

1 kilogram (kg)	$= 10^3$ g $= 1000$ g
1 hectogram (hg)	$= 10^2$ g $= 100$ g
1 dekagram (dag)	$= 10^1$ g $= 10$ g
1 gram (g)	$= 10^0$ g $= 1$ g
1 decigram (dg)	$= 10^{-1}$ g $= 0.1$g
1 centigram (cg))	$= 10^{-2}$ g $= 0.01$g
1 milligram (mg)	$= 10^{-3}$ g $= 0.001$g

Mass

1 lb = 16 oz
1 ton = 2000 lb
1 long ton = 2240 lb

1 kg = 2.2 lb
1 lb = 454 g
1 oz = 28.4 g = 16 drams
1 ton = 0.9072 metric ton

Some "**bridge**s" for converting from one system to the other

1 kiloliter (kl)	$= 10^3$ l $= 1000$ l
1 hectoliter (hl)	$= 10^2$ l $= 100$ l
1 dekaliter (dal)	$= 10^1$ l $= 10$ l
1 liter (l)	$= 10^0$ l $= 1$ l
1 deciliter (dl)	$= 10^{-1}$ l $= 0.1$l
1 centiliter (cl)	$= 10^{-2}$ l $= 0.01$l
1 milliliter (ml)	$= 10^{-3}$ l $= 0.001$ l

Volume

16 fluid oz (fl-oz)= 1 pint (pt)
2 pints (pt) = 1 quart (qt)
4 quarts = 1 gallon (gal)

1 liter (l) = 1.057 qt
1 gal = 3.785 l
1 liter = 2.1 pt
1 pt = .473 l

Must remember the following (metric system:)

100 cm = 1 m	
1000 m = 1 km	

1000 mg = 1 g
1000 g = 1 kg

1000 ml = 1 l
1 ml = 1 cc = 1cm^3
1000 cc = 1 l

Mnemonic device (metric system)

k – ilo – 10^3
h – ecto – 10^2
d – eka – 10^1
d – eci – 10^{-1}
c – enti – 10^{-2}
m – illi – 10^{-3}

Say the following aloud:
Step 1: First go down vertically as kei-eitch-dii-dii-see-em, then Step 2
Step 2: Kilo-hecto-deka-deci-centi-milli, and then note how the powers
    decrease vertically downwards.
  Examples: 1 Kilometer = 10^3meter; 1 milligram =10^{-3}gram;
  1 centimeter =10^{-2} meter =$\frac{1}{100}$ meter --->100 centimeters = 1 meter.

Mixed Conversion Factors

Units of length

1 cm = .3937 in = .0328ft = .01094 yd

1 m = 100 cm = 39.3701 in = 3.2808 ft.= 1.0936 yd

1 in. = 2.54 cm = .0833 ft = .0254 m

1 ft = 12 in. = 30.48 cm = .3048 m = .3333 yd

1 mile = 1760 yd = 5280 ft = 1.6093 km

1 yd = 3 ft = 36 in.= .9144 m = 91.44 cm

1 km = .62137 mile = 1000 m = 100,000 cm

Units of mass

1 kg = 1000 g = 2.2046 lb =35.274 oz

1 lb (avdp) = 453.592 g = 16 oz

1 metric ton = 1000 kg = 10^6 g = 1.1023 ton

1 ton = 907.1847 kg = 2000 lb = .9072 metric ton

1 gm = .03527 oz = 1000 mg =.0022046 lb

1 oz = 28.3495 g = .0625 lb = 16 drams

1 long ton (British) = 2240 lb

Units of volume

1 liter = 1000 cm^3 = 61.0237 $in.^3$= .26417 gal = 1.0567 qt =.03531 ft^3= 2.113 pt

1 gal (U.S.) = 4 qt = 3.7854 liter = 8 pt = 231 $in.^3$ = .13368 ft^3

1 qt = 2 pt = .946353 liter =946.353 cm^3 = 57.75 $in.^3$ = .25 gal =.034201 ft^3

1 cord = 128 ft^3

1 pt = .473 liter

1 ft^3 = 1728 $in.^3$

1 yd^3 = 27 ft^3 = 46656 $in.^3$

Units of area

1 ft^2 (sq. ft) = 144 $in.^2$ (sq. in.)

1 yd^2 (sq. yd) = 9 ft^2 (sq. ft) = 1296 sq. in.

1 $mile^2$ (sq. mile) = 640 acres

1 m^2 = $10^4 cm^2$

1 acre = 4840 yd^2

Symbols for units

cm = centimeter	yd = yard	gal = gallon	g = gram
m = meter	km = kilometer	qt = quart	kg = kilogram
in. = inch	mi = mile	pt = pint	lb = pound
ft = foot		oz = ounce	mg = milligram

Some prefixes (International System)

Prefix	Power
tera	10^{12}
giga	10^9
mega	10^6
kilo	10^3
hecto	10^2
deka	10^1
deci	10^{-1}
centi	10^{-2}
milli	10^{-3}
micro	10^{-6}
nano	10^{-9}
pico	10^{-12}

Lesson 27 Exercises

1.Convert 84 in. to m ;   **2.** Convert 1200 cm to yd;   **3.** Convert 4300 m to hm;
4. Convert 6 kg to oz;   **5.** Convert 1760 yd to km;   **6.** Convert 60 pt to ℓ (liters)

Answers: **1.** 2.13 m;   **2.** 13.12 yd;   **3.** 43 hm;   **4.** 211.64;   **5.** 1.61;   **6.** 28.38 ℓ

Test # 2- Student's Self-Test (Always, Test yourself before you are tested)

Attempt all questions on clean sheets of paper, **Do not write in the book** Show all necessary work.

---

1. Write as decimals:

(a) $\frac{56}{1000}$; (b) $\frac{60}{100}$.

2. Add: $53.050 + 73.2 + 9 + .0504 + 26$

---

3. Subtract: $58 - 14.603$

4. What must be added to 2.53 to obtain 8?

---

5. Simplify: (a) 62.8×100; (b) 4.206×10^2.
(*c*)

6. Simplify: (a) $54.06 \div 10^4$; (b) $\frac{364.8}{10000}$

---

7. Determine the smallest decimal:

(a) .028, .0098, .12706

8. Write in scientific notation:

(a) 7846.3; (c) .000832;

---

9. In the bar graph below, find the ratio of the 1980 population to the 1982 population.

10. Simplify and leave answer in scientific notation:

$$[9.4 \times 10^5] \times [5.0 \times 10^{-8}]$$

---

11. Change $\frac{2}{7}$ to a decimal, rounding off quotient to the nearest tenth.

12. Change each decimal to a fraction in its lowest terms

(a) 0.048; (b) 0.0725

---

13. (a) Find 60% of 35;

(b) Find $6\frac{2}{9}\%$ of 2700.

14. 18% of what number is 72?

---

15. The monthly rent for Maria's apartment is $600. If Maria spends 20% of her monthly income on this rent, what is her monthly income?

16. What rate% of 16 is 64?

17. Last year, a homeowner borrowed money from a bank at interest rate of 10% per year. If the homeowner is to pay the bank an interest of $5,000 after one year, how much money did the homeowner borrow?

18. A new math textbook sells for $30.00. However, a used edition of this book is sold at a 5% discount. What is the selling price of the used edition?

19. Last year, the president of a corporation was earning $180,000 per year. This year, because of financial problems, the annual salary is to be reduced to $150,000. What is the percent decrease in salary?

20. A student's grades for class tests were 85, 79, 0, 83, and 78. What was the student's average class grade?

21. Simplify: $58 + 5 \times (2 + 4)^2$

22. The ratio of male students to female students at a certain college is 3:5. if there are 2,400 students at this college, how many of the students are females?

23. (a) Find the area of the figure below.
   (b) Find the perimeter of the figure.

5
4
12
16

24. (a). Find the area of the figure below.
   (b). Find the perimeter of the figure below.

$2\frac{4}{5}$
$3\frac{3}{4}$
$7\frac{1}{2}$
11

25. Simplify and leave answer in scientific

notation: $\dfrac{[2.1 \times 10^4] \times [2.7 \times 10^{-8}]}{[1.4 \times 10^{-7}] \times [9.0 \times 10^2]}$

Bonus:

James and Mary are to divide $84 in the ratio 1:2. How much does each receive?

.

Answers: 1 (a) .056 ; (b) .60 ; **2.** 161.3004 ; **3.** 43.397 ; **4.** 5.47 ; **5.** (a) 6280; (b) 420.6;

   6. (a) .005406; (b) .03648; **7.** .0098 ;   **8.** (a) 7.8463×10^3; (b) 8.32×10^{-4}

9. Ratio 5:6 ; **10.** 4.7×10^{-2} ; **11.** 0.3 ; **12.** (a) $\dfrac{6}{125}$; (b) $\dfrac{29}{400}$; **13.** (a) 21; (b) 168 ; **14.** 400 ;

15. $3,000   ; **16.** 400% ; **17.** $50,000 ; **18.** $28.50 ; **19.** $16\frac{2}{3}\%$; **20.** 65 ; **21.** 238 ;

22. 1,500 ; **23.** (a) 172 sq. units; (b) 56. units.; **24.** (a) 72 sq. units; (b) 37 units ;

25. 4.5×10^0 ; Bonus: James receives $28, and Mary receives $ 56.

Test # 3 - Class Test

Student's Name......................................

Student's ID.................................

Attempt all questions on the spaces provided. Show all necessary work

---

1. Simplify: $\frac{1}{2} \div \frac{1}{4} \times \frac{4}{5}$

2. Evaluate: $7 + 4 \div 2 \times 3^2$

---

3. Write using digits:
    One hundred seventy-two and nine hundredths.

4. Simplify: $43\frac{1}{4} - 18\frac{5}{8}$

---

5. Simplify: $16\frac{3}{4} + 28\frac{2}{3}$

6. Simplify: $\frac{2}{3} + \frac{5}{6} + \frac{7}{8}$

7. Find (a) the perimeter and
(b) the area of a rectangle of
length 8 inches and width 4 inches.

8. Round off 7,893,200 to the nearest
thousand.

9. Write 420.05 in words.

10. Change $\frac{6}{7}$ to a decimal rounded off to
the nearest hundredth.

11. Add the following:
68.03;   142; 7.875; and .0832

12. Find the smallest and the largest.:
.08; .3; 1.004; .009;   .08001

13. 24 is 30% of what number?

14. Change $4\frac{2}{3}\%$ to a fraction

15. Solve for *x:* $\frac{x}{6} = \frac{5}{2}$

16. A bag which originally sold for $24.50 is now being sold at a 20% discount. What is the present price?

17. Divide: $20.24 \div 4.6$

18. Multiply: .37 by 2.8

19. Maria got grades of 86, 92, 100, 88, and 94 on her tests. What was her average grade.?

20. Simplify: $\left(\frac{3}{4} \div \frac{1}{2}\right) \div 1\frac{1}{5}$

Answers: **1.** $1\frac{3}{5}$; **2.** 25 ; **3.** $172\frac{9}{100}$ or 172.09; **4.** $24\frac{5}{8}$; **5.** $45\frac{5}{12}$; **6.** $2\frac{3}{8}$; **7.** (a) 24 in.; (b) 32 sq. in. ; **8.** 7,893,000; **9.** Four hundred twenty and five hundredths ; **10.** 0.86 ; **11.** 217.9882; **12.** Smallest: .009; largest: 1.004; **13.** 80 ; **14.** $\frac{7}{150}$; **15.** $x = 15$; **16.** \$19.60 ; **17.** 4.4; **18.** 1.036 ; **19.** 92 ; **20.** $1\frac{1}{4}$.

Test # 4 - Class Test

Student's Name.....................................
Student's ID..................................

Attempt all questions on the spaces provided. Show all necessary work

---

1. Find x from the proportion: $\dfrac{x}{5} = \dfrac{7}{15}$

2. If 80 pounds of a material cost \$1.60, what is the cost of 320 pounds of the same material?

---

3. Express 65% as a fraction in its lowest terms.

4. Change $\dfrac{7}{25}$ to a percent.

---

5. Find 8% of 2400.

6. Simplify: $7 + 6 \div 3 \times 2$

7. Find $6\frac{2}{9}\%$ of 3600

8. Divide: $324.07 \div 10^2$

9. Change .64 to a fraction in its lowest terms.

10. Change 4.8% to a fraction in its lowest terms

11. 56 is 25% of what number?

12. Multiply: 23.46×10^3

13. Simplify: .07894 × 100

14. Change $\frac{68.29}{1000}$ to a decimal.

15. Show whether or not the following is a
proportion: $\frac{6}{9} = \frac{8}{12}$?

16 Change .35 to a fraction in its lowest terms.

17. Multiply: 6.23 × .47

18. Divide: .56 ÷ .7

19. What rate% of 36 is 108?

20. Subtract 7.146 from 34

Answers: **1.** $2\frac{1}{3}$; **2.** $6.40 ; **3.** $\frac{13}{20}$; **4.** 28% ; **5.** 192 ; **6.** 11 ; **7.** 224 ; **8.** 3.2407 ;

9. $\frac{16}{25}$; **10.** $\frac{6}{125}$; **11.** 224 ; **12.** 23460 ; **13.** 7.894 ; **14.** .06829 ; **15.** Yes ; **16.** $\frac{7}{20}$;

17. 2.9281; **18.** 0.8 ; **19.** 300% ; **20.** 26.854 .

Test # 5 Student's Self-Test (Always, Test yourself before you are tested)

Attempt all questions. Show all necessary work **Word Problems**

1. Maria has $2048 in her savings account and $900 in her checking account. Both accounts are at the same bank. She wants to write a check against her checking account for $1,500.

(a) At least, how much money should Maria transfer from the savings account to the checking account in order to have sufficient funds in the checking account?

(b) After this transfer, what are the balances in the savings and checking accounts?

(c) After this transfer, what percent (approximately) of the total deposit at this bank is in her savings account.

(d) Before the above transfer, what percent (approximately) of the total deposit was in her checking account.

2. 300 ml of a 10% solution of hydrochloric acid are available for a chemistry class experiment. If each student needs 15 ml of this 10% solution for his/her experiment, how many students can perform the experiment.

3. The population of a city is 3,500,000. Of this population, 1,200,000 are males.

(a) How many females live in this city?

(b) What is the ratio of males to females?

(c) Approximately, what percent of the population are females?

4. Maria has $200 to purchase her books. She buys 5 notebooks at $1.20 each, a math textbook at $55, a chemistry textbook at $60 and a biology textbook at $50.00. How much money is left over after these purchases?

5. Janet bought 4 cans of soda for $2, and a pound of sugar for $1. If she paid for these items with a $5.00 bill, how much change does she expect?

6. An engineer's monthly salary a year ago was $3,000. This year, the engineer's salary is $3,150 ` per month.

(a) What is the percent increase in monthly salary.

(b) How much more money will the engineer earn this year than last year?

7. Albert weighs 6 kilograms and Joana weighs $8\frac{1}{4}$ kilograms. By how much does Joana weigh more than Albert?

8. If 4 drams of a drug are given to a person weighing 20 pounds, what will be the weight of a person who is given 12 drams of this drug?

9. How many subway tokens at $1.50 can one buy for $13.50?

10. How many subway tokens at $1.50 each can one buy if one has $10?

11. A rectangular room is 12 feet long, 10 feet wide and 9 feet high.
    (a) Find the area of the floor.
    (b) Find the total area of the walls of the room
    (c) If carpet costs $5.00 per square foot, what is the cost of carpet needed to cover the floor of
       this room?

12. What is the area of a square of side 4.5 feet?

13. A box contains 90 marbles in two colors, namely, red and black. 20% of these marbles are red
    (a) What is the ratio of the number of red marbles to the number of black marbles?

    (b) How many red marbles must be added so that 70% of the marbles are red?

    (c) How many red marbles must be added so that the ratio of red marbles to black marbles is 2:3?

14. James deposited $2,000 in his savings account for one year at interest rate of $5\frac{1}{4}\%$ per year.
    What is the balance in this savings account after one year?

Answers: 1. (a) $600; (b) Savings: $1448. Checking: $1500; (c) 49%; (d) 31% ;

 2. 20 students ; **3**. (a) 2,300,000 females ; (b) 12:23; (c) 66%; **4**. $29 ; **5**. $2; **6**. (a) 5%;

(b) $1800 ; **7**. $2\frac{1}{4}$ kilograms ; **8**. 60 pounds ; **9**. 9 tokens ; **10**. 6 tokens ;

11. (a) 120 sq. ft. ; (b) 396 sq. ft.; (c) $600; **12**. 20.25 sq. ft ; **13**. (a) 1:4 ; (b) 150 red marbles
(c) 30 red marbles ; **14**. $2,105.

CHAPTER 9

Signed Numbers and Real Number Operations
Lesson 28: **Addition and Subtraction of Signed Numbers**
Lesson 29: **Multiplication and Division of Signed Numbers**
Lesson 30: **Operations Involving Zero, Powers and Roots of Numbers**

Lesson 28
Addition and Subtraction of Signed Numbers

Signed Numbers

A signed number is a number with either a plus sign " +" or a minus sign "-" preceding it
(in front of it). If there is no sign in front of a number , we will assume that the number has a
plus sign. We call a number with a plus sign a positive number, and we call a number with
a minus sign a negative number. For a positive number, we sometimes do not write the plus
 sign, but for a negative number we **must** always write the minus sign.

The Real Number Line and Real numbers

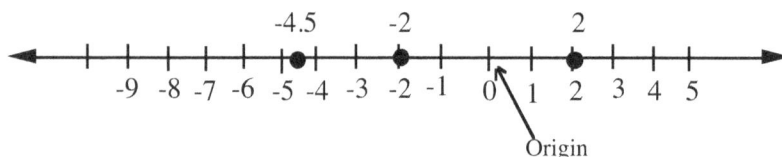

Figure 1

The real number line is a horizontal straight line with equally spaced intervals as in Figure 1 above.
We label a point called the origin, 0 (zero). Points to the right of the origin are labeled positive and
points to the left of the origin are labeled negative. The numbers increase as one moves from the left
to the right on the real number line. Roughly speaking, a real number is a number that can be
represented by a point on the real number line. The real numbers consists of the integers, fractions,
 mixed numbers, decimals, and radicals. In Figure 1, if the real numbers, -4.5, -2, and 2 are of
interest, we can represent them by the dots shown. Every point on this line is associated with
 a real number; and every real number is associated with a point on this line. We can also say t
hat the set of real numbers consists of the signed numbers and zero.

Absolute Value

The absolute value of a signed number may be defined as the number without its sign.

Examples: **1**. The absolute value of -4 , symbolized $|-4| = 4$.

   2. The absolute value of 6, symbolized $|6| = 6$.

   3.   The absolute value of $+9$ is 9.

   4. The absolute value of 0 is 0.

Note: The absolute value of a signed number is also its distance from zero on the number line.

Addition of Two Signed Numbers

To add any two signed numbers, first determine which of the following two cases is involved.

Case 1: The two numbers have the same sign

Step 1: Add the absolute values of the numbers (the unsigned parts)
Step 2: Write the common sign in front of the sum from Step 1.

Examples

1. $(+2) + (+3) = +5$ or 5

2. $(-3) + (-4) = -7$

3. $(-.5) + (-.25) = (-.75)$

Case 2: The two numbers have different signs

Step 1 : Subtract the smaller absolute value from the larger absolute value
(This subtraction is exactly like the subtraction in arithmetic, where we always subtract a smaller number from a larger number)

Step 2. Write the sign of the number with the larger absolute value in front of the difference (result) from Step 1.

Examples

1. $(+5) + (-7) = -2$

2. $(-6) + (+9) = +3$ or 3

3. $(-8) + (+6) = -2$

Note that

1. $-8 + 6$ implies $(-8) + (+6) = -2$

2. $-6 + 9$ implies $(-6) + (9) = +3$ or 3

3. $-8 + 9 - 7$ implies that we are to add the numbers, with each number carrying the sign in front of it, and thus
$-8 + 9 - 7 = (-8) + (+9) + (-7) = -6$

Absolute value defined more formally:

The absolute value of a real number x is x if x is a positive number or zero, but it is $-x$ if x is a negative number . (i.e. the negative of a negative number).

Subtraction of Signed Numbers

Procedure:

Change the sign of the number being subtracted, **and add** the changed number to the number from which you are subtracting (i.e., add the negative of the subtrahend to the minuend).

Examples

1.   (+5) - (+7)   (subtract)   (Here, +5 is the minuend and +7 is the subtrahend)

Step 1: We change +7 to -7

Step 2: We add +5 and -7   (-7 is the changed number; it is the negative of +7)

(+5) + (-7)   (add)

= - 2

2.   (-6) - (-9)   (subtract)

Step 1: We change -9 to +9

Step 2: We add -6 and +9   (+9 is the changed number)

(-6) + (+9)   (add)

= +3

3.   -7 - (-10)   (subtract)

Step 1: We change -10 to + 10

Step 2: We add -7 and + 10   (+10 is the changed number)

-7 + (+10)   (add)

= +3

4.   8 - (12)   (subtract)

Step 1: We change 12 to -12

Step 2: We add 8 and -12   (-12 is the changed number)

8 + (-12)   (add)

= -4

Note the following about the plus "+" and minus "-" signs.

The plus sign "+" may imply either the sign of a positive number, for example, +2, positive 2; or it may imply the operation of addition, for example, (+2) + (-8).　　　　　Add

The minus sign "-" may imply the sign of a negative number, for example, -6, negative 6, or the operation of subtraction, for example, (-6) - (+7).
　　　　　　　　　　　　　　Subtract

Note also that whenever we are to "add" we must use the rules for the addition of signed numbers and to subtract we must use the subtraction rule(s) for signed numbers.

Subtraction defined more formally:

To subtract a signed number y from another signed number x means add the negative of y to x.

　　　that is, $x - y = x + (-y)$.

Lesson 28 Exercises

A. Add:　**1.** (-3) + (-9);　**2.** (+5) + (+2);　**3.** (-7) + (-6)

Answers 1. -12;　**2.** +7 or 7 ;　**3.** -13

B. Add:　**1.** (-8) + (3);　**2.** (5) + (-9); **3.** (7) + (-4)

Answers　1. -5; **2.** -4 ;　**3.** +3 or 3

C. Simplify:　**1.** 5 - 7 ;　**2.** - 4 - 7;　**3.** -8 + 3 - 2;　**4.** -3 + 8; **5.** -2 + 9 + 3;

Answers 1. -2 ;　**2.** -11 ;　**3.** -7;　**4.** 5 ;　**5.** 10.

D. Perform the indicated operations:

1. (+7) + (+8) ;　　**2**. $-3\frac{2}{5} + 4$;　　**3**. (-6) + (-8) ;　　**4**. 12 + (-15)

5 . (+9) + (7) ;　　**6**. (-3) + (4) ;　　**7.** 9 + (8) ;　　**8.** -3 + 1

9. $(-2\frac{1}{2}) + (\frac{1}{4})$;　**10**. - 5 + (8) ;　　**11**. (4.3) + (-.21);　**12**. 9 - 12

13. (-6) + (-5);　**14**. - 6 + 5 ;　　**15**. 8 - 0 ;　　**16**. (-3) + (-2)

17. - 4 + 8;　　**18**. -1 - 1;　　**19**. $-\frac{11}{2} + 3$;　　**20**. 5.32 - 7.45

Also find the following:

21. $|-2| =$　　**22**. $|+3| =$　　**23**. $|-1\frac{1}{2}| =$　　**24**. $|-5| =$

Answers: **1.** 15 ; **2.** $\frac{3}{5}$; **3.** -14 ; **4.** -3 ; **5.** 16; **6.** 1; **7.** 17 ; **8.** -2; **9.** - $2\frac{1}{4}$; **10.** 3; **11.** 4.09 ; **12.** -3 ;

13. -11 ; **14.** -1 ; **15.** 8 ; **16.** -5 ; **17.** 4 ; **18.** -2 ; **19.** $-\frac{5}{2}$ or $-2\frac{1}{2}$; **20.** -2.13 ; **21.** 2 ; **22.** 3 ;

23 $1\frac{1}{2}$; **24.** 5

E.

Simplify the following:

1. (-5) - (-7) =　　**2.** (-6) - (+7) =　　**3.** -6 - (2)　　**4.** 6 - (9)

5. -1 - (-1) =　　**6.** $-2\frac{1}{2} - (2\frac{3}{4})$　　**7.** (8) - (-10)

8. Subtract 6 from 9　　**9.** Subtract 9 from 6.　　**10.** From 4, subtract 6

11. $\frac{3}{4} - (-\frac{1}{2}))$　　**12.** (-4) - (7)　　**13.** (-6) - (-8)　　**14.** - 8 - (6)

Answers: **1.** 2 ;　**2.** -13;　**3.** -8 ;　**4.** -3 ;　**5.** 0 ; **6.** $-5\frac{1}{4}$; **7.** 18 ;　**8.** 3;　**9.** -3;　**10.** -2; **11.** $1\frac{1}{4}$;

12. -11 ;　**13.** 2 ;　**14.** -14

Lesson 29
Multiplication and Division of Signed Numbers

Multiplication of Two Signed Numbers

Implication of Multiplication   Example: $2 \times 3 = 2(3) = (2)(3) = (2) \cdot (3) = (2)3 = 2 \cdot 3 = 6$

Step 1: Multiply the absolute values.
Step 2: The product is positive (that is, use a plus sign or no sign) if the two numbers have the same sign; but the product is negative (that is, use a minus sign) if the two numbers have different signs.

Examples. The two numbers having the same sign

    1. $(+2)(+3) = +6$ or 6

    2. $(-2)(-3) = +6$ or 6

    3. $-4(-2) = + 8$

The two numbers having different signs

    1. $(-4)(+2) = -8$

    2. $(+5)(-3) = - 15$

    3. $-6(4) = -24$

    4. $(3)(-7) = - 21$

Note the following distinctions:

    1. $-7 - 8 = (-7) + (-8) = -15$ (Addition)

    2. $-7 - (-8) = -7 + (+8) = 1$ (Subtraction)

    3. $-7(-8) = 56$ (Multiplication)

Note also that $(-3)(-2)(-4) = (+6)(-4) = -24$ (The multiplication rules are for **two** numbers at a time.)

Division of Signed Numbers

The rules of signs for division are the same as those for multiplication.
Procedure:
Step 1: Divide the absolute values of the numbers.

Step 2: The quotient is positive (that is, use a plus sign or no sign) if the two numbers have the same sign; but the quotient is negative (that is, use a minus sign) if the two numbers have different signs.

Examples: **The two numbers have the same sign**

1. $\frac{+8}{+4} = +2$ or 2;   2. $\frac{-8}{-4} = +2$ or 2;   3. $\frac{-12}{-9} = +\frac{4}{3})$ or $\frac{4}{3}$

The two numbers have different signs

1. $\frac{-6}{+2} = -3$;   2. $\frac{8}{-2} = -4$;   3. $\frac{-13}{3} = -\frac{13}{3}$

Reciprocals: The reciprocal of a real number A is $\frac{1}{A}$. Example : The reciprocal of $\frac{2}{3}$ is $\frac{3}{2}$.

Thus, to find the reciprocal of a number, invert the number (or interchange the numerator and the denominator).

The reciprocal of number is also known as the multiplicative inverse of that number.

The product of a number and its reciprocal is 1. Example $\frac{1}{4} \times \frac{4}{1} = 1$

Application: In $\frac{3}{7} \div \frac{9}{28}$ ($\frac{3}{7}$ is the dividend; $\frac{9}{28}$ is the divisor)

Procedure : Multiply the dividend by the **reciprocal** of the divisor. (Same as invert the divisor and multiply)

$$\frac{3}{7} \div \frac{9}{28} = \frac{3}{7} \times \frac{28}{9} = \frac{4}{3}$$

Lesson 29 Exercises

A. Simplify: **1.** (-5)(-4); **2.** -8(-9) ; **3.** 6(7) ; **4.** +5(4); **5.** (-6)(+7); **6.** -4(9)

  7. -6(+3); **8.** 2(-8)

Answers **1.** 20; **2.** 72; **3.** 42; **4.** 20 ; **5.** -42; **6.** -36; **7.** -18; **8.** -16

B. Perform the indicated operations:

  1. -8 - 8; **2.** -8(-8); **3.** -6 - (8); **4.** -6 - (-8); **5.** 9 - 7 + 3 - 8

Answers **1.** -16 ; **2.** 64; **3.** -14 **4.** +2; **5.** -3

C. Perform the indicated operations:

1. (-4)(-5) =   **2.** (-8)(6) =   **3.** 8(-7) =

4. -8(-7) =   **5.** -1(-5) =   **6.** +6(+4) =

7. 5(6) =   **8.** $4\frac{1}{3}\left(\frac{1}{2}\right)$ =   **9.** (2.3)(.5) =

10. (-4)(+2) =   **11.** (-4)(2) =   **12.** (-3)(-2)(-4) =

Review

13. - 3 - 4 =   **14.** 7 - (-10) =   **15.** -8 - 9 =

16. - 4 + (3) =   **17.** 3 - (-8) =   **18.** 4 - (9) =

19. $5\frac{3}{4} - 8\frac{1}{4}$ =   **20.** (-10) - (-12) =

Answers: **1.** 20 ; **2.** -48 ; **3.** -56 ; **4.** 56 ; **5.** 5 ; **6.** 24 ; **7.** 30; **8.** $2\frac{1}{6}$; **9.** 1.15 ; **10.** -8 ;
11. -8 ; **12.** -24 ; **13** -7 ; **14.** 17; **15.** -17 ; **16.** -1 ; **17.** 11 ; **18.** -5 ; **19.** $-2\frac{1}{2}$; **20.** 2

D. Simplify: **1.** (-6) ÷ (-3); **2.** $\frac{-12}{-4}$; **3.** $\frac{-4}{-6}$; **4.** (-10) ÷ (2); **5.** $\frac{24}{-6}$; **6.** $\frac{-15}{3}$

Answers **1.** 2 ; **2.** 3 ; **3.** $\frac{2}{3}$; **4.** -5 ; **5.** -4; **6.** -5

E. Carry out the indicated operations:

1. (+12) ÷ (+2);   **2.** (-6) ÷ (-2);   **3.** $\frac{14}{-7}$;   **4.** (6) ÷ (-2);   **5.** $-\frac{1}{2} \div \frac{1}{4}$;

6. Divide -9 by 2;   **7.** Divide $2\frac{1}{4}$ by $-\frac{1}{4}$;   **8.** Simplify: $\frac{4\frac{1}{5}}{2\frac{1}{5}}$;   **9.** Simplify: $\frac{-7.5}{.3}$;

10. Simplify: $\frac{4.28}{-100}$;   **11.** $\frac{+12}{-16}$ **12.** $\frac{-16}{12}$;   **13.** Find the reciprocal of $\frac{4}{5}$.

Review: **14.** -4 - 2 =   **15.** 8 - 3 =   **16.** 3 - 8 =

     17. -1 - 1 =   **18.** 3 - (-4) =   **19.** 4 - (12) =

Answers: **1.** 6 ; **2.** 3 ; **3.** -2 ; **4.** -3 ; **5.** -2 ; **6.** $-\frac{9}{2}$; **7.** -9; **8.** $1\frac{10}{11}$; **9.** -25 ; **10.** -.0428; **11.** $-\frac{3}{4}$;
     12. $-\frac{4}{3}$ or $-1\frac{1}{3}$;   **13.** $\frac{5}{4}$;**14.** -6 ; **15.** 5 ; **16.** -5 ; **17.** -2 ; **18.** 7 ; **19.** -8.

Lesson 30

Operations Involving Zero, Powers and Roots of Numbers

Operations Involving Zero

Addition

Examples **1**. $6 + 0 = 6$;   **2.** $0 + 4 = 4$;   **3.** $-7 + 0 = -7$;   **4**. $0 - 5 = -5$;   **5.** $b + 0 = b$

Multiplication

Examples **1**. $5(0) = 0$;   **2.** $(7)(9)(0)(14) = 0$

    Note: **3.** $0 \times 0 = 0$

Division

Examples   **1**. $\dfrac{0}{5} = 0$ (because $5 \times 0 = 0$)

        2. $\dfrac{0}{-6} = 0$ (because $-6 \times 0 = 0$

But   **3.** $\dfrac{5}{0}$ is undefined.(There is **no** number such that $0 \times$ "that number" = 5. Do not divide by 0)

       4. $\dfrac{0}{0}$ is indeterminate. (Any number will do since $0 \times$ "any number" = 0)

Square root: $\sqrt{0} = 0$

Powers of Signed Numbers

Examples **1.** $2^5 = (2)(2)(2)(2)(2) = 32$

          2^5 is read as 2 to the fifth power. (or 2 raised to the fifth power)
         The exponent "5" indicates how many times the base "2" is being used as a factor.

       2. $(-2)^3 = (-2)(-2)(-2)$

             $= -8$

       3. $(-3)^2 = (-3)(-3)$

            $= 9$

Zero as an Exponent

       Any (nonzero) number raised the to power zero is 1.

Examples 1. $b^0 = 1$;   **2.** $4^0 = 1$;   **3.** $(7x^2 dz)^0 = 1$

Note that 0^0 is indeterminate. Example: $0^0 = \dfrac{0^2}{0^2} = \dfrac{0}{0}$ which is indeterminate (see above)

Roots of Signed Numbers

Root finding and Power finding are inverse operations (in much the same way as multiplication and division are inverse operations).

Square Root: Symbol " $\sqrt{}$ " or " $\sqrt[2]{}$ "

Finding the square roots of perfect squares by inspection or by factoring

1. We may cheaply define the square root (principal square root) of a nonzero number as one of the two equal **positive** factors of that number. We exclude negative roots.

2. The square root of 0 is 0.

Examples

1. The square root of 9 is symbolized $\sqrt{9}$, and

$$\sqrt{9} = 3$$
because $3^2 = 9$

Note that , according to the above definition,
$$\sqrt{9} = \sqrt{(3)(3)} = 3$$

2. The square root of 64 is written $\sqrt{64}$, and

$\sqrt{64} = 8$, because $8^2 = 64$.

Also, $\sqrt{64} = \sqrt{(8)(8)} = 8$

3. **Note:** $\sqrt{0} = 0$, because $0^2 = 0$

If the given number is not a perfect square, we will use tables or a calculator to find the square root.

The **Cube Root** : Symbol" $\sqrt[3]{}$ "

From the above square root definition, we can similarly define the cube root of a number as one of the three equal factors of that number. (In this definition, we do **not** specify **positive root,** since the cube root may be positive or negative, depending on whether the given number is positive or negative.)

Examples

1. $\sqrt[3]{8} = 2$; because $2^3 = 8$.

Note that $\sqrt[3]{8} = \sqrt[3]{(2)(2)(2)} = 2$

2. $\sqrt[3]{-8} = -2$; because $(-2)^3 = -8$.

Note that $\sqrt[3]{(-2)(-2)(-2)} = -2$

More formal definition of square root

1. $\sqrt{A} = r$ if $r^2 = A$, where A, and r are both real and positive.

2. The square root of zero is zero (i.e. $\sqrt{0} = 0$,).

Lesson 30 Exercises

A.. Simplify: **1.** $9 + 0$;   **2.** $3 - 0$;   **3.** $-6 - 0$;   **4.** $(6)(0)$;   **5.** $(0)(-7)$;

6. $\frac{0}{4}$;   **7.** $\frac{0}{-5}$;   **8.** $\frac{9}{0}$;   **9.** $0(0)$;   **10.** $(78)(5)(0)(32)$;   **11.** $\frac{0}{+6}$;   **12.** $0(-8)$

Answers **1.** 9; **2.** 3 ; **3.** -6 ; **4.** 0; **5.** 0 ; **6.** 0; **7.** 0 ; **8.** undefined; **9.** 0 ; **10.** 0 ; **11.** 0 ; **12.** 0

B. Evaluate the following:

1. 3^3 **2.** 2^4 **3.** 8^2 **4.** $(-5)^3$

5. $(-1)^3$ **6.** $(-1)^4$ **7.** $(-1)^5$ **8.** $(-\frac{1}{2})^3$

9. 4^0 **10.** 0^0 **11.** 0^3 **12.** $(\frac{3}{2})^2$

13 $(-5)^0$ **14.** $(-2)^3$ **15.** $(-2)^4$

Review:

16. - 6 - 6 ; **17.** -7(-5); **18.** 9 - (-8); **19.** 9 - 8 **20** - 9 - 8

21 8 - 9; **22** $\frac{-16}{4}$ **23.** $\frac{0}{-3}$; **24.** $\frac{6}{0}$; **25.** (.6)(-.7)

Answers: **1.** 27; **2.** 16 ; **3.** 64 ; **4.** -125 ; **5.** -1 ; **6.** 1 ; **7.** -1 ; **8** $-\frac{1}{8}$; **9.** 1 ; **10.** indeterminate; **11.** 0 ; **12.** $\frac{9}{4}$; **13.** 1 ; **14.** -8 ; **15.** 16; **16.** -12; **17.** 35 ; **18.** 17 ; **19.** 1 ; **20.** -17 ; **21.** -1 ; **22.** -4 ; **23.** 0 ; **24.** undefined ; **25.** -.42

C. Evaluate the following : **1.** 4^3; **2.** $(-3)^4$; **3.** $\sqrt{64}$; **4.** $(-5)^3$; **5.** $\sqrt[3]{-64}$.

Answers **1.** 64 ; **2.** 81 ; **3.** 8 ; **4.** -125 ; **5.** -4

D.

Simplify the following:

1. $\sqrt{25}$ **2.** $\sqrt{81}$ **3.** $\sqrt{36}$ **4.** $\sqrt{256}$

5. $\sqrt{1024}$ **6.** $\sqrt[3]{64}$ **7.** $\sqrt[4]{81}$ **8.** $\sqrt{121}$

9 $\sqrt{\frac{25}{49}}$ **10.** $\sqrt[3]{-8}$

Review

11. - 2 - 2 **12.** -3(-2) **13.** - 4 - (-7) **14.** -2 ÷ 6

15. 3(0) **16.** 3^4 **17.** $\sqrt{100}$ **18.** 6 - (-14)

Answers: **1.** 5 ; **2.** 9 ; **3.** 6 ; **4.** 16 ; **5.** 32 ; **6.** 4 ; **7.** 3 ; **8.** 11 ; **9.** $\frac{5}{7}$; **10.** -2; **11** -4; **12.** 6 ; **13.** 3 ; **14.** $-\frac{1}{3}$; **15.** 0 ; **16.** 81 ; **17.** 10 ; **18.** 20.

Test # 1 - Student's Self-Test (Always, Test yourself before you are tested)

Attempt all questions on clean sheets of paper, **Do not write in the book** Show all necessary work.

1. (a) $(-4) + (-7) =$

   (b) $(-6) + (2) =$

2. $-6\frac{2}{5} + 4 =$

3. (a) $(4.3) + (-.27) =$
   (b) $-6 - 8 =$

4. (a) $(-6) - (-7) =$

   (b) $5 - 20 =$

5. (a) Subtract 4 from 7.

   (b) $-5 - 12 =$

6. (a) $-4 - (7) =$

   (b) $-6(-6) =$

7. (a) $(-7)(-3) =$
   (b) $(-2)(+6) =$

8. (a) $\left| -4\frac{1}{2} \right| =$
   (b) $-7 - (-3) =$

9. (a) $8 - 5 + 7 - 9 =$

   (b) $(-20) \div (5) =$

10. (a) $4\frac{1}{3}\left(\frac{1}{2}\right) =$

    (b) $\frac{-14}{21} =$

11. (a) $6\frac{3}{4} - 8\frac{1}{4} =$
    (b) $-5(-7) =$

12. (a) $\frac{0}{-7} =$

    (b) $-1 - 1 - 1 =$

13. (a) $\frac{8}{0} =$
    (b) $(-5) + (-4) =$

14. (a) $8^0 =$

    (b) $-3(-5) =$

15. (a) $\dfrac{0}{-4} =$

    (b) $-1 - 5 =$

16. (a) $\sqrt{49} =$

    (b) $-4 - (2) =$

17. (a) $\sqrt{\dfrac{16}{25}} =$

    (b) $4(0) =$

18. (a) $\sqrt{512} =$

    (b) $7 - 3 - 6 =$

19. (a) $-3 - 1 + 9 - 8 =$

    (b) $\dfrac{-8}{12} =$

20. (a) $8 - 8 =$

    (b) $\dfrac{40}{-16} =$

21. (a) $(-12) \div (4) =$

    (b) $(-7) + (-8) =$

22. (a) $-1 - 5 + 6 - 8 =$

    (b) $-9 - 2 =$

23. (a) $8 - (-5) =$

    (b) $(-6) + (-9) =$

24. (a) $-2(-6)(-2) =$

    (b) $-2(-6)(-2)(-3) =$

25. (a) $-6 - (-8) =$

    (b) $(+3) + (-7) =$

Bonus: (a) $8 - 7 - 5 =$

    (b) $\dfrac{0}{0} =$

Answers: **1.** (a) -11; (b) -4 ; **2.** $-2\frac{2}{5}$; **3.** (a) 4.03; (b) -14 ; **4.** (a) 1; (b) -15 ; **5.** (a) 3; (b) -17; **6.** (a) -11; (b) 36 ; **7.** (a) 21; (b) -12 ; **8.** (a) $4\frac{1}{2}$; (b) -4 ; **9.** (a) 1; (b) -4 ; **10.** (a) $2\frac{1}{6}$; (b) $-\frac{2}{3}$; **11.** (a) $-1\frac{1}{2}$; (b) 35; **12.** (a) 0; (b) -3 ; **13.** (a) undefined; (b) -9 ; **14.** (a) 1; (b) 15 ; **15.** (a) 0; (b) -6 ; **16.** (a) 7; (b) -6 ; **17.** (a) $\frac{4}{5}$; (b) 0; **18.** (a) $16\sqrt{2}$; (b) -2; **19** (a) -3; (b) $-\frac{2}{3}$; **20.** (a) 0; (b) $-\frac{5}{2}$; **21.** (a) -3; (b) -15; **22.** (a) -8; (b) -11; **23.** (a) 13; (b) -15 ; **24** (a) -24; (b) 72 ; **25.** (a) 2; (b) -4; **Bonus:** (a) -4; (b) indeterminate.

CHAPTER 10
Order of Operations and Evaluation of Algebraic Expressions
Lesson 31: **Order of Operations in Performing Algebraic operations**

Lesson 32: **Evaluation of Algebraic Expressions Containing Variables**

Lesson 31
Order of Operations in Performing Algebraic Operations

We repeat the order of operations covered previously in Chapter 2; and in the examples covered below, we include evaluations involving negative numbers.

Perform the operations according to the following order
1. Grouping Symbols: If there are grouping symbols, evaluate within the grouping symbols first;
2. Then, evaluate the powers and the roots in any order ; followed by
3. Division and Multiplication from left to right (or invert the divisors and multiply in any order); and then
4. Addition and Subtraction from left to right.

Note: If grouping symbols occur within other grouping symbols, evaluate within the innermost grouping symbols first, followed by the next innermost symbols and so on.

Example 1

$5 + 2(4 - 7) - 9$

$= 5 + 2(-3) - 9$ Evaluating within the parentheses first

$= 5 - 6 - 9$ Multiplying next

$= -1 - 9$ Adding from left to right

$= -10$

Example 2

$2^3 - 9(4 + 3) + 8$ Evaluating within the parentheses first

$= 2^3 - 9(7) + 8$

$= 8 - 9(7) + 8$ **Scrapwork:** $2^3 = (2)(2)(2) = 8$

$= 8 - 63 + 8$

$= -47$

Example 3

$18 \div 6 \times 3$ From left to right, we encounter division first and

$= 3 \times 3$ therefore, we divide first, according to

$= 9$ the order of operations.

Note in Example 3 above that, if we had multiplied first, we would **not** obtain the correct answer of 9. In multiplying first, $18 \div 6 \times 3 = 18 \div 18 = 1$ <-----**This is the wrong solution**.

Example 4

$6 \times 3 \div 18$

$= 18 \div 18$ (From left to right, we encounter multiplication

$= 1$ first, therefore, we multiply first and then divide)

Note however that, if we divide first, we will still get the correct answer.

$$\text{Thus}, 6 \times 3 \div 18 = \frac{6}{1} \times \frac{3}{18} = \frac{6}{1} \times \frac{1}{6} = 1$$

We conclude, from Examples 3 and 4, that if we always divide first (whether or not division comes first) , we will obtain the correct answer; but if multiplication does not come first in going from left to right, then we should **not** multiply first. Thus if we always perform division first, we will obtain the correct answer. Note that Example 3 and Example 4 are different problems.

EXTRA. Simplify: $20 \div \frac{1}{2}$ of 10 (Hint: Perform the "of" operation first before dividing.)

Solution $20 \div \frac{1}{2}$ of 10

$= 20 \div (\frac{1}{2} \times 10)$ (The notion of changing "of" to multiplication and performing the operations from

$= 20 \div 5 = 4$ left to right will lead to the **wrong result** in this problem: "of" precedes $\times$ and $\div$.)

Example 5
$$10 - [7 - 3(5 - 1)]$$
$$= 10 - [7 - 3(4)]$$
$$= 10 - [7 - 12]$$
$$= 10 - [- 5]$$
$$= 10 + [+5]$$
$$= 15$$

(Add the terms in the numerator, then

Example 6 $\dfrac{-18 - 6}{9 - 3} = \dfrac{-24}{6}$ add the terms in the denominator before dividing)

$$= -4$$

Example 7 Evaluate.

$0 - 4^2$ $= 0 - (4)(4)$

$= 0 - 16$

$= - 16$

Thus, $-4^2 = -16$

Example 8 Evaluate. $0 + (-4)^2$

Solution: $0 + (-4)^2 = 0 + (-4)(-4)$

$= 0 + 16$

$= 16$

Thus, $(-4)^2 = 16$; but note that $- 4^2 = - 16$ (From Examples 7 and 8)

Example 9 Evaluate $(-3)^2 + 7(-3)$

Procedure: Apply the order of operations.
$(-3)^2 + 7(-3)$

$= 9 - 21$

$= -12$

Scrapwork: 1. $(-3)^2 = (-3)(-3) = 9$
2. $9 - 21 = (+9) + (-21) = -12$

Example 10

$(-2)^3 - 9(-3)$

$= - 8 + 27$

$= 19$

Scrapwork: 1. $(-2)^3 = (-2)(-2)(-2) = - 8$
2. $-9(-3) = +27$

Example 11

$3\sqrt{16} - 2(-6) \div 3$

$= 3(4) + 12 \div 3$

$= 12 + 4$

$= 16$

Lesson 31 Exercises

A.

Evaluate the following: **1.** $7 - 5(3 + 4)$; **2.** $8 + 3(5 + 1)^2$; **3.** $6\sqrt{81} - 5(4)(-2)$

4. $-3[9 - (- 4 - 5) + 8]$; **5.** $15 \div 3 \times 5$ **6.** $12 \times 8 \div 4$

7. $\dfrac{4 - 7}{-8 + 6}$; **8.** $(-6)^2$; **9.** -6^2 ; **10.** $5(-2)^3 - 7(-4)$

Answers: **1.** -28 ; **2.** 116; **3.** 94 ; **4.** -78 ; **5.** 25 ; **6.** 24 ; **7.** $\dfrac{3}{2}$; **8.** 36 ; **9.** -36 ; **10.** -12

B.

Simplify the following:

1. $6 + 2(3 - 8)$ **2.** $4^2 - 3(2 + 5) - 9$ **3.** $12 \div 6 \times 2$

4. $6 \times 2 \div 12$ **5.** $12 - 4[3 - (8 - 2) - 5] + 3$ **6.** $4 -1[5 + 3]^2 - 8$

7. $0 - 3^2$ **8.** $0 - (-3)^2$ **9.** -3^2 **10.** $(-3)^2$

11. $-2(3) - (-4)$ **12.** $5(-4) + (-3)(2)$ **13.** $-7(-6) - 9(-4)$

14. $(-6)2(0)(8)$ **15.** $3 - 4\{ - (6)(-2) + 2[4 - 7]\}$ **16** $\dfrac{- 6 - 3}{-2 - 3}$

17. $-3(2 - 8) - (-3)$ **18.** $\sqrt{4^2 + 3^2}$ **19.** $(\tfrac{2}{3})^3 + \dfrac{1}{9}$

Review

20. $-8^2 + 3(0)$ **21.** $(-4)^2 - (-3)^2$ **22.** $\dfrac{(-8) - (-2)}{-2}$

Answers: **1.** -4 ; **2.** -14 ; **3.** 4 ; **4.** 1 ; **5.** 47 ; **6.** -68 ; **7.** -9 ; **8.** -9 ; **9.** -9 ; **10.** 9 ; **11.** -2 ; **12.** -26 ; **13.** 78 ; **14.** 0; **15.** -21 ; **16.** $\dfrac{9}{5}$ or $1\tfrac{4}{5}$; **17.** 21 ; **18.** 5 ; **19.** $\dfrac{11}{27}$; **20.** -64 ; **21.** 7 ; **22.** 3.

C.

1. At 7 AM. today, the temperature was 10 degrees below zero. By 3 PM, today, the temperature had increased by 12 degrees. What was the temperature at 3 PM.

2. If an original temperature of 30 degrees above zero decreases by 50 degrees, what is the new temperature?

3. Maria's score on a math test was 4 points above average, while John's score was 3 points below average. What is the difference between Maria's score and John's score?

Answers: **1.** The temperature is 2 degrees above zero; **2.** The temperature is 20 degrees below zero. **3.** The difference is 7 points.

Lesson 32
Evaluation of Algebraic Expressions Containing Variables and Numbers, Given or Knowing the Numerical Values of the Variables

Procedure:
Step 1: Replace each letter by its given numerical value in the expression, (enclosing each value in parentheses during the replacement).
Step 2: Evaluate the resulting expressions using the order of operations learned previously (see p.147).

Example 1   Given that $x = -3$, evaluate $x^2 + 7x + 2$

Solution: $(-3)^2 + 7(-3) + 2$
$= 9 - 21 + 2$ Scrapwork: $(-3)^2 = (-3)(-3)$
$= -12 + 2$ $= 9$
$= -10$

Example 2   Find the value of $5 - 4[3x - 1]$, if $x = 2$.

Solution   Substituting 2 for x, we obtain
$$5 - 4[3(2) - 1]$$
$$= 5 - 4[6 - 1]$$
$$= 5 - 4[5]$$
$$= 5 - 20$$
$$= -15$$

Example 3   (a) Evaluate: $200 (2^n)$ when $n = 3$

Procedure: Substituting n = 3 in $200(2^n)$ and applying the order of operations, then we obtain:

$200 (2^3)$ Scrapwork
$= 200 (8)$ $2^3 = (2)(2)(2) = 8$
$= 1600$

(b) Evaluate: $1000 (4)^n$ if $n = 2$

Procedure: Substitute $n = 2$ in $1000 (4)^n$

$1000 (4)^2$ Scrapwork
$= 1000 (16)$ $(4)^2 = (4)(4) = 16$
$= 16000$

Lesson 32 Exercises

A. 1. Find the value of $b^2 - 6b + 7$ when $b = -3$

2. Given that $A = \frac{1}{2} bh$, find the value of A if $b = 6, h = 4$

3. Find the value of F if $C = -60$, given that $F = \frac{9}{5}C + 32$

4. Given that $A = L \times W$, find the value of A when $L = 8, W = 6$

Answers 1. 34 ; **2.** 12 ; **3.** -76; **4.** 48

B. Evaluate the following if $a = 2$, $b = 1$, $c = 3$, $x = -5$, $y = -1$

1. $4x + 5y$ **2.** $a^2 - b^2$ **3.** $abcxy$

4. $2ab - xy$ **5.** $-ax - by + 10$ **6.** $8 - a^2 + bc$

7. $(-5)^2 - by$ **8.** $\dfrac{3a - 2}{ax + 2}$

9. Find the value of F if $C = -20$, given that $F = \frac{9}{5}C + 32$

10. Given that $A = \frac{1}{2}bh$, find the value of A when $b = 10$ and $h = 4$.

Answers: 1. -25 ; **2.** 3 ; **3.** 30 ; **4.** -1 ; **5.** 21 ; **6.** 7 ; **7.** 26 ; **8.** $-\frac{1}{2}$; **9.** -4 ; **10.** 20

Test # 2 - Student's Self-Test (Always, Test yourself before you are tested)

Attempt all questions on clean sheets of paper, **Do not write in the book** Show all necessary work.

1. (a) $(-2)^4 =$

   (b) $(-3)^3 =$

2. (a) $\sqrt{81}$

   (b) $\sqrt[3]{-8}$

3. Evaluate (a) $7 - 2(3 + 4)$

   (b) $16 \times 4 \div 4$

4. (a) $5\sqrt{49} - 3(4)(-2)$

   (b) $-3[8 - (-4 - 3) + 8]$

5. $\dfrac{5 - 3}{-10 + 6}$

6. (a) $5 + 3(4 + 1)^2$

   (b) $2(-3)^3 - 7(-5)$

7. (a) $(-5)^2 =$

   (b) $-5^2 =$

8. (a) $6 \times 12 \div 2$

9. (a) $2 - 4\{ - (4)(-2) + 2[4 - 7]$

10. $\dfrac{-6 - 3}{-2 - 3}$

11. (a) $0 - 4^2$
    (b) $0 - (-4)^2$
    (c) $0 + (-4)^2$

12. (a) $\sqrt{4^2 + 3^2}$

    (b) $\left(\dfrac{2}{3}\right)^3 + \dfrac{1}{9}$

13. (a) $(-4)2(0)(9) =$

    (b) $-7^2 + 2(0) =$

14. (a) $-3(2 - 7) - (-3)$

    (b) $(-5)^2 - (-3)^2$

15. (a) $6 - 2[4 + 3\}^2 - 8$

16. (a) Find the value of $b^2 - 6b + 7$ when $b = -2$

Lesson 32: Evaluation of Algebraic Expressions Containing Variables

17. Given that $A = \frac{1}{2}bh$ find the value of A if $b = 10, h = 3$

18. Find the value of F if $C = -20$, given that $F = \frac{9}{5}C + 32$

19. Given that $A = L \times W$, find the value of A when $L = 7, W = 4$.

20. $\frac{(-4) - (-1)}{-1}$

Questions 21-24
Evaluate if $a = -2, b = 1, c = 3, x = -5, y = -1$
21. $6x + 5y$

22. $abcxy$

23. $\frac{4a - 2}{ax + 2}$

24. $a^2 - b^2$

25 Find the value of F if $C = -40$, given that $F = \frac{9}{5}C + 32$

.

Bonus:

Answers: **1.** (a) 16; (b) -27 ; **2.** (a) 9; (b) -2 ; **3.** (a) -7; (b) 16 ; **4.** (a) 59; (b) -69 ; **5.** $-\frac{1}{2}$; **6.** (a) 80; (b) -19 ; **7.** (a) 25; (b) -25 ; **8.** 36 ; **9.** -6 ; **10.** $\frac{9}{5}$; **11.** (a) -16; (b) -16; (c) 16 ; **12.** (a) 5 ; (b) $\frac{11}{27}$; **13.** (a) 0; (b) -49 ; **14.** (a) 18; (b) 16 ; **15.** -100 ; **16.** 23 ; **17.** 15 ; **18.** -4 ; **19.** 28 ; **20.** 3; **21.** -35 ; **22.** -30 ; **23.** $-\frac{5}{6}$; **24.** 3 ; **25.** -40.

CHAPTER 11
EXPONENTS
Lesson 33
Rules of Exponents and Applications

Example 1. $2^5 = 2 \cdot 2 \cdot 2 \cdot 2 \cdot 2 = 32$

2^5 is read as 2 raised to the fifth power.(or 2 to the fifth power)
The exponent "5" indicates how many times the base "2" is being used as a factor.

Example 2. $x^6 = x \cdot x \cdot x \cdot x \cdot x \cdot x$

Mnemonic Examples for the Rules of Exponents
(Match the following examples with the rules of exponents which are presented later, below.)

1. $x^2 x^3 = x^5$

2. $\dfrac{x^7}{x^4} = x^{7-4} = x^3$

3. $(x^4)^2 = x^8$

4. $(xy)^5 = x^5 y^5$

5. $(\dfrac{x}{y})^6 = \dfrac{x^6}{y^6}$

6. $x^{-2} = \dfrac{1}{x^2}$

7. $x^0 = 1$

8. $\dfrac{x^4}{x^7} = x^{4-7} = x^{-3} = \dfrac{1}{x^3}$

Rules of Exponents

1. $x^a x^b = x^{a+b}$ (Product of powers of the same base rule)   Example $x^2 x^3 = x^5$

2. $(x^a)^b = x^{ab}$ (Power to a power rule)   Example $(x^4)^2 = x^8$

3. $x^{-a} = \dfrac{1}{x^a}$; also $\dfrac{1}{x^{-a}} = x^a$ (where a is positive) Negative exponent rule .Ex.1: $x^{-2} = \dfrac{1}{x^2}$; Ex.2: $\dfrac{1}{x^{-2}} = x^2$

4. $\dfrac{x^a}{x^b} = x^{a-b}$ (Quotient of powers of the same base rule) Example $\dfrac{x^7}{x^4} = x^{7-4} = x^3$

5. $(xy)^n = x^n y^n$ (Product to a power rule)   Example $(xy)^3 = x^3 y^3$

6.. $\left(\dfrac{x}{y}\right)^n = \dfrac{x^n}{y^n}$ ($y \neq 0$)   (Quotient to a power rule) Example $(\dfrac{x}{y})^5 = \dfrac{x^5}{y^5}$

Zero as an exponent: $x^0 = 1$ ($x \neq 0$) Example: $5^0 = 1$

Applications of the rules of exponents

Example 1   Simplify: $(3x^2 y)^2$

$$= 3^2 x^4 y^2 \qquad\qquad \text{(Using the rules of exponents)}$$

$$= 9x^4 y^2$$

(**Note**: also that $(3x^2 y)^2 = (3x^2 y)(3x^2 y) = 9x^4 y^2)$

Example 2 Simplify: $(2x^4y^6)^4$

$$= 2^4x^{16}y^{24}$$

$$= 16x^{16}y^{24}$$

Scrapwork: $2^4 = (2)(2)(2)(2)$
$= 16$

Lesson 33 Exercises

A. Simplify the following: **1.** x^6x^3 ;   **2.** $(y^7)^3$;   **3.** $(bc^4)^6$; **4.** $\dfrac{x^9}{x^5}$;

5. $\dfrac{(y^3)^5x^9}{(y^7)^2x^4}$; **6.** $(x^3y^6)^4$;   **7.** $(-\dfrac{2}{3})^3$

Answers: **1.** x^9 ;   **2.** y^{21} ; **3.** b^6c^{24} ; **4.** x^4 ; **5.** x^5y ;   **6.** $x^{12}y^{24}$; **7.** $-\dfrac{8}{27}$

B. Simplify the following: **1.** $(3x^5y)^3$; **2.** $(-2x^2y^7)^5$; **3.** Multiply: $\left(\dfrac{-8b^2c}{c}\right)\left(\dfrac{9d}{3bcd}\right)\left(\dfrac{-15}{27c}\right)$;

4. $(-2x^2y^7)^4$

Answers: **1.** $27x^{15}y^3$; **2.** $-32x^{10}y^{35}$; **3.** $\dfrac{40b}{3c^2}$; **4.** $16x^8y^{28}$

C. Simplify, leaving answers with positive exponents.

1. $x^8 \cdot x^2$

2. $x^{-4} \cdot x^6$

3. $x(x^2)$;

4. $\dfrac{x^9}{x^6}$

5. $\dfrac{x^6}{x^9}$

6. $(a^2b^3)^5$

7. $(xy)^0$;

8. $(x^{-2})^4$

9. $10^{-2} \cdot 10^6$

10. $\dfrac{m^2}{n^{-4}}$

11. $\dfrac{10^{-3} \cdot 10^8}{10^3}$;

12. $(x^2y^{-3}z)^4$

13. x^5x^7

14. $(x^8)^4$

15. $(-3x)^3$;

16. $(2x^2y^3)^4$

17. $(-3ab^2c^3)^3$

18. $\dfrac{x^2y^4}{xy}$

19. $\dfrac{xy}{x^2y^5}$;

20. $2x(x)$

21. $\dfrac{(x^3y^4)^2}{x^2y}$

22. $(a^3)^{12}(a^2)^5$

23. $\dfrac{b^5(b^2)^3}{b^8(b^3)^4}$;

24. $\dfrac{(x^3)^5(x^4)^2}{(x^5)^2x^9}$

Answers: **1.** x^{10} ; **2.** x^2 ; **3.** x^3 ; **4** x^3 ; **5.** $\dfrac{1}{x^3}$; **6.** $a^{10}b^{15}$;   **7.** 1 ;   **8.** $\dfrac{1}{x^8}$; **9.** 10^4 ; **10.** m^2n^4 ;

11. 10^2 or 100; **12.** $\dfrac{x^8z^4}{y^{12}}$; **13.** x^{12} ; **14.** x^{32} ; **15.** $-27x^3$; **16.** $16x^8y^{12}$; **17.** $-27a^3b^6c^9$; **18.** xy^3 ;

19. $\dfrac{1}{xy^4}$; **20.** $2x^2$; **21.** x^4y^7 ; **22.** a^{46} ; **23.** $\dfrac{1}{b^9}$; **24.** x^4

Test # 3 - Student's Self-Test (**Always**, **Test yourself before you are tested**)

Attempt all questions on clean sheets of paper, **Do not write in the book** Show all necessary work.

Simplify the following and leave answers with positive exponents

1. (a) $x^4 x^5$

   (b) $(y^5)^4$

2. (a) $(bc^4)^6$

   (b) $\dfrac{x^8}{x^5}$

3. (a) $\left(-\dfrac{2}{3}\right)^3$

   (b) $\dfrac{(y^3)^5 x^9}{(y^7)^2 x^4}$

4. (a) $(x^9 y^6)^4$

   (b) $(-3x^2 y^7)^3$

5. $(2x^5 y)^3$

6. Multiply: $\left(\dfrac{-8b^2 c}{c}\right)\left(\dfrac{9d}{3bcd}\right)\left(\dfrac{-15}{27c}\right)$

7. (a) $x^8 \cdot x^2$

   (b) $x^{-4} \cdot x^6$

8. (a) $x(x^2)$

   (b) $\dfrac{x^9}{x^6}$

9. (a) $\dfrac{x^6}{x^9}$

   (b) $(a^2 b^3)^5$

10. $(-3x)^5$

11. (a) $x^5 x^7$

    (b) $(x^8)^4$

12. $\dfrac{10^{-4} \cdot 10^9}{10^3}$

13. $10^{-2} \cdot 10^6$

14. (a) $(xy)^0$

    (b) $(x^{-2})^4$

15. $\dfrac{(x^3 y^4)^2}{x^2 y}$

16. $(a^3)^{12} (a^2)^5$

17. $\dfrac{x^2 y^4}{xy}$

18. $\dfrac{m^2}{n^{-4}}$

19. $(-3ab^2c^3)^3$

20. (a) $\dfrac{xy}{x^2 y^5}$

 (b) $2x(x)$

21. $(2x^2y^3)^4$

22. $(-2x^2y^7)^5$

23. $\dfrac{(x^3)^4(x^4)^2}{(x^5)^2 x^9}$

24. $\dfrac{b^5(b^2)^3}{b^8(b^3)^4}$

25. $(x^2y^{-3}z)^4$

Bonus: (a) What is your last name?
 (b) What is the fourth power of the
  number of letters in your last name?

Answers: **1.** (a) x^9; (b) y^{20}; **2.** (a) b^6c^{24}; (b) x^3; **3.** (a) $-\dfrac{8}{27}$; (b) x^5y;

4. (a) $x^{36}y^{24}$; (b) $-27x^6y^{21}$; **5.** $8x^{15}y^3$; **6.** $\dfrac{40b}{3c^2}$; **7.** (a) x^{10}; (b) x^2; **8.** (a) x^3; (b) x^3;

9. (a) $\dfrac{1}{x^3}$; (b) $a^{10}b^{15}$ **10.** $-243x^5$; **11.** (a) x^{12}; (b) x^{32}; **12.** 100 or 10^2; **13.** 10^4 or 10,000;

14. (a) 1; (b) $\dfrac{1}{x^8}$; **15.** x^4y^7; **16.** a^{46}; **17.** xy^3; **18.** m^2n^4; **19.** $-27a^3b^6c^9$;

20. (a) $\dfrac{1}{xy^4}$; (b) $2x^2$; **21.** $16x^8y^{12}$; **22.** $-32x^{10}y^{35}$; **23.** x; **24.** $\dfrac{1}{b^9}$; **25.** $\dfrac{x^8z^4}{y^{12}}$

Bonus: Varies

CHAPTER 12

Algebraic Expressions and Polynomials

Lesson 34: **Basic Definitions; Addition of Like terms**
Lesson 35: **Removing Grouping Symbols; Addition,** Subtraction of Polynomials
Lesson 36: **Multiplication and Division of Polynomials**

Lesson 34
Basic Definitions; Addition of Like terms

Definitions

Constant: a symbol representing a quantity that does not change in value in a certain discussion or operation. We usually represent a constant by a number.
Example: In $4x^2 - 3xy$, the constants are 4, 2 and -3.
We may also represent a constant by a letter, such as one of the first few letters of the alphabet. In a discussion in which letters are used to represent constants, it is good practice to state which letters represent constants.
Example: A quadratic trinomial is of the form $ax^2 + bx + c$, where a, b, and c are **constants**.

Variable: a variable is a symbol representing a quantity that changes in value. We usually represent a variable by a letter.
Example: In $4x^2 - 3xy$, the variables are x and y.

Algebraic Expression

An algebraic expression is any variable (letter) or constant (number) or any combination of variables and constants related by signs of operation (such as $+, -, \times$, etc.) and grouping symbols.
Examples 1. $5x^2 + 8x - 9$.
  2. $-2x(x - y) + 8$.
The plus and the minus signs separate an algebraic expression into parts. Each part is called a **term**.
$5x^2 + 8x - 9$ has three terms : The first term is $5x^2$; the second term is $+ 8x$, and the third term is $- 9$.
$-2x(x - y) + 8$ has two terms : The first term is $-2x(x - y)$ and the second term is $+8$.

Monomial: A monomial in x is a one-term algebraic expression of the form ax^n, where a is a real number and n is a whole number. Examples: **1.** $4x^3$ **2.** $5x$; **3.** $6x^3$; **4.** 8

A monomial in x and y is a one-term algebraic expression of the form $ax^m y^n$, where a is a real number, and m and n are whole numbers.

Polynomial: A polynomial is an algebraic expression consisting of two or (a finite number of) more unlike monomials.
Examples: **1.** $7x^4 - 5x^3 - 4x + 7$ **2.** $4x + 8$; **4.** $\frac{2}{3}x^5 + 7x + 4$; **5.** $3x^5 y^2 + 7x^3 y$.

An example of an expression which is **not** a polynomial: $3x^2 + 6x^4 + \frac{2}{x} - 9$ is **not** a polynomial because of the x in the denominator of the third term. Note also that $\frac{2}{x} = 2x^{-1}$ (has a negative exponent).

Another definition

A polynomial in x is an algebraic expression consisting of a finite number of terms with each term being of the form ax^n, where a is a real number and n is a whole number. (In a polynomial, there are **no** variables in the denominator or **no** negative exponents; no variables with fractional exponents or the radical equivalents.)

Types of polynomials

Binomial : a polynomial of exactly two unlike monomials. Examples: **1.** $5x^2 - 3x^4$;   **2.** $a + b$

Trinomial: a polynomial of exactly three unlike monomials. Examples: **1.** $2x^3 - 5xy - 8$; **2.** $x^2 + 7x + 12$.

Note that a monomial is a special type of term. All monomials are terms but not all terms are monomials.

A polynomial may also be in terms of two or more variables. It may also be in terms of any letter(s).
Definition: A polynomial in x and y is an algebraic expression in which all the terms are of the form $ax^m y^n$, where a is any real number and m and n are whole numbers.

The **degree** of a **term** is the sum of the exponents of the variables of that term.

Example: Find the degree of the following: **1.** $3x^5 y^2$, **2.** $7x^4$, **3.** x, **4.** 8
Solution:

1. degree = 5 + 2 =7, **2.** degree = 4;   **3.** degree = 1 ;   **4.** degree = 0 (since $8 = 8x^0$ and $x^0 = 1$)

The **degree** of a **polynomial** is the degree of the term with the highest degree.

Example: Find the degree of the following polynomials

1. $7x^4 - 5x^3 - 4x + 7$, **2.** $4x + 8$, **3.** $\frac{2}{3}x^5 + 7x + 4$, **4.** $3x^5 y^2 + 7x^3 y$

Solution **1.** Degree = 4 (degree of $7x^4$);   2. degree = 1 (degree of $4x$)
3. degree =5 (degree of $\frac{2}{3}x^5$) ; **4.** degree = 7 (degree of $3x^5 y^2$).

Like Terms

Like terms have exactly the same literal (letter) factors, with each letter raised to same power.
Examples $3x$ and $-7x$ are like terms.
             $5x^2 y$ and $6x^2 y$ are like terms.
             $2bc$ and $-5cb$ are like terms.
However, $3x^2 y^2$ and $-8x^2 y$ are **unlike** terms.

Note: Like terms **can** be added or subtracted,   Unlike terms **cannot** be added or subtracted.

Numerical Coefficient of a Term

Examples   The numerical coefficient of $8x$ is 8
             The numerical coefficient of $-8x$ is -8
             The numerical coefficient of $-7xy$ is -7
             The coefficient of $6x^2$ is 6.

The Distributive Property $\boxed{a(b + c) = ab + ac}$

Also, $a(b + c + d) = ab + ac + ad$
     $a(b - c - d) = ab - ac - ad$
       $a(b + c) = (b + c)a = ba + ca = ab + ac$

Examples   **1.** $5(2x + y)$ **2.** $3(a - b)$
             $= 10x + 5y$ $= 3a - 3b$

Addition of Like Terms

To add like terms, add the numerical coefficients and keep the literal parts.

You may skip writing this step

Example 1 $4x + 2y + 6x + 3y$ **Scrapwork**

$= (4 + 6)x + (2 + 3)y$ <---------   For the x-terms: $4 + 6 = 10$

$= 10x + 5y$ For the y-terms: $2 + 3 = 5$

Example 2 $-8x + 5x + 2y$

$= (-8 + 5)x + 2y$ <----------------------------You may skip this step

$= -3x + 2y$ For the x-terms: $-8 + 5 = -3$

Lesson 34 Exercises

A. Add the like terms: **1**. $-7x + 5y + 4x - 3y$; **2**. $7a + 3b - 8b - 9a$; **3**. $6y - 6x + 3y - 9z$

Answers **1**. $-3x + 2y$; **2**. $-2a - 5b$; **3**. $-6x + 9y - 9z$

B. Add the like terms:

1. $3x - 6y + 8x - 4y$ **2**. $3a - 4b - 2a - 5b$ **3**. $-5x + 6y - 3x + 4$

4. $2a + 3b - 5a + 4b - 9$ **5**. $-4a^2b + 5ab^2 + 3a^2b + ab^2$ **6**. $-2x^2y - 5xy + 2x^2y + y^2$

7. $6ab - 8ba + 3ab + 6b$ **8**. $7a^2 - 6ab^2 - 4a^2 + 9$ **9**. $5x - 7y + 3y - 4x$

Answers: 1. $11x - 10y$; **2**. $a - 9b$; **3**. $-8x + 6y + 4$; **4**. $-3a + 7b - 9$; **5**. $-a^2b + 6ab^2$;
      6. $-5xy + y^2$; **7**. $ab + 6b$; **8**. $3a^2 - 6ab^2 + 9$; **9**. $x - 4y$.

Lesson 35

Grouping Symbols; Addition and Subtraction of Polynomials

Grouping Symbols

1. The parentheses : ( ) , example: $4(x + y)$

2. The brackets : [ ] , example : $4[x + y]$

3. The braces : { }, example : $4\{x + y\}$

* **4.** The bar : " ——— " example: $6\sqrt{9 + 16}$ (also see note below.)

**Note that the grouping symbols can be used interchangeably, and thus
$$3(4 + 2) = 3[4 + 2] = 3\{ 4 + 2\}$$

*Note also that $\dfrac{2 + 4}{3 + 5} = \dfrac{(2 + 4)}{(3 + 5)}$

How to Remove Grouping Symbols

We will consider four cases.

Case 1: If there is a plus sign or no sign preceding the grouping symbols, delete the plus sign, delete the grouping symbols, and copy the expression within the grouping symbols.

Example 1: $+ (2x + 4y - 6)$
$= 2x + 4y - 6$

Example 2 $(2x + 4y - 6)$
$= 2x + 4y - 6$

Case 2: If a minus sign precedes the grouping symbols, delete the minus sign, delete the grouping symbols, and change the sign of every term within the grouping symbols.

Example: $- (3x + 4y - 7)$ (Note that changing the sign of every term within the parentheses is
$= - 3x - 4y + 7$ equivalent to multiplying every term within the parentheses by -1)

Case 3: If a factor precedes the grouping symbols, multiply each term within the grouping symbols by this factor. This case is the application of the distributive property:
$$a(b + c) = ab + ac$$

Example 1 $2(5x - 6y + 3)$
$= 10x - 12y + 6$

Example 2 $- 3(4a + 2b - 5c)$
$= -12a - 6b + 15c$

Case 4: If there are grouping symbols within other grouping symbols, remove the innermost grouping symbols first, followed by the next innermost, and so on.

Example: Remove the grouping symbols.
$6[- 3(4x - 1) + 11] - 12x$
$= 6 [- 12x + 3 + 11] - 12x$ (Removing the parentheses first)
$= - 72x + 18 + 66 - 12x$
$= - 84x + 84$ (Adding the like terms)

Note that this interchangeability applies only to the present topics. As you learn more mathematics, you will cover topics such as ordered pairs and sets where these symbols **cannot** be used interchangeably.

Addition of Polynomials

Procedure: To add polynomials, add the like terms.

Example 1   Find the sum of $-2b + 3c$ and $5b - 9c$

Procedure: Add the like terms.
   $-2b + 3c + 5b - 9c$
   $= 3b - 6c$

Scrapwork: For the b- terms:
$-2 + 5 = 3$
For the c-terms:
$3 - 9 = -6$

Example 2   Add $b^2 + 4c$ and $-3b^2 - 9c$

Procedure: Add the like terms: $b^2 + 4c - 3b^2 - 9c$

$$= -2b^2 - 5c$$

Subtraction of Polynomials

Example   Subtract $5x^2 - 3$ from $2x - 4$

Procedure: Write the "from" expression first,
          followed by the subtracted expression in parentheses with
          a minus sign in front of the parentheses.   **or**
                    $2x - 4 - (5x^2 - 3)$
          Now simplify by removing the parentheses   Change the signs of all the terms of the
          and adding any like terms.   subtracted expression and add to the
                    $2x - 4 - (5x^2 - 3)$ expression from which you are subtracting.
          $= 2x - 4 - 5x^2 + 3$
          $= -5x^2 + 2x - 1$

Note: The subtracted expression (the subtrahend) follows the word subtract
          and the "from" expression (the minuend) follows the word "from".

Lesson 35 Exercises

A. Remove the grouping symbols: **1.** $-(5x - 8y - 6)$;   **2.** $-4(3x - 2y - 9)$;

   3. $-(3x + 4y - 7)$;   **4.** $7[-5(3y - 2) + 9] - 8x$;   **5.** $-6(2x - 3) - (4 - 5x) - 2x$

Answers: **1.** $-5x + 8y + 6$;   **2.** $-12x + 8y + 36$;   **3.** $-3x - 4y + 7$;   **4.** $-8x - 105y + 133$;   **5.** $-9x + 14$

B. Add the following:

1. $3x - 6y - 5$ and $2y - 4x + 2$ **2.** $4x^2y - 9xy^2$ and $5x^2y + 3xy^2$

3. $2a - 7b - 8$ and $a - 36$ **4.** $-a^2b + 3ab^2$ and $-a^2b + 5a^2b - 4a^2b$

5. Simplify: $\frac{1}{2}a^2b + 9 + \frac{1}{4}a^2b$

Answers:**1.** $-x - 4y - 3$;   **2.** $9x^2y - 6xy^2$;   **3.** $3a - 7b - 44$;   **4.** $-a^2b + 3ab^2$; **5.** $\frac{3}{4}a^2b + 9$

Lesson 35: Removing Grouping Symbols; Addition, Subtraction of Polynomials

C.
1. Subtract $-6x^3 + 4x - 5$ from $7x - 2$;   **2**. From $7x - 2$, subtract $-6x^3 + 4x - 5$

3. Simplify : $4x^2y^3 - (x^2y^3 + x^2y) + xy^2 - 5x^2y$;   **4. Simplify**: $5a^4b^5 + 3ab - (7a^4b^5 + 3ab)$

Answers: **1**. $6x^3 + 3x + 3$;   **2**. $6x^3 + 3x + 3$;   **3**. $3x^2y^3 - 6x^2y + xy^2$; ;   **4**. $-2a^4b^5$

D.
1. Subtract 6 from 2.   **2**. Subtract 2 from 6

3. Subtract $2y - 4x + 2$ from $3x - 6y - 5$ **4**. Subtract $2a - 7b - 8$ from $a - 36$

5. Simplify: $2a^2 - 4ab + b^2 - (5a^2 + 2ab - 3b^2)$ **6**. Subtract $a - 36$ from $2a - 7b - 8$

7. Simplify: $(6x - 8y + 7) - (-3x + 5y - 1)$

Answers: **1**. -4 ;   **2**. 4 ;   **3**. $7x - 8y - 7$;   **4**. $-a + 7b - 28$;   **5**. $-3a^2 - 6ab + 4b^2$; **6**. $a - 7b + 28$;
       7. $9x - 13y + 8$.

Lesson 36
Multiplication and Division of Polynomials

Multiplication of Polynomials

Example 1 Find the product of $4xy^2$ and $-3y^2$

Multiply: $(4xy^2)(-3y^2) = -12\,xy^4$

Example 2 $(-2x^3y)(3xy^5)$
$$= (-2\cdot3)x^3\cdot x\cdot y\cdot y^5$$
$$= -6x^4\,y^6$$

Example 3 $2x(x + 4)$
$$= 2x^2 + 8x$$

Example 4 $(x + 2)(x + 5)$ <------------ Multiplication of two binomials
$x(x + 5) + 2(x + 5)$ (Distribute x over $(x + 5)$, followed by distribution of $+2$ over $(x + 5)$.
                    You may skip writing this step, and write the next step.
$$= x^2 + 5x + 2x + 10$$
$$= x^2 + 7x + 10$$

Example 5 $(3x + 2)(4x - 3)$
$$= 12x^2 - 9x + 8x - 6$$
$$= 12x^2 - x - 6$$

Special Products: **1.** $(x + y)^2 = (x + y)(x + y)$
$$= x^2 + xy + xy + y^2$$ (Always read the letters alphabetically)
$$= x^2 + 2xy + y^2$$

2. $(x - y)^2 = (x - y)(x - y)$
$$= x^2 - xy - xy + y^2$$
$$= x^2 - 2xy + y^2$$

3. $(a + b)(a - b) = a^2 - ab + ab - b^2$
$$= a^2 - b^2$$ $(- ab + ab = 0)$

The expression $a^2 - b^2$ is known as the difference between two squares. We will meet this expression again when we factor polynomials in the future (Lesson 47).

What is meant by to simplify an algebraic expression?

The meanings given below will be sufficient for the topics in this course.

1. It may mean carry out the indicated operations, which may be addition , subtraction, multiplication, division, root extraction, or power finding; or

2. It may mean remove the grouping symbols and add any like terms, if there are any like terms; or

3. It may mean reduce the fractions to lowest terms.

Examples   Simplify each of the following:

    1. $3b + 4(b + c)$
        $= 3b + 4b + 4c$
        $= 7b + 4c$

    2. $5(2x - 3y) - 6(2x - 4y)$
        $= 10x - 15y - 12x + 24y$ (Removing the parentheses)
        $= -2x + 9y$ **Scrapwork:** For the x-terms
                                      $10 - 12 = -2$
                                For the y-terms
                                $-15 + 24 = 9$

    3. $11x - 12y + 6x + 4$ **Scrapwork**: For the x-terms
      $=17x - 12y + 4$ $11 + 6 = 17$

  4. $- (5b + 8c) - 3(4b - 20 + c) - 8$
      $= - 5b - 8c - 12b + 60 - 3c - 8$ (Removing the parentheses)
      $= -17b - 11c + 52$ (Adding the like terms)

5. $3xy(2x^2y - 5y) - 2xy^2$

    Procedure: Remove the parentheses and add any like terms.

      $3xy(2x^2y - 5y) - 2xy^2$

    $= 3xy(2x^2y) + 3xy(-5y) - 2xy^2$ <------You may skip this step
                                                  Scrapwork

    $= 6x^3y^2 - 15xy^2 - 2xy^2$ $- 15 - 2 = -17$

    $= 6x^3y^2 - 17xy^2$

    6.   Simplify: $6 + 8x^2y - 4x(xy^2 - 3xy)$
                         $= 6 + 8x^2y - 4x^2y^2 + 12x^2y$

                         $= 6 + 20x^2y - 4x^2y^2$

                         $= 20x^2y - 4x^2y^2 + 6$ (Rewriting)

Division of Polynomials

Division of a monomial by a monomial

Example 1

$$\frac{8x^3y}{2xy}$$

$$= \frac{8}{2} \cdot \frac{x^3}{x} \cdot \frac{y}{y}$$

$$= \frac{4}{1} \cdot \frac{x^2}{1} \cdot 1$$

$$= 4x^2$$

Example 2

$$\frac{-6x^3y^4}{-2x^8y}$$

$$= \frac{-6}{-2} \cdot \frac{x^3}{x^8} \cdot \frac{y^4}{y}$$

$$= \frac{3}{1} \cdot \frac{1}{x^5} \cdot \frac{y^3}{1}$$

$$= \frac{3y^3}{x^5}$$

(Leaving answer with positive exponents only)

Division of a polynomial by a monomial

Example 1 Simplify: $\dfrac{12x^2 - 8x}{4x}$

Step 1: $\dfrac{12x^2}{4x} + \dfrac{-8x}{4x}$

Step 2: $3x + (-2)$

Step 3: $3x - 2$

Example 2 Simplify: $\dfrac{15y^3 + 20y^2}{-5y}$

Step 1: $\dfrac{15y^3}{-5y} + \dfrac{20y^2}{-5y}$

Step 2: $-3y^2 + (-4y)$

Step 3: $-3y^2 - 4y$

Note that in Examples 1 and 2 (above), you may skip writing Steps 1 and 2 if you **make sure** t hat you divide **every term** (taking into account its sign) in the **numerator** by the denominator. The author suggests skipping Steps 1 and 2, provided attention is paid to the above **note**.

Division of a polynomial by a polynomial (The long division method)

We will consider four cases.

Case 1: The divisor divides the dividend exactly (i.e. without a remainder).

Example: Divide $7x^2 + x^3 + 6 + 13x$ by $x + 2$

Note : $7x^2 + x^3 + 6 + 13x$ is the dividend; $x + 2$ is the divisor.

We will use a long division method which is similar to the long division in arithmetic.

Step 1: Check the dividend and the divisor to see if the terms of each expression are arranged in descending powers of the variable, x. Descending powers of x means the term with the highest power of x is written first, followed in decreasing orders by terms with lower powers of x.
The divisor $x + 2$ is already arranged in descending powers of x.
The dividend $7x^2 + x^3 + 6 + 13x$ is not arranged in descending powers of x. In descending powers of x, the dividend becomes $x^3 + 7x^2 + 13x + 6$.

Step 2: Arrange the divisor and the dividend as done in arithmetic long division, as shown below.

$$x + 2 \overline{\smash{\big)}\ x^3 + 7x^2 + 13x + 6}.$$

Step 3: Divide the first term (leading term) of the dividend by the first term (leading term) of the divisor, and write quotient x^2 above the x^3 (as we do in arithmetic).

$$
\begin{array}{r}
x^2 \\
x + 2 \overline{\smash{\big)}\ x^3 + 7x^2 + 13x + 6}
\end{array}
\qquad \text{Scrapwork: } \frac{x^3}{x} = x^2
$$

Step 4: As done in arithmetic, multiply the divisor, $x + 2$ by the x^2, and write the product under the terms of the dividend, lining up the like terms vertically.

$$
\begin{array}{r}
x^2 \\
x + 2 \overline{\smash{\big)}\ x^3 + 7x^2 + 13x + 6} \\
+x^3 + 2x^2
\end{array}
\qquad \text{Scrapwork: } x^2(x + 2) = x^3 + 2x^2
$$

Step 5: Subtract the $x^3 + 2x^2$ from the like terms above them; but recall that to subtract we change the signs of the terms being subtracted and add the changed terms to the terms from which we are subtracting.

$$
\begin{array}{r}
x^2 \\
x + 2 \overline{\smash{\big)}\ x^3 + 7x^2 + 13x + 6} \\
\end{array}
$$

Subtract: $x^3 + 2x^2$

                        $\overline{5x^2}$

Scrapwork:

$$
\begin{array}{cc}
x^3 & 7x^2 \\
-x^3 & -2x^2 \\
\hline
0 & 5x^2
\end{array}
$$

Step 6: Bring down the other terms; divide, multiply and subtract as done in Steps 3, 4, and 5 above. Remember, always the first term of divisor divides the first term of dividend.

$$
\begin{array}{r}
x^2 + 5x \\
x + 2 \,\overline{)\, x^3 + 7x^2 + 13x + 6} \\
+ x^3 + 2x^2 \\
\hline
+ 5x^2 + 13x + 6
\end{array}
$$

Scrapwork: $\dfrac{5x^2}{x} = 5x$

Subtract: $\qquad\qquad + 5x^2 + 10x$

Scrapwork: $5x(x + 2) = 5x^2 + 10x$

$$\overline{ + 3x}$$

Step 7: Bring down the 6, divide, multiply, and subtract.

$$
\begin{array}{r}
x^2 + 5x + 3 \\
x + 2 \,\overline{)\, x^3 + 7x^2 + 13x + 6} \\
+ x^3 + 2x^2 \\
\hline
+ 5x^2 + 13x + 6
\end{array}
$$

Subtract: $\qquad + 5x^2 + 10x$

$$
\begin{array}{r}
\hline
+ 3x + 6 \\
+ 3x + 6 \\
\hline
0 + 0
\end{array}
$$

Scrapwork: $\dfrac{3x}{x} = 3$

$\therefore \qquad \dfrac{x^3 + 7x^2 + 13x + 6}{x + 2} = x^2 + 5x + 3$

Normally, when we perform the division process, we show only Step 7 above.

Note: In the above division process, we divide, multiply, subtract, bring down the remaining terms of the dividend and repeat the process. Note also that always, the **first term** of divisor divides the **first term** of the dividend.

Lesson 36: Multiplication and Division of Polynomials

Case 2: The division process leaves a remainder.

Solution: The steps are exactly like those of Example 1; however we will show only the compact form in Step 7 of Example 1.

Divide $6x^4 + x^2 - x^3 + 30x + 7$ by $3 + 2x$

Step 1 : Arrange the divisor and dividend in descending powers of x; and set up the problem as in arithmetic long division.

Step 2:

$$\begin{array}{r} 3x^3 - 5x^2 + 8x + 3 \text{ R } -2 \\ 2x + 3 \overline{)\, 6x^4 - x^3 + x^2 + 30x + 7} \\ + 6x^4 + 9x^3 \\ \hline -10x^3 + x^2 + 30x + 7 \\ -10x^3 - 15x^2 \\ \hline 16x^2 + 30x + 7 \\ +16x^2 + 24x \\ \hline +6x + 7 \\ +6x + 9 \\ \hline -2 \end{array}$$

Scrapwork: $\dfrac{6x^4}{2x} = 3x^3$

-2 <---------Remainder

There is a remainder of -2. The division process ends when the degree of the divisor is larger than the degree of the dividend.

We may write the answer as $3x^3 - 5x^2 + 8x + 3$ R -2 or $3x^3 - 5x^2 + 8x + 3 + \dfrac{-2}{2x + 3}$

$$= 3x^3 - 5x^2 + 8x + 3 - \dfrac{2}{2x + 3}$$

To check the answer: Multiply $3x^3 - 5x^2 + 8x + 3$ by $2x + 3$ and add the remainder -2. If you obtain the original dividend the answer is correct.

Case 3: The dividend has missing powers

Example   Divide $x^5 - 2x^3 - 10x^2 - 30$ by $x - 3$.

In setting up the problem vertically, we may either leave a column space for the missing powers or we may write zero-coefficient terms for the powers. We choose to write zero coefficient powers.

Then the dividend becomes $x^5 + 0x^4 - 2x^3 - 10x^2 + 0x - 30$

Solution   Arrange the divisor and dividend in descending powers of x and set up the problem as in arithmetic long division, and divide.

$$x - 3 \overline{)\begin{array}{l} x^4 + 3x^3 + 7x^2 + 11x + 33 \\ x^5 + 0x^4 - 2x^3 - 10x^2 + 0x - 30 \end{array}}$$

$$\begin{array}{l} +x^5 - 3x^4 \\ \hline + 3x^4 - 2x^3 - 10x^2 + 0x - 30 \\ + 3x^4 - 9x^3 \\ \hline + 7x^3 - 10x^2 + 0x - 30 \\ + 7x^3 - 21x^2 \\ \hline + 11x^2 - 0x - 30 \\ + 11x^2 - 33x \\ \hline +33x - 30 \\ + 33x - 99 \\ \hline 69 \end{array}$$

Scrapwork

Divide: $\dfrac{x^5}{x} = x^4$

Subtract: $0x^4 - (-3x^4) = 3x^4$

Divide $\dfrac{3x^4}{x} = 3x^3$

Subtract: $-2x^3 - (-9x^3) = 7x^3$

Subtract: $-10x^3 - (-21x^2) = 11x^2$

Subtract: $-30 - (-99) = -30 + 99 = 69$
<---------Remainder

In this problem, the division process ends when the degree of the first term of the divisor is larger than the degree of the first term of the dividend (if the terms are in descending powers of x).

$$\therefore \frac{x^5 - 2x^3 - 10x^2 - 30}{x - 3} = x^4 + 3x^3 + 7x^2 + 11x + 33 \text{ R } 69$$

or

$$= x^4 + 3x^3 + 7x^2 + 11x + 33 + \frac{69}{x - 3}$$

Case 4: Division of polynomials involving two or more variables.
     In this case, we may base the descending order arrangement on one of the variables.
Example   Divide $- 3y^3 + 2x^2y - 5xy^2$ by $2x + y$

We base the order of arrangement of the terms on x.

$$
\begin{array}{r}
xy - 3y^2 \\
2x + y \; | \; \overline{\;2x^2y - 5xy^2 - 3y^3\;} \\
+\, 2x^2y + xy^2 \\
\hline
-6xy^2 - 3y^3 \\
-6xy^2 - 3y^3 \\
\hline
0 + 0
\end{array}
$$

Note above that if we base the order of arrangement of the terms on y, we will obtain the same quotient, (except for the order of the terms) $-3y^2 + xy$.

Lesson 36 Exercises

A. 1. Find the product of $-5ab^2$ and $(6b^4)$;   **2**. Simplify: $-4x(x - 5)$;   **3**. $(x - 3)(x + 7)$.

4. Find the product: $(4x + 2)(3x - 6)$;   **5**. Simplify: $(x - y)(x + y)$

Answers: **1**.$-30ab^6$;   **2**. $- 4x^2 + 20x$;   **3**. $x^2 + 4x - 21$;   **4**. $12x^2 - 18x - 12$;   **5**. $x^2 - y^2$

B. Carry out the indicated operations:

1. $2x(bx)$ **2**. $-3x(-7x)$ **3**. $x(2x + 6y)$

4. $2(3a - b)$ **5**. $(x + 2)(x + 3)$ **6**. $(x - 2)(x + 3)$

7. $(2a + 3b)(a - b)$ **8**. $(a + b + c)(a - b + c)$ **9**. $(3x - 2)(2x + 5)$

10. $(2x - 3)^2$ **11**. $(a + b)^2$ **12**. $2a^2(5a - 5b - c)$

13. Simplify: $(x + 4)(x - 5) + (x - 6)(x + 3)$

Answers: **1**. $2bx^2$;   **2**. $21x^2$;   **3**. $2x^2 + 6xy$;   **4**. $6a - 2b$;   **5**. $x^2 + 5x + 6$;   **6**. $x^2 + x - 6$;   **7**. $2a^2 + ab - 3b^2$;   **8**. $a^2 + 2ac - b^2 + c^2$;   **9**. $6x^2 + 11x - 10$;   **10**. $4x^2 - 12x + 9$;   **11**. $a^2 + 2ab + b^2$;   **12**. $10a^3 - 10a^2b - 2a^2c$;   **13**. $2x^2 - 4x - 38$.

C. Simplify the following **1**. $- (-4x + 6y + 6)$;   **2**. $2 + 4(3x - 5y - 2)$;   **3**. $-4(3a + c) - 3(5a - 4)$

4. $+(6x + 4y) + 4(7x + 8)$;   **5**. $4xy(5x^2y - 3y) - 7xy^2$

Answers: **1**. $4x - 6y - 6$; **2**. $12x - 20y - 6$; **3**. $- 27a - 4c + 12$;   **4**. $34x + 4y + 32$;   **5**. $20x^3y^2 - 19xy^2$

D. Simplify and leave answers with positive exponents:

1. $\dfrac{10a^7}{2a^2}$ **2.** $\dfrac{12x^4y^6}{-2x^3y^2}$ **3.** $\dfrac{-18a^5b^7}{-3a^2b^4}$;   **4.** $\dfrac{16a^2bc}{14a^3bc^2}$

5. $\dfrac{-10x^5y^7}{20x^6y^6}$ **6.** $\dfrac{16c^3d}{-12cd^2}$ **7.** $\dfrac{.75a^4bc}{2.5abc}$

Answers: 1. $5a^5$; **2.** $- 6xy^4$; **3.** $6a^3b^3$; **4.** $\dfrac{8}{7ac}$; **5.** $-\dfrac{y}{2x}$; **6.** $-\dfrac{4c^2}{3d}$; **7.** $.3a^3$

E. Simplify the following:

1. $\dfrac{-8x^5y^3}{2x^2y}$; **2.** $\dfrac{15a^4b^2 - 20a^3b^2 - 5ab}{5ab}$;   **3.** $\dfrac{14x^3y^2 - 21x^2y}{7xy}$

Answers: 1. $-4x^3y^2$;   **2.** $3a^3b - 4a^2b - 1$;   **3.** $2x^2y - 3x$

F. Divide:

1. $\dfrac{16x^2y - 8x}{2x}$ **2.** $\dfrac{16x^2y - 8x}{-2x}$ **3.** $\dfrac{14c^5d^2 - 21cd}{-7cd}$

4. $\dfrac{20x^4yz - 5xy}{5xy}$ **5.** $\dfrac{12a^2bc - 14ab^3c}{6}$

Answers: 1. $8xy - 4$;   **2.** $-8xy + 4$;   **3.** $-2c^4d + 3$;   **4.** $4x^3z - 1$;   **5.** $2a^2bc - \dfrac{7ab^3c}{3}$

G. Carry out the indicated operation.

1. Divide $6x^2 + x^3 + 6 + 11x$ by $x + 2$ **2.** Divide $8x^4 + 9x^2 + 14x^3 + 19x + 12$ by $3 + 2x$

3. $(x^4 + 4x^2 + 5x + 2) \div (x + 3)$ **4.** $(4x^3y - 6x^2y^2 + 5y^4 + 6xy^3) \div (2x + y)$

Answers   **1.** $x^2 + 4x + 3$;   **2.** $4x^5 + x^2 + 3x + 5$ R $- 3$;
3. $x^3 - 3x^2 + 13x - 34$ R 104; **4.** $2x^2y - 4xy^2 + 5y^3$

H. 1. Divide $x^3 + 2x^2 - 5x - 6$ by $x - 2$ **2.** Divide $6x^2 - 8x + 5$ by $x + 3$

3. Divide $x^3 + 10x^2 + 33x + 36$ by $x + 4$ **4.** Divide $4x^4 - 2x^2 - 3$ by $x - 3$

5. Find the remainder when $x^3 + 6x^2 + 6x - 12$ is divided by $x + 2$

Answers: **1.** $x^2 + 4x + 3$;   **2.** $6x - 26$ R 83 ;   **3.** $x^2 + 6x + 9$;
4. $4x^3 + 12x^2 + 34x + 102$ R 303;   **5.** Remainder $= -8$

Test # 4 - Student's Self-Test (**Always**, **Test yourself before you are tested**)

Attempt all questions on clean sheets of paper, **Do not write in the book** Show all necessary work.

Add the like terms:

1. (a) $4x - 6y + 7x - 4y$

  (b) $2a - 4b - 2a - 5b$

Simplify:

2. (a) $-7x + 6y - 3x + 4$

  (b) $5a + 3b - 5a + 4b - 9$

---

3. (a) Subtract $-8x^3 + 2x - 5$ from $6x - 2$.

  (b) From $9x - 2$, subtract $-6x^3 + 4x - 4$

4. (a) Subtract 5 from 3.

  (b) Subtract 3 from 7.

---

5. (a) Subtract $5y - 4x + 2$ from $3x - 6y - 5$

  (b) Subtract $2a - 7b - 8$ from $a - 36$

6. (a) Simplify: $3a^2 - 4ab + b^2 - (5a^2 + 2ab - 3b^2)$

  (b) Subtract $a - 32$ from $2a - 7b - 8$

---

7. (a) Find the product of $-4ab^2$ and $(5b^4)$.

  (b) Find the product: $(3x + 2)(5x - 6)$.

8. Simplify :

  (a) $6x^2y^3 - (x^2y^3 + x^2y) + xy^2 - 5x^2y$;

  (b) $5a^4b^5 + 3ab - (5a^4b^5 + 3ab)$

---

9. Simplify:

  (a) $2(3a - b)$

  (b) $(x + 2)(x + 3)$

10. Simplify:

  (a) $(6x - 8y + 7) - (-3x + 5y - 1)$

  (b) $-4x(x - 5)$;

---

11. Simplify:

  (a) $(x - 4)(x + 6)$.

  (b) $(x - y)(x + y)$

12. Simplify: (a) $(a + b + c)(a - b + c)$

  (b) $(3x - 2)(2x + 5)$

---

13. Simplify: (a) $(4x - 3)^2$

  (b) $(a + b)^2$

14. Simplify:

  (a) $2a^2(5a - 5b - c)$

  (b) $(x + 2)(x - 5) + (x - 6)(x + 3)$

15. Carry out the indicated operations:

(a) $(x - 2)(x + 5)$

(b) $(4a + 3b)(a - b)$

16. Simplify: (a) $2x(bx)$

(b) $-4x(-7x)$

---

17. Simplify the following
(a) $- (-4x + 6y + 6)$

(b) $2 + 5(3x - 5y - 2).$

18. Simplify: (a) $-4(3a + c) - 3(5a - 4)$

(b) $+(2x + 4y) + 3(7x + 8);$

---

19. Simplify: (a) $3xy(5x^2y - 3y) -7xy^2$

(b) $7ab - 8ba + 2ab + 6b$

20 Simplify:
(a) $6x - 7y + 3y - 4x$

(b) $\frac{1}{2}a^2b + 8 + \frac{1}{4}a^2b$

---

21. Add the following:

(a) $2x - 6y - 5$ and $2y - 4x + 3$

(b) $3x^2y - 9xy^2$ and $5x^2y + 3xy^2$

22. Simplify: (a) $-3a^2b + 5ab^2 + 3a^2b + ab^2$

(b) $- 4x^2y - 5xy + 2x^2y + y^2$

Simplify:

23. $\dfrac{.75a^4bc}{2.5abc}$

Simplify:

24.   (a) $\dfrac{10a^7}{2a^2}$; (b) $\dfrac{12x^4y^6}{-2x^3y^2}$

Simplify:

25. (a) $\dfrac{-18a^5b^7}{-3a^2b^4}$;   (b) $\dfrac{16a^2bc}{14a^3bc^2}$

Simplify:

Bonus: (a) $\dfrac{-10x^5y^7}{20x^6y^6}$;   (b) $\dfrac{16c^3d}{-12cd^2}$

Answers: **1**. (a) $11x - 10y$; (b) $-9b$;   **2**. (a) $-10x + 6y + 4$; (b) $7b - 9$;
3. (a) $8x^3 + 4x + 3$; (b) $6x^3 + 5x + 2$; **4**. (a) -2; (b) 4; ; **5**. (a) $7x - 11y - 7$; (b) $-a + 7b - 28$;
6. (a) $-2a^2 - 6ab + 4b^2$; (b) $a - 7b + 24$; **7**.(a) $-20ab^6$; (b) $15x^2 - 8x - 12$;
8 (a) $5x^2y^3 - 6x^2y + xy^2$; (b) 0; **9**. (a) $6a - 2b$; (b) $x^2 + 5x + 6$; **10**. (a) $9x - 13y + 8$;
(b) $-4x^2 + 20x$; **11**. (a) $x^2 + 2x - 24$; (b) $x^2 - y^2$; **12**. (a) $a^2 + 2ac - b^2 + c^2$; (b) $6x^2 + 11x - 10$;
13. (a) $16x^2 - 24x + 9$; (b) $a^2 + 2ab + b^2$;   **14**. (a) $10a^3 - 10a^2b - 2a^2c$; (b) $2x^2 - 6x - 28$;
15. (a) $x^2 + 3x - 10$; (b) $4a^2 - ab - 3b^2$; **16**. (a) $2bx^2$; (b) $28x^2$; **17** (a) $4x - 6y - 6$;
(b) $15x - 25y - 8$; **18**. (a) $-27a - 4c + 12$; (b) $23x + 4y + 24$; **19**. (a) $15x^3y^2 - 16xy^2$;
(b) $ab + 6b$; **20**. (a) $2x - 4y$; (b) $\frac{3}{4}a^2b + 8$; **21**. (a) $-2x - 4y - 2$; (b) $8x^2y - 6xy^2$;
22. (a) $6ab^2$; (b) $-2x^2y - 5xy + y^2$; **23**. $.3a^3$ or $\frac{3}{10}a^3$; **24.** (a) $5a^5$; (b) $-6xy^4$;
25. (a) $6a^3b^3$; (b) $\dfrac{8}{7ac}$; **Bonus:** (a) $-\dfrac{y}{2x}$; (b) $-\dfrac{4c^2}{3d}$.

Test # 5 - Student's Self-Test (**Always, Test yourself before you are tested**)

Attempt all questions on clean sheets of paper, **Do not write in the book** Show all necessary work.

1. Simplify: $\dfrac{42x^3y^2 - 21x^2y}{7xy}$

2. Simplify: $\dfrac{30a^4b^2 - 20a^3b^2 - 5ab}{5ab}$

---

3. **Divide**:

(a) $\dfrac{24x^2y - 8x}{2x}$;   (b) $\dfrac{14x^2y - 8x}{-2x}$

Divide:

4. (a) $\dfrac{14c^5d^2 - 21cd}{-7cd}$;   (b) $\dfrac{20x^4yz - 5xy}{5xy}$

---

5. Divide $6x^2 + x^3 + 6 + 11x$ by $x + 3$

6. Simplify: $(x^4 + 4x^2 + 5x + 2) \div (x + 3)$

---

7. Divide $6x^2 - 8x + 5$ by $x - 3$

8. Divide $x^3 + 10x^2 + 33x + 36$ by $x + 4$

---

9. Divide $4x^4 - 2x^2 - 3$ by $x - 2$

10. $(4x^3y - 6x^2y^2 + 5y^4 + 6xy^3) \div (2x + y)$

11.
Divide $8x^4 + 9x^2 + 14x^3 + 19x + 12$ by $3 + 2x$

12.   Divide $x^3 + 2x^2 - 5x - 6$ by $x - 2$

Answers: **1.** $6x^2y - 3x$; **2.** $6a^3b - 4a^2b - 1$; **3.** (a) $12xy - 4$; (b) $-7xy + 4$; **4.** (a) $-2c^4d + 3$;
(b) $4x^3z - 1$; **5.** $x^2 + 3x + 2$; **6.** $x^3 - 3x^2 + 13x - 34 + \dfrac{104}{x+3}$; **7.** $6x + 10 + \dfrac{35}{x-3}$; **8.** $x^2 + 6x + 9$;
9. $4x^3 + 8x^2 + 14x + 28 + \dfrac{53}{x-2}$; **10.** $2x^2y - 4xy^2 + 5y^3$; **11.** $4x^3 + x^2 + 3x + 5 - \dfrac{3}{2x+3}$;
12. $x^2 + 4x + 3$.

CHAPTER 13

First Degree Equations Containing One Variable
(Simple Linear Equations)

Lesson 37: **Axioms for Solving Equations**
Lesson 38: **Solving First Degree Equations**
Lesson 39: **Solving Decimal and Literal Equations**

Lesson 37
Axioms for Solving Equations

Axioms are general mathematical statements that we accept as true, without any proof, in order to deduce other less obvious statements. The following are very useful in constructing proofs and solving equations.

1. A quantity is equal to itself. (reflexive property of equality, also identity principle)
   $a = a$

2. An equality may be reversed. (symmetric property of equality)
   If $a = b$, then
   $b = a$

3. Quantities equal to the same quantity are equal to each other.(transitive property of equality)
   If $a = b$, and $b = c$, then
   $a = c$.

4. A quantity may be substituted for its equal in any expression or equation. (substitution axiom)

5. A whole equals the sum of all its parts. (partition axiom)

6. If equal quantities are added to equal quantities, the sums are equal (addition axiom)

If $a = b$,		(If $a = b$, and
then $a + c = b + c$	**ALSO**	$c = d$, then
		$a + c = b + d$)

7. If equal quantities are subtracted from equal quantities, the differences are equal. (subtraction axiom)

If $a = b$,		(If $a = b$, and
then $a - c = b - c$	**ALSO**	$c = d$, then
		$a - c = b - d$)

8. If equal quantities are multiplied by equal quantities, the products are equal. (multiplication axiom)

If $a = b$,		(If $a = b$, and
then $ac = bc$	**ALSO**	$c = d$, then
		$ac = bd$.)

9. If equal quantities are divided by equal quantities (not zero), the quotients are equal. (division axiom)

If $a = b$,

then $\dfrac{a}{c} = \dfrac{b}{c}$

ALSO

(If $a = b$, and

$c = d$, then

$\dfrac{a}{c} = \dfrac{b}{d}$)

10. Like powers of equals are equal . (powers axiom)

Example: If $a = 3$, then

$a^2 = 3^2$ or $a^2 = 9$.

11. Like roots of equals are equal. (roots axiom)

Example: if $a^3 = 8$, then

$\sqrt[3]{a^3} = \sqrt[3]{8}$

$a = 2$.

Lesson 38
Solving First Degree Equations

An equation is a statement of the equality between two expressions.

Example: $2x + 3 = x - 10$

The equality symbol "=" breaks up an equation into two sides or members, namely the left-hand side and the right-hand side of the equation. To solve an equation involving a single variable, say x, means we are to find values of x which satisfy the equation. **A value of x is said to satisfy an equation if this value when substituted in the equation makes both the left-hand side and the right-hand side of the equation equal to each other.**
A value of x which satisfies a given equation is said to be a solution or a root of the given equation. **To obtain a value for the variable, x, we will get x by itself on one side of the equation** .
We agree that a value of x has been obtained if we have x by itself on one side of the equation and all the other quantities on the other side of the equation do **not** involve x. To get x by itself, we will use inverse operations. Addition and subtraction are inverse operations. Multiplication and division are inverse operations. Inverse operations "undo" each other. For example, to undo multiplication by a number, we will use division by the same the number. We will keep the above discussion in mind when we solve linear equations.

Example 1 Solve for x:

1. $x - 5 = 11$

To get x by itself on the left-hand side of the equation, we will remove the "-5" by adding its opposite to both sides of the equation.(This is the same as adding $+ 5$ to both sides of the equation)

$$x - 5 = 11$$
$$\underline{\quad + 5 \quad + 5 \quad}$$
$$x = 16$$

The solution is 16.

Example 2 Solve for x: $x + 8 = -11$

To get x by itself on the left-hand side of the equation, we remove the "+8" by adding -8 to (that is, subtracting 8 from) both sides of the equation.

$$x + 8 = -11$$
$$\underline{\quad - 8 \quad - 8 \quad}.$$
$$x = -19$$

(by stressing on **addition** of - 8, the right-hand side of the equation is done without confusion)

Note that to remove (undo) any number, use the inverse (opposite) operation.

Example 3   Solve for x: $5x = 20$

To obtain x by itself, we will undo (remove) the "5". We use division by 5   (since the "5" is multiplying the x)

$$\frac{5x}{5} = \frac{20}{5}$$
$$x = 4$$

The solution is 4.

Example 4 Solve for x:

$$- 6x = 24$$
$$\frac{-6x}{-6} = \frac{24}{-6}$$
$$x = -4$$

The solution is -4.

Example 5   Solve for x: $\dfrac{x}{6} = 4$

To remove the "6" which is a divisor, we will multiply both sides of the equation by 6

$$(6)\dfrac{x}{6} = (6)4$$

$$\cancel{(6)}\dfrac{x}{\cancel{6}}_1 = (6)4$$

$x = 24$. The solution is 24.

Example 6 Solve for x: $\dfrac{-x}{5} = 2$ (or $\dfrac{x}{5} = -2 \text{---}>x = 5(-2) = -10$)

$$\cancel{(5)}\dfrac{(-x)}{\cancel{5}}_1 = (5)2$$

$-x = 10$ <- -------------------------

$(-1)\,(-1x) = (-1)10$ **Note:** $-x = -1x$

$x = -10$ OR   from $\boxed{\text{this step}}$, just change

the sign of the left-hand side and

The solution is -10.   change the sign of the right-hand side.

Example 7   Solve for x

$2x - 6 = 10$

We will remove the "- 6 " first, and then remove the "2" (i.e., we will always undo multiplication and division last)

$$\begin{array}{r} 2x - 6 = 10 \\ +6 \quad +6 \\ \hline 2x = 16 \end{array}$$

$$\dfrac{2x}{2} = \dfrac{16}{2}$$

$x = 8$.   The solution is 8.

Example 8   Solve for x: $5x - 6 = x + 2$

In this case, we will collect all the x's on one side by removing either the x-term on the right-hand side or the x-term on the left-hand side.

Let us remove the x-term on the right-hand side.

$$\begin{array}{r} 5x - 6 = x + 2 \\ -x \qquad -x \quad \\ \hline 4x - 6 = +2 \\ +6 \quad +6 \\ \hline 4x = 8 \end{array}$$

$$\dfrac{4x}{4} = \dfrac{8}{4}$$

$x = 2$.   The solution is 2.

Example 9   Solve for x

$7x + 2 - 3x = 2x + 18 + 3x$

In this case, first, simplify each side of the equation, and then proceed.

Step 1: $7x + 2 - 3x = 2x + 18 + 3x$

$4x + 2 = 5x + 18$

Step 2: $\begin{array}{r} -5x \qquad -5x \quad \\ \hline -x + 2 = 18 \\ -2 \quad -2 \\ \hline -x = 16 \end{array}$

$x = -16$.   The solution is -16.

Example 10 Solve for *x*

$$3(x + 2) = 5(x - 6)$$

First, remove the parentheses on both sides of the equation

$$3(x + 2) = 5(x - 6)$$
$$3x + 6 = 5x - 30$$
$$\underline{-5x \qquad\ -5x}$$
$$-2x + 6 = -30$$
$$\underline{\ - 6 \quad - 6}$$
$$- 2x = -36$$
$$\frac{-2x}{-2} = \frac{-36}{-2}$$
$$x = 18$$

The solution is 18.

Checking Solutions of Equations (see also Lesson 54)

Example (a)   Determine if - 5 is a solution of the equation

$$4x + 18 - x = 8x + 32 + 2x.$$

- 5 will be a solution of the given equation if this value when substituted in the given equation makes the left-hand side of the equation equal to the right-hand of the equation.

Checking: Replace *x* by -5 in the equation.

$$\text{Then, } 4(-5) + 18 - (-5) \overset{?}{=} 8(-5) + 32 + 2(-5)$$
$$-20 + 18 + 5 \overset{?}{=} - 40 + 32\ - 10$$
$$-2 + 5 \overset{?}{=} - 8 - 10$$
$$3 = -18 \quad \text{False}$$

The left-hand side of the equation and the right-hand side of the equation are **not** equal. Therefore, -5 is **not** a solution of the given equation.

(b) Determine if -2 is a solution of $4x + 18 - x = 8x + 32 + 2x.$

Checking: Replace *x* by -2 in the equation.

$$\text{Then, } 4(-2) + 18 -(-2) \overset{?}{=} 8(-2) + 32 + 2(-2)$$
$$-8 + 18 + 2 \overset{?}{=} -16 + 32 - 4$$
$$10 + 2 \overset{?}{=} 16 - 4$$
$$12 = 12 \quad \text{True}$$

Since the left-hand side of the equation equals the right-hand side of the equation, -2 is a solution of the given equation.

More Examples on Solutions of Linear Equations

In the following examples, attempt the problems before reading the solutions:

Example 1   Solve for x: $5x - 9 = 6$

Solution

$$5x - 9 = 6$$
$$\underline{ + 9 \ +9}$$
$$5x + 0 = 15$$
$$\frac{5x}{5} = \frac{15}{5}$$

Check: $5(3) - 9 \overset{?}{=} 6$

$$x = 3$$

$15 - 9 \overset{?}{=} 6$

$6 = 6$ True

Example 2   Solve for x: $6x - 3 = 2x + 8$

Solution :

$$6x - 3 = 2x + 8$$
$$\underline{-2x -2x }$$
$$4x - 3 = 0 + 8$$
$$\underline{ +3 + 3}$$
$$4x = 11$$
$$\frac{4x}{4} = \frac{11}{4}$$

$$x = \frac{11}{4}$$

Check: $6(\frac{11}{4}) - 3 \overset{?}{=} 2(\frac{11}{4}) + 8$

$33 - 6 \overset{?}{=} 11 + 16$

$27 = 27$ True

Example 3   Solve for x:

$$\frac{x}{6} = \frac{11}{12}$$

Method 1   To get x by itself on the left-hand side of the equation, we undo the "6" (which is a divisor) by multiplying each side of the equation by 6.

$$(6)\frac{x}{6} = (6)\frac{11}{12}_2$$

$$(6)\frac{x}{6} = {}^1(6)\frac{11}{12}_2$$

$$x = \frac{11}{2}$$

Method 2   Multiply each term of the equation by the LCD of the fractions. The LCD is 12.

$$(12)\frac{x}{6} = (12)\frac{11}{12}$$

$$2\,(12)\frac{x}{6} = (12)\frac{11}{12}$$

$$2x = 11$$
$$\frac{2x}{2} = \frac{11}{2}$$
$$x = \frac{11}{2}$$

Method 3   **Cross-multiplication**

$$\frac{x}{6} = \frac{11}{12}$$

$$12x = 6(11)$$

$$\frac{{}^1\cancel{12}x}{\cancel{12}_1} = \frac{{}^1\cancel{6}(11)}{\cancel{12}_2}$$

$$x = \frac{11}{2}$$

Note: We can cross-multiply if there is only a single fraction on the left-hand side of the equation, and only a single fraction on the right-hand side of the equation. For instance, in the following example, we cannot cross-multiply immediately, but rather, we will have to combine the right-hand side into a single fraction before cross-multiplying.

Example 4   Solve for *x*: $\dfrac{x}{4} = \dfrac{x}{20} + \dfrac{2}{5}$

Method 1   **The LCD method**. Multiply each term of the equation by the LCD of the fractions. The LCD is 20.

$$(20)\frac{x}{4} = (20)\frac{x}{20} + (20)\frac{2}{5}$$

$$^5\cancel{(20)}\frac{x}{\cancel{4}_1} = {}^1\cancel{(20)}\frac{x}{\cancel{20}_1} + {}^4\cancel{(20)}\frac{2}{\cancel{5}_1}$$

$$5x = x + 8$$

$$\underline{-x \quad -x \qquad}$$

$$4x = 8$$

$$\frac{4x}{4} = \frac{8}{4}$$

$$x = 2$$

The solution is 2.

Method 2   By cross-multiplication

   To apply cross-multiplication to the above problem, first, combine the two terms on the right-hand side into a single fraction. (We have a single fraction if we have a single denominator.)

continuing, $\dfrac{x}{4} = \dfrac{x}{20} + \dfrac{2}{5}$

$$\frac{x}{4} = \frac{x}{20} + \frac{2(4)}{5(4)}$$

$$\frac{x}{4} = \frac{x}{20} + \frac{8}{20}$$

$$\frac{x}{4} = \frac{(x + 8)}{20}$$

Now, we cross-multiply:

$$20x = 4(x + 8) \;\longleftarrow\text{------ The parentheses are important}$$

$$20x = 4x + 32$$

$$\underline{- 4x \; - 4x \qquad}$$

$$16x = 0 + 32$$

$$\frac{16x}{16} = \frac{32}{16}$$

$$x = 2$$

Method 3   You may undo one denominator at a time.

$$\frac{x}{4} = \frac{x}{20} + \frac{2}{5}$$

Step 1:   Undo the "4" : $\dfrac{1\,(4)x}{4_1} = \dfrac{1\,(4)x}{20} + \dfrac{(4)2}{5}$ (Multiply each term by 4)

$$x = \frac{x}{5} + \frac{8}{5}$$

Step 2:   To undo the "5" multiply by 5:

$$(5)x = \frac{1\,(5)x}{5_1} + \frac{(5)8}{5_1}$$

$$5x = x + 8$$
$$4x = 8$$
$$x = 2$$

Note: In method 3, multiplying the equation first by 4 and then by 5 is equivalent to multiplying the equation by 20 (as was done in Method 1).

Lesson 38 Exercises

A

Solve for x: **1.** $x - 7 = 12$;   **2.** $x + 6 = 9$; **3.** $4x = 20$;   **4.** $\frac{x}{6} = 8$; **5.** $4x + 3 = 35$

6. $7x - 6 = 2x + 24$;   **7.** $5x - 7 + 3x + 9 = 6x + 5 + 8x$;   **8.** $\frac{x}{6} + 4 = 9$;

9. $3(2x - 5) + 4 = -4(x + 7)$

Answers: **1.** $x = 19$; **2.** $x = 3$; **3.** $x = 5$; **4.** $x = 48$; **5.** $x = 8$; **6.** $x = 6$; **7.** $x = -\frac{1}{2}$

8. $x = 30$;   **9.** $x = -\frac{17}{10}$

B   Solve and check:

1. $x + 4 = 6$;   2. $x + 5 = -7$ 3. $x - 3 = 9$

4. $x - 6 = -11$ 5. $x - \frac{1}{3} = \frac{1}{4}$ 6. $\frac{x}{2} = 8$

7. $\frac{x}{-2} = 7$ 8. $\frac{-x}{4} = 9$ 9. $3x = 18$

10. $-4x = 20$ 11. $6x = 22$ 12. $7x = -28$

13. $\frac{x}{4} - 6 = 14$ 14. $\frac{-x}{3} + 5 = 7$ 15. $2x + 4 = 9$

16. $-3x - 5 = 10$ 17. $5x - 4 + x = 4x + 6 - 3x$

18. $2x - 5 + 3x + 2 = 7x - 5 - 3x$ 19. $2(x - 4) + 3 = 15$;   20. $-3(x + 5) + 1 = 5(x + 4) - 6$

Answers: **1.** 2 ; **2.** -12 ; **3.** 12 ; **4.** -5 ; **5.** $\frac{7}{12}$; **6.** 16 ; **7.** -14 ; **8.** -36 ; **9.** 6 ; **10.** -5 ; **11.** $\frac{11}{3}$;

12. -4 ; **13.** 80 ; **14.** -6 ; **15.** $\frac{5}{2}$; **16.** -5 ; **17.** 2 ; **18.** -2 ; **19.** 10 ; **20.** -$\frac{7}{2}$

C

1. Is 2 a solution of $3x + 1 = 7$?   **2.** Is -3 a solution of $4x - 1 = 15$?

3. Is 2 a solution of $5x + 2 = x - 4$?

Answers: **1.** Yes ; **2.** No ; **3.** No.

Solve for x: **1.** $\dfrac{x}{6} + \dfrac{x}{3} = \dfrac{3}{4}$ **2.** $\dfrac{x}{8} = \dfrac{x}{2} - 6$;   **3.** $\dfrac{x}{2} + 1 = \dfrac{x}{3}$;   **4.** $\dfrac{x-2}{4} + \dfrac{2}{3} = \dfrac{1}{2}$

Answers : **1.** $x = \dfrac{3}{2}$; **2.** $x = 16$; **3.** $x = -6$; **4.** $x = \dfrac{4}{3}$

E

Solve for x: **1.** $\dfrac{x+3}{4} + \dfrac{x-2}{8} = 2$ **2.** $\dfrac{x-2}{3} + 1 = \dfrac{x}{5}$ **3.** $\dfrac{x}{4} + \dfrac{1}{2} = \dfrac{x}{5} + 1$

4. $\dfrac{x}{3} = \dfrac{x}{6} + \dfrac{3}{5}$ **5.** $\dfrac{x}{2} + \dfrac{2}{3} = \dfrac{x}{4}$

Answers: **1.** 4 ; **2.** $-\dfrac{5}{2}$; **3.** 10 ; **4.** $\dfrac{18}{5}$; **5.** $-\dfrac{8}{3}$

Lesson 39
Solving Decimal and Literal Equations

Solving Decimal Equations

Discussion: There are a number of methods for solving decimal equations.

Method **1:** We can solve a decimal equation by strictly using decimals.

Method **2:** We can also change a decimal equation to an equivalent equation in which all the constants are integers, by multiplying every term of the equation by a power of 10. The power of 10 to use is based on the constant with the most number of decimal places or decimal digits.

Method **3:** We can make a partial conversion by multiplying the equation by a power of 10 so that only the numerical coefficients of the variables are necessarily changed to integers; some of the constant terms may consequently be changed partially or wholly, since we have to multiply every term of the equation by a power of 10. In multiplying by a power of 10, just move the decimal point accordingly.(see p.44.)

Example 1 Solve for x:
$$.5x = 2$$

Solution: **Using Method 2** above, multiply both sides of the equation by 10.
$$5x = 20$$
$$x = 4$$

Example 2 Solve for x:
$$.02x - 6 = .1$$

Solution; Multiply every term by 100. (**Using Method 2**)
$$2x - 600 = 10$$
$$2x = 610$$
$$x = 305$$

Example 3 Solve for x:
$$.7x + .002 + .04 = .01x$$

 Solution: **Method 3** , multiply every term 100.

$$70x + .2 + 4 = x$$
$$70x + 4.2 = x$$
$$69x = -4.2$$
$$x = \frac{-4.2}{69}$$
$$x = -.06 \text{ (to the nearest hundredth)}$$

Note that in Example 3, if we used **Method 2**. we would have
$$700x + 2 + 40 = 10x$$
$$700x + 42 = 10x$$
$$690x = -42$$
$$x = -\frac{42}{690} = x = -.06$$

Or if we used **Method 1** (strictly decimals) we would have
$$.7x - .01x = -.042$$
$$.69x = -.042 ----> x = \frac{-.042}{.69} = -.06$$

Solving Literal Equations

Here, note that the letters represent numbers and therefore the principles and techniques we
have used, so far, in solving linear equations are applicable.

Example 1   Solve for x: $4x + 3y = 7$

We want to get x by itself on one side of the equation. Let us use some common sense here. We will undo
(remove) the $3y$ and the 4 on the left-hand side of the equation.

$$4x + 3y = 7$$
$$\underline{\quad -3y \qquad -3y}$$
$$4x + 0 = 7 - 3y$$

$$\frac{4}{4}x = \frac{7 - 3y}{4}$$

$$x = \frac{7 - 3y}{4} = \frac{-3y + 7}{4} = \frac{-3y}{4} + \frac{7}{4} = -\frac{3}{4}y + \frac{7}{4}$$

If the question is of the multiple-choice type, then, be on the look out for any of these equivalent forms of answers.

Example 2   Solve Example 1 for y.

i.e., Solve $4x + 3y = 7$ for y.

Note in this case that, we want y by itself one side of the equation.

$$4x + 3y = 7$$
$$\underline{-4x \qquad\qquad -4x}$$
$$0 + 3y = 7 - 4x$$

$$\frac{3y}{3} = \frac{7 - 4x}{3}$$

$$y = \frac{-4x + 7}{3} = \frac{-4}{3}x + \frac{7}{3}$$

$$y = -\frac{4}{3}x + \frac{7}{3}$$

Example 3 Solve for A: $B = AC + CE$

We will obtain A by itself on one side of the equation.

$$B = AC + CE$$
$$B - CE = AC + CE - CE \qquad\qquad \text{(subtracting } CE \text{ from both sides)}$$
$$B - CE = AC$$

$$\frac{B - CE}{C} = \frac{AC}{C} \qquad\qquad \text{(dividing both sides by } C\text{)}$$

$$\frac{B - CE}{C} = A$$

$$\therefore A = \frac{B - CE}{C} \text{ or } \frac{B}{C} - \frac{CE}{C} = \frac{B}{C} - E$$

Lesson 39 Exercises

A   Solve for x:

1. $2x + .6 = 1$ **2.** $5x + .04 = .2$ **3.** $4x + .1 = 2.5$

Answers: **1.** $.2$;   **2.** $.032$;   **3.** $.6$

B

Solve for y: **1.** $3x + 8y = 32$;   **2.** $2y - 6x = 21$;   **3.** Solve for C: $F = \frac{9}{5}C + 32$

Answers: **1.** $y = -\frac{3}{8}x + 4$ **2.** $y = 3x + \frac{21}{2}$;   **3.** $C = \frac{5}{9}F - \frac{160}{9}$ or $C = \frac{5}{9}(F - 32)$

C

1. Solve for x: $5x + 4y = 20$ **2.** Solve for y: $5x + 4y = 20$

3. Solve for x: $2x - 3y = 12$ **4.** Solve for b: $abc = x$

5. Solve for P $nRT = PV$ **6.** Solve for R: $nRT = PV$

Answers: **1.** $x = -\frac{4}{5}y + 4$;   **2.** $y = -\frac{5}{4}x + 5$;   **3.** $x = \frac{3}{2}y + 6$;   **4.** $b = \frac{x}{ac}$;   **5.** $P = \frac{nRT}{V}$;   **6.** $R = \frac{PV}{nT}$

Test # 6 - Student's Self-Test (Always, Test yourself before you are tested)

Attempt all questions on clean sheets of paper, **Do not write in the book**
Show all necessary work.

Problems 1- 23: Solve the equations unless
  otherwise instructed:

1. (a) $x - 5 = 12$

  (b) $x + 4 = -9$

2. (a) $4x + 5 = 35$

  (b) $2x + 3 = 9$

3. (a) $5x = 20$

  (b) $-2x = 20$

4. (a) $8x = -32$

  (b) $-8x = 32$

5. (a) $7x - 8 = 3x + 24$

6. (a) $\frac{x}{6} + 3 = 7$

  (b) $8x = 22$

7. (a) $\frac{x}{6} = 8$ (b) $3x = 18$

8. (a) $5x - 7 + 3x + 9 = 6x + 5 + 8x$;

  (b) $\frac{x}{4} - 6 = 14$

9. (a) $2(x - 4) + 3 = 15$

  (b) $3(x + 5) + 1 = 5(x + 4) - 6$

10. (a) $3(2x - 5) + 4 = -4(x + 7)$

  (b) $5x - 4 + x = 4x + 6 - 3x$

11. $2x - 5 + 3x + 2 = 7x - 5 - 3x$

12. $\frac{-x}{3} + 5 = 7$

13. Is 2 a solution of $5x + 2 = x - 4$?

14. Is -3 a solution of $4x - 1 = 15$?

15. Is 2 a solution of $3x + 1 = 7$?

16. (a) $x - \frac{1}{3} = \frac{1}{4}$

  (b) $\frac{x}{3} = \frac{x}{6} + \frac{3}{5}$

17. (a) $\frac{x}{-2} = 7$ (b) $\frac{-x}{4} = 9$

18. (a) $\frac{x}{6} + \frac{x}{3} = \frac{3}{4}$

    (b) $\frac{x}{8} = \frac{x}{2} - 6$

---

19. (a) $\frac{x}{2} + 1 = \frac{x}{3}$

   (b) $\frac{x-2}{4} + \frac{2}{3} = \frac{1}{2}$

20. (a) $\frac{x+3}{4} + \frac{x-2}{8} = 2$

    (b) $\frac{x}{4} + \frac{1}{2} = \frac{x}{5} + 1$

---

21. (a) $\frac{x-2}{3} + 1 = \frac{x}{5}$

   (b) $\frac{x}{2} + \frac{2}{3} = \frac{x}{4}$

22. (a) $2x + .6 = 1$

    (b) $5x + .04 = .2$

---

23. (a) Solve for x $4x + .1 = 2.$
    (b) Solve for x: $2x - 3y = 12$

24. (a) Solve for C: $F = \frac{9}{5}C + 32$
    (b) Solve for y: $5x + 4y = 20$

---

25. (a) Solve for y: $3x + 8y = 32$

    (b) Solve for x: $2y - 6x = 21$

Bonus: (a) Solve for P $nRT = PV$

    (b) Solve for R: $nRT = PV$

    (c) Solve for b: $abc = x$

    (d) Solve for x: $5x + 4y = 20$

Answers: **1.** (a) $x = 17$; (b) $x = -13$; **2.** (a) $x = \frac{15}{2}$; (b) $x = 3$; **3.** (a) $x = 4$; (b) $x = -10$;

4. (a) $x = -4$; (b) $x = -4$; **5.** $x = 8$; **6.** (a) $x = 24$; (b) $x = \frac{11}{4}$; **7.** (a) $x = 48$; (b) $x = 6$;

8. (a) $x = -\frac{1}{2}$; (b) $x = 80$; **9.** (a) $x = 10$; (b) $x = 1$; **10.** (a) $x = -\frac{17}{10}$; (b; $x = 2$; **11.** $x = -2$;

12. $x = -6$; **13.** No ; **14.** No; **15.** Yes ; **16.** (a) $x = \frac{7}{12}$; (b) $x = \frac{18}{5}$; **17.** (a) $x = -14$;

(b) $x = -36$; **18.** (a) $x = \frac{3}{2}$; (b) $x = 16$; **19.** (a) $x = -6$; (b) $x = \frac{4}{3}$; **20.** (a) $x = 4$; (b) $x = 10$;

21. (a) $x = -\frac{5}{2}$; (b) $x = -\frac{8}{3}$; **22.** (a) $x = .2$ or $\frac{1}{5}$; (b) $x = .032$ or $\frac{4}{125}$; **23.** (a) $x = .475$ or $\frac{19}{40}$;

(b) $x = \frac{3}{2}y + 6$; **24.** (a) $C = \frac{5F - 160}{9}$ or $C = \frac{5}{9}(F - 32)$; (b) $y = -\frac{5}{4}x + 5$; **25.** (a) $y = -\frac{3}{8}x + 4$;

(b) $x = \frac{1}{3}y - \frac{7}{2}$; **Bonus:** (a) $p = \frac{nRT}{V}$; (b); $R = \frac{PV}{nT}$;   (c) $b = \frac{x}{ac}$; (d) $x = -\frac{4}{5}y + 4$.

CHAPTER 14
WORD PROBLEMS

Lesson 40: **Words and Phrases for Translating Word Problems into Equations**
Lesson 41: **Solving Number Word Problems**
Lesson 42: **Perimeter Problems; Work-Time Problems**
Lesson 43: **Motion Problems** (Distance-Time-Speed Problems)
Lesson 44: **Consecutive Integer Problems; Age Problems; Percent Problems**
Lesson 45: **Mixture and Mixture-Like Problems**

To translate a word problem into an equation, we need an algebraic expression for the left-hand
 side of the equation and an algebraic expression for the right-hand side of the equation.
The equality symbol "=" is placed between these two expressions. **Note** that an expression could
consist of only a single term such as the integer 3.

Lesson 40
Words and Phrases for Translating Word Problems into Equations

Implication of Equality   "="

1.   Equals
2.   Is equal to
3.   Result is
4.   Equivalent to   } All these mean "="
5.   One gets
6.   Answer is
7.   One obtains

Begin each of the following with " **the number which is**"
Example: **The number which is** the sum of 6 and 2 is: 6 + 2 = 8.

Implication of Addition:
1. the sum of 6 and 2 is **6+2=8**
2. 2 added to 6 is   **6+2**
3. 2 more than 6 is **6+2**
4. 6 increased by 2 is **6+2**
5. 2 greater than 6 is **6+2**

Implication of Subtraction
1. the difference between 6 and 2 is **6 - 2**=4; but
2. the difference between 2 and 6 is **2 - 6 = -** 4
3. the difference of 6 and 2 is **6 - 2**= 4 ; but   (Note that in subtraction, order is important)
5. 6 decreased by 2 is **6 - 2**
6. 6 less 2 is 6 - 2; **but also note that**
7. 2 less than 6 also is **6 - 2**
Note the difference between #1 and #2

Implication of Multiplication

1. the product of 6 and 2 is $\mathbf{6 \times 2}$ (or $\mathbf{2 \times 6}$)=12
2. 6 times 2 is $\mathbf{6 \times 2}$
3. 6 times as much as 2 is $\mathbf{6 \times 2}$
4. twice 6 is $\mathbf{2 \times 6}$
5. double 6 is $\mathbf{2 \times 6}$

Implication of Division

1. the quotient of 6 (divided) by 2 is $\dfrac{6}{2}$;**but**

2. the quotient of 2 (divided) by 6 is $\dfrac{2}{6}$ (Note that in division, order is important)

3. the ratio 6:2 is $\dfrac{6}{2}$

4. the ratio 2:6 is $\dfrac{2}{6}$

Note also that in the future, if you come across any words or phrases, you may add them to the above list.

Lesson 41

Solving Number Word Problems (Using a Single Variable)

1. Read the word problem twice, first quickly and then slowly and determine what is unknown (or are unknown).
2. Represent the unknown (or one of the unknowns) by a letter and express any other unknowns in terms of the same letter.
3. Read the whole problem again, and break it up into "parts" with the help of some of the phrases or words in the above table (or from any other source).
4. Put the parts together to form an equation.
5. Read the translated problem (equation), in your own words if possible, to see if it is equivalent to the original word problem.
6. Solve the equation for the unknown (variable) and check the solution in the original word problem.

Number Problems

Example 1   If five less than twice a number is 11, what is the number?

Solution   Let the number be x. Then, "twice the number" translates to $2x$.
"Five less than twice the number" translates to $2x - 5$.

"Five less than twice x is 11" translates to: $2x - 5 = 11$

$$\text{Solve for } x:$$
$$2x - 5 = 11$$
$$\underline{ + 5 \quad + 5}$$
$$2x = 16$$
$$\frac{2x}{2} = \frac{16}{2}$$
$$x = 8$$

The number is 8.

Example 2   If 6 is subtracted from five times a number, the result is 14
Find the number.

Solution   Let the number be x
Then, translating, $5x - 6 = 14$
Solve for x: $5x - 6 = 14$
$$\underline{ +6 \quad +6}$$
$$5x = 20$$
$$x = 4$$

The number is 4.

Example 3 If the product of four and a number is subtracted from 14, the result is equal the number diminished by 1. Find the number.

Solution   Let the number be x
Then, $14 - 4x = x - 1$
Solve for x: $14 - 5x = -1$
$$-5x = -1 - 14$$
$$-5x = -15$$
$$x = 3$$
The number is 3.

Example 4

The sum of two numbers is 72. One of the numbers is 12 more than five times the other number. Find these numbers.

Solution Let "the other number" be x.

Then "one of the numbers" $= 5x + 12$

"The sum of the two numbers is 72" translates to: $x + (5x + 12) = 72$

Solve for x:

$$x + (5x + 12) = 72$$
$$x + 5x + 12 = 72$$
$$6x + 12 = 72$$
$$6x = 60$$
$$x = 10$$

Therefore, "the other number" is 10 and

"one of the numbers" $= 5(10) + 12$

$$= 50 + 12 = 62$$

The numbers are 62 and 10.

Note: The use of the phrases "the other number" and "one of the numbers" was to help direct translation. This technique is useful in problems in which it is not easy or we do not want to define the smaller or the larger number at the beginning of solution of the problem. If it is easy to define the smaller or the larger number at the beginning of the problem, do so. In the above problem, we could have done this: Let the smaller number be x. Then the larger number $= (5x + 12)$. The equation needed : $x + (5x + 12) = 72$, which is the same equation as before.

Example 5 A number is three more than another number. Twice their sum is 38. Find these numbers.

Let the smaller number $= x$

Then the larger number $= x + 3$

The sum of the numbers $= x + x + 3$

Twice their sum $= 2(x + x + 3) = 2(2x + 3)$

The equation needed: $2(2x + 3) = 38$

Solve for x:

$$4x + 6 = 38$$
$$4x = 32$$
$$x = 8$$

The smaller number is 8 and the larger number is 11.

Lesson 41 Exercises

A. If twice the difference of a number and six is increased by four, the result is 10. Find the number.

Answer: 9

B

1. If six more than twice a number is 48, find this number.

2. If five times a number is subtracted from 8, the result is 68. Find this number.

3. A number is 6 less than twice another number. If the sum of these numbers is 42, find them.

4 In a chemistry class, there are 32 students. The number of females is 4 more than three times the number of males. How many males and how many females are in this class?

Answers: **1.** 21 ; **2.** -12 ; **3.** 26 and 16; **4.** There are 7 males and 25 females in this class.

Lesson 42

Perimeter Problems; Work-Time Problems

Perimeter Problems

The **perimeter** of a two-dimensional figure is the total distance around the figure.
(That is, the sum of the lengths of all the sides enclosing the figure)

We will approach perimeter problems in much the same way as we approached the "number" problems above. However, here ,we need to know some geometric facts which will not usually be stated in the problem. For example, for a rectangle, we need to know how the length and the width are related to the perimeter.

Perimeter of a Rectangle

Let the length of a rectangle = L; and let its width (or altitude) = W
Then the perimeter, P, of a rectangle is given by the formula, $P = 2(L + W)$ or $2L + 2W$

Example 6 We will reword Example 5 as a perimeter problem.

The length of a rectangle is three units more than its width. If the perimeter of this rectangle is 38 units, find the dimensions (length and width) of this rectangle.

Solution
Let the width of this rectangle = x
Then the length = $x + 3$
The perimeter formula is $P = 2(L + W)$ or $2L + 2W$

The equation needed: $2(x + x + 3) = 38$ or $2x + 2(x + 3) = 38$
Solve for x:
$$4x + 6 = 38$$
$$4x = 32$$
$$x = 8$$
The width is 8 units and the length is 11 units.

Work and Time Problems

Example 7 (This problem was solved using arithmetic in Lesson 24.)

Working alone, Maria can paint an apartment in 4 hours, while James can paint the same apartment in 6 hours, working alone. How long it will take Maria and James to paint the same apartment if they both paint together?

Method 1: Using Algebra

Let the painting work $= w$.
Let the rate (per hr.) of doing the work be r.   (i.e., speed of working or how fast the work is done)
Let the time taken to do the work $= t$.
These three quantities are related as follows:

1. $r = \dfrac{w}{t}$;   **2.** $t = \dfrac{w}{r}$;   **3.** $w = r \times t$

The rate (per hr.) at which Maria can do the work $= \dfrac{w}{4}$.

The rate (per hr.) at which James can do the work $= \dfrac{w}{6}$.

Working together, the rate (per hr.) at which Maria and James do the work $= \dfrac{w}{4} + \dfrac{w}{6} = \dfrac{5w}{12}$.

Now, r (for Maria and James) $= \dfrac{5w}{12}$; $t = \dfrac{w}{r}$ or $w \div r$ and

$\therefore$ t (for Maria and James) $= w \div \dfrac{5w}{12} = \dfrac{w}{1} \times \dfrac{12}{5w} = \dfrac{12}{5} = 2\dfrac{2}{5}$ hrs.

Method 2: Using Algebra

Let the time taken when both Maria and James work together $= t$.
Let the painting work $= w$.

Part of the work done by Maria $= r \times t = \dfrac{w}{4} \times t$

Part of the work done by James $= r \times t = \dfrac{w}{6} \times t$

Total work done by Maria and James in t hours when working together $= (\dfrac{w}{4} \times t) + (\dfrac{w}{6} \times t) = W$ .....(1)

Dividing through equation (1) by w, we obtain $\boxed{\dfrac{t}{4} + \dfrac{t}{6} = 1}$ ....(2)

Multiplying equation (2) by 12 (the LCM of the denominators), we obtain
$$3t + 2t = 12$$
$$5t = 12$$
$$t = \dfrac{12}{5} = 2\dfrac{2}{5} \text{ hours.}$$

Method 3: Compact formula (Derived from the general form equation (2) of Method 2)

Let the time Maria takes to do the work alone = t_1

Let the time James takes to do the work alone = t_2

Let the time both Maria and James take working together = t

Then $\boxed{t = \dfrac{t_1 t_2}{t_1 + t_2}}$ (i.e., t = **Product** of the individual times **divided** by the **sum** of the individual times.)

Substituting $t_1 = 4$, $t_2 = 6$ in the above compact formula,

$$t = \frac{4 \times 6}{4 + 6}$$

$$t = \frac{24}{10} = \frac{12}{5} = 2\frac{2}{5} \text{ hours.}$$

Note above that in Example 7, three quantities are involved (t, t_1, and t_2). Therefore, given any two of these quantities, we can find the third quantity by substituting the given quantities in the compact formula (or in the general form of equation (2) of Method 2, given below *) and solving for the unknown quantity.

* The general form of $\dfrac{t}{4} + \dfrac{t}{6} = 1$ is $\boxed{\dfrac{t}{t_1} + \dfrac{t}{t_2} = 1}$ OR $\boxed{\dfrac{1}{t} = \dfrac{1}{t_1} + \dfrac{1}{t_2}}$

where t is the "working-together" time; and t_1, and t_2 are the individual times.

Lesson 42 Exercises

A

1. The width of a rectangle is four units less than its length. If the perimeter of this rectangle is 20 units, find the dimensions of this rectangle.

2. Maria fenced a rectangular farm with 600 ft of wire. If this farm is twice as long as it is wide, what are the dimensions of this farm?

Answers: **1.** Length = 7 units, width = 3 units; **2**. Length = 200 ft, width = 100 ft

B

The length of a rectangle is 3 units more than the width. If the perimeter of this rectangle is 66, find the length and width of this rectangle.

Answer: $w = 15$; $L = 18$

C

Working alone, John can sort a quantity of mail in 5 hours, while Betty working alone can sort the same quantity of mail in 2 hours. How long will it take John and Betty to sort the same quantity of mail if they work together?

Answer : $1\frac{3}{7}$ hours.

Lesson 43
Motion Problems
(Distance-Time-Speed-Problems)

Let the distance traveled $= d$. Let the time taken to travel $d = t$. Let the constant speed at which d was traveled $= s$. These three quantities are related as follows:

$$d = st \quad \text{(distance = speed × time)} ; \quad s = \frac{d}{t} \ \text{(i.e., speed} = \frac{\text{distance}}{\text{time}}; \quad t = \frac{d}{s} \ \text{(i.e., time} = \frac{\text{distance}}{\text{speed}})$$

In the problems covered below, we may base our **main equation** on the quantity (time, speed, or distance) which is the same (sometimes with some minor adjustments) for say, both car A and car B. The main equation may also be based on the following: sum of distances; equality of distances; sum or difference of elapsed times or equality of elapsed times or equality of speeds. We will consider five main cases.

Case 1   Two Cars Start at the Same Time and Travel Towards Each Other

Example 1a

Two cars A and B, initially d miles apart, travel towards each other in opposite directions along a straight road. They both start at the same time. Car A's speed is a mph and Car B's speed is b mph Obtain an expression for where they meet. (Note: mph means miles per hour.)

Solution

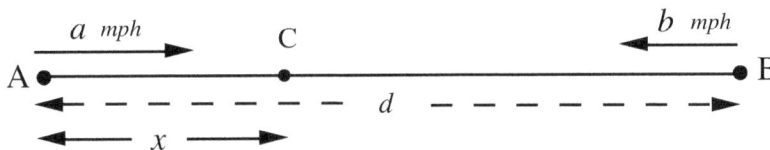

Step 1: In this problem, the time they take to travel to meet is the same for A and B since they both start at the same time. Therefore, our **main equation** is $t_A = t_B$. (t_A is A's time and t_B is B's time)

Step 2: Let Car A and Car B meet at point C which is x miles from Car A's starting point (Point A). Then the distance Car A covered is x miles and the distance Car B covered $= (d - x)$ miles. Now, we will obtain expressions for t_A and t_B in terms of a, b, d and x.

Step 3: $t_A = \dfrac{x}{a} \qquad \left(\dfrac{\text{distance A traveled}}{\text{A's speed}} \right)$

$t_B = \dfrac{d - x}{b} \qquad \left(\dfrac{\text{distance B traveled}}{\text{B's speed}} \right)$

Step 4: Since $t_A = t_B$ (since they both started traveling towards each other at the same time)

$$\frac{x}{a} = \frac{d - x}{b}$$

Step 5: Solve for x: $bx = ad - ax$

$$ax + bx = ad$$
$$x(a + b) = ad$$
$$x = \frac{ad}{a + b}$$

Step 6: The two cars meet at a distance $\dfrac{ad}{a + b}$ miles from A.

(Similarly, if they meet y miles from B, then $y = \dfrac{bd}{a + b}$ miles)

Example 1b   (Numerical Example)

Two cars A and B, initially 60 miles apart, travel towards each other in opposite directions
along a straight road. They both start at the same time. Car A's speed is 25 mph and Car B's
speed is 15 mph Obtain an expression for where they meet. (Note: mph means miles per hour.)

Solution

Step 1: In this problem, the time they take to travel to meet is the same for A and B since they both
start at the same time. Therefore, our **main equation** is $t_A = t_B$. (t_A is A's time and t_B is B's time)

Step 2: Let Car A and Car B meet at point C which is x miles from Car A's starting point (Point A).
Then the distance Car A covered is x miles and the distance Car B covered = (60 - x) miles.
Now, we will obtain expressions for t_A and t_B.

Step 3: t_A, A's travel time = $\dfrac{x}{25}$ $\left(\dfrac{\text{distance A traveled}}{\text{A's speed}} \right)$

  t_B, B's travel time = $\dfrac{60 - x}{15}$ $\left(\dfrac{\text{distance B traveled}}{\text{B's speed}} \right)$

Step 4: Since A's travel time = B's travel time (They started traveling towards each other at the same time)
$$\frac{x}{25} = \frac{60 - x}{15}$$

Step 5: Solve for x: $15x = 1500 - 25x$
$$x = \frac{1500}{40} = 37.5$$

Step 6: The two cars meet at a distance 37.5 miles from A.

(Similarly, if they meet y miles from B, then $y = \dfrac{15(60)}{25 + 15} = 22.5$ miles)

Example 2a   Two Cars Start at Different Times and Travel Towards Each Other

Two cars A and B, initially d miles apart, travel towards each other (in opposite directions) along a straight road. Car A starts first and travels at a mph; and e hours later, Car B starts traveling at b mph. Obtain an expression for where they meet.

Solution

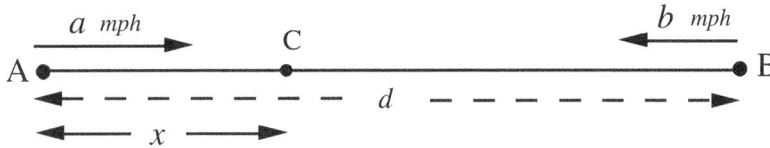

Step 1. In this problem, Car A's travel time before meeting Car B is given by:

$t_A = t_B + e$ (where t_A is Car A's time and t_B is Car B's time), since Car A started e hours before Car B.

Step 2: Let Car A and Car B meet at point C which is x miles from Car A's starting point (Point A). Then the distance Car A covered is x miles and the distance Car B covered $= (d - x)$ miles. Now, we obtain expressions for t_A and t_B in terms of $a, b, d, x,$ and e.

Step 3: $t_A = \dfrac{x}{a}$ $\left(\dfrac{\text{distance A traveled}}{\text{A's speed}}\right)$

$t_B = \dfrac{d - x}{b}$ $\left(\dfrac{\text{distance B traveled}}{\text{B's speed}}\right)$

Step 4: Since $t_A = t_B + e$

$\dfrac{x}{a} = \dfrac{d - x}{b} + e$ (Substituting for t_A, and t_B, from Step 3)

Step 5: Solve for x:

$\dfrac{x}{a} = \dfrac{d - x + be}{b}$

$bx = a(d - x + be)$ (by cross-multiplication)

$bx = ad - ax + abe$ <----Now, solve this equation for x.

$ax + bx = ad + abe$

$x(a + b) = ad + abe$

$x = \dfrac{ad + abe}{a + b}$

Step 6 . The two cars meet $\dfrac{ad + abe}{a + b}$ miles from Car A's starting point.

(Similarly, if they meet y miles from Car B's starting point then $y = ?$)

Example 2b (Numerical Example)

Two cars A and B, initially 60 miles apart, travel towards each other (in opposite directions) along a straight road. Car A starts first and travels at 30 mph; and one hour later, Car B starts traveling at 20 mph. Obtain an expression for where they meet.

Solution

Step 1. Let Car B's travel time $= t$.
Then Car A's travel time $= t + 1$ (Since Car A started one hour before Car B.

Step 2: Let Car A and Car B meet at point C which is x miles from Car A's starting point (Point A).
Then the distance Car A covered is x miles and the distance Car B covered $= (60 - x)$ miles.

Step 3:   Car A's travel time $= \dfrac{x}{30}$ $\left(\dfrac{\text{distance A traveled}}{\text{A's speed}}\right)$

   Car B's travel time$= \dfrac{60 - x}{20}$ $\left(\dfrac{\text{distance B traveled}}{\text{B's speed}}\right)$

Step 4: Since Car A's travel time = Car B's travel time + 1

$$\frac{x}{30} = \frac{60 - x}{20} + 1$$

Step 5: Solve for x:

$$\frac{x}{30} = \frac{60 - x + 20}{20}$$
$$20x = 30(60) - 30x + 30(20)$$
$$20x = 1800 - 30x + 600$$
$$50x = 2400$$
$$x = \frac{2400}{50}$$
$$x = 48$$

Step 6. The two cars meet 48 miles from Car A's starting point.

Case 3: **Two cars start at different times from the same point and travel in the same direction**

Example 3a

Two cars A and B embark on a journey in the same direction and from the same point. Car A leaves first and travels at a mph; and e hours later, Car B follows at b mph ($b > a$) and travels to catch-up with Car A. After how many hours, and where, did Car B catch-up with Car A.?

Solution

A $\xrightarrow{a \ mph}$

B $\xrightarrow{b \ mph}$

Step 1: Car B will catch-up with Car A when both Car A and Car B have traveled the same distance from the starting point. Therefore, the quantity which is the same for Car A and Car B is the distance traveled.

Therefore, $d_A = d_B$ (equality of distances)

(d_A and d_B are the distances traveled by Car A and Car B respectively.)

Step 2: Let the travel time for Car A = t hours.

Then the travel time for Car B = $t - e$ (since Car A left e hours earlier)

$d_A = at$ ($d = st$)

$d_B = b(t - e)$

Step 3: Since $d_A = d_B$

$at = b(t - e)$

Step 4: Solve for t:

$at = bt - be$

$at - bt = -be$

$t(a - b) = -be$

$t = -\dfrac{be}{a - b} = \dfrac{be}{b - a}$

Answer: Car B caught-up with Car A $\dfrac{be}{b - a}$ hours from the time Car A started the journey.

Car B caught-up with Car A $\quad a \cdot \dfrac{be}{b - a} = \dfrac{abe}{b - a}$ miles from the starting point. ($d = st$)

Note that the time Car B took to catch-up with Car A $= \dfrac{be}{b - a} - e = \dfrac{be - be + ae}{b - a} = \dfrac{ae}{b - a}$ hours, which we could also obtain directly as follows:

If Car B's travel time is t , then $bt = ae + at$ (distance Car B traveled = distance Car A traveled), which on solving yields $t = \dfrac{ae}{b - a}$, as before.

Example 3b (Numerical Example)

Two cars A and B embark on a journey in the same direction and from the same point Car A leaves first and travels at 20 mph; and 2 hours later, Car B follows at 50 mph and travels to catch-up with Car A. After how many hours, and where, did Car B catch-up with Car A.?

Solution

Step 1: Car B will catch-up with Car A when both Car A and Car B have traveled the same distance from the starting point. Therefore, the quantity which is the same for Car A and Car B is the distance traveled.

Step 2: Let the travel time for Car A $= t$ hours.

Then the travel time for Car B $= t - 2$ (Car B's time is 2 hours less than Car A's time)
  Car A's travel distance $= 20t$ ($d = st$)
  Car B's travel distance $= 50(t - 2)$

Step 3: Since Car A's travel distance = Car B's travel distance
$$20t = 50(t - 2)$$

Step 4: Solve for t:
$$20t = 50t - 100$$
$$-30t = -100$$
$$t = \frac{-100}{-30}$$
$$t = 3\frac{1}{3}$$

Distance from starting point at which Car B caught-up with car A $= 20 \times \dfrac{10}{3} = 66\frac{2}{3}$ ($S = 20$; $t = 3\frac{1}{3}$; $d = st$)

Answer: Car B caught-up with Car A $3\frac{1}{3}$ hours from the time Car A started the journey.

  Car B caught-up with Car A at a distance of $66\frac{2}{3}$ miles from the starting point.

Note: If we were asked to find Car B travel time, t, then we could obtain t directly from
$$50t = 20(2) + 20t$$ (distance Car B traveled = distance Car A traveled)
$$30t = 40$$
$$t = 1\frac{1}{3} \text{ hours}$$

Case 4 **Two cars start at the same time and travel from the same point in opposite directions**

Example 4a Car A and Car B leave point C at the same time and travel in opposite directions . Car A's speed is a mph and Car B's speed is b mph. After how many hours are they d miles apart.?

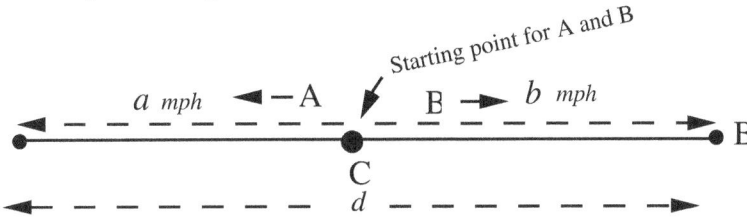

Solution

Let the time taken for the cars to be d miles apart be t hours.
Distance Car A traveled at a mph in t hours is at, and the distance Car B traveled at b mph in t hours is bt miles (Note: distance = speed × time)
The equation needed is based on the total distance traveled by Car A and Car B.
Therefore, $at + bt = d$ <---------Solve for t.

$$t(a + b) = d$$

$$t = \frac{d}{a + b} \text{ hours}$$

Answer: Car A and Car B are d miles apart after $\dfrac{d}{a + b}$ hours.

Example 4b (Numerical Example)

Car A and Car B leave point C at the same time and travel in opposite directions .
Car A's speed is 40 mph and Car B's speed is 30 mph. After how many hours are they 210 miles apart.?

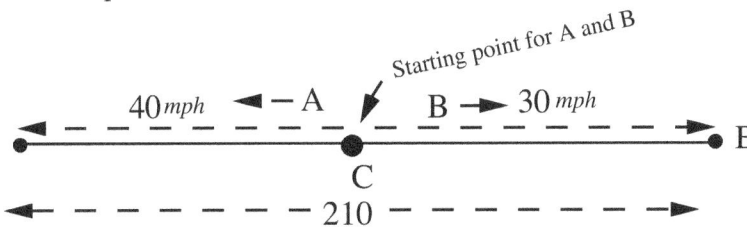

Solution

Let the time taken for the cars to be 210 miles apart be t hours.
Distance Car A traveled at 40 mph in t hours is $40t$; and the distance Car B traveled at 30 mph in t hours is $30t$ miles (Note: distance = speed × time)
The equation needed is based on the total distance traveled by Car A and Car B.
Therefore, $40t + 30t = 210$ <---------Solve for t.

$$70t = 210$$

$$t = \frac{210}{70}$$

$$t = 3$$

Answer: Car A and Car B are 210 miles apart after **3** hours.

Case 5 **Two cars start at different times and travel from the same point in opposite directions**

Example 5a Car A and Car B leave point C and travel in opposite directions.
Car A departs 2 hours before Car B. Car A's speed is a mph and Car B's speed is b mph. After how many hours from the time Car A departed are they d miles apart.?

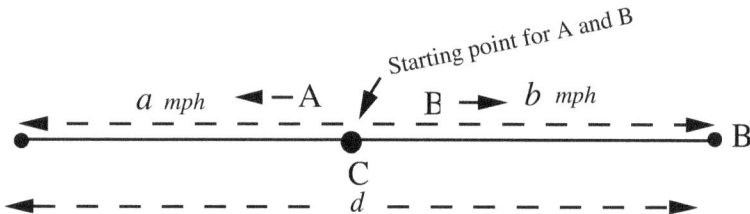

Solution Let the travel time for Car B = t hours.
Then the travel time for Car A = $(t + 2)$ hours   (Since Car A started 2 hours before Car B)
Distance Car A traveled = $a(t + 2) = at + 2a$ ($2a$ is the distance A traveled before B started)
Distance Car B traveled = bt.

The equation needed is based on the total distance traveled.

$\quad (at + 2a) + bt = d$ <---------Solve for t.

$\quad at + 2a + bt = d$

$\quad\quad at + bt = d - 2a$

$\quad\quad t(a + b) = d - 2a$

$\quad\quad\quad t = \dfrac{d - 2a}{a + b}$

Answer: The two cars are d miles apart after $\left(\dfrac{d - 2a}{a + b} + 2 \right)$ hours from the time Car A started.

Example 5b **Two cars start at different times and travel from the same point in opposite directions** (Numerical Example)

Car A and Car B depart point C and travel in opposite directions.
Car A departs 2 hours before Car B. Car A's speed is 50 mph and Car B's speed is 40 mph. After how many hours from the time Car A departed are they 280 miles apart?

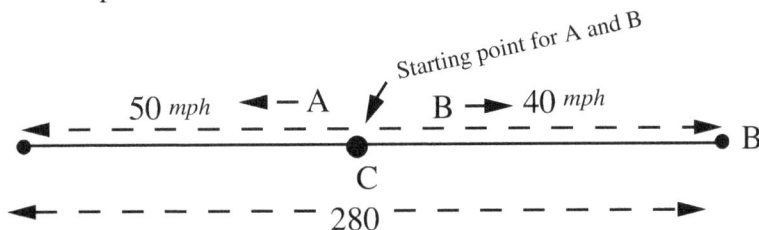

Solution Let the travel time for Car B = t hours.
Then the travel time for Car A = $(t + 2)$ hours   (Since Car A started 2 hours before Car B)
Distance Car A traveled = $50(t + 2) =$
Distance Car B traveled = $40t$

The equation needed is based on the total distance traveled.
Car A's distance + Car B's distance = 280

$$50(t + 2) + 40t = 280 \longleftarrow \text{Solve for } t.$$
$$50t + 100 + 40t = 280$$
$$90t + 100 = 280$$
$$90t = 180$$
$$t = \frac{180}{90}$$
$$t = 2$$

The two cars are 280 miles apart after $(t + 2) = 2 + 2$ or 4 hours from the time Car A started.

Answer: The two cars are 280 miles apart after **4** hours from the time Car A started.

EXTRA

Example 6a

Maria rode her bicycle at b mph from point A to point B, and returned by car at c mph. If her total traveling time was t hours, find the distance between point A and point B

We will base our (main) equation on: **car's time + bicycle's time = total time.** Also, $t = \dfrac{d}{s}$.

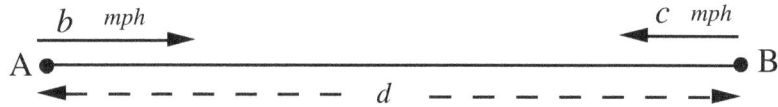

Solution: Step 1: $\dfrac{d}{b} + \dfrac{d}{c} = t$ (where d is the distance between point A and point B; b is the bicycle's speed;

c is the car's speed and t is the total time for the bicycle and the car.)

Step 2: Solve for d:

$$\frac{cd + bd}{bc} = t$$

$$cd + bd = bct$$

$$d(c + b) = bct$$

$$d = \frac{bct}{b + c}.$$

The distance between point A and point B is $\dfrac{bct}{b + c}$ miles.

Example 6b (Numerical Example)

Maria rode her bicycle at 5 mph from point A to point B, and returned by car at 40 mph. If her total traveling time was 3 hours, find the distance between point A and point B

(We will base our (main) equation on: **car's time + bicycle's time = total time.** Also, $t = \dfrac{d}{s}$)

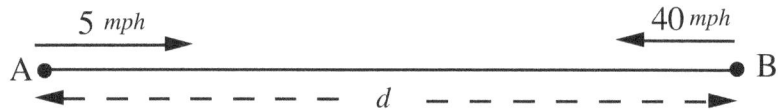

Solution: Step 1: car's time + bicycle's time = total time.

$$\frac{d}{5} + \frac{d}{40} = 3 \qquad \text{(where } d \text{ is the distance between point A and point B)}$$

Step 2: Solve for d:

$$8d + d = 120$$

$$9d = 120$$

$$d = \frac{120}{9}$$

$$d = 13\tfrac{1}{3}$$

The distance between point A and point B is $13\tfrac{1}{3}$ miles.

Distance-Time-Speed Problems Involving Travel by Water or by Air

In problems of travel by water (such as **boat rowing**) or by air (such as **aircraft fligh**t or **bird flight**) in which we have to consider the effect of the direction of the flow of water or air, there is an additional variable involved, namely, the water (river) current or the wind current. We can solve these problems using a single variable. However, because of the additional variable, we will prefer to solve these problems using two variables in **Chapter 19.**

Lesson 43 Exercises

A

1. Two cars A and B, initially 100 miles apart, travel towards each other in opposite directions along a straight road. They both start at the same time. Car A's speed is 20 mph and Car B's speed is 30 mph Where do they meet?

Answer: They meet 40 miles from point A;

B

Two cars A and B, initially 100 miles apart, travel towards each other (in opposite directions) along a straight road. Car A starts first and travels at 40 mph; and one hour later, Car B starts traveling at 30 mph. Where do they meet?

Answers: They meet $74\frac{2}{7}$ miles from point A.

C

Two cars A and B embark on a journey in the same direction and from the same point Car A leaves first and travels at 30 mph; and 3 hours later, Car B follows at 50 mph and travels to catch-up with Car A. After how many hours, and where, did Car B catch-up with Car A.?

Ans: Car B caught-up with Car A $4\frac{1}{2}$ hours after Car B started, and at a point 225 miles from the starting poin**t.**

D

1. Car A and Car B leave point C at the same time and travel in opposite directions .Car A's speed is 60 mph and Car B's speed is 50 mph. After how many hours are they 440 miles apart.?

Answer: After 4 hours;

E

Car A and Car B depart point C and travel in opposite directions. Car A departs 1 hour before Car B. Car A's speed is 50 mph and Car B's speed is 30 mph. After how many hours from the time Car A departed are they 290 miles apart?

Answer: After 4 hours.

F

Maria rode her bicycle at 10 mph from point A to point B, and returned by car at 60 mph. If her total traveling time was 4 hours, find the distance between point A and point B.

Answer: The distance between point A and point B is $34\frac{2}{7}$ miles.

Lesson 44
Consecutive Integer Problems; Age Problems; Percent Problems

Consecutive Integer Problems

Consecutive integers are integers written in increasing or decreasing order such that each integer differs from the integer immediately before it or immediately after it by 1. Example: 5, 6, and 7 are consecutive integers.

Example   The sum of three consecutive integers is 21. Find the numbers.

Solution   Let the first consecutive integer be x.
Then the second consecutive integer is $(x + 1)$
and the third consecutive integer is $(x + 1 + 1)$.
"Their sum is 21" translates to:
$$x + (x + 1) + (x + 1 + 1) = 21$$
$$x + x + 1 + x + 2 = 21$$

Solve for x.
$$3x + 3 = 21$$
$$\underline{ - 3 \quad -3}$$
$$3x = 18$$
$$\frac{3x}{3} = \frac{18}{3}$$
$$x = 6$$

∴   The first consecutive integer is 6.
The second consecutive integer is $6 + 1 = 7$.
The third consecutive integer $6 + 1 + 1 = 8$.

Check: $6 + 7 + 8 \overset{?}{=} 21$
$21 = 21$ True

∴   The three consecutive integers are 6, 7 and 8 (**note** that the integers are consecutive and their sum is 21)

Note also that in the above problem, if we know a given integer, to obtain next consecutive integer we just add 1.

Consecutive Even Integer Problems

Consecutive **even** integers are even integers written in increasing or decreasing order such that each integer differs from the integer immediately before it or immediately after it by 2. Example: 6, 8, 10, and 12 are consecutive even integers.

Here, if we know the first consecutive even integer, to obtain
the next consecutive integer we add 2.

Example 2   The sum of three consecutive even integers is 36. What are these even integers?
Solution:

Step 1: Let the first consecutive even integer be x.
Then the second consecutive even integer is $(x + 2)$
and the third consecutive even integer is $(x + 2 + 2)$
"Their sum is 36" translates to:
$$x + (x + 2) + (x + 2 + 2) = 36$$
$$x + x + 2 + x + 4 = 36$$
$$3x + 6 = 36$$
$$\underline{ - 6 \quad -6}$$
$$3x + 0 = 30$$
$$\frac{3x}{3} = \frac{30}{3}$$
$$x = 10$$

Step 2:
The first consecutive even integer is 10
The second consecutive even integer is 12
The third consecutive even integer 14
Check: $10 + 12 + 14 = 36$
$36 = 36$ True
The three consecutive even integers are
10 12 and14

Consecutive Odd Integer Problems

Consecutive **odd** integers are odd integers written in increasing or decreasing order such that each integer differs from the integer immediately before it or immediately after it by 2. Example: 5, 7, 9 and 11 are consecutive odd integers.

Here **also**, as in the case of even integer problems, if we know the first odd integer, to obtain the next odd integer, we add 2. For example, if the first odd integer is 5, then the next odd integer is $5 + 2 = 7$, and the next odd integer is $7 + 2 = 9$.

Example 3 The sum of three consecutive odd integers is 33. Find the integers.

Solution Let the first consecutive odd integer be x.

Then the second consecutive odd integer is $(x + 2)$
and the third consecutive odd integer is $(x + 2 + 2)$

"Their sum is 33" translates to:
$$x + (x + 2) + (x + 2 + 2) = 33$$
$$x + x + 2 + x + 4 = 33$$
$$3x + 6 = 33$$
$$\underline{\quad - 6 \quad -3}$$
$$3x = 27$$
$$\frac{3x}{3} = \frac{27}{3}$$
$$x = 9$$

∴ The first consecutive odd integer is 9
The second consecutive odd integer is $9 + 2 = 11$
The third consecutive odd integer $11 + 2 = 13$

Check: $9 + 11 + 13 \overset{?}{=} 33$
$33 = 33$ True

∴ The three consecutive odd integers are 9, 11 and 13.

A note about consecutive even and odd integer problems. Sometimes, because of the use of the word "odd" in consecutive odd integer problems, some students think that if given an odd integer, to obtain the next odd integer, they would add odd integers 1, 3, 5 etc.
This notion is incorrect. For example, if the first odd integer is 5, the next odd integer 7 is obtained by adding 2; and similarly, the next odd integer, 9, is obtained by adding 2 to 7.
 Also $5 + 1 = 6, 5 + 3 = 8$ are all even.
Therefore, note that whether an integer is odd or even, to obtain the next consecutive odd or even integer we **add 2.** The difference in these types of problems will be that if the problem states that the integers are even, then the answers we obtain must be even integers, otherwise, there is no solution. On the other hand, if the problem states that the integers are odd, then the solutions we obtain should be odd integers, otherwise there is no solution.

Age Problems

Example 1 A father is six times as old as his daughter. Last year he was seven times as old as his daughter. How old are they?

Solution: Let the daughter's present age be x
Then the father's present age is $6x$.
The father's age last year = $(6x - 1)$
The daughter's age last year = $(x - 1)$
 Last year, their ages were related as follows :

Father's age = seven times daughter's age.

$$(6x - 1) = 7(x - 1)$$
$$6x - 1 = 7x - 7$$
$$\underline{\quad +1 \qquad + 1\quad}$$
$$6x = 7x - 6$$
$$\underline{-7x \quad -7x\quad}.$$
$$-x = - 6$$
$$x = 6$$

The daughter is 6 years old.
The father is $6(6) =$ 36 years old.

Extra: How old was the father when the daughter was born?

Percent Problems

In Chapter 1, problems involving percent were solved by both arithmetic and algebraic methods.
The algebraic methods are repeated below.

Finding the Original Number (Finding the base)

Example If 20% of a number is 64, what is the number?

Solution

Method 1 Let the number be x

Then, "20% of the number is 64" translates to $\dfrac{20x}{100} = 64$

i.e. $\dfrac{20x}{100} = 64$

Solve for x: $\dfrac{\overset{1}{(\cancel{100})}}{\underset{1}{(\cancel{20})}} \dfrac{\overset{1}{\cancel{20}x}}{\cancel{100}} = 64 \dfrac{\overset{5}{(\cancel{100})}}{\underset{1}{(\cancel{20})}}$

$$x = 320$$

The number is 320 .
Note also that since 20% of the number is 64, the number must be greater than 64.

Method 2:   **Using Proportions**

20% is to 64 as 100% is to x, where 100% of the
original number (the base) is x.
Translating the proportion,

$$\frac{20\%}{64} = \frac{100\%}{x} \qquad\qquad (\text{ or } \frac{.20}{64} = \frac{1}{x})$$

Solve for x: $20\%\, x = 100\%\, (64)$ (or $.20x = 64$)

$$x = \frac{100\%}{20\%}(64) \qquad (\text{or } x = \frac{64}{.20})$$

$$x = 320$$

Note that 20% as a fraction is $\frac{20}{100} = \frac{1}{5}$

Therefore, in the above problem, instead of asking "if 20% of a number is 64, what is the number?",
we could have asked "if $\frac{1}{5}$ of a number is 64, what is the number? "

Example   If $\frac{1}{5}$ of a number is 64, what is the number? We can solve this by any of the
methods discussed above. Let us use the algebraic method.

    Solving:

Let the number be x

Then $\frac{x}{5} = 64$ (Note that $\frac{1}{5}x = \frac{x}{5}$)

$$(5)\frac{x}{5} = 64(5)$$

$$x = 320$$

The number is 320.

Example   Mary spends 20% of her weekly income on food. If she spends $64 on
food every week, what is her weekly income?

Solution : We could reword this problem in the familiar form as " If 20% of a number is 64,
what is the number?"; and then use exactly the same method as in the example, above.

Answer: $320 (numerically, the same answer as for the last example).

Finding the Rate %

Example 1 What rate % of 24 is 15?

Solution

Let $x\%$ be the rate %. We will write an equation in terms of x, and solve for x.
From the wording, "$x\%$ of 24 is 15" translates to:

$$\frac{x}{100} \times \frac{24}{1} = 15$$

$$\frac{24x}{100} = \frac{15}{1}$$

$24x = 100 \times 15$ (by cross-multiplication or by multiplying both sides of the equation by 100).

$$x = \frac{100 \times 15}{24}$$ (by dividing both sides of the equation by 24)

$$x = 62\frac{1}{2}$$

$\therefore$ the required rate $= 62\frac{1}{2}\%$.

Example 2 What rate % of 18 is 44?

Solution

Let the required rate be $x\%$.
Then, "$x\%$ of 18 is 44" translates to:

$$\frac{x}{100} \times \frac{18}{1} = 44$$

$$\frac{18x}{100} = \frac{44}{1}$$

$18x = 100 \times 44$ (by cross-multiplication or by multiplying both sides of the equation by 100)

$$x = \frac{100 \times 44}{18}$$ (by dividing both sides of the equation by 18)

$$x = 244\frac{4}{9}$$

$\therefore$ the required rate $= 244\frac{4}{9}\%$.

Example 3 A family's annual income last year was \$20,000. This year, the income is \$33,000. What is the percent increase in the annual income?

Step 1: The increase in income $= \$33,000 - \$20,000$
$$= \$13,000$$

Step 2: The percent increase in income $= \frac{13000}{20000} \times 100\%$ (Finding the rate percent)
$$= 65\%$$

Note: In Step 2 , the question could have been posed as: What percent of 20,000 is 13,000 ?

More Applications of Base Finding and Percentage Finding

In these applications, the questions are **not** worded in forms such as "find 20% of a number; "if 30% of a number is 45, what is the number?". A good approach is to reword the problem in any of these familiar forms and then proceed accordingly.

Note 1. The fraction involved in the problem may very likely be the **rate** (but it may **not** be the rate).

Note for example that $\frac{3}{5} = 60\%$

Example 1   3 out of 5 students at a certain college study Biology.

(a) If 600 students at this college study Biology, how many students are there at this college?

(b) If one were to collect a sample of 200 students at this college, how many of these students would study Biology?

Solution:

Note that 3 out of 5 means $\frac{3}{5}$

Part (a): We can reword this part of the question as: if $\frac{3}{5}$ of a number is 600, what is the number?

Solution   Let the number be x

Then $\frac{3x}{5} = 600$

$3x = 3000$

$x = 1000$

There are 1000 students at this college.

We solve part(b) using arithmetic, since after rewording the calculation becomes straightforward.

Part (b)   We can reword this part of the question as: Find $\frac{3}{5}$ of 200.

$$\frac{3}{5} \text{ of } 200 = \frac{3}{5} \times \frac{200}{1} = 120$$

That is, of a sample of 200 students, 120 students would study Biology.

Note: In finding the base and the rate% , the use of algebra facilitates the solution process.

In finding the percentage as in part (b) of the last problem, perform a straightforward arithmetic calculation.

Lesson 44 Exercises

A

1. The sum of three consecutive integers is 48.   Find these integers.

2. The sum of two consecutive integers is 35 less than four times the larger. Find these integers.

Answers: **1**.15,16 and 17 : **2**.   16 and 17

B   The sum of three consecutive even integers is 60 . Find these integers.

Answer: 18, 20, 22.

C
Find three consecutive even integers such that the sum of the second and third integers is 18 less than three times the first integer.

Ans: 24, 26 and 28.

D   Find three consecutive odd integers whose sum is 63.

Answer: 19,21 and 23

E
Find two consecutive integers such that the difference of twice the larger integer and the smaller integer is 51

Answers : 49 and 50

F
1. James is 3 years older than Mary. Five years ago, James was twice as old as Mary .
  Find their ages.

2. Four years from now, Paul's age will be $3x + 3$. Express his age five years ago in terms of x.

3. A father is 20 years older than his daughter. In 3 years, the sum of their ages will be 40 years.
    Find their present ages.

1. Answers: Mary is 8 years old;   James is 11 years old.;   **2** Five years ago, Paul's age was $3x - 6$ years.

3. The daughter's age is 7 years and the father's age is 27 years.

G
1. If 20% of a number is 72, what is the number?

2. 53 is 25% of what number?

3. 16% of what number is 82?

Answers **1.**360 ; **2.** 212; **3.** 512.5 or $512\frac{1}{2}$

H
    The monthly rent for Maria's apartment is $800. If Maria spends 25% of her monthly income on this rent, what is her monthly income?

Answer: $3,200

I
1.   What rate percent of 32 is 12?
2.   What rate% of 20 is 80?
3.   On the last CUNY test, there were 40 questions. James answered 32 questions correctly. What was his grade in percent?
4. Repeat Step 2 of Example 3, above using algebra.

Answers   **1.** 37.5% or $37\frac{1}{2}$% **2**. 400% ; **3.** 80%

Lesson 45

Mixture and Mixture-Like Problems, Problems on Averages

We will call a problem a mixture problem if the solution method conforms to the methods to be presented below. There are problems which may not appear to be mixture problems, but which on close examination of the solution method will reveal that they are mixture problems

Example 1 A publishing company has 120 workers of whom 40% are women. How many new women workers must be hired such that 50% of the workers are women? Assume that no workers are fired.

Solution
Method 1   **Using Arithmetic**

Step 1: Find the present number of women workers.

The present number of women workers is 40% of 120.

$$\text{i.e.} \frac{40}{100} \times \frac{120}{1} = 48$$

Therefore, there are at present 48 women and (120 - 48) = 72 men

Step 2: Now, we want 50% or ($\frac{1}{2}$) of the workers to be women.

This is equivalent to saying that we want the number of women to be equal to the number of men.
Since there are at present 72 men and 48 women, we need 72 - 48 = 24 women.
Therefore 24 new women must be hired.

Note: If 24 new women are hired, the number of men would be 72; the number of women would be 72, and the total number of workers (both men and women) would be 144.

Method 2: Using Algebra

The present number of women = 40% of 120 = 48
Let the number of the new women workers hired be x.
Then after this hiring, the total number of women would be 48 + x, and
the total number of all the workers would be 120 + x
From the problem, "the number of women should be 50% of the number of workers" translates to:

$$48 + x = \frac{50}{100}(120 + x)$$

Solve for x: $48 + x = \frac{1}{2}(120 + x)$

$2(48 + x) = 120 + x$ (multiplying both sides of the equation by 2)
$96 + 2x = 120 + x$
$\underline{-96 \qquad\quad -96}$
$\qquad\qquad 2x = 24 + x$
$\underline{\qquad\quad -x \qquad\; -x}$
$\qquad\qquad\quad x = 24$

Again, the number of new women to be hired is 24.

Example 2   A chemist has 120 liters of a 40% solution of alcohol. How many liters of
pure alcohol should be added to obtain a 50% solution of alcohol?
(Assume the solution is composed of only pure alcohol and water)
Solution: Use exactly the same procedure as in Example 1, above).
Answer: 24 liters (Numerically, same answer as in Example 1)

A rather more involved problem is the next example.

Example 3   A hospital has 120 doctors of whom 40% are surgeons. Assuming that no doctors
are fired, how many new surgeons must be hired such that 75% of the doctors are
surgeons?

We will solve this problem both algebraically and arithmetically.

Algebraic Method

The present number of surgeons = 40% of $120 = \dfrac{40}{100} \times 120 = 48$

Thus, there are at present 48 surgeons. Let the number of new surgeons be x. Then after hiring
x new surgeons the total number of surgeons would be $(48 + x)$ and the total number of all the
doctors would be $120 + x$.

From the problem, the number of surgeons is to be equal to 75% of all the doctors.

$$48 + x = \frac{75}{100}\ (120 + x) \qquad\qquad \textbf{Note: } 75\% = \frac{75}{100} = \frac{3}{4}$$

Now solve this equation for x.

$$48 + x = \frac{75}{100}\ (120 + x) \qquad \text{or } 48 + x = \frac{3}{4}(120 + x)$$
$$4800 + 100x = 9000 + 75x \qquad \text{or } 192 + 4x = 360 + 3x$$
$$25x = 4200 \qquad\qquad\qquad x = 168$$
$$x = 168$$

$\therefore$ 168 new surgeons must be hired.

After this new hiring ,the total number of surgeons would be $48 + 168 = 216$, and the total
number of all the doctors would be $168 + 120 = 288$.

Arithmetic Method

Step 1: The present number of surgeons = 40% of $120 = \dfrac{40}{100} \times 120 = 48$

Step 2:   The present number of non-surgeons = 120 - 48 = 72.
Since, we want 75% of the doctors to be surgeons,
we want 25% of the doctors to be non-surgeons.

Step 3:   If 25% of the number of doctors is (and is to be) 72,

then , the total number (100%) of all the doctors $= \dfrac{72}{1} \times \dfrac{100}{25} = 288$

(Step 3 could have been posed as: if 25% of a number is 72, what is the number?; the calculation is
based on the quantity that remains unchanged in the process).

Since, the total number of all the doctors is to be 288,
the total number of the surgeons is to be 288 -72 = 216.
At present, there are 48 surgeons; and
therefore, the number of new surgeons to be hired = 216 - 48 = 168
Again, we obtain the same answer as by the algebraic method.

Generalization of Mixture Problems

Example: Mixing solutions of alcohol

Consider mixing two kinds of alcohol solutions. Let q_1 = volume of alcohol #1 and let its concentration = $r_1\%$. Similarly, let q_2 = volume of alcohol #2, and let its concentration = $r_2\%$. Let the volume of the mixture (made up of alcohol #1 and alcohol #2) = V, and let its concentration = $r\%$

Then we may write a general equation as:

$$q_1 \, r_1\% \; + q_2 \, r_2\% \; = qr\%$$

Note #1: A 20% alcohol solution, for example, means 100 units of the alcohol solution contains 20 units of pure alcohol and 80 units of water, assuming the solution contains alcohol and water only.

Note #2: 100% alcohol (or pure alcohol) means $r\% = 100\% = \dfrac{100}{100} = 1$

Note: Adding pure water means $r\%$ for alcohol = 0, and therefore, $q_{alcohol} r_{alcohol}\% = q_{alcohol} \times 0 = 0$

General Formula for mixture Problems

A modified form of the above equation, (given below) may be applied to other mixture problems.

$$q_1 \, r_1 \; + q_2 \, r_2 = qr$$

where q_1, q_2, q (for quantity) may be mass, weight, volume or a number ; r_1, r_2, r may be a fraction, a rate %, a price or cost per unit, or a value per item, etc. Note that sometimes the product qr on the right-hand side of the equation may be given and you may not have to perform the operation $q \times r$.

The author found this formula to be surprisingly applicable to most (if not all) of the popular mixture problems. Moreover, it is easy to memorize and recall. The applications include the hiring of men and women to obtain a certain percent of workers who are men or women; mixing of alcohol solutions; mixing of ingredients of different values (cost, price, etc.) to obtain a desired mixture of a specified value (cost, price, etc.); investing money at different rates in different accounts; and coin problems. (See also Chapter 19.)

Problems on Averages

Example 1   During the semester, James scored 85, 70, 75 on the first three tests. He wants his class average for the term to be 80. What is the score he needs on the next and last test?

Solution   We will cover three methods.

Method 1: Using algebra.

Let x be the next and last score.

Then, the total score for the four tests will be $85 + 70 + 75 + x$

$$\text{Average score} = \frac{\text{Total score}}{\text{Number of tests}} = \frac{85 + 70 + 75 + x}{4}$$

Now we want this average score to be 80

$$\frac{85 + 70 + 75 + x}{4} = 80 \qquad \text{Scrapwork: } 85+70+75 =230$$

$$\frac{230 + x}{4} = 80$$

We solve for x. $\qquad \dfrac{4(230 + x)}{4} = 80\,(4)$

$$230 + x = 320$$
$$\underline{-230 \qquad\quad -230}$$
$$x = 90$$

∴ James needs a score of 90 on the next and last test.

Method 2: Arithmetic approach.

This method can follow from Method 1 (the algebraic approach).

Step 1:   For an average of 80 on 4 tests, total score must be $80 \times 4 = 320$

Step 2:   Total score on the first three tests is $75 + 70 + 75 = 230$

Step 3:   Score needed on the 4th test to make total score 320 is
$320 - 230 = 90$

∴   James needs a score of 90 on the next test.

Method 3: Layperson's approach

Step 1:

1st **test**	2nd **est**	3rd **test**	4th **test**
85	**70**	**75**	**?**
Here, there are an extra 5 points to spare	Here, we need 10 points	Here, we need 5 points	Here, we need 80 points

Step 2: If we add the 5 extra points from the first test to the 3rd test, we would have

1st **test**	2nd **test**	3rd **test**	4th **test**
80	**70**	**80**	**?**
This is O.K	Here, we need 10 more points	This is O.K	Here, we need 80 points

Since there is only one more test to be taken, James must make up the 10 points needed for the 2nd test, and obtain 80 points needed for the 4th test, all on the last test.

Hence, James needs a score of $10 + 80 = 90$ on the last test.

Lesson 45 Exercises

A. A chemist has 120 liters of a 40% solution of alcohol. How much pure alcohol must be added in order to obtain a 75% solution of alcohol? (Assume the solution is composed of only pure alcohol and water)

Answer: 168 liters

B. 1. An engineering company has 260 workers of whom 30% are women. How many new women must be hired such that 50% of the workers are women? Assume that no workers are fired.

2. A school system has 400 teachers of whom 70% are males Assuming that no teachers are fired, approximately, how many new female teachers must be hired such that 60% of the teachers are males?

3. A company has 240 workers of whom 100 are males Assuming that no workers are fired, how many new male workers must be hired such that $\frac{3}{4}$ of the workers are males?

4. Betty deposited a total of $20,000 in two bank accounts. The interest rate for one of the accounts was 6.5%, and the rate for the other account was 4%. If the total interest Betty was paid, at the end of last year, was $1100, how much money was invested at 6.5% and at 4% ?

Ans: **1.** 104 women; **2.** 67 females; **3.** 320 males; **4.** $12,000 at 6.5% and $8,000 at 4%.

C. Betty has $850, $700, $750 in three different bank accounts. This morning, she has decided to open a fourth bank account, today. What is the total amount of money Betty needs to deposit so that at the end of the day, she would have exactly $800 in each of the four bank accounts?

Answer: $900

D. During the semester, Maria scored 92 and 84 on the first two math tests. If there is one more test to take, what is the score she needs on the third test so that her class average is 90? Ans: 94

Test # 7-Student's Self-Test (Always, Test yourself before you are tested)

Attempt all questions on clean sheets of paper, **Do not write in the book**
Show all necessary work.

1. If three tines the difference of a number and two is increased by eight, the result is 17. Find the number.

2. If five more than twice a number is 48, find this number.

3. If four times a number is subtracted from 20, the result is 28. Find this number.

4 A number is 4 less than twice another number. If the sum of these numbers is 32, find them.

5. The length of a rectangle is 4 units more than the width. If the perimeter of this rectangle is 64, find the length and width of this rectangle.

6. Maria fenced a rectangular farm with 1200 ft of wire. If this farm is twice as long as it is wide, what are the dimensions of this farm?

7. The width of a rectangle is three units less than its length. If the perimeter of this rectangle is 30 units, find the dimensions of this rectangle.

8. During the semester, Maria scored 82, 78 and 84 on the first three math tests. If there is one more test to take, what is the score she needs on the fourth test so that her class average is 85?

9. Working alone, John can sort a quantity of mail in 2 hours, while Betty working alone can sort the same quantity of mail in 3 hours. How long will it take John and Betty to sort the same quantity of mail if they work together?

10. The sum of three consecutive integers is 99. Find these integers.

11. The sum of two consecutive integers is 49 less than five times the larger. Find these integers.

12. The sum of three consecutive even integers is 84 . Find these integers.

13. Find three consecutive even integers such that the sum of the first and third integers is 12 less than the second integer.

14. Find three consecutive odd integers whose sum is 135.

15. Find two consecutive integers such that the difference of three times the larger integer and the smaller integer is 39.

16. James is 4 years older than Mary. Three years ago, James was twice as old as Mary . Find their present ages.

17. If 30% of a number is 45, what is the number?

18. 60 is 30% of what number?

19. 18% of what number is 126?

20. What rate percent of 16 is 12?

21 The monthly rent for Maria's apartment is $600. If Maria spends 30% of her monthly income on this rent, what is her monthly income?

22. What rate% of 80 is 240?

23. On the last CUNY test, there were 40 questions. James answered 4 questions incorrectly. What was his grade in percent?

24. Betty has $820, $780, $840 in three different bank accounts. This morning, she has decided to open a fourth bank account, today. What is the total amount of money Betty needs to deposit so that at the end of the day, she would have exactly $850 in each of the four bank accounts?

25. The difference between two positive numbers is 40. The larger of the two numbers is 6 times the smaller. Find these two numbers.

Bonus:
In a chemistry class, there are 35 students. The number of females is 5 less than three times the number of males. How many males and how many females are in this class?

Answers: **1**. The number is 5 ; **2**. The number is $\frac{43}{2}$; **3**. The number is -2 ;

4. The numbers are 12 and 20 ; **5**. The width is 14 units and the length is 18 units;
6. The length is 400 ft and the width is 200 ft ; **7**. The width is 6 units and the length is 9 units ;
8. Maria needs a score of 96; **9**. $1\frac{1}{5}$ hours; **10**. The numbers are 32, 33, and 34 ;
11. The numbers are 15 and 16; **12**. The numbers are 26, 28 and 30;
13. The numbers are -14, -12 and -10 ; **14**. The numbers are 43, 45 and 47 ;
15. The numbers are 18 and 19; **16**. James is 11 years old and Mary is 7 years old;
17. The number is 150; **18**; The number is 200; **19**. The number is 700; **20**. 75% ;
21. Maria's monthly income is $2,000 ; **22**. 300%;
23. James' grade was 90%; **24**. Betty needs to deposit $960;
25. The smaller number is 8, and the larger number is 48;
   Bonus: There are 25 females and 10 males

CHAPTER 15
RATIO and PROPORTION (Variation)

Lesson: **Definition; Reduction of Ratios to Lowest Terms**
Lesson: **Using Ratios to Compare Quantities**
Lesson: **Using Ratios to Divide a Quantity into Parts**

Lesson**: Introduction to Direct and Indirect Proportion**
Lesson: **Methods for Solving Direct Proportion Problems**
Lesson: **Methods for Solving Inverse Proportion Problems**
Lesson: **Compound Proportion (Compound Variation) Problems**

Use Chapters 5 & 6

The above was designed to avoid duplication and to reduce the volume of this book, and also, such a design is good for the environment.

CHAPTER 16A
Factoring Polynomials

Lesson 46: **Greatest Common Factor and Common Monomial Factoring**
Lesson 47: **Difference between two Squares; Factoring by Grouping**
Lesson 48: **Factoring Monic Quadratic Trinomials** (Coefficient of the x^2-term is 1)
Lesson 49: **Non-monic Quadratic Trinomials** (Coefficient of the x^2-term is **not** 1)
Lesson 50: **Factoring trinomials quadratic in form (Substitution method)**
Lesson 51: **Solving Quadratic Equations by Factoring and Applications**

You are responsible for three main types of factoring:

1. Common monomial factoring. Example: $2x + 8 = \mathbf{2(x+4)}$
2. Factoring quadratic trinomials. Example: $x^2 + 7x + 12 = \mathbf{(x+3)(x+4)}$
3. Factoring the difference between two squares. Example: $x^2 - y^2 = \mathbf{(x+y)(x-y)}.$

Factoring Completely:

Note: Step 1. In any factoring, perform common monomial factoring first, if this is possible. (That is, factor out the greatest common divisor.)

Step 2. Look out for the difference between squares or quadratic trinomials and factor accordingly.

Lesson 46
Greatest Common Factor and Common Monomial Factoring

The Greatest Common Factor (GCF)

Definition: The **greatest common factor** of two or more given numbers is the largest **common** divisor of the given numbers.

We will cover two methods which are similar to the methods (see page 24) used in finding the LCM of numbers. Note that the greatest common factor is also the greatest common divisor (GCD) or the highest common factor (HCF).

Example   Find the greatest common factor (GCF) of 36 and 48.

Method 1

We will use a direct method which is similar to the prime factorization technique discussed previously (see page 23). We will use **prime numbers** successively as divisors. At each step of the division process, each prime number used **must divide all** the dividends. The division process ends when there are no more common divisors to be used. The GCF is the product of the divisors used.

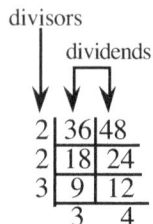

```
divisors
   |    dividends
   |   ↓  ↓
   ↓  ↓  ↓
  2 | 36|48
  2 | 18| 24
  3 |  9 | 12
        3    4
```

Answer: The GCF of 36 and 48 = $2 \times 2 \times 3$
$$= 12.$$

Method 2

Solution Step 1 : Factor each number into primes, and write repeated factors in power form.

$$36 = 2 \times 2 \times 3 \times 3 = 2^2 \times 3^2$$
$$48 = 2 \times 2 \times 2 \times 2 \times 3 = 2^4 \times 3$$

Step 2 :Consider each prime factor which is common to both 36 and 48 (i.e., consider 2 and 3) and choose the lowest power of each prime factor.

Step 3: The GCF = the product of the **lowest powers** of the prime factors considered.

$$= 2^2 \times 3$$
$$= 4 \times 3$$

The GCF of 36 and 48 = 12

Note : 2×2 or 2^2 is common to both 36 and 48; and also 3 is common to both 36 and 48.

Also, if a prime factor is **not common** to all the given numbers, we do not "consider" it.

Common Monomial Factoring

Example   Factor completely: $9x^3 - 6x$

$$9x^3 - 6x$$

Step 1: $= 3x\left(\dfrac{9x^3}{3x} - \dfrac{6x}{3x}\right)$ <---You may skip Step 1. ($3x$ is the greatest common factor (divisor) of $9x^3$ and $-6x$)

↓   ↓   ↓   Scrapwork:

Step 2: $= 3x(3x^2 - 2)$ $\left(\mathbf{1.}\ \dfrac{9x^3}{3x} = 3x^2;\ \ \mathbf{2.}\ \dfrac{-6x}{3x} = -2\right)$

Step 3:   We perform two checks (The checking may be done by mere inspection)

Check # 1:   Multiply the factors:
$$3x\,(3x^2 - 2)$$
$$= 9x^3 - 6x \text{ <------------ This must agree with the original expression.}$$

Check # 2: Inspect $3x^2 - 2$ (the expression within the parentheses to make sure the terms do not have anymore common factors (divisors).

This checking is usually done mentally (by inspection).

$\therefore\ 9x^3 - 6x\ =\ \boxed{3x(3x^2 - 2)}$ <----**answer**

Incomplete Factorization:

Example: $9x^3 - 6x$

$= 3(3x^3 - 2x)$ <------------This is incomplete factorization because
$3x^3$ and $- 2x$ still have common divisors (factors).

To complete the factorization:
$$= 3(3x^3 - 2x)$$
$$= 3x(3x^2 - 2)$$
and then , $9x^3 - 6x = 3x(3x^2 - 2)$ as obtained previously.

Lesson 46 Exercises

A.

Find the GCF of the following: **1.** 24 and 30   **2.** 56, 70 and 105   **3.** 12, 64 and 72.

Answers: **1.** 6 ;   **2.** 7 ;   **3.** 4

B.

Factor completely: **1.** $4x^4y^6 - 20xy^2$ **2.** $2a^5b^3c^7 + a^4b^2c^6 - a^2b^2c^5$

3. $4x^4y^2 - 2x^2y$;   **4.** $6a^3bc - 2x^2y$;   **5.** $7x^5y^2z - 21x^4y$;   **6.** $8a^2b^2 - ab$;   **7.** $4x^2 - 3xy$

Answers   **1.** $4xy^2(x^3y^4 - 5)$;   **2.** $a^2b^2c^5(2a^3bc^2 + a^2c - 1)$;   **3.** $2x^2y(2x^2y - 1)$;
     4. $2(3a^3bc - x^2y)$;   **5.** $7x^4y(xyz - 3)$;   **6.** $ab(8ab - 1)$;   **7.** $x(4x - 3y)$

Lesson 47

Difference between two Squares; Factoring by Grouping

Factoring the Difference Between Two Squares

Examples **1.** $a^2 - b^2 = (a + b)(a - b);$

    2. $c^2 - d^2 = (c + d)(c - d)$

    3. $16 - x^2 = (4 + x)(4 - x)$

    4. $9x^2 - 1 = (3x + 1)(3x - 1)$

    5. $6x^4 - 6 = 6(x^4 - 1) = 6(x^2 + 1)(x^2 - 1) = 6(x^2 + 1)(x + 1)\ (x - 1)$

    6. $a^4 - b^4 = (a^2 + b^2)(a^2 - b^2) = (a^2 + b^2)(a + b)(a - b)$

    7. $3x^2 - 75 = 3(x^2 - 25) = 3(x + 5)(x - 5)$

    8. $9y^2 - 16 = (3y + 4)(3y - 4)$

Factoring by Grouping

Sometimes, a seemingly non-factorable expression can be factored after a proper grouping of the terms of the expression. Look out for this type of factoring when there are four, six, or an even number of terms.

Example 1:   Factor by grouping: $ab + cd + ac + bd$

Step 1: Pick the first term and then pick another term (of the expression)
           which has a common factor with the first term picked.
          $\underline{ab + ac}\ \ + \underline{cd + bd}$

Step 2: Perform common monomial factoring on the first two terms, and then on the
        other terms.
          $a(b + c) + d(c + b)$
         $= a(b + c) + d(b + c)$ (Note that $c + b = b + c$)

Step 3:   Perform "common binomial" factoring. (This is similar to common monomial factoring)
         Since $(b + c)$ is common , we pick it first.
         Then we obtain $(b + c)(a + d)$

    Note : Step 3 is a common monomial factoring if we consider $(b + c)$ as a single quantity.

Example 2: Factor by grouping (without adding any like terms). 228

$$6x^2 + 15x - 4x - 10$$

Step 1: Perform common monomial factoring on the first two terms and then on the other terms.

$3x(2x + 5) - 2(2x + 5)$ <------- Adjust the signs here so that we obtain
$- 4x - 10x$ when the parentheses are removed.

Step 2: Pick the common factor $2x + 5$ and factor as done in common monomial factoring.

$(2x + 5)(3x - 2)$

You can check this factoring by multiplying the factors.
Note: We will obtain the same factors if we group the first and the third terms.

i.e. $6x^2 - 4x + 15x - 10$
     $2x(3x - 2) + 5(3x - 2)$

Remove or factor out the common factor $(3x - 2)$, and then we obtain

$(3x - 2)(2x + 5)$
Again, we obtain the same binomial factors.

Lesson 47 Exercises

A

Factor completely: **1.** $y^2 - 16$;   **2.** $4x^2 - 36$;   **3.** $4x^2 - 9$;   **4.** $25 - 4y^2$

Answers: **1.** $(y + 4)(y - 4)$; **2.** $4(x + 3)(x - 3)$; **3.** $(2x + 3)(2x - 3)$; **4.** $(5 + 2y)(5 - 2y)$ or $-(2y + 5)(2y - 5)$

B.

Factor completely : **1.** $x^2 + 2y + xy + 2x$;   **2.** $15x^2 - 9x + 20x - 12$

3. $ac - bc + ad - bd$ **4.** $ax + by + bx + ay$ **5.** $8x^2 + 12x + 2x + 3$

Answers: **1.** $(x + y)(x + 2)$;   **2.** $(3x + 4)(5x - 3)$; **3.** $(a - b)(c + d)$; **4.** $(a + b)(x + y)$; **5.** $(2x + 3)(4x + 1)$

Lesson 48
*Factoring Monic Quadratic Trinomials
(Coefficient of the x^2-term is 1)

Case 1: All the terms are positive

Factor completely:

Example 1: $x^2 + 7x + 12$

Step 1: $(x \quad)(x \quad)$

Step 2: $(x + 3)(x + 4)$

Scrapwork:
Find two numbers whose product is 12 and whose sum is 7
$12 = 6 \times 2$
$12 = \boxed{4 \times 3}$

Always check:
$(x + 3)(x + 4) = x^2 + 4x + 3x + 12$ (multiplying the factors)

$\qquad\qquad = x^2 + 7x + 12$ <------------------ This agrees with the original expression.

$\therefore x^2 + 7x + 12 = (x + 3)\ (x + 4)$

Case 2: The last term is positive and the middle term is negative

Example 2: Factor completely: $x^2 - 5x + 6$

Step 1: $(x \quad)(x \quad)$

$\boxed{\text{These two terms}}$ must have the same sign because $\boxed{\text{this}}$ is positive.

Step 2: $(x - 2)(x - 3)$

Scrapwork : $6 \times 1 = 6$
$\boxed{2 \times 3} = 6$

Always check by multiplying.

$\qquad (x - 2)(x - 3)$ (We want two numbers whose
$\qquad = x^2 - 3x - 2x + 6$ product is +6, but whose sum is -5)
$\qquad = x^2 - 5x + 6$
$\therefore\ \ x^2 - 5x + 6 = (x - 2)(x - 3)$

*The coefficient of the x^2-term of the quadratic trinomial is also called the leading coefficient of the quadratic trinomial. Also, a quadratic trinomial in which the coefficient of the x^2-term is 1 is called a **monic** quadratic trinomial.

Case 3: The last term is negative and the middle term is negative or positive

Example 3 Factor completely:

$$x^2 - 7x - 18$$

Step 1: $(x \quad)(x \quad)$ | These two terms | must have different signs because | this term | is negative

Step 2: $(x + 2)(x - 9)$

Scrapwork:

$9 \times 2 = 18$

$3 \times 6 = 18$

We want two numbers whose product is -18 and whose sum is -7
(Give the sign of the middle term to the larger of 2 and 9 (absolute values))

After checking, $x^2 - 7x - 18 = (x + 2)(x - 9)$.

Note that $x^2 + 7x - 18$ would be factored as $(x - 2)(x + 9)$

Example 4   Factor completely: $2x^2 - 12x + 16$

$$2x^2 - 12x + 16$$

Step 1: $2(x^2 - 6x + 8)$ (Factoring out the greatest common divisor)

Step 2: $2(x - 2)(x - 4)$ (Factoring the quadratic trinomial)

After checking, $2x^2 - 12x + 16 = 2(x - 2)(x - 4)$

Lesson 48 Exercises

A. Factor completely: **1.** $x^2 + 9x + 20$;   **2.** $x^2 + 16x + 28$;   **3.** $x^2 + 11x + 30$

Answers **1.** $(x + 4)(x + 5)$; **3.** $(x + 2)(x + 14)$;   **4.** $(x + 5)(x + 6)$

B. Factor completely: **1.** $x^2 - 9x + 18$;   **2.** $x^2 - 15x + 36$;   **3.** $x^2 - 13x + 40$.

Answers **1.** $(x - 3)(x - 6)$ **2.** $(x - 3)(x - 12)$ **3.** $(x - 5)(x - 8)$

C. Factor completely: **1.** $x^2 + 5x - 24$;   **2.** $x^2 - 5x - 24$;   **3.** $x^2 - 13x - 48$

Answers: **1.** $(x - 3)(x + 8)$;   **2.** $(x + 3)(x - 8)$;   **3.** $(x + 3)(x - 16)$

D. Factor completely: **1.** $3x^2 - 24x + 36$;   **2.** $2x^2 + 2x - 84$

Answers **1.** $3(x - 2)(x - 6)$;   **2.** $2(x - 6)(x + 7)$

E. Factor completely

1. $x^2 + 9x + 18$

2. $x^2 + 8x + 15$

3. $x^2 + 17x + 72$

4. $x^2 + 3x + 2$

5. $x^2 + 14x + 33$

6. $x^2 + 14x + 45$

7. $x^2 + 15x + 36$

8. $x^2 - 4x - 21$

9. $x^2 + 4x - 21$

10. $x^2 - 11x + 18$

11. $x^2 - 10x + 21$

12. $x^2 - 9x + 18$

13. $x^2 + 8x - 33$

14. $4x^2 + 44x + 72$

15. $2x^2 + 28x + 90$

Answers: 1. $(x + 3)(x + 6)$; **2**. $(x + 3)(x + 5)$; **3**. $(x + 8)(x + 9)$; **4**. $(x + 1)(x + 2)$; **5**. $(x + 3)(x + 11)$;
6. $(x + 5)(x + 9)$; **7**. $(x + 3)(x + 12)$; **8**. $(x + 3)(x - 7)$; **9**. $(x - 3)(x + 7)$; **10**. $(x - 2)(x - 9)$;
11. $(x - 3)(x - 7)$; **12**. $(x - 3)(x - 6)$; **13**. $(x - 3)(x + 11)$; **14**. $4(x + 2)(x + 9)$; **15**. $2(x + 5)(x + 9)$.

F. Factor completely:

1. $x^2 - 9x + 20$;

2. $x^2 + 3x - 18$

3. $6x^3y - 14x^2y$;

4. $x^2 - 11x + 24$;

5. $2x^2 - 12x - 54$;

6. $a^2 + 3a - 28$;

7. $b^2 - 2b - 48$;

8. $x^2 - 49$;

9. $2y^2 - 32$;

10. $4x^2y^2 - 9b^2$;

11. $b^4 - a^4$;

12. $9y^2 - 36$;

13. $9 - x^2$

14. $-x^2 - 7x - 12$.

Answers: 1. $(x - 5)(x - 4)$; **2**. $(x + 6)(x - 3)$; **3**. $2x^2y(3x - 7)$ **4**. $(x - 8)(x - 3)$; **5**. $2(x + 3)(x - 9)$; **6**. $(a + 7)(a - 4)$;

7.$(b + 6)(b - 8)$; **8**.$(x + 7)(x - 7)$; **9**. $2(y + 4)(y - 4)$; **10**. $(2xy + 3b)(2xy - 3b)$; **11**. $(b^2 + a^2)(b + a)(b - a)$

12. $9(y + 2)(y - 2)$; **13**. $(3 + x)(3 - x)$ or $-(x + 3)(x - 3)$; **14**. $-(x + 3)(x + 4)$

Test # 8 - Student's Self-Test (**Always**, **Test yourself before you are tested**)

Attempt all questions on clean sheets of paper, **Do not write in the book** Show all necessary work.

1. Find the GCF of the following:

(a). 32 and 36; (b) 14, 70 and 182 ;

(c). 12, 56 and 72.

2. Factor completely: (a) $x^2 + 11x + 18$;

(b) $x^2 - 11x + 18$.

3. Factor completely: (a) $x^2 + 9x + 20$;

(b) $x^2 + 13x + 42$.

4. Factor completely: (a) $x^2 + 12x + 20$;

(b) $x^2 - 17x + 72$.

5. Factor completely: (a) $x^2 - 10x + 24$;

(b) $2x^2 + 22x + 60$.

6. Factor completely: (a) $x^2 - 15x + 36$;

(b) $4x^2 + 44x + 72$.

7. Factor completely: (a) $x^2 - 12x + 32$;
(b) $x^2 + 4x - 77$.

8. Factor completely: (a) $x^2 + 5x - 24$;
(b) $x^2 - 9x + 18$.

9. Factor completely: (a) $x^2 - 5x - 24$;

(b) $x^2 - 10x + 9$.

10. Factor completely: (a) $x^2 - 13x - 48$;

(b) $x^2 + 4x - 21$.

11. Factor completely: (a) $10x^2 + 4x - 6$;

(b) $x^2 + 14x + 33$

12. Factor completely: (a) $2x^2 - 4x - 96$;

(b) $x^2 + 14x + 45$.

13. Factor completely: (a) $x^2 - 5x - 36$;

(b) $x^2 + 3x + 2$.

14. Factor completely: (a) $x^2 + 9x + 18$;

(b) $x^2 + 8x + 15$.

15. Factor completely: (a) $4x^2 - 16$;

(b) $4 - 9x^2$.

16. Factor completely: (a) $y^2 - 16$;

(b) $25 - 4y^2$.

17. Factor completely: (a) $9y^2 - 36$;

(b) $-x^2 - 7x - 12$.

18. Factor completely: (a) $4x^2y^2 - 9b^2$;

(b) $b^4 - a^4$.

19. Factor completely: (a) $x^2 - 49$;

(b) $2y^2 - 32$.

20. Factor completely: (a) $a^2 + 3a - 28$;

(b) $b^2 - 2b - 48$.

21. Factor completely: (a) $x^2 - 11x + 24$;

(b) $2x^2 - 12x - 54$.

22. Factor completely: (a) $-x^2 - 3x + 18$

(b) $x^2 + 3xy + 2y^2$

23. Factor completely: (a) $x^2 - 17x + 72$;
(b) $2x^2 + x - 15$

Factor completely:

24.   (a) $ad - bc - ac + bd$

(b) $ay + bx + by + ax$

25.   Factor completely: $a^2 + 2x + ax + 2a$

Bonus: Write down your first and last names, and place a plus sign between these names. Assuming that each letter in your name represents a real number, factor completely, if possible, the expression obtained.

Answers: 1 (a) 4; (b) 14 ; (c) 4;   **2.** (a) $(x + 2)(x + 9)$; (b) $(x - 2)(x - 9)$; **3.** (a) $(x + 4)(x + 5)$;
 (b) $(x + 6)(x + 7)$;   **4.** (a) $(x + 2)(x + 10)$; (b) $(x - 8)(x - 9)$;   **5.** (a) $(x - 6)(x - 4)$;
 (b) $2(x + 5)(x + 6)$; **6.** (a) $(x - 3)(x - 12)$; (b) $4(x + 2)(x + 9)$;   **7.** (a) $(x - 4)(x - 8)$;
 (b) $(x - 7)(x + 11)$; **8.** (a) $(x - 3)(x + 8)$; (b) $(x - 3)(x - 6)$; **9.**(a) $(x - 8)(x + 3)$; (b) $(x - 1)(x - 9)$;
 10. (a) $(x + 3)(x - 16)$; (b) $(x + 7)(x - 3)$;   **11.** (a) $2(5x - 3)(x + 1)$; (b) $(x + 3)(x + 11)$;
 12. (a) $2(x - 8)(x + 6)$; (b) $(x + 9)(x + 5)$;   **13.** (a) $(x - 9)(x + 4)$; (b) $(x + 1)(x + 2)$;
 14. (a) $(x + 3)(x + 6)$; (b) $(x + 3)(x + 5)$;   **15.** (a) $4(x + 2)(x - 2)$; (b) $(2 + 3x)(2 - 3x)$;
 16. (a) $(y + 4)(y - 4)$; (b) $(5 + 2y)(5 - 2y)$;   **17.** (a) $9(y + 2)(y - 2)$; (b) $-(x + 3)(x + 4)$;
 18. (a) $(2xy + 3b)(2xy - 3b)$; (b) $(b^2 + a^2)(b + a)(b - a)$;   **19.** (a) $(x + 7)(x - 7)$;
 (b) $2(y + 4)(y - 4)$; **20.** (a) $(a - 4)(a + 7)$; (b) $(b + 6)(b - 8)$; **21.** (a) $(x - 3)(x - 8)$;
 (b) $2(x + 3)(x - 9)$;   **22.** (a) $-(x - 3)(x + 6)$; (b) $(x + y)(x + 2y)$; **23.** (a) $(x - 8)(x - 9)$;
 (b) $(2x - 5)(x + 3)$; **24** (a) $(d - c)(a + b)$; (b) $(a + b)(x + y)$;   **25.** $(a + x)(a + 2)$.
 Bonus: Answer varies.

Lesson 49
Factoring Non-monic Quadratic Trinomials

(Coefficient of the x^2-term is **not** 1)

Example   Factor completely: $6x^2 + 11x - 10$

Solution

There are a number of methods for factoring non-monic (the coefficient of the x^2-term is not 1) quadratic trinomials, namely

   1. A general method (Method 1)
   2. A Substitution Method (Method 2) and
   3. A third method (Method 3, the **ac**-method), which involves factoring by grouping.

Method 1   The General Method

This method involves considering the products of the various possible binomial factors to determine which factors yield the original quadratic trinomial.

We will now factor $6x^2 + 11x - 10$
We consider the possible binomials
(a) $(3x + 2)(2x - 5)$
(b) $(3x - 2)(2x + 5)$
(c) $(3x + 5)(2x - 2)$
(d) $(3x - 5)(2x + 2)$
(e) $(6x + 5)(x - 2)$
(f) $(6x - 5)(x + 2)$
(g) $(6x + 2)(x - 5)$
(h) $(6x - 2)(x + 5)$

Multiply the above factors. The pair of binomial factors which yield the original quadratic expression are the correct factors. **Note** that since the original quadratic trinomial did not have any common factors, each possible binomial factor will not have any common factors; and therefore, we could have eliminated the factors in (c),(d) and (h) without having to multiply them out, and then check only the products for cases (a), (b), (e) and (f).
In the above example, the correct factors are $(3x - 2)$ and $(2x + 5)$.

$\therefore$ $6x^2 + 11x - 10 = (3x - 2)(2x + 5)$

Factorability of a quadratic trinomial: $ax^2 + bx + c$

A quadratic trinomial is factorable if the discriminant $b^2 - 4ac$ is a perfect square.

Example Is $6x^2 + 11x - 10$ factorable?
     Check: $a = 6, b = 11, c = -10$
     $b^2 - 4ac = 11^2 - 4(6)(-10)$
            $= 121 + 240$
            $= 361$
      361 is a perfect square because $\sqrt{361} = 19$, a rational number.
Answer: Yes. $6x^2 + 11x - 10$ is factorable.

Method 2:   The Substitution Method

In this method, we change the quadratic trinomial to a monic trinomial which is then easily factored.

Example   Factor $6x^2 + 11x - 10$

Step 1: Multiply the expression by the coefficient of the x^2-term.

$6(6x^2) + 6(11x) - 6(10)$

Step 2: Write the first term as a square and interchange the 6 and the 11 in the second term.

$36x^2 + 11(6x) - 60$

$(6x)^2 + 11(6x) - 60$ ....(A)

Step 3: Let $6x = s$ (That is, replace $6x$ by s in expression (A))

Then, we obtain $s^2 + 11s - 60$

Now, since this is a monic trinomial (coefficient of s^2 is 1),
    factor easily.

$(s - 4)(s + 15)$ ....(B)

Scrapwork
$60 \times 1 = 60$
$30 \times 2 = 60$
$\boxed{15 \times 4 = 60}$

Step 4:   Replace s by $6x$,and then, expression (B) becomes

$(6x - 4)(6x + 15)$.............................(C)

Since we multiplied the original trinomial by 6, we must divide expression (C) by 6 (that is we must undo the "6" we introduced in Step 1.)

In order to divide (C) by 6, we perform common monomial factoring on the two binomial factors (in some cases, this factoring is performed only on one of the binomial factors).

$(6x - 4)(6x + 15)$
$2(3x - 2)\ 3(2x + 5)$
$2(3)(3x - 2)(2x + 5)$
$6(3x - 2)(2x + 5)$

Now, we divide by 6: $\dfrac{6(3x - 2)(2x + 5)}{6}$

and then the complete factorization of

$6x^2 + 11x - 10$ is $(3x - 2)(2x + 5)$

Method 3 Factoring by the ac-Method

The "a" is the coefficient of the x^2-term, and the "c" is the constant term of the general quadratic trinomial $ax^2 + bx + c.$

Example   Factor completely. $6x^2 + 11x - 10$
$a = 6, \ c = -10.$

Step 1: Multiply the coefficient of the x^2-term by the last term (the constant term).

$$6x^2 + 11x - 10 \ ....(A)$$

$$6(-10) = -60$$

Step 2: Replace the $+11x$ (the middle-term) of expression (A) by two terms whose coefficients add up to 11 and whose product is -60.

Then, we obtain

$$6x^2 + 15x - 4x - 10$$

Scrapwork

$60 \times 1 = 60$
$30 \times 2 = 60$
$\boxed{15 \times (-4) = -60}$
$20 \times 3 = 60$

Step 3: Factor the expression by grouping. (Break up the expression into two pairs and factor)
$$6x^2 + 15x - 4x - 10$$
$$3x(2x + 5) - 2(2x + 5)$$
$$(2x + 5)(3x - 2)$$
Therefore, $6x^2 + 11x - 10 = (2x + 5)(3x - 2)$

Factoring by any method

Example   Factor by any method
$$3x^2 - 7x - 10$$
We will use Method 2 and Method 3.

Method 2

Step 1:   Multiply every term by 3 (the coefficient of the x^2-term)

$$3(3x^2) - 7x(3) - 30$$
$$9x^2 - 7(3x) - 30$$
$$(3x)^2 - 7(3x) - 30 \ ....(A)$$

Step 2: Let $3x = s$ in expression (A) and factor the resulting monic trinomial.
          Then $s^2 - 7s - 30$
          $(s + 3)(s - 10)$

Step 3:   Replace s by $3x$.
          $(3x + 3)(3x - 10)$

Step 4:   Factor the first binomial and divide by 3

$$\frac{3(x + 1)(3x - 10)}{3}$$ This "3" was used as a multiplier in Step 1

$\therefore 3x^2 - 7x - 10 = (x + 1)(3x - 10)$

Method 3   Factor $3x^2 - 7x - 10$.

Step 1: Multiply the coefficient of the x^2-term by the last term.

$$3x^2 - 7x - 10$$

$$3(-10) = -30$$

Step 2: Replace $-7x$ by two terms whose coefficients add up to -7 and whose product is -30.

$$3x^2 - 10x + 3x - 10$$

Step 3:   Factor by grouping

$$3x^2 - 10x + 3x - 10$$
$$= x(3x - 10) + 1(3x - 10)$$

$$= (3x - 10)(x + 1)$$

$\therefore$ $3x^2 - 7x - 10 = \boldsymbol{(x + 1)(3x - 10)}$

Lesson 49 Exercises

A. Factor completely: **1.** $12x^2 - x - 6$;   **2.** $18x^2 - 3x - 10$;   **3.** $10x^2 - 29x + 21$

Answers: **1**. $(3x + 2)(4x - 3)$;   **2.** $(6x - 5)(3x + 2)$;   **3.** $(2x - 3)(5x - 7)$

B. Factor completely: **1.** $12x^2 - x - 6$;   **2.** $18x^2 - 3x - 10$;   **3.** $10x^2 - 29x + 21$

Answers: **1.** $(3x + 2)(4x - 3)$;   **2.** $(6x - 5)(3x + 2)$;   **3.** $(2x - 3)(5x - 7)$

C Factor completely:

1. $8x^2 - 2x - 15$ **2.** $4x^2 - 12x + 9$ **3.** $6x^2 - 19x + 10$

4. $15x^2 - 14x - 8$ **5.** $12x^2 + x - 6$

Answers: **1** . $(2x - 3)(4x + 5)$ **2.** $(2x - 3)(2x - 3)$ or $(2x - 3)^2$ **3.** $(2x - 5)(3x - 2)$
4. $(3x - 4)(5x + 2)$ **5.** $(3x - 2)(4x + 3)$

Lesson 50

Factoring trinomials quadratic in form (Substitution method)

Example (a) Factor completely: $x^4 + 7x^2 + 12$ ....(1)

Solution Step 1: Let $x^2 = u$. Then, expression (1) becomes $u^2 + 7u + 12$...........(2)

Step 2 : Factoring (2), we obtain $(u + 3)(u + 4)$.................................(3)

Step 3: Replacing u by x^2 in expression (3), we obtain $(x^2 + 3)(x^2 + 4)$
$\therefore x^4 + 7x^2 + 12 = (x^2 + 3)(x^2 + 4)$.

Example (b) Factor completely: $2a^4 - 5a^2 - 12$ ....(1)

Solution Let $a^2 = u$. Then, we obtain $2u^2 - 5u - 12$ ....(2)

Factoring (2), we obtain $(2u + 3)(u - 4)$ ....(3)

Replacing u by a^2 in expression (3), we obtain $(2a^2 + 3)(a^2 - 4) = (2a^2 + 3)(a + 2)(a - 2)$

$\therefore 2a^4 - 5a^2 - 12 = (2a^2 + 3)(a + 2)(a - 2)$.

Lesson 50 Exercises

Factor completely: **1**. $x^4 - 7x^2 + 12$;   **2**. $2a^4 + 5a^2 - 12$

Answers. **1**. $(x^2 - 3)(x + 2)(x - 2)$;   **2**. $(2a^2 - 3)(a^2 + 4)$

Lesson 51
Solving Quadratic Equations by Factoring and Applications

Standard form of the quadratic equation: $ax^2 + bx + c = 0$, where a, b, and c are constants and $a \neq 0$

Examples 1. $x^2 - 3x - 28 = 0$ <---------in standard form
     2. $x^2 - 2x = 0$ <--------- in standard form
     3. $x^2 = 5x + 14$ <---------**not** in standard form
     4. $5x^2 = 8x$ <---------**not** in standard form

Principle of zero products: If $ab = 0$ then either $a = 0$, or $b = 0$ (or both $= 0$).

To solve by factoring, the quadratic equation must be in standard form before factoring

Example 1   Solve by factoring
      $x^2 - 3x - 28 = 0$

Step 1:   Factor the quadratic trinomial.
        $(x + 4)(x - 7) = 0$
Step 2:   Set each factor equal to zero and solve for x.

     $x + 4 = 0$ $x - 7 = 0$
       $\underline{-4 \ \ -4}$ $\underline{+ 7 \ \ +7}$
        $x = -4$ $x = 7$
            $\therefore \ x = -4, x = 7$

The solutions are -4 and 7.

Example 2   Solve for x by factoring.
        $3x^2 - 4x = 0$

Step 1: Factor. $3x^2 - 4x = 0$

          $x(3x - 4) = 0$ (performing common monomial factoring)

Step 2: Set each factor equal to zero and solve for x.   (We set only factors having a **variable** to zero)

      $x = 0$ or $3x - 4 = 0$
                  $\underline{+ 4 \ \ +4}$
                  $\dfrac{3}{3}x = \dfrac{4}{3}$
                   $x = \dfrac{4}{3}$

        $\therefore \ x = 0, \ \ x = \dfrac{4}{3}$ **Note** that 0 is also a solution .

The solutions are 0 and $\dfrac{4}{3}$.

Note: A common **wrong** approach in the above problem:
     $3x^2 - 4x = 0$
       $3x^2 = 4x$
   $3x = 4$ (Canceling, that is dividing out an x on both sides of the equation ; such a cancellation excludes
         $x = 0$ as a solution.)

   $x = \dfrac{4}{3}$ (Of course, you still obtain one of the solutions but lose the other solution).

Example 3 Solve by factoring: $9x^2 = 36$

Solution $9x^2 = 36$
Step 1: $9x^2 - 36 = 0$

Step 2:   Factor the left-hand side completely.
        $9(x^2 - 4) = 0$
        $9(x + 2)(x - 2) = 0$
Step 3:   Set each factor equal zero and solve for x: (Note that $9 \neq 0$; only factors with a variable = 0)

$$x + 2 = 0 \quad \text{or} \quad x - 2 = 0$$
$$\underline{\quad -2 \quad -2} \qquad \underline{+2 \quad +2}$$
$$x = -2 \qquad\qquad x = 2$$

The solutions are 2 and -2.

Another method (The square root method)
$$9x^2 = 36$$
$$\frac{9x^2}{9} = \frac{36}{9}$$
$$x^2 = 4$$
$$x = \pm\sqrt{4}$$
$$x = \pm 2 \ (+2 \text{ or } -2)$$
Again, the solutions are 2 and -2.

Applications of the quadratic equation

Example   The width of a rectangle is 3 units less than the length. If the area of the rectangle is 28 sq. units, determine the dimensions of this rectangle.

Solution
Let the length of the rectangle $= x$.
Then, the width of this rectangle $= (x - 3)$
Area of a rectangle is given by LW (Where L is the length and W is the width).
The required equation: $x(x - 3) = 28$.
$$x^2 - 3x - 28 = 0$$
$$(x - 7)(x + 4) = 0$$
$$x = 7 \ \text{ or } x = -4 \qquad\qquad \text{(Solving by factoring)}$$
We reject the negative value, -4, since the dimension of a rectangle cannot be negative.
When the length is 7 units, the width is 7 - 3 = 4 units.

(**Check**: If the length is 7 and the width is 4, the area is 7(4) = 28 sq. units; and also the width, 4, is 3 less than the length, 7. Therefore, the conditions in the original word problem have been satisfied.)
The length is 7 units and the width is 4 units.

Lesson 51 Exercises

A. Solve for x by factoring: **1.** $x^2 - 3x - 10 = 0$;   **2.** $2x^2 - 6x = 0$ **3.** $10x^2 = 15x$

Answers: **1.** - 2 and 5;   **2.** 0 and 3 ;   **3.** 0 and $\frac{3}{2}$

B. 1. Solve for x: $2x^2 = 32$;   **2.** Solve for x: $3x^2 - 75 = 0$.

3. The sum of two numbers is 20, and their product is 96. Find these numbers.

4. The length of a rectangle is 5 units more than the width. If the area of this rectangle is 126 sq, units, find the length and width of this rectangle.

Ans: **1.** 4 and -4;   **2.** 5 and -5;   **3.** 8 and 12;   **4.** Length = 14 units, width = 9 units.

Test # 9 -Student's Self-Test (Always, Test yourself before you are tested)

Attempt all questions on clean sheets of paper, **Do not write in the book**
Show all necessary work.

---

1 Factor completely: $6x^2 - 13x + 6$.

2. Factor completely: $10x^2 - 21x + 2$.

---

3. Factor completely: $3x^2 + 13x + 4$.

4. Factor completely: $10x^2 - 13x$.

---

5. Factor completely: $2x^2 + 20x + 18$.

6. Factor completely: $4x^2 + 44x + 72$.

---

7. Factor completely: $8x^2 - 2x - 1$.

8. Factor completely: $8x^2 - 26x + 15$.

---

9. Factor completely: $18x^2 - 3x - 10$.

10. Factor completely: $15x^2 - 14x - 8$.

---

11. Factor completely: $3x^2 - 24x + 36$.

12. Factor completely: $2x^2 + 2x - 84$.

---

13. Factor completely: $7x^2 - 23x + 6$.

14. Solve for x by factoring: $4x^2 - 9 = 0$.

---

15. Solve for x by factoring: $4x^2 - 36 = 0$.

16. Factor completely: $24x^2 + 2x - 12$.

17. Solve for x by factoring: $-x^2 - 7x - 12 = 0$

18. Factor completely: $6x^2 - 19x + 10$.

19. Solve for x by factoring: $3x^2 - 6x = 0$.

20. Factor completely: $10x^2 - 29x + 21$.

21. Solve for x by factoring: $4x^2 = 36$.

22. Solve for x by factoring: $3x^2 - 75 = 0$.

23. Solve for x by factoring:
$x^2 - 3x - 10 = 0$.

24. The sum of two numbers is 24, and their product is 108. Find these numbers.

25. The length of a rectangle is 6 units more than the width. If the area of this rectangle is 112 sq. units, find the length and width of this rectangle.

Bonus:

$a^4 - 7a^2 + 12$;

$2y^4 + 5y^2 - 12$.

Answers: 1. $(2x - 3)(3x - 2)$; **2**. $(x - 2)(10x - 1)$; **3**. $(x + 4)(3x + 1)$; **4**. $x(10x - 13)$;
5. $2(x + 1)(x + 9)$; **6**. $4(x + 2)(x + 9)$; **7**. $(2x - 1)(4x + 1)$; **8**. $(2x - 5)(4x - 3)$;
9. $(6x - 5)(3x + 2)$; **10**. $(3x - 4)(5x + 2)$; **11**. $3(x - 6)(x - 2)$; **12**. $2(x - 6)(x + 7)$;
13. $(x - 3)(7x - 2)$; **14**. $x = \pm\frac{3}{2}$; **15**. $x = \pm3$; **16**. $2(3x - 2)(4x + 3)$; **17**. $x = -3$, $x = -4$;
18. $(2x - 5)(3x - 2)$; **19**. $x = 0$, $x = 2$; **20**. $(5x - 7)(2x - 3)$; **21**. $x = \pm3$; **22** $x = \pm5$;
23. $x = 5$, $x = -2$; **24**. The numbers are 6 and 18;
25. The length is 14 units and the width is 8 units.

Bonus: (a) $(a^2 - 3)(a + 2)(a - 2)$; (b) $(2y^2 - 3)(y^2 + 4)$.

CHAPTER 16B
More Factoring Polynomials
Lesson 52

Factoring the Sum and Difference between two Cubes;

Factoring the Sum of Two Cubes

$$\boxed{a^3 + b^3 = (a + b)(a^2 - ab + b^2)} \dots\dots\dots\dots\dots(1)$$

Similarly, $x^3 + y^3 = (x + y)(x^2 - xy + y^2)$

Example   Factor completely: $8x^3 + y^3$

$$8x^3 + y^3$$
$$= (2x)^3 + y^3$$
$$= (2x + y)((2x)^2 - 2x(y) + y^2) \qquad \text{(Replace } a \text{ by } 2x; b \text{ by } y \text{ in the factoring formula (1) above.}$$
$$= (2x + y)(4x^2 - 2xy + y^2)$$

Factoring the Difference between (of) Two Cubes

$$\boxed{a^3 - b^3 = (a - b)(a^2 + ab + b^2)} \dots\dots\dots\dots(2)$$

Similarly, $x^3 - y^3 = (x - y)(x^2 + xy + y^2)$

Example   Factor completely: $x^3 - 8y^3$

$$x^3 - 8y^3$$
$$= x^3 - (2y)^3$$
$$= (x - 2y)(x^2 + x(2y) + (2y)^2) \qquad \text{(Replace } a \text{ by } x; \text{ and } b \text{ by } 2y \text{ in the factoring formula (2) above)}$$
$$= (x - 2y)(x^2 + 2xy + 4y^2)$$

It is good practice to check the factoring by multiplying the factors.
In the future, imitate the substitution (replacement) technique when factoring the sum of or the difference
between two cubes.
Mnemonic device:
Find a way to remember and distinguish between the basic factoring formulas as in (1) and (2) above: You may
observe that in the factors on the right-hand side, there is only one minus, "-", involved; and the other signs
 are all "+" signs.

$a^3 + b^3 = \boxed{(a + b)(a^2 - ab + b^2)}$. <-- the minus sign precedes the ab-term. Also there is only one minus sign.

$a^3 - b^3 = \boxed{(a - b)(a^2 + ab + b^2)}$ <--The minus sign precedes the b-term . Also there is only one minus sign

To make sure you have the correct factorization, check by multiplying the binomial and the trinomial.

Lesson 52 Exercises

A Factor completely: $y^3 + 8$;   **2.** $x^3 + 27$;   **3.** $x^3 - 27$;   **4.** $8x^3 - 27$;   **5.** $3x^3 - 81$

Answers: **1.** $(y + 2)(y^2 - 2y + 4)$; **2.** $(x + 3)(x^2 - 3x + 9)$; **3.** $(x - 3)(x^2 + 3x + 9)$; **4.** $(2x - 3)(4x^2 + 6x + 9)$
     5. $3(x - 3)(x^2 + 3x + 9)$

Introductory Theme for Chapter 17A
(Next Chapter)
Straight Line
Theme: Two points

1. Why two points? Two points, because given or knowing two points, a straight line can be drawn by connecting the two points, using a straight edge and pencil.

2. Why two points? Two points, because given (knowing) two points; the slope, m, of the line segment connecting the two points $P_1(x_1, y_1)$ and $P_2(x_2, y_2)$ can be found by applying $m = \dfrac{y_2 - y_1}{x_2 - x_1}$

3a. Why two points? Two points, because if we know the **slope, m,** and the **y-intercept, b,** of the line, we can obtain two points and draw the graph of the line

3b Note: y–intercept,b implies the point $(0, b)$, By choosing a point (x, y) on a line, we have two points, and the slope (as well as an equation) of the line connecting the two pints $(0, b)$ and (x, y)

is given by $m = \dfrac{y - b}{x - 0}$ <--**slope = slope**

$mx = y - b$ or $\boxed{y = mx + b}$ <------**slope-intercept form** of the equation of a line

4. Why two points? Two points, because given or knowing two points an equation of the line segment connecting the two points $P_1(x_1, y_1)$ and $P_2(x_2, y_2)$ can be found

by applying $\boxed{y - y_1 = \left(\dfrac{y_2 - y_1}{x_2 - x_1}\right)(x - x_1)}$ (from $\dfrac{y - y_1}{x - x_1} = \dfrac{y_2 - y_1}{x_2 - x_1}$ <-- **is slope = slope**

or $\boxed{y - y_1 = m(x - x_1)}$ <------**point-slope form,** where $m = \dfrac{y_2 - y_1}{x_2 - x_1}$

5. Why two points? Two points, because given or knowing the two-intercept points $(a, 0)$, $(0, b)$ an equation of the line segment connecting the two points $P_1(a, 0)$ and $P_2(0, b)$ can be found by

applying $y - y_1 = \left(\dfrac{y_2 - y_1}{x_2 - x_1}\right)(x - x_1)$ to obtain $y - 0 = \dfrac{b - 0}{0 - a}(x - a)$; from $\dfrac{y - 0}{x - a} = \dfrac{b - 0}{0 - a}$)

or $y = \dfrac{b}{-a}(x - a)$ or $y = \dfrac{b}{-a}x + (\dfrac{b}{-a})(-a)$ or $\boxed{y = -\dfrac{b}{a}x + b}$ also $\dfrac{x}{a} + \dfrac{y}{b} = 1$ (Two intercept form 6.

Why two points? Two points, because given the graph (picture) of a line , we are given infinitely many points from which we can read the coordinates of any two points on the line and write an equation of a line by applying **3, 4** or **5** above. A picture is worth a thousand words

7. Why two points? Two points, because given or knowing two points; the **midpoint** of the line segment connecting the points $P_1(x_1, y_1)$ and $P_2(x_2, y_2)$ is given by x-coordinate, $x_m = \dfrac{x_1 + x_2}{2}$,

and the y-coordinate, $y_m = \dfrac{y_1 + y_2}{2}$

8. Why two points? Two points, because given or knowing) two points; $P_1(x_1, y_1)$, $P_2(x_2, y_2)$, the distance, d, between the two points on a line **in a plane** is given by

$$d = \sqrt{(x_2 - x_1)^2 + (y_2 - y_1)^2}$$

9. Why two points? Two points, because given or knowing two points, $P_1(x_1, y_1)$, $P_2(x_2, y_2)$ on each of two lines, the slopes m_1, m_2 as well as parallelism or perpendicularity can be determined. **The above theme** summarizes next two chapters.

CHAPTER 17A

Lesson 53A: **Rectangular Coordinate System**
Lesson 53B: **Drawing the Graph of a Straight Line**
Lesson 54: **Points on a Line; Slopes of lines; Intercepts**
Lesson 55: **Equations of Straight Lines**

Lesson 53A

Graphing in the Rectangular Coordinate System of Axes

We begin by means of an example. Let us consider the equation $y = 3x + 2$ (1)

If we let $x = 1$ in equation (1) then $y = 3(1) + 2$
$$= 5$$

Thus, when $x = 1$, $y = 5$

If we let $x = 2$ in equation (1) m then $y = 3(2) + 2$
$$= 8$$

Similarly, when $x = 2$, $y = 8$.

From the above example, we can say that for each value of x we choose, there is a corresponding value of y. By agreement, we can write the x– and y–values values as an ordered pair (x, y), where x is the first component and y is the second component.

Then for the solutions to $y = 3x + 2$, when $x = 1$, $y = 5$, we can represent the solution as the ordered pair $(1, 5)$.

For $x = 2$, $y = 8$, we can write the ordered pair $(2, 8)$

Similarly, when $x = 4$, $y = 14$, giving us the ordered pair $(4, 14)$.

The equation $y = 3x + 2$ has infinitely many solutions. The solution set consists of all ordered pairs.

If an ordered pair is a solution of a given equation, then these numbers when substituted in the equation for the unknowns, should make both sides of the equation equal. We should note that we could have chosen values for y and then calculate the corresponding values of x. In fact, in some instances, it may be more convenient to choose y and calculate x

Rectangular Coordinate System of Axes (Cartesian Coordinate System)

Consider a horizontal number line (scale) labeled as in Figure. On this line, all numbers to the right of the origin are positive, and all numbers to the left of the origin are negative.

Figure 1: Horizontal real number line

Consider also a vertical line (**Figure 2**). On this line, all numbers above the origin are positive,
and all numbers below the origin are negative.

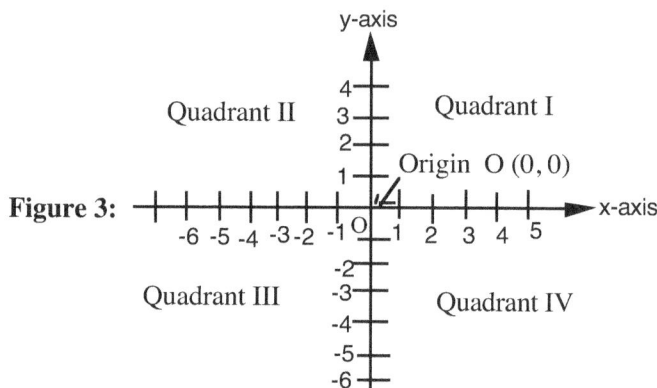

Figure 2:
Vertical line

Figure 3:

If we combine the horizontal and vertical lines at right angles so that the zero points (the origins) coincide, then we obtain what is called the rectangular coordinate system of axes. (**Figure 3**). We call the intersection of these lines, the origin and we label it O(0, 0). We call the horizontal number scale the x–axis, and we call the vertical number scale the y–axis. The two axes divide the plane (area) into four quadrants. These quadrants are numbered counterclockwise as quadrants I, II, III, and IV (i.e., first quadrant, second quadrant, third quadrant and fourth quadrant).

Graphing Ordered Pairs in a Rectangular Coordinate Plane

Geometrically, the first component (element) of each ordered pair is called the x-coordinate or the abscissa of the point. The second component (element) of the ordered pair is called the y-coordinate or ordinate of the point. The x-coordinate is the directed (positive or negative) distance from the y–axis and the y-coordinate is the directed distance from the x–axis. In quadrant I, both the x- and y-coordinates are positive. In quadrant II, the x–coordinate is negative but the y–coordinate is positive, In quadrant III, both coordinates are negative, In quadrant IV, the x–coordinate is positive, but the y–coordinate is negative.

Geometrically, each ordered pair represents a **point** in an x–y coordinate system of axes.

Plotting Points

In plotting points, we will be guided as follows:
1. We always count from the origin $O(0, 0)$.
2. If the x–coordinate is positive, we count horizontally to the right from the origin, but if the x–coordinate is negative, we count horizontally to the left from the origin.
3. If the y–coordinate is positive, we count vertically upwards from the origin, but if the y–coordinate is negative, we count vertically downwards from the origin. The counting is done visually, and we do not make any marks on the graph paper as we count.

Example .Graph the following ordered pairs (points) in a rectangular coordinate system of axes. $A(1,5)$. $B(-2,4)$, $P_1(-3,-4)$. $P_2(1,-2)$.

Solution

For $A(1,5)$: From the origin, count horizontally one unit to the right ($x=1$) and stop. From this point, count 5 units ($y=5$) vertically upwards and stop. Place a dot here and label this point as $A(1,5)$.

For $B(-2,4)$: From the origin, count horizontally 2 units to the left ($x=-2$) and stop, From this point, count 4 units ($y=4$) vertically upwards and stop. Place a dot here and label this point as $B(-2,4)$.

For $P_1(-3,-4)$: From the origin, count horizontally 3 units to the left ($x=-3$) and stop. From this point, count 4 units vertically downwards ($y=-4$) and stop. Place a dot here and label this point as $P_1(-3,-4)$.

Similarly, for $P_2(1,-2)$, count horizontally 1 unit to the right ($x=1$) and 2 units vertically downwards ($y=-2$) and stop. Place a dot here and label this point as $P_1(-3,-4)$.

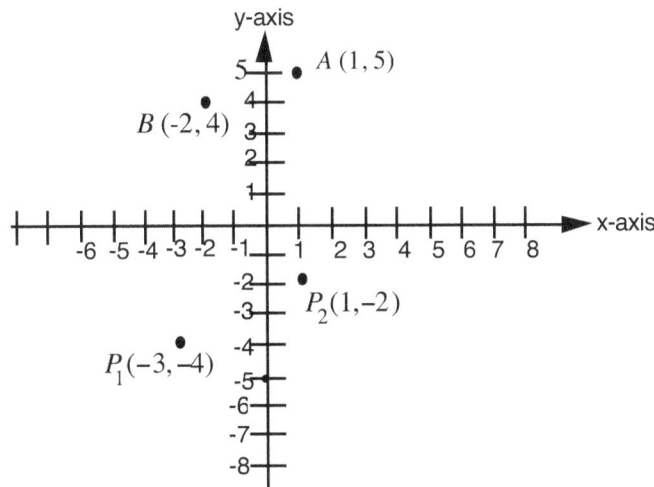

Figure 4

Lesson 53A Exercises

Graph the following ordered pairs in a system of rectangular axes.

$A(2,3)$, $B(-2,-3)$, $C(-4,5)$, $D(5,-4)$

Lesson 53B
Drawing the Graph of a Straight Line

Example Draw the graph of the line whose equation is given by : $y = 3x - 5$

We will cover three methods, namely a general method; the x- and y-intercepts method; and the intercept-slope method.

Method 1: General method

Step 1: Choose three convenient x-values and calculate the corresponding y-values to obtain ordered pairs.
(Actually, two ordered pairs will be sufficient; the third pair is used as a check: all **three** points must be in line)

choosing $x = 1$, $y = 3(1) - 5$ ordered pairs

$$y = 3 - 5$$
$$y = -2$$
$\Big\}$ ------> $(1,-2)$

choosing $x = 0$, $y = 3(0) - 5$ $\Big\}$ ------> $(0,-5)$
$$y = -5$$

choosing $x = -2$, $y = 3(-2) - 5$

$$y = -6 - 5$$
$$y = -11$$
$\Big\}$ ------> $(-2,-11)$

Step 2: Plot the points $(1,-2)$, $(0,-5)$ and $(-2,-11)$ on a rectangular coordinate system of axes
(on graph paper) and connect the points by a straight line.

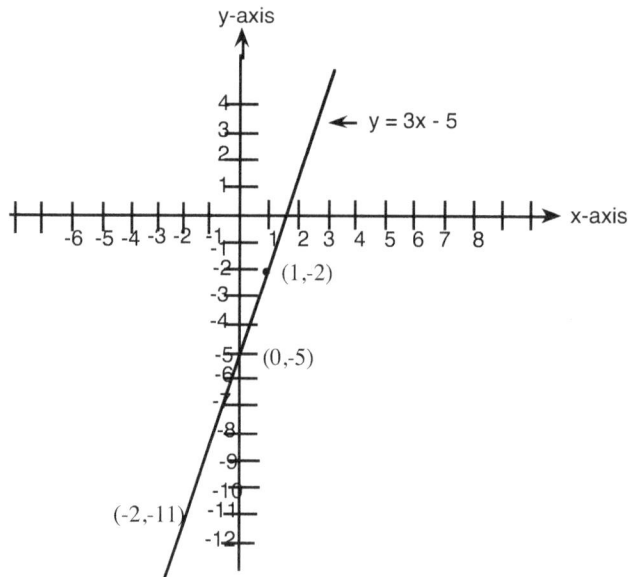

Note that from the given equation, the line has a positive slope and therefore the direction (see also page 260) of
the line (it **leans to the right**) in the figure is appropriate.

Note also that the above general method can be extended to draw the graphs of nonlinear equations such as $y = x^2$.

Method 2: The *x*- and *y*-intercepts method

We need two points to plot and connect by a straight line. The coordinates of the *x*-intercept yield one point and the coordinates of the *y*-intercept yield a second point..

Step 1: Find the *x*-intercept by letting $y = 0$ in the equation, $y = 3x$ -5, and solving for *x*.

Then $0 = 3x - 5$ and from which $x = \frac{5}{3}$. This step yields the point $(\frac{5}{3}, 0)$.

Step 2: Find the *y*-intercept by letting $x = 0$ in the equation and solving for *y*.

$$\left.\begin{array}{l} y = 3(0) - 5 \\ y = -5 \end{array}\right\} \text{This yields the point } (0, -5)$$

Step 3: Plot these two points from Steps 1 and 2 and connect them by a straight line.

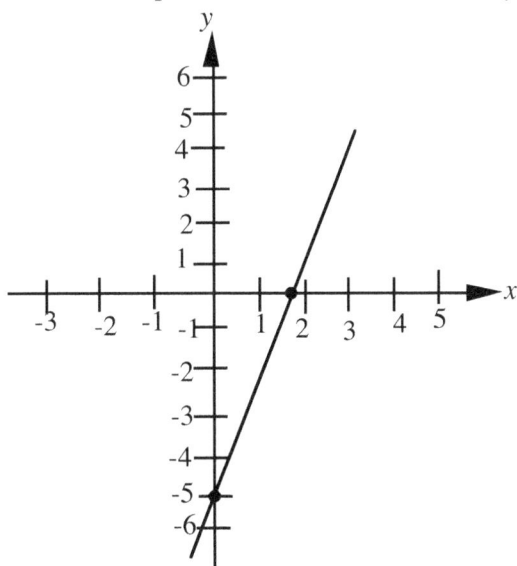

Figure: Graph of $y = 3x - 5$

Note that in Method 2, if the *y*-intercept is 0 (i.e., the equation is of the form $y = mx$), then we need to obtain another point by choosing another *x*-value (other than 0), calculate the corresponding *y*-value to obtain an ordered pair which we plot to locate a second point. An example of such a case is given by the line $y = 3x$. Note also that if the *y*-intercept is 0, the line passes through the origin.

Method 3: Slope-Intercept method

Here also, we need two points to plot and connect by a straight line. The *y*-intercept locates one point, and we use the slope to locate a second point.

Step 1: Solve the equation for *y*, and read the *y*-intercept.
  Since the equation has already been solved for *y*, we read the intercept, -5.
   This step locates the point $(0, -5)$.

Step 2: Graph the point $(0, -5)$.

Step 3: Since the slope $= +3 = \frac{3}{1}$, the vertical change is +3 and the horizontal change is +1.

  From the point $(0, -5)$, the *y*-intercept, count 3 units vertically upwards and 1 unit horizontally to the right, stop and place a dot here; this is a second point.
  Connect the two points by a straight line.

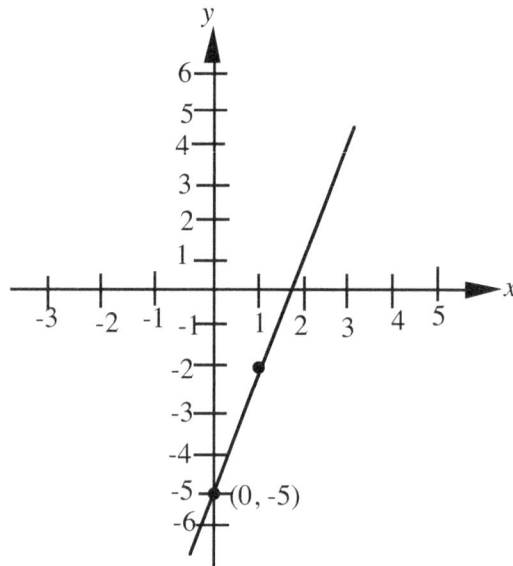

Figure: Graph of $y = 3x - 5$

Note In Step 3, we adopt the following convention.

1. Write the slope as a fraction if it is not a fraction. and if the slope is negative, give the minus sign to the numerator.

   For example, a slope of -3 is expressed as $\frac{-3}{1}$; $-\frac{4}{5}$ is expressed as $\frac{-4}{5}$, $-\frac{1}{2}$ is expressed as $\frac{-1}{2}$

2. In counting from the *y*-intercept, if the numerator (change in *y*) is positive, we count vertically upwards, but if it is negative, we count vertically downwards.
   For the denominator (change in *x*) we always count to the right since we have agreed to give the minus sign to the numerator.

   Examples: For a slope of $\frac{-2}{3}$, starting from the *y*-intercept, we count 2 units vertically downwards (negative numerator) and 3 units horizontally to the right. For a slope of $\frac{-1}{2}$, starting from the *y*-intercept, we count

   1 unit vertically downwards (negative numerator) and 2 units horizontally to the right. For a slope of $\frac{4}{5}$,

   starting from the *y*-intercept, we count 4 units upwards (positive numerator) and 5 units to the right

Lesson 53B Exercises

Draw the graphs of the following lines whose equations are given:

1. $y = -3x + 5$ **2**. $6y = 18 - 3x$ **3**. $2y - 6x = -3$

Answers: Graphs of the lines **1.** $y = -3x + 5$ **2.** $6y = 18 - 3x$ **3.** $2y - 6x = -3$

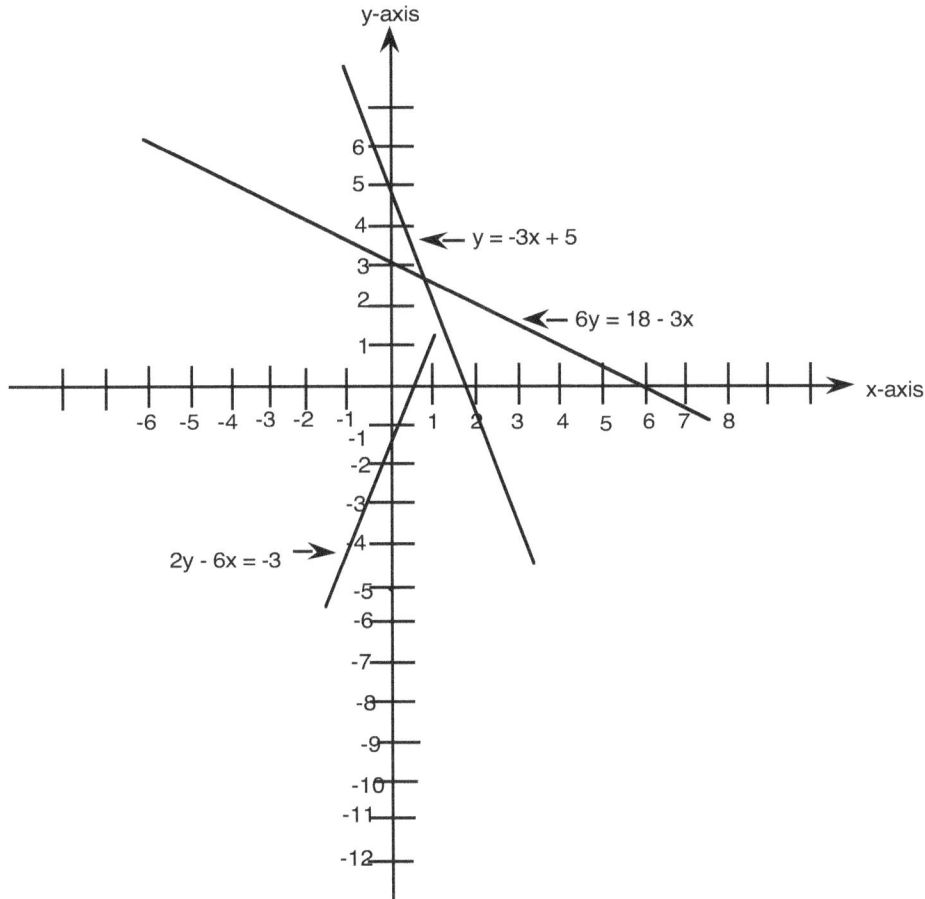

Lesson 54

Points on a Line; Slopes of Lines; x- and y-intercepts

Determining if a point is on a line whose equation is given (See also p.180)

Example   Which of the points $(5, 2)$ and $(3, 11)$ is on the line whose equation
is given by $y = 3x + 2$.

We will substitute each ordered pair in turn in the given equation. The ordered pair whose
substitution makes the left-hand side of the equation **equal** to the right-hand side **is on** the line.

Step 1:   Checking for the point (5,2)

Substitute the ordered pair (5,2) in

$y = 3x + 2$ $\qquad$ $(x = 5, y = 2)$

The question mark "?" above the equality symbol

then, $2 \overset{?}{=} 3(5) + 2$ $\qquad$ is there to show that at this step, we do not yet know
if the left-hand side and the right-hand side of the

$2 \overset{?}{=} 15 + 2$ $\qquad$ equation are equal. In checking solutions, it is a good
practice to use the question mark.

$2 \overset{?}{=} 17$ No   (or $2 = 17$ is False)

The left-hand side is **not** equal to the right-hand side of the equation and therefore the
point (5,2) is **not** on the given line.

We also say (algebraically) that the ordered pair (5,2) is **not** a solution of the
the equation $y = 3x + 2$

Step 2: Checking for the point (3,11)

Substitute (3, 11) in $y = 3x + 2$

then, $11 \overset{?}{=} 3(3) + 2$

$11 \overset{?}{=} 9 + 2$

$11 = 11$ True (We no longer use the question mark since it is obvious that
the left- hand side equals the right-hand side of the equation)

Since the left-hand side **is equal to** the right-hand side of the equation,
the point (3,11) **is on** the line $y = 3x + 2$.

Algebraically, we also say that
the ordered pair (3,11) **is a solution** of the equation $y = 3x + 2$.

In fact, the last example (question) could have been posed as : Which of the following ordered
pairs is a solution of the equation $y = 3x + 2$? (a) (5,2), (b) (3,11)

Solution: Proceed exactly as steps 1 and 2 of example above.

In calculations, the following have the same interchangeable implications:

1. A line passes through a given point.

2. A line contains a given point.

3. A given point is on a line.

4. The coordinates of a given point satisfy an equation of a line.

5. The coordinate pair of a given point is a solution of an equation of a line.

Example: The problem " find an equation of the line **passing through** the points.$(2, 3)$ and
$(6, 8)$" is **equivalent to** "find an equation of the line **containing** the points $(2, 3)$ and $(6, 8)$.

Finding the slope of a line given the coordinates of two points on the line

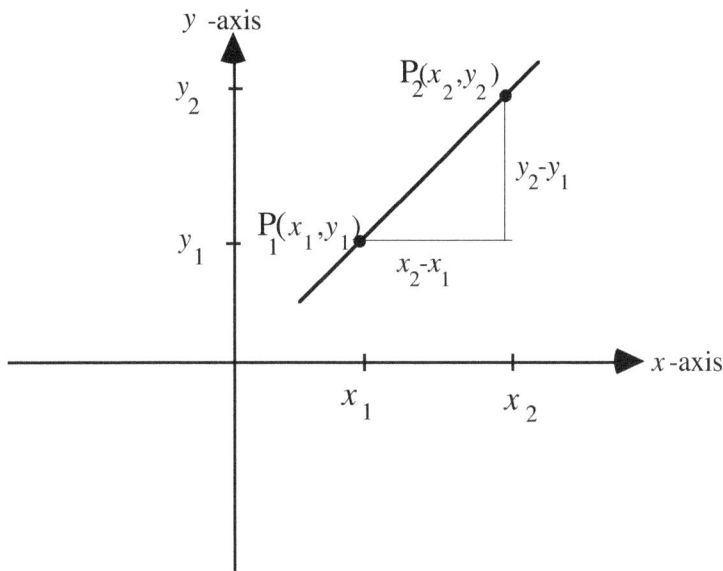

The slope, m, of the line segment connecting the points $P_1(x_1, y_1)$ and $P_2(x_2, y_2)$ is given by

$$m = \frac{y_2 - y_1}{x_2 - x_1}$$

Example   Find the slope of the line passing through the points (2,3) and (6,8).

Solution :   Identify the first point as $P_1(2,3)$ and the second point as $P_2(6,8)$

Then $x_1 = 2, y_1 = 3;$ and $x_2 = 6, y_2 = 8$

Applying the slope formula, $m = \frac{y_2 - y_1}{x_2 - x_1}$

$$m = \frac{8 - 3}{6 - 2}$$

$$m = \frac{5}{4}$$

The slope is $\frac{5}{4}$.

Special Cases of the Slopes of Lines
The special cases are for horizontal and vertical lines

Slope of a horizontal Line
The **slope** of a **horizontal** line is **zero**, since the vertical change is zero. For example, the
slope, m, of the horizontal line in Figure **1**, below, by the slope formula is $m = \frac{3-3}{5-2} = \frac{0}{3} = 0.$

$\left(\text{Note that } m = \frac{\text{vertical change}}{\text{horizontal change}} = \frac{\text{change in } y}{\text{change in } x}\right)$

Slope of a Vertical Line

The **slope** of a **vertical** line is **undefined** since the horizontal change is zero. For example, the slope, m, of the vertical line in Figure **2**, below, by the slope formula is $m = \frac{2-(-3)}{4-4} = \frac{5}{0}$ is undefined.

Figure 1

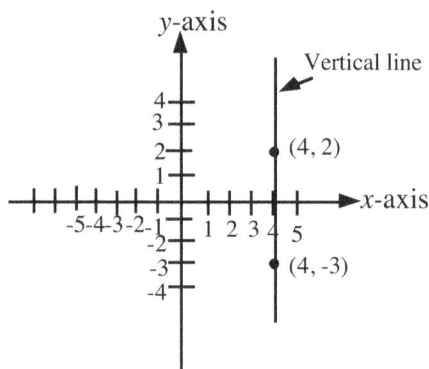

Figure 2

The *x*- and *y*-intercepts of a line

The ***x*-intercept** of a line is the x-coordinate of the point where the line crosses or meets the x-axis. At this point, $y = 0$. (Geometrically, we should also say that the x-intercept is the point where the line intersects the x-axis, and in which case, we must specify two coordinates, with the y-coordinate always being zero.) See Figure 1 below.

The *y*-intercept of a line is the y-coordinate of the point where the line crosses or meets the y-axis. At this point, $x = 0$. (Geometrically, we should also say that the y-intercept is the point where the line intersects the y-axis, and in which case, we must specify two coordinates, with the x-coordinate always being zero.) See Figure 1 below.

Example

In Figure 1 below, the x-intercept is -3; and the y-intercept is 2. The x- and y--intercepts are at (-3,0) and (0,2) respectively.

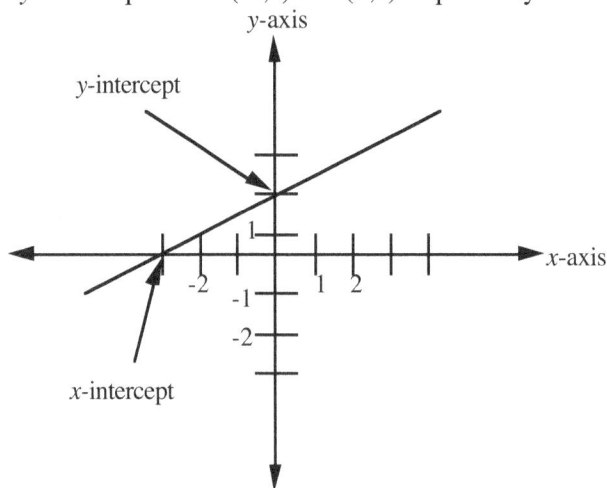

Figure 1

Example   Find the x- and y-intercepts for the graph of $3x + 2y = 12$

For the x-intercept: Let $y = 0$ in $3x + 2y = 12$.
$\qquad\qquad\qquad$ Then $3x + 2(0) = 12$; $3x = 12$; and from which $x = 4$.
$\qquad\qquad$ The x-intercept is 4.

For the y-intercept: Let $x = 0$ in $3x + 2y = 12$.
$\qquad\qquad\qquad$ Then $3(0) + 2y = 12$; $2y = 12$; and from which $y = 6$.
$\qquad\qquad$ The y-intercept is 6.

Finding the slope and the y-intercept of a line, given the equation of the line

Example 1 Find the slope and the y-intercept of the line whose equation is given by $y = -5x + 6$.

The slope-intercept form of the equation of a straight line is given by $y = mx + b$, where m is the slope and b is the y-intercept.

$\qquad$ Compare $y = -5x + 6$ to $\qquad$ **Note:** (The slope is the coefficient of the x- term
$\qquad\qquad\quad$ $y = mx + b$, $\qquad\qquad\qquad\qquad$ if the equation is in slope-intercept form.)

$\qquad$ then, the slope, $m = -5$,
$\qquad$ and the y-intercept $= 6$

Example 2   Find the slope and the y-intercept of the line whose equation is given by
$\qquad\qquad\qquad$ $5y + 4x = 20$

Step 1: Solve the equation for y (that is, write the equation in slope-intercept form).
$$5y + 4x = 20$$
$$\underline{\quad -4x \qquad\qquad -4x \quad}$$
$$5y = -4x + 20$$
$$\frac{5}{5}y = \frac{-4}{5}x + \frac{20}{5}$$
$$y = -\frac{4}{5}x + 4$$

Step 2: By comparison with $y = mx + b$, the slope is $-\frac{4}{5}$ $\qquad$ (the coefficient of the x-term).

$\qquad\qquad\qquad$ and the y-intercept is $+4$ or 4.

Lesson 54 Exercises

A Determine which of the points (2,4), (-2,5), (-2,-16) are on the line whose equation is given by
$$y = 5x - 6$$

Answer: The points (2,4), and (-2,-16) are on the given line

B Find the slope of each line passing through the given points:

1. (2, 3) and (6, 4) **2.** (3, -2) and (2, -4) **3.** (5, 2) and (1, -1)

4. (2, 2) and (-3, 3) **5.** (-4, 2) and (5, 2) **6.** (-3, -5) and (-3, -6)

7. Find c so that the slope of the line passing through the points $(c, 2c)$ and $(-6, 8)$ is 1.

Answers: 1. $\frac{1}{4}$; **2.** 2 ; **3.** $\frac{3}{4}$; **4.** $-\frac{1}{5}$; **5.** 0 ; **6.** undefined; **7.** $c = 14$

C **1..** Find the slope of the line passing through the points (3,2) and (5,9).

  2. The points (2,2) and (-5,6) are on a line whose slope we want to determine. What is the slope of this line?

  3. Find the slope of the line containing the points (1,6) and (2,8).

  4. Find the slope of the line passing through the points (1,-1), (3,3) and (-2,-7).

Answers: 1. slope = $\frac{7}{2}$ **2.** slope = $-\frac{4}{7}$ **3.** slope = 2 ; **4.** slope = 2

D Find the slope and the y-intercept of the following:

  1.. $y = -6x + 4$; **2.** $-7 + 5x = y$; **3.** $y = -\frac{x}{2} + 8$

Answers: 1. slope = - 6, y-intercept = 4; **2.** slope = 5, y-intercept = -7; **3.** slope = $-\frac{1}{2}$, y-intercept = 8

E Find the slope and y-intercept: **1.** $3x + 8y = 32$ **2.** $2y - 6x = 21$; **3.** $y = -3x + 4$

  4. $y = \frac{1}{2}x - 3$; **5.** $3y = 6x + 12$; **6.** $y = \frac{x}{3} + 5$; **7.** $2y + 5x = 8$; **8.** $4x + 3y = 12$

Ans 1. slope = $-\frac{3}{8}$, y-intercept = 4; **2.** slope = 3, y-intercept = $\frac{21}{2}$; **3.** slope = - 3 ; y-intercept = 4;

  4. slope = $\frac{1}{2}$, y-intercept = -3 ; **5.** slope = 2, y-intercept = 4; **6.** slope = $\frac{1}{3}$, y-intercept = 5;

  7. slope = $-\frac{5}{2}$, y-intercept = 4; **8.** slope = $-\frac{4}{3}$, y-intercept = 4;

Lesson 55
Equations of Straight Lines

There are a number of approaches that we can use in **finding equations** of straight lines. Each approach depends upon what we are given.

Generally, we can easily write down an equation of a line if we know any of the pairs of properties or characteristics of (information about) a line in the formulas presented below.

These formulas are based on finding two different expressions for the slope of a line and equating these expressions to each other.

1. If we know the **slope,** m, and the **y-intercept, b,** of the line, we can apply the slope-intercept form, $y = mx + b.$

2. If we know the slope, m, and the coordinates (x_1, y_1) of one point on the line, we can apply the point-slope form, $(y - y_1) = m(x - x_1).$

3. If we know the coordinates (x_1, y_1) and (x_2, y_2) of two points on the line, we can apply $(y - y_1) = \dfrac{y_2 - y_1}{x_2 - x_1}(x - x_1).$

4. If we are given the graph (picture), we can determine any of the above pairs of properties (information) from the graph, and then apply the formulas in the above cases; however, if we want to memorize one more formula, we can apply the equation

$y = -\dfrac{b}{a}x + b,$ where a is the x-intercept and b is the y-intercept. (If $b = 0$ or $a = 0$ use the other methods.)

(This equation is another form of the two-intercept form: $\dfrac{x}{a} + \dfrac{y}{b} = 1$)

We will now cover the above **four** cases in detail with examples.

Case 1: Finding an equation of a line given the slope and the y-intercept of the line

Example 1   Find an equation of the line with slope 3 and y-intercept of 4.

Solution   We will apply the slope-intercept form of the equation
of a straight line, $y = mx + b.$
Substituting the slope, $m = 3$, and the y-intercept, $b = 4$ in this equation,
$$y = 3x + 4$$

Example 2   Find an equation of the line with slope - 2 and y-intercept of - 5.

Solution: Substituting $m = -2, b = -5$ in $y = mx + b$, we obtain
the equation $y = -2x - 5$

Case 2: **Finding an equation of a line given the slope of the line and the coordinates of one point on the line**

Example 1 Find an equation of the line passing through the point $(3,-2)$ and having a slope 4.

Solution. We will cover two methods.

Method 1

Step 1: Find the y-intercept, b, by substituting $x_1 = 3$, $y_1 = -2$, $m = 4$ in

$$y = mx + b$$
$$\text{then, } -2 = 4(3) + b$$
$$-2 = 12 + b$$
$$-14 = b$$

Step 2: Now, since we know that $m = 4$, $b = -14$, we can apply $y = mx + b$ (as in Case 1, above) and then, $y = 4x - 14$

Method 2

We will use the point-slope form of the equation of a straight line which is given by

$$y - y_1 = m(x - x_1) \text{ <-------point-slope form.} \tag{1}$$

Substituting the coordinates, $x_1 = 3$, $y_1 = -2$ and slope, $m = 4$ in equation (1), we obtain

$$y - (-2) = 4(x - 3)$$

$$y + 2 = 4(x - 3) \tag{2}$$

Equation (2) is the point-slope form of the required equation.

By solving equation (2) for y, we obtain

$$y = 4x - 14 \quad \text{<----------- slope-intercept form} \tag{3}$$

Equation (3) is the slope-intercept form of the required equation.
In this form, we can, by inspection, determine the slope and the y-intercept.

The author recommends the slope-intercept form (for this course), unless otherwise specified.

Case 3: **Finding an equation of a line given the coordinates of two different points on the line**

Example Find an equation of the line passing through the points (2,1) and (-3,-4).

Solution
    Step 1: Find the slope, m, with $x_1 = 2, y_1 = 1, x_2 = -3 \ y_2 = -4$

$$m = \frac{y_2 - y_1}{x_2 - x_1}$$ Scrapwork

$$m = \frac{-4 - 1}{-3 - 2}$$ $\frac{-4-1}{-3-2} = \frac{-5}{-5} = 1$

$$m = 1$$

Now, $m = 1, x_1 = 2, y_1 = 1, x_2 = -3, y_2 = -4$, and we can apply the procedure in Case 2.

    Step 2: Applying Method 2 of Case 2 and
          Substituting $m = 1, x_1 = 2, y_1 = 1$ in $y - y_1 = m(x - x_1)$, we obtain
$$y - 1 = 1(x - 2)$$

$$y = x - 1$$

Also, If we substitute $m = 1, x_2 = -3, y_2 = -4$ in $y - y_2 = m(x - x_2)$, we obtain

$$y - (-4) = 1(x - (-3))$$

$$y + 4 = x + 3$$
$$y = x - 1$$

Again, we obtain the same equation. An equation of the line is $y = x - 1$.
We conclude also that any of the two given points can be used in finding an equation of the straight line. From the above solution, we can state a general formula for Case 3 as

$$(y - y_1) = \frac{y_2 - y_1}{x_2 - x_1}(x - x_1)$$

Case 4: Finding an equation of a line given the graph (picture) of the line

If we are given the graph (picture), we can determine any of the above pairs of properties (information) from the graph (see p. 257, case 4).

It is also useful to be able to tell immediately from the graph if the line has a positive slope, a negative slope, a zero slope, or an undefined slope.

The signs of the slopes of lines

The lines in Figure 1 have positive slopes.   The lines in Figure 2 have negative slopes.

Figure 1

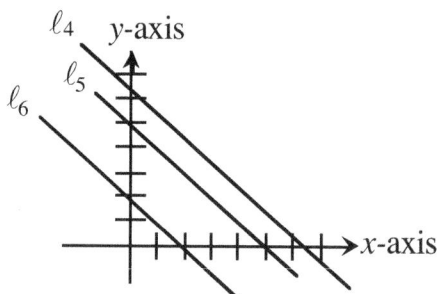

Figure 2

Lines ℓ_1, ℓ_2, and ℓ_3 have positive slopes. (Simply, these lines **lean** to the **right** in the page; or these lines rise as one moves ones head from the left to the right in the page.)

Lines ℓ_4, ℓ_5 and ℓ_6 have negative slopes. (Simply, these lines **lean** to the **left** in the page; or these lines fall as one moves ones head from the left to the right in the page.)

Note: The slope of a **horizontal line** is zero. The slope of a **vertical line** is undefined. (For the equations of horizontal and vertical lines, see p.263-264)

Example 1   Find an equation of the line whose graph is given below.

Method 1: Apply $y = -\dfrac{b}{a}x + b,$

where a is the x-intercept and b is the y-intercept. (see also p.254)
Step 1: From the graph, we read the values of a and b: $.a = 3, b = -4$
Step 2: Substitute $a = 3, b = -4$ in

$$y = -\frac{b}{a}x + b.$$

Then $y = -\dfrac{-4}{3}x + (-4)$

(Make sure you take into account the **minus sign** that comes with the formula.)

$y = +\dfrac{4}{3}x - 4$ (Two minus signs make

the x-term positive)

An equation of the line is $y = \dfrac{4}{3}x - 4$.

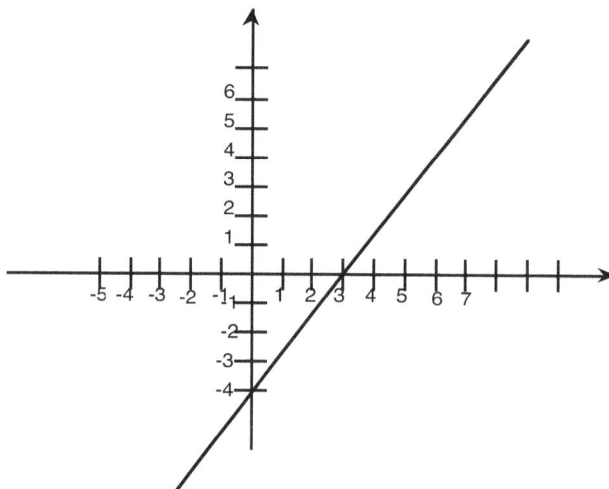

Method 2

Step 1: By inspection, the line has a positive slope. (See figure 1 above.)

Step 2: Slope, $m = \dfrac{\text{vertical change}}{\text{horizontal change}}$

$\qquad = \dfrac{\text{change in } y}{\text{change in } x}$

Pick any two convenient points on the line, say, $P_1(0,-4)$ and $P_2(3,0)$ (see figure 2, below).

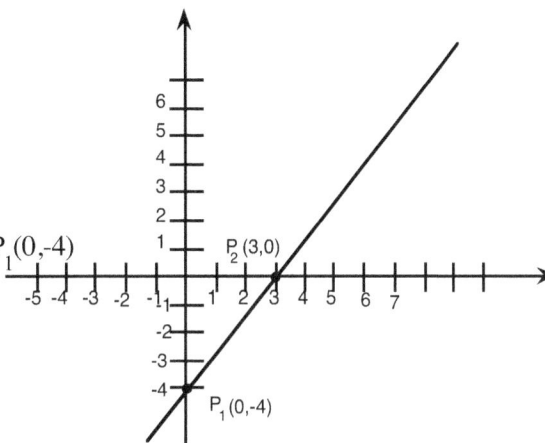

Figure 2

Finding the vertical and horizontal changes by counting.

We will assume that we want to go from the point P_1 to the point P_2, and count the number of equal intervals (units) we will travel vertically, and the number of equal intervals (units) we will travel horizontally. Then, we will go up vertically 4 equal intervals and horizontally to the right 3 equal intervals. Divide the vertical change, 4, by the horizontal change, 3 , to obtain the slope $\dfrac{4}{3}$

Note that the slope is positive. (see p.260, Figure 1).

Step 3: From the graph , the y-intercept, $b = -4$

Step 4: With $m = \dfrac{4}{3},\ b = -4,$ apply $y = mx + b$

$\qquad$ then, $y = \dfrac{4}{3}x - 4$

Therefore, an equation of the line is

$y = \dfrac{4}{3}x - 4$

In fact, once we have been able to specify the points $P_1(0,-4)$, $P_2(3,0)$, we

can apply m $= \dfrac{y_2 - y_1}{x_2 - x_1} = \dfrac{0 - (-4)}{3 - 0} = \dfrac{+4}{3}$

to find the slope $m = \dfrac{4}{3}$,

and then apply $y - y_1 = m(x - x_1)$ to obtain

$y + 4 = \dfrac{4}{3}(x - 0) \qquad (x_1 = 0, y_1 = -4)$

$y + 4 = \dfrac{4}{3}x - 0$

$y = \dfrac{4}{3}x - 4$ <------- slope-intercept form of the equation of a straight line.

Example 2 Find an equation of the line whose graph is given below.

Method 1 Apply $y = -\dfrac{b}{a}x + b$ where a is the x-intercept and b is the y-intercept.

Step 1: From the graph, $a = 3$, $b = 6$

Step 2: Substitute $a = 3$, $b = 6$ in $y = -\dfrac{b}{a}x + b$

Then $y = -\dfrac{6}{3}x + 6$

$y = -2x + 6$ < ---slope-intercept form of the equation of a straight line.

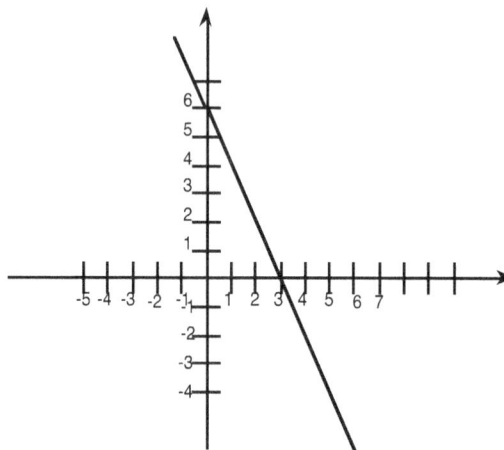

Figure

Method 2

Step 1: Note that the slope is negative.
(See p.260, Figure 2.)

Step 2: Pick any two convenient points on the line, say P_1 and P_2

Step 3: Assume that we want to travel (vertically and horizontally only) from P_1 to P_2.
Then, we will go down 6 units and then to the right 3 units.

The slope, $m = -\dfrac{6}{3} = -2$

(The slope is negative since the line leans to the left. See page 260, Fig..2)

Step 4: Read the y-intercept, $b = 6$

Step 5: Apply $y = mx + b$, with $m = -2$, $b = 6$

Then, an equation of the given graph is
$y = -2x + 6$

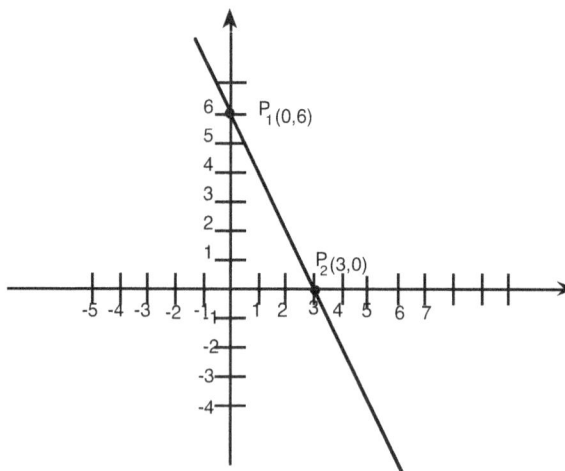

Figure

Method 3   From $P_1(0,6)$ and $P_2(3,0)$

Step 1: $m = \dfrac{y_2 - y_1}{x_2 - x_1}$ $(x_1 = 0, y_1 = 6)$

$= \dfrac{0 - 6}{3 - 0} = \dfrac{-6}{3} = -2$

$m = -2$

Step 2: Apply $y - y_1 = m(x - x_1)$

$y - 6 = -2(x - 0)$

$y - 6 = -2x + 0$

$y = -2x + 6$

An equation of the line is $y = -2x + 6$

Note that, in Method 3, the negative sign results solely from the calculation and we do not have to know in advance the sign of the slope)

In **Case 4**, the author recommends **Method 1.**

Special Cases of the Equations of Straight Lines and their Graphs

The equation $y = mx + b$, (with m defined and, $m \neq 0$) geometrically represents oblique lines (i.e., lines which are neither vertical nor horizontal). The special cases of the equation of a straight line are for horizontal and vertical lines. In these cases, either the x- or the y-term is missing.

The Equation and Graph of a Horizontal Line

The equation of a horizontal line is of the form $y = b$, where b is the y-intercept.

The slope, $m = 0$, since the vertical change is zero.

Substituting $m = 0$, in $y = mx + b$,

$$y = 0x + b$$

$$y = b \qquad\qquad (1)$$

Equation (1) means that as x varies, y remains unchanged.

Examples: Sketch the graphs of the following lines:

1. $y = 3$, **2.** $y = -4$. **3.** $y = 0$.

Solution: The line $y = 3$ is the horizontal line passing through $(0,3)$. See Figure 1, below.

The line $y = -4$ is the horizontal line passing through $(0,-4)$.

The line $y = 0$ is the horizontal line along the x-axis (i.e., the line $y = 0$ is the x-axis)

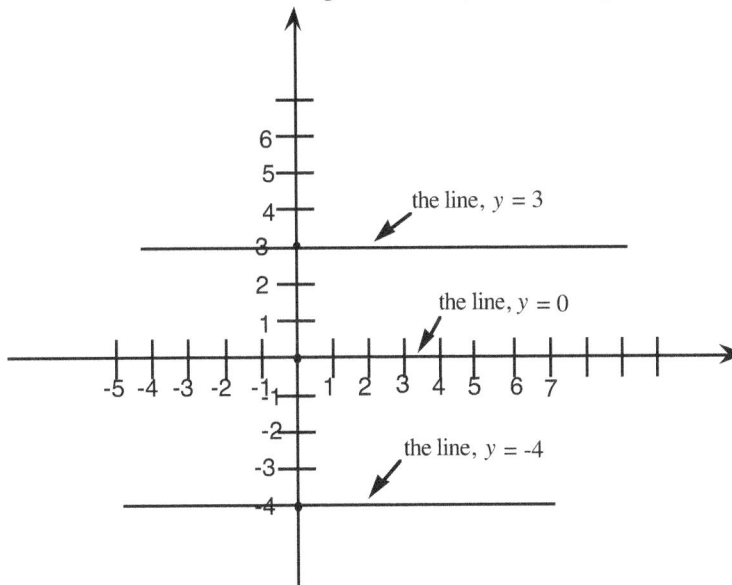

Figure 1

The Equation and Graph of a Vertical Line

The slope of a vertical line is undefined since the horizontal change is zero.
The equation of a vertical line is of the form $x = a$, where a is the x-intercept.
This form of the equation means that as y varies, x remains unchanged.

Examples   Sketch the lines with the following equations:

      1. $x = 2$,   **2.** $x = 0$,   **3.** $x = -4$

Solution: See Figure 2, below.

   1. The line $x = 2$ is the vertical line passing the point (2,0).

   2. The line $x = 0$ is the vertical line along the y-axis (i.e., the line $x = 0$ is the y-axis).

   3. The line $x = -4$ is the vertical line passing through the point (-4,0).

   4. The line $x = -2$ is the vertical line passing through the point (-2,0).

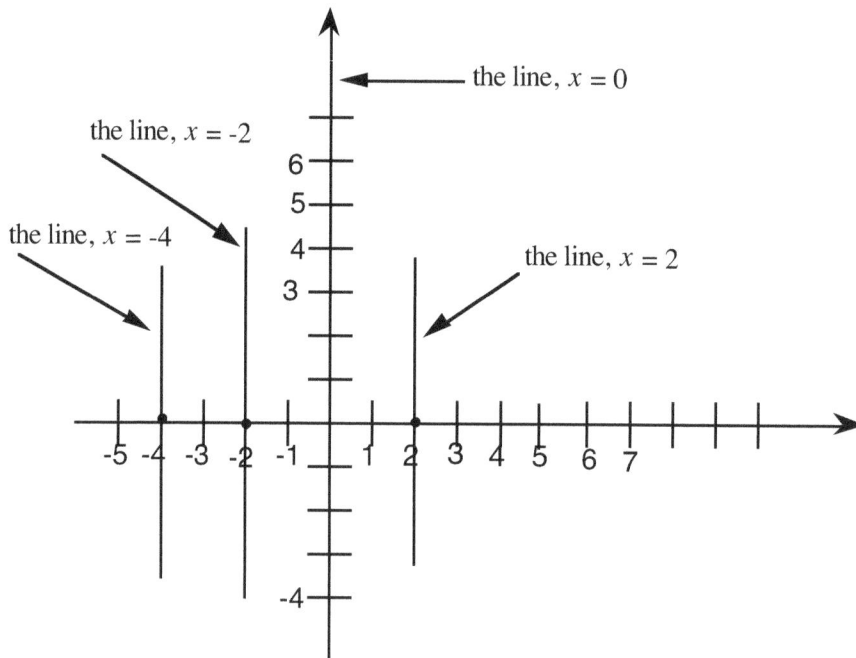

Figure 2

Lesson 55 Exercises

A 1. Find an equation of the line with slope 7 and *y*-intercept -2.

  2. Find an equation of the line with slope $\frac{2}{3}$ and *y*-intercept 5.

  3. A line has a *y*-intercept -3 and a slope of 6. Find an equation for this line.

Answers: 1. $y = 7x - 2$;   2. $y = \frac{2}{3}x + 5$;   3. $y = 6x - 3$

B 1. Find an equation of the line with slope -4 and passing through the point (3,-2).

2. A line passes through the point (-1,-7) and has a slope of 5. Find an equation for this line.

Answers: **1** .$y = - 4x + 10$;   **2** . $y = 5x - 2$

C **1.** Find an equation of the line passing through the points (2,2) and (-5,6)

  2. If the points (2,-5) and (-3,1) are on a certain line, find an equation for this line.

  3. Find an equation of the line with slope -3 and y-intercept 8.

  4. Find an equation of the line with slope 2 and passing through the point (1,6)

Answers: **1.** $y = -\frac{4}{7}x + \frac{22}{7}$;   **2.** $y = -\frac{6}{5}x - \frac{13}{5}$;   **3.** $y = -3x + 8$;   **4.** $y = 2x + 4$.

D Find the equations (slope-intercept forms) of the lines ℓ_1 , ℓ_2 , ℓ_3, ℓ_4

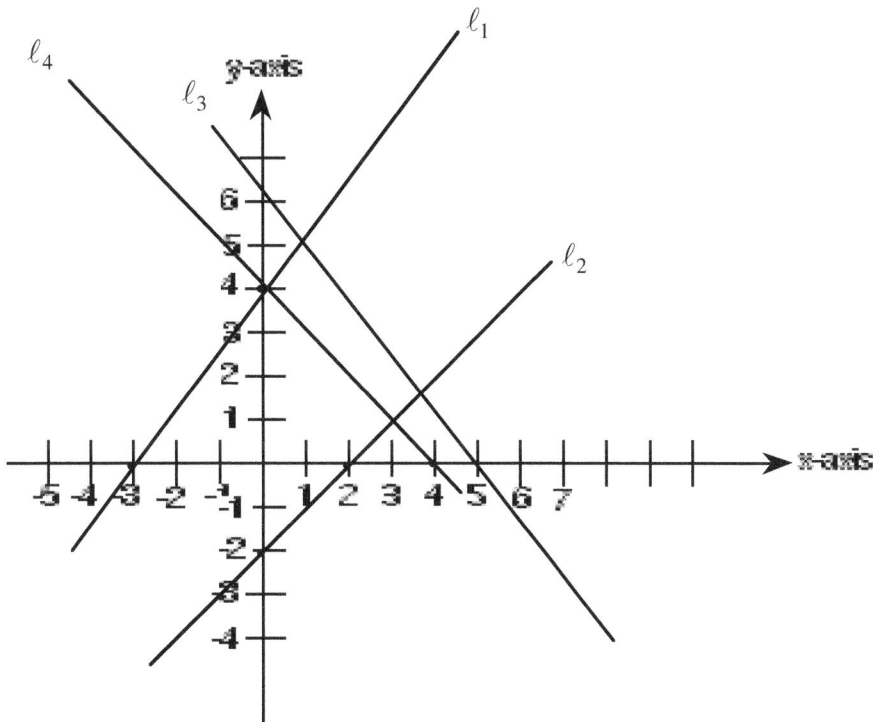

Answers: For ℓ_1: $y = \frac{4}{3}x + 4$. For ℓ_2 : $y = x - 2$.   For ℓ_3: $y = -\frac{5}{4}x + \frac{25}{4}$ For ℓ_4: $y = -x + 4$

E 1. Find an equation of the horizontal line passing through the point (3,-4).

2. Sketch the graph the line in Problem 1.

3. Does the line in Problem 1 have a y-intercept ? If yes, find it.

4. Sketch the graph of the line $y = -5$.

5. Sketch the graph of the line $y = 2$.

Answers: **1.** $y = -4$; **2.** See Fig 1, above and imitate. ;**3.** Yes. y-intercept $= -4$; **4. & 5.** Imitate Fig. 1 above

F 1. Find an equation of the vertical line passing through the point (2,-3).

2. Sketch the graph the line in Problem **1**.

3. Does the line in Problem **1** have a y-intercept ? If yes, find it.

Answers: **1.** $x = 2$;   **3.** No.

G 1. Draw the graph of the line $x = 4$.   **2.** Draw the graph of the line $x = -1$.

H **1.** Find an equation of the line with slope -5 and y-intercept 2.

2. Find an equation of the line whose y-intercept is -3 and whose slope is $\frac{1}{2}$.

3. Find an equation of the line whose slope is -2 and which passes through the point (4, -5).

4. A line passes through the point (-3, 2) and has slope 4. Find an equation for this line.

5. Find an equation of the line passing through the points (5, 4) and (8, 6).

Answers: **1.** $y = -5x + 2$; **2.** $y = \frac{1}{2}x - 3$; **3.** $y = -2x + 3$; **4.** $y = 4x + 14$; **5.** $y = \frac{2}{3}x + \frac{2}{3}$

CHAPTER 17B

Lesson 56A: **Equations of Parallel lines and Perpendicular Lines**
Lesson 56B: **Midpoint of a Line; Distance between Points;** Distance Formula
Lesson 56C: **Applications: 1. Distance from a point to a line whose equation is given**
2. Perpendicular Bisector of a Line

Lesson 56A

Parallel Lines: Slopes and Equations

Note: If two lines are parallel, then they have the same slope.

If the slope of one line is m_1, then the slope of the other parallel line is also m_1.

Example Find an equation of the line passing through the point (2,-5), and parallel to the
line $y = -3x + 2$.

Solution

The slope of the line whose equation is given = -3. (comparing $y = -3x + 2$. with $y = mx + b$.
$m = -3$) The slope of the line whose equation we want to find = -3 (since two parallel lines have the
same slope).

Applying $y - y_1 = m(x - x_1)$; with $m = -3, x_1 = 2, y_1 = -5$

$$y - (-5) = -3(x - 2)$$
$$y + 5 = -3x + 6$$
$$y = -3x + 1 \; \text{<------slope-intercept form.}$$

Perpendicular Lines: Slopes and Equations

If two lines are perpendicular, then their slopes are negative reciprocals of each other; or simply,

if m_1 and m_2 are their slopes, then $m_1 = -\dfrac{1}{m_2}$ or $m_2 = -\dfrac{1}{m_1}$ or $m_1 \cdot m_2 = -1$

Example Find an equation of the line passing through the point (-2,4) and
perpendicular to the line $y = 3x - 5$.

Solution

Step 1: Determine the slope of the line whose equation we want to find.

The slope of $y = 3x - 5$ is 3

The slope of the line whose equation we are to find is $-\dfrac{1}{3}$ (Since the two lines are perpendicular).

Step 2: Apply $y - y_1 = m(x - x_1)$

with $m = -\dfrac{1}{3}$ and $x_1 = -2, y_1 = 4$

Then, $y - 4 = -\dfrac{1}{3}(x - (-2))$

$$y - 4 = -\dfrac{1}{3}(x + 2)$$
$$y - 4 = -\dfrac{1}{3}x - \dfrac{2}{3}$$
$$y = -\dfrac{1}{3}x - \dfrac{2}{3} + 4$$
$$y = -\dfrac{1}{3}x + \dfrac{10}{3} \; \text{<--------slope-intercept form.}$$

A **note** about finding the slope in the above problem:

The slope of the given line is 3
To find the slope of the other (perpendicular line) line,

Step 1. Invert 3 to obtain $\dfrac{1}{3}$

Step 2. Change the sign. Since $\dfrac{1}{3}$ has a plus sign, after the change, we obtain $-\dfrac{1}{3}$

Similarly, if the slope of one line were -5, the slope of the other perpendicular line would be $+\dfrac{1}{5}$.

If the slope of one line were $-\dfrac{1}{4}$, the slope of the other perpendicular line would be +4.

Summary: Invert and change the sign, or change the sign and invert.

Distinction between the point $x = a$ and the line $x = a$

We must distinguish between, for example, the point $x = 2$ and the line $x = 2$. We use a dot to locate the point $x = 2$ on a number line.

For the graph of the line $x = 2$, we draw a vertical line through the point $(2, 0)$

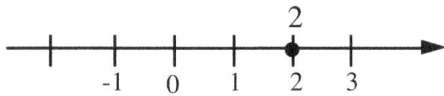

Figure: The graph of the point $x = 2$.

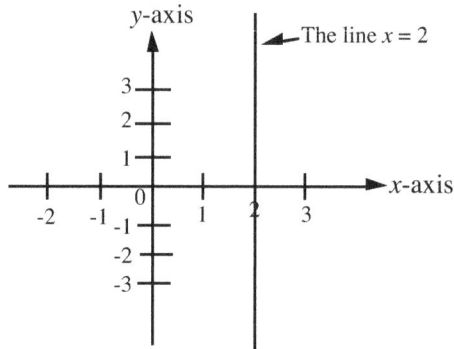

Figure: The graph of the line $x = 2$.

Similarly, the point $y = 2$ and the line (horizontal line) $y = 2$ are shown in Figures .. and respectively.

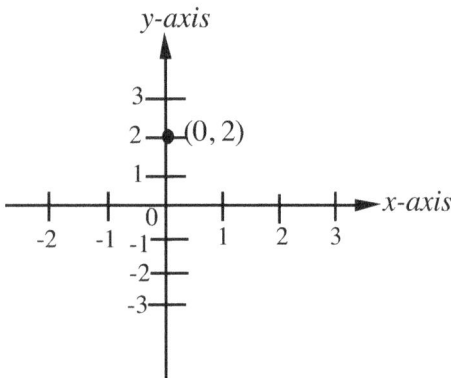

Figure: The graph of the point $y = 2$.

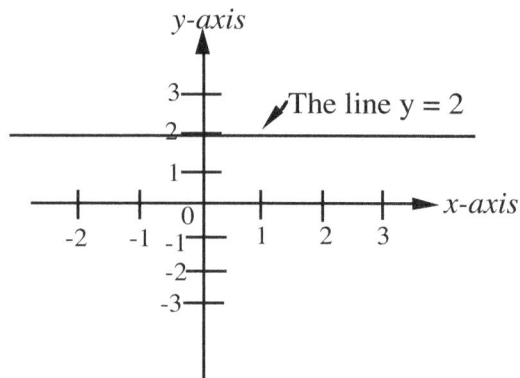

Figure: The graph of the line $y = 2$.

Lesson 56A Exercises

A 1. Find an equation of the line through the point $(3, -4)$ and perpendicular to the line $y = 2x + 5$.

2. Find an equation of the line with the same y–intercept as the line $2y = 3x + 8$ and parallel to the line $y = -4x + 2$.

Answers: **1.** $y = -\frac{1}{2}x - \frac{5}{2}$; **2.** $y = -4x + 4$

B 1. **By** graphing, distinguish between the point $x = 2$ on the number line and the line $x = 2$.

2. By graphing, distinguish between the point $y = 2$ on the y-axis and the line $y = 2$.

1.

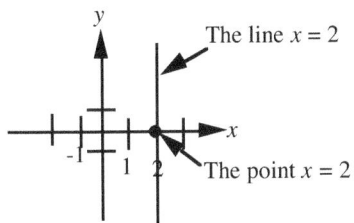

The line $x = 2$

The point $x = 2$

2.

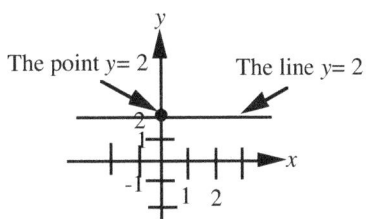

The point $y = 2$

The line $y = 2$

Lesson 56B
Midpoint of a Line; Distance between Points; Distance Formula

Midpoint of a Line

The **midpoint** of (Figure below), the line, P_1P_2, connecting the points $P_1(x_1, y_1)$ and $P_2(x_2, y_2)$ has the coordinates given by the following formulas:

The x-coordinate, x_m, of the midpoint is given by $x_m = \dfrac{x_1 + x_2}{2}$

The y-coordinate, y_m of the midpoint is given by $y_m = \dfrac{y_1 + y_2}{2}$

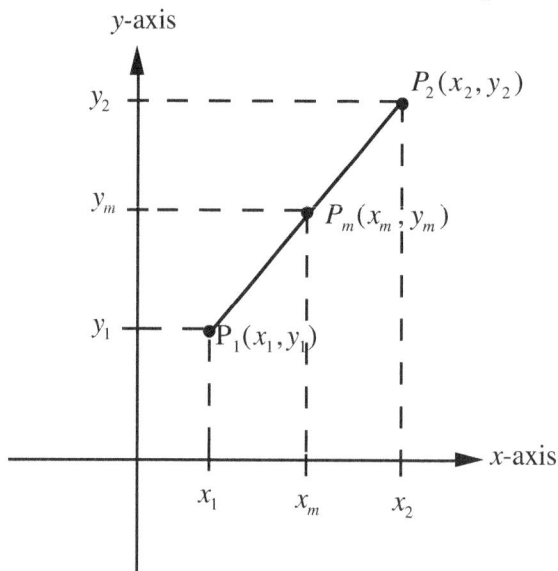

Example Find the coordinates of the mid-point of the line connecting the points $(3, 2)$ and $(-9, 4)$.
Solution

$$\text{The } x\text{-coordinate of the mid-point} = \frac{x_1 + x_2}{2} = \frac{3 + (-9)}{2} \qquad (x_1 = 3, x_2 = -9)$$
$$= \frac{-6}{2}$$
$$= -3$$

$$\text{The } y\text{-coordinate of the mid-point} = \frac{y_1 + y_2}{2} = \frac{2 + 4}{2} \qquad (y_1 = 2, y_2 = 4)$$
$$= \frac{6}{2}$$
$$= 3$$

Therefore $(x_m, y_m) = (-3, 3)$ where x_m and y_m are the coordinates of the mid-point.
The mid-point is at $(-3, 3)$.

Distance Between two Points on a Line

The distance between two points A and B on a line is equal to the number of units (equal intervals) between A and B. If we denote the distance between A and B by $d(A, B)$. Then, we can write

$$d(A, B) = |A - B|$$

that is, the distance between A and B is the absolute value of the difference between A and B.

Figure: Distance on a horizontal line

Figure: Distance on a vertical line

Example

$$d(-3, 1) = |-3 - 1|$$
$$= |-4|$$
$$= 4$$

We conclude also that the distance between two points on the same **horizontal line** or the same **vertical line** can be found algebraically.

However, the distance between two points in a plane cannot be found algebraically, but must be found geometrically as discussed in the next section.

Distance Between two Points on a Line in a Plane: Distance Formula

The distance, d, between the points $P_1(x_1, y_1)$ and $P_2(x_2, y_2)$ on a line **in a plane** (Fig.1) is given by

$d = \sqrt{(x_2 - x_1)^2 + (y_2 - y_1)^2}$ (By applying the Pythagorean theorem to the right triangle, and solving for d)

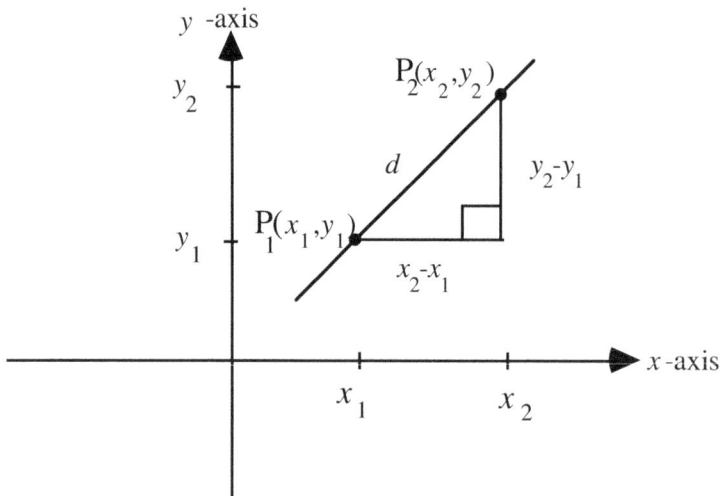

Figure 1

Example Find the distance between the points (-2,3) and (4, -5) .

Solution Apply the distance formula:

$$d = \sqrt{(x_2 - x_1)^2 + (y_2 - y_1)^2} \qquad (1)$$

(where d is the distance between the points $P_1(x_1, y_1)$ and $P_2(x_2, y_2)$).

Substituting $x_1 = -2, y_1 = 3, x_2 = 4, y_2 = -5$ in equation (1) above,

$d = \sqrt{(4 - (-2))^2 + (-5 - 3)^2}$

$d = \sqrt{(4 + 2)^2 + (-8)^2}$

$= \sqrt{(6)^2 + (-8)^2}$

$= \sqrt{36 + 64}$

$= \sqrt{100}$

$= 10$

$\therefore$ the distance between the given points is 10 units.

Lesson 56B Exercises

A   **1.** Find the distance between the points (3,4) and (5, -1) .

2. Find the distance between the points (-4, 2) and (-6, -3)

Answers: **1**. $\sqrt{29}$; **2**. $\sqrt{29}$

B   Find the distance between each given pair of points.

1. (2, 3) and (−7, 4);   **2**. (−4, 3) and (6, 2);   **3.** (−3, 5) and (7, 10)

4. (a, b), and (c, d);   **5**. (−3, 5), and (4, 7)

6. (a) Find the directed distance from (4,−5) to (4,−7);

(b) Find the distance from (4,−5) to (4,−7)

Find the coordinates of the mid–point of each of the following points :

7. (2,−4), and (4, 0);   **8.** (3, 2), and (4, 6).

9. (−4, − 4), and (5, 3);   **10.** (5, 12), and (0, 0).

Answers: **1**. $\sqrt{82}$; **2**. $\sqrt{101}$; **3**. $5\sqrt{5}$; **4**. $\sqrt{a^2 + b^2 + c^2 + d^2 - 2ac - 2bd}$; **5**. $\sqrt{53}$; **6**. (a) −2 ; (b) 2.
7. $(3, -2)$; **8**. $(\frac{7}{2}, 4)$; **9**. $(\frac{1}{2}, -\frac{1}{2})$; **10**. $(\frac{5}{2}, 6)$.

Lesson 56C
Applications
1. Distance from a point to a line whose equation is given
2. Perpendicular Bisector of a Line

Distance from a point to a line whose equation is given

Find a formula for the distance, d, from the point $A(x_0, y_0)$ to the line $y = mx + b$.

Solution

Method 1

Given: The point $A(x_0, y_0)$ and the line $y = mx + b$.

Required: To find a formula for the distance, d, between the point $A(x_0, y_0)$ and the line $y = mx + b$.

Construction: Draw the perpendicular line from $A(x_0, y_0)$ to meet the line $y = mx + b$ (line ℓ_2) at E.

Also, draw $\overline{AC} \perp$ to the x-axis.

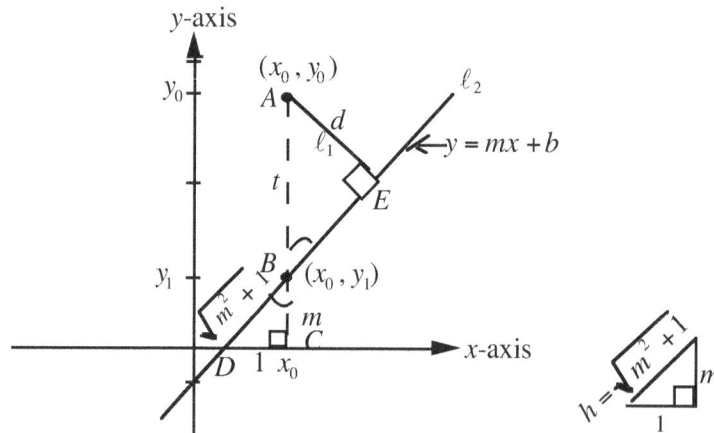

Step 1: Let $h = DB$. If the slope of $y = mx + b$ is $m = \frac{m}{1}$, then $BC = m$, $DC = 1$, and $h = \sqrt{m^2 + 1}$.

$\triangle ABE$ and $\triangle DBC$ are similar: (Two angles of $\triangle ABE$ are congruent to two angles of $\triangle DBC$)

$\therefore \dfrac{AE}{DC} = \dfrac{AB}{DB}$ (Corresponding sides of similar triangles are in proportion)

$\dfrac{d}{1} = \dfrac{t}{\sqrt{m^2 + 1}}$ (1)   ($AE = d$, $AB = t$, $DC = 1$, $DB = \sqrt{m^2 + 1}$)

Step 2: $t = y_0 - y_1$

$t = y_0 - (mx_0 + b)$ ($y_1 = mx_0 + b$ is obtained by substituting x_0 in $y = mx + b$)

$t = y_0 - mx_0 - b$

Step 3: Substitute for t in equation (1)

$\dfrac{d}{1} = \dfrac{y_0 - mx_0 - b}{\sqrt{1 + m^2}}$

$d = \dfrac{y_0 - mx_0 - b}{\sqrt{1 + m^2}}$

$\therefore d = \dfrac{|y_0 - mx_0 - b|}{\sqrt{m^2 + 1}}$

Note: If given $Ax + By + C = 0$, $y = -\frac{A}{B}x - \frac{C}{B}$

Then $d = \dfrac{|Ax_0 + By_0 + C|}{\sqrt{A^2 + B^2}}$ by substitution.

(We take the absolute value of the numerator in case the numerator is negative)

Method 2

Given: The point $A(x_0, y_0)$ and the line $y = mx + b$.

Required: To find a formula for the distance, d, between the point $A(x_0, y_0)$ and the line $y = mx + b$.

Construction: Draw the perpendicular line from the point $A(x_0, y_0)$ to meet the line $y = mx + b$ ℓ_2 at E.

Step 1: Let lines ℓ_1, and ℓ_2 meet at (r, s). (Figure below)

Let the distance between the points (x_0, y_0) and $(r, s) = d$.

Then $d^2 = (x_0 - r)^2 + (y_0 - r)^2$ or $d = \sqrt{(x_0 - r)^2 + (y_0 - r)^2}$

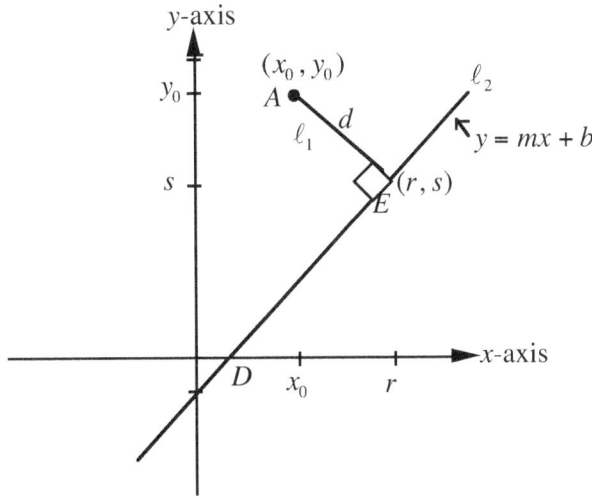

Step 2: We shall now express d in terms of x_0, y_0, m, and b only.

(That is we shall eliminate r, and s from the formula)

The slope of line $\ell_1 = \dfrac{y_0 - s}{x_0 - r}$. (1)

The slope of line $\ell_2 = m$. (from $y = mx + b$)

Therefore slope of $\ell_1 = -\dfrac{1}{m}$ (2) (Since ℓ_1 and ℓ_2 meet at right angles at E, the slopes are negative reciprocals of each other)

Equating right-hand-sides of (1) and (2) to each other, we obtain

$\dfrac{y_0 - s}{x_0 - r} = -\dfrac{1}{m}$

$my_0 - ms = -x_0 + r$ (cross-multiplying)

$r = my_0 + x_0 - ms$ (3) (solving for r)

Since (r, s) is on the line $y = mx + b$

$s = mr + b$ (4) (substituting r for x and s for y.)

Step 3: We now solve equations (3) and (4) simultaneously for r and s.

Substitute for s from (4) in (3): $my_0 + x_0 - m(mr + b) = r$

Now, we solve for r: $my_0 + x_0 - m^2 r - mb = r$

$my_0 + x_0 - mb = m^2 r + r$

$my_0 + x_0 - mb = r(m^2 + 1)$

$$r = \frac{my_0 + x_0 - mb}{m^2 + 1} \qquad (5)$$

Substitute for r from (5) in (4). Then we obtain

$$s = m\left[\frac{my_0 + x_0 - mb}{m^2 + 1}\right] + b$$

$$= \frac{m^2 y_0 + mx_0 - m^2 b + m^2 b + b}{m^2 + 1}$$

$$s = \frac{m^2 y_0 + mx_0 + b}{m^2 + 1} \qquad (6)$$

Step 4: We now substitute right-hand sides of equations (5) and (6) for r and s respectively in the formula $d^2 = (x_0 - r)^2 + (y_0 - r)^2$. Then we obtain

$$d^2 = \left\{x_0 - \left[\frac{my_0 + x_0 - mb}{m^2 + 1}\right]\right\}^2 + \left\{y_0 - \left[\frac{m^2 y_0 + mx_0 + b}{m^2 + 1}\right]\right\}$$

$$= \frac{\{m^2 x_0 + x_0 - my_0 - x_0 + mb\}^2}{(m^2 + 1)^2} + \frac{\{m^2 y_0 + y_0 - m^2 y_0 - mx_0 - b\}^2}{(m^2 + 1)^2}$$

$$= \frac{\{m^2 x_0 - my_0 + mb\}^2}{(m^2 + 1)^2} + \frac{\{y_0 - mx_0 - b\}^2}{(m^2 + 1)^2}$$

$$= \frac{\{-m(-mx_0 + y_0 - b)\}^2}{(m^2 + 1)^2} + \frac{\{y_0 - mx_0 - b\}^2}{(m^2 + 1)^2}$$

$$= \frac{\{(-m)^2(-mx_0 + y_0 - b)\}^2}{(m^2 + 1)^2} + \frac{\{y_0 - mx_0 - b\}^2}{(m^2 + 1)^2}$$

$$= \frac{m^2(y_0 - mx_0 - b)^2}{(m^2 + 1)^2} + \frac{(y_0 - mx_0 - b)^2}{(m^2 + 1)^2}$$

$$= \frac{m^2(y_0 - mx_0 - b)^2 + (y_0 - mx_0 - b)^2}{(m^2 + 1)^2}$$

$$= \frac{(y_0 - mx_0 - b)^2(m^2 + 1)}{(m^2 + 1)^2}$$

$$= \frac{(y_0 - mx_0 - b)^2(m^2 + 1)}{(m^2 + 1)^2}$$

$$d^2 = \frac{(y_0 - mx_0 - b)^2}{(m^2 + 1)}$$

$$d = \sqrt{\frac{(y_0 - mx_0 - b)^2}{(m^2 + 1)}}$$

$$= \frac{\sqrt{(y_0 - mx_0 - b)^2}}{\sqrt{m^2 + 1}}$$

$$\therefore d = \frac{|y_0 - mx_0 - b|}{\sqrt{m^2 + 1}} \qquad \text{(noting that } \sqrt{a^2} = |a|\text{)}$$

and the derivation is complete.

Application 2: Perpendicular Bisector of a Line

Example

Find the slope-intercept form of the equation of line L_1 (Figure below) which passes through the mid-point C of line L_2 given that line L_2 passes through the two points A (1 ,4) and B (7, 8)

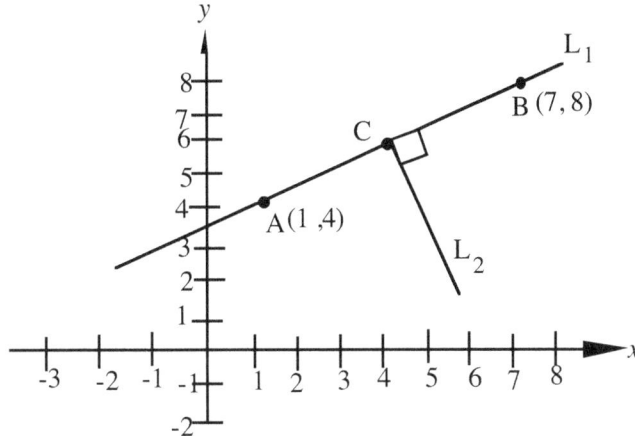

Solution

Step 1 Find the slope of line L_2

$$m = \frac{y_2 - y_1}{x_2 - x_1}$$

$$= \frac{8 - 4}{7 - 1}$$

$$= \frac{4}{6}$$

$$m = \frac{2}{3}$$

Step 2: Find the slope of L_1

Since L_1 is perpendicular to L_2, the slope of L_2 is $-\frac{3}{2}$ ($m_2 = -\frac{1}{m_1}$. The lines are perpendicular)

Step 3 : Find the coordinates of the midpoint C

$$x_m = \frac{1 + 7}{2} = 4 \quad y_m = \frac{4 + 8}{2} = 6$$

The mid-point is at $(4, 6)$

Step 4: now find the slope-intercept form of the equation of the line with slope $-\frac{3}{2}$ and passing through the point $(4, 6)$.

Applying $y - y_1 = m(x - x_1)$, we obtain

$y - 6 = -\frac{3}{2}(x - 4)$ <-------- point-slope form

$y - 6 = -\frac{3}{2}x + 6$

$y = -\frac{3}{2}x + 6 + 6$

$y = -\frac{3}{2}x + 12$ <-----slope-intercept form

Lesson 56C: 1. Distance from a point to a line. 2. Perpendicular Bisector of a Line

Test # 10 -Student's Self-Test (**Always**, **Test yourself before you are tested**)

Attempt all questions on clean sheets of paper, **Do not write in the book** Show all necessary work.

---

1. Draw the graphs of the following lines whose equations are given:

    (a) $y = -2x + 5$

    (b) $3y = 18 - 3x$

    (c) $2y - 6x = -8$

2. Determine which of the points $(2, 3)$, $(-2, 5)$, $(-2, -16)$ are on the line whose equation is given by $y = 5x - 6$.

---

3. Find the slope of each line passing through the given points:

    (a) $(5, 6)$ and $(3, 2)$;

    (b) $(2, -4)$ and $(3, -2)$.

4. Find the slope of each line passing through the given points:

    (a) $(1, -1)$; and $(5, 2)$.

    (b) $(-3, 3)$ and $(2, 2)$.

---

5. Find the slope of each line passing through the given points:

    (a) $(5, 2)$ and $(-4, 2)$;

    (b) $(-3, -6)$ and $(-3, -5)$.

6. Find the slope of the line passing through the points $(5, 9)$ and $(3, 2)$.

---

7. The points $(-5, 6)$ and $(2, 2)$ are on a line whose slope we want to determine. What is the slope of this line?

8. Find the slope of the line containing the points $(2, 8)$ and $(1, 6)$.

---

9. Find the slope of the line passing through the points $(1, -1)$, $(3, 3)$ and $(-2, -7)$.

10. Find the slope and the y-intercept of the following:

    (a) $y = -6x + 4$;   (b) $-7 + 5x = y$;

    (c) $y = -\frac{x}{2} + 8$

---

11. Find the slope and the y-intercept:

    (a) $4x + 8y = 32$

    (b) $5y - 6x = 21$

12. Find the slope and the y-intercept:

    (a) $y = -2x + 4$

    (b) $y = \frac{1}{5}x - 3$

---

13. Find the slope and the y-intercept:

    (a) $4y = 6x + 12$

    (b) $y = \frac{x}{3} + 5$

15. Find an equation of the line with slope 5 and y-intercept -3.

14. Find the slope and the y-intercept:

    (a) $3y + 5x = 8$

    (b) $5x + 4y = 12$

16. Find an equation of the line with slope $\frac{2}{3}$ and y-intercept 5.

17. A line has a y-intercept -2 and a slope of 6. Find an equation for this line.

18. Find an equation of the line with slope -3 and passing through the point (5, -2).

19. A line passes through the point (-1, -7) and has a slope of 5. Find an equation for this line.

20. Find an equation of the line passing through the points (2, 2) and (-5, 6)

21 (a) If the points (2, -5) and (-3, 1) are on a certain line, find an equation for this line.

(b) Find an equation of the line with slope -3 and y-intercept 8.

22. Find equations for lines m and l

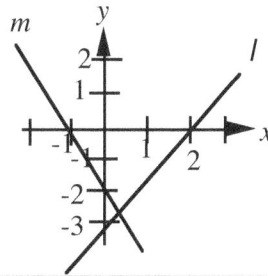

23. (a) Find an equation of the line with slope 2 and passing through the point (1, 6)

(b) Sketch the graph of the line $y = 2$

(c) Draw the graph of the line $x = 4$.

(d) Draw the graph of the line $x = -1$.

24. (a) Find an equation of the horizontal line passing through the point (3, -4).

(b) Find an equation of the vertical line passing through the point (2, -3).

(c) Does the line in (b) have a y-intercept ? If yes, find it.

25. (a) Does the line in Problem **23** (b) have a y-intercept ? If yes, find it.

(b) Sketch the graph of the line $y = -5$.

Bonus:

(a) Sketch the graph of the line $y = -4$.
(b) Sketch the graph of the line $y = 1$.

Answers: **2.** The point (-2, -16) ; **3.** (a) 2 ; (b) 2; **4.** (a) $\frac{3}{4}$; (b) $-\frac{1}{5}$; **5.** (a) 0 ;

(b) undefined; **6.** $\frac{7}{2}$; **7.** $-\frac{4}{7}$; **8.** 2 ; **9.** 2 ; **10.** (a) Slope:-6, y-intercept: 4;

(b) Slope: 5, y-intercept:: -7; (c) Slope: $-\frac{1}{2}$, y-intercept : 8 ; **11.** (a) slope: $-\frac{1}{2}$, y-intercept: 4;

(b) Slope: $\frac{6}{5}$, y-intercept: $\frac{21}{5}$; **12.** (a) Slope: -2, y-intercept : 4 ; (b) Slope: $\frac{1}{5}$, y-intercept: -3;

13. (a) Slope: $\frac{3}{2}$, y-intercept: 3 ; (b) Slope: $\frac{1}{3}$, y-intercept: 5; **14.** (a) Slope: $-\frac{5}{3}$, y-intercept: $\frac{8}{3}$;

(b) Slope: $-\frac{5}{4}$, y-intercept: 3; **15.** $y = 5x - 3$; **16.** $y = \frac{2}{3}x + 5$; **17.** $y = 6x - 2$;

18. $y = -3x + 13$; **19.** $y = 5x - 2$; **20.** $y = -\frac{4}{7}x + \frac{22}{7}$ or $4x + 7y = 22$;

21. (a) $y = -\frac{6}{5}x - \frac{13}{5}$ or $6x + 5y = -13$; (b) $y = -3x + 8$; **22.** For the line m: $y = -2x - 2$,

for the line l: $y = \frac{3}{2}x - 3$; **23.** (a) $y = 2x + 4$; **24.** (a) $y = -4$; (b) $x = 2$; (c) No;

25. (a) Yes. y-intercept: 2.

CHAPTER 18

Solving Systems of Linear Equations (Simultaneous Equations)

Lesson 57: **Addition Method; Substitution Method; Equality Method**

Lesson 58: **Graphical Method**

Lesson 57

Solving by the Addition (or Subtraction) Method

Example 1   Solve for x and y by the elimination method.

$$\begin{cases} 2x + 4y = 6 ....(A) \\ 3x + 5y = 7 ....(B) \end{cases}$$

You may read the following discussion first, or you may skip it and read it after studying the solution.

Discussion: The above system is an example of a general form of a linear system of two equations involving two variables.

We will combine (**add** or subtract) the two equations or their equivalent equations so that either the x-term or the y-term drops out (**is eliminated**). We will then have an equation of a single variable; this equation is then solved for this variable. We then substitute the value obtained in one of the original equations to find the value of the other variable. Finally, we will check the x- and y-values in the original system of equations to determine if they satisfy the system of equations. If they do, then, these values are accepted as solutions of the system of equations, otherwise they are rejected as solutions.

First, let us eliminate x (or the x-term) by adding the equivalent forms of the two equations. We would like the coefficients of the x-terms to be the same, but of opposite signs. More preferably, we want the coefficients of the x-terms to be the LCM of the coefficients of the x-terms of the original equations. We will multiply equation (A) by 3 and equation (B) by -2. Note that 3 is the coefficient of the x-term of (B) and 2 is the coefficient of the x-term of equation (A). We use a minus sign for the 2, so that the new coefficients obtained have different signs. We could also give the minus to the 3 and then, we would multiply equation (A) by -3 and equation (B) by 2. We use different signs for the "3" and "2" because the given x-terms have the same sign. If the x-terms had different signs, then the multipliers 3 and 2 must have the same sign, and then we would have multiplied equation (A) by 3 and equation (B) by 2.

Solution: Now, let us apply the above discussion to solve the given system of linear equations.

$$\begin{cases} 2x + 4y = 6 ....(A) \\ 3x + 5y = 7 ....(B) \end{cases}$$

(A) × 3: $3(2x + 4y = 6)$
(B) × -2: $-2(3x + 5y = 7)$

$$6x + 12y = 18 .... (C)$$
$$\underline{-6x - 10y = -14 ....(D)}$$

adding (C) and (D) $2y = 4$ (The x-term has been **eliminated**)
$$y = 2$$

Next, we find the value of x:

Substitute $y = 2$ in (A) or (B) and solve for x.

We substitute in (A): $2x + 4(2) = 6$
$$2x + 8 = 6$$
$$\underline{\quad - 8 \quad -8 \quad}$$
$$2x = -2$$
$$x = -1$$
$$\therefore \ x = -1, \ y = 2.$$

We will check the values $x = -1$, $y = 2$ in equations (A) and (B) to determine if this ordered pair (-1,2) is a solution of the system of equations.

Check #1: For $x = -1$ and $y = 2$ in (A):

we have $2(-1) + 4(2) \overset{?}{=} 6$

$-2 + 8 \overset{?}{=} 6$

$6 = 6$ True

.: The ordered pair (-1, 2) is a solution of (A).

Check #2: For $x = -1$ and $y = 2$ in (B):

we have $3(-1) + 5(2) \overset{?}{=} 7$

$-3 + 10 \overset{?}{=} 7$

$7 = 7$ True

.: The ordered pair (-1, 2) is a solution of (B).

Since the ordered pair (-1, 2) is a solution of both equation (A) and (B).
the ordered pair (-1, 2) is the solution of the given system of equations.

Let us solve the same system of equations by eliminating y first (instead of x).

$$\begin{cases} 2x + 4y = 6 ....(A) \\ 3x + 5y = 7 ....(B) \end{cases}$$

(A) × 5: $5(2x + 4y = 6)$ ---------> $10x + 20y = 30$ ....(C)

(B) × –4: $-4(3x + 5y = 7)$ --------> $\underline{-12x - 20y = -28}$ .... (D)

(C) + (D): $-2x = 2$

$$\frac{2x}{2} = \frac{2}{-2}$$

$$x = -1$$

To find y, substitute $x = -1$ in (A)

$$2(-1) + 4y = 6$$
$$-2 + 4y = 6$$
$$\underline{+2 \qquad +2}$$
$$4y = 8$$
$$y = 2$$
$$\therefore \ x = -1, \ y = 2.$$

Again, we obtain the same solution as in the case in which we eliminated x first.
The above problem was a most general example. Sometimes we may not have to multiply any of the given equations by any (nonzero constant) number, but rather, we may just add the two equations to eliminate one of the variables, as the next example shows.

Example 2   Solve for x and y: $\begin{cases} 2x + 5y = 4 ....(A) \\ 3x - 5y = 11 ....(B) \end{cases}$

We can see that the coefficients of the y-terms are the same (absolute values) but of opposite signs.

Step 1:   Add (A) and (B)   (The objective here is to combine the two equations so that

$2x + 5y = 4$ either the x-term or the y-term drops out)

$\underline{3x - 5y = 11}$

$5x + 0 = 15$

Step 2:   Solve for x:

$$5x = 15$$
$$x = 3$$

To find y, substitute $x = 3$ in (A) and solve for y

$$2(3) + 5y = 4$$
$$6 + 5y = 4$$
$$\underline{-6 \qquad -6}$$
$$5y = -2$$
$$\frac{5y}{5} = \frac{-2}{5}$$

Thus $x = 3$, $y = -\frac{2}{5}$

Checking in (A):   Substitute $x = 3$, $y = -\frac{2}{5}$

$$2(3) + 5\frac{(-2)}{5} \overset{?}{=} 4$$
$$6 - 2 \overset{?}{=} 4$$
$$4 = 4 \qquad \text{True}$$

Checking in (B):   Substitute $x = 3$, $y = -\frac{2}{5}$

$$3(3) - 5\frac{(-2)}{5} \overset{?}{=} 11$$
$$9 + 2 \overset{?}{=} 11$$
$$11 = 11 \qquad \text{True}$$

$\therefore$ The solution is $x = 3$ and $y = -\frac{2}{5}$

Example 3   If $2x + 5y = 4$ and $3x - 5y = 11$, then $x = ?$ and $y = ?$ The solution is the same as that of **Example 2**, above. $x = 3$ and $y = -\frac{2}{5}$.

Note: Each of the above systems of linear equations is called a consistent and independent system with a unique solution. Two other systems follow.

Special Cases of Systems of Linear Equations

System with infinitely many solutions. Consistent but dependent system	**System with no solution. Inconsistent System**
Example	**Example**
Solve for x and y	Solve for x and y
$\begin{cases} x + 2y = 4 & (A) \\ 2x + 4y = 8 & (B) \end{cases}$	$\begin{cases} 2x + 4y = 2 & (D) \\ x + 2y = -3 & (E) \end{cases}$
Solution	**Solution**
$(A) \times -2:\ -2x - 4y = -8 \qquad (C)$	$(E) \times -2:\ -2x - 4y = 6 \qquad (F)$
$(C) + (B):\ \underline{2x + 4y = 8} \qquad (B)$	$\underline{\qquad 2x + 4y = 2} \qquad (D)$
$\qquad\qquad 0 = 0 \quad$ True	$(F) + (D):\qquad 0 = 8 \quad$ False
Such a result, $0 = 0$, an identity, implies that the system is consistent, and there are infinitely **many solutions**. For example, $x = 1$ in (A), $1 + 2y = 4$ and from which $y = \frac{3}{2}$.	Such a false statement, $0 = 8$, a contradiction, implies that the original system of equations is inconsistent, and there are **no** solutions.
Note: If the graphs of (A) and (B) were drawn, they would coincide since they have the same slope **and** the same y–intercept.	**Note**: If the graphs of (D) and (E) were drawn, they would be parallel, since they have the same slope ($m = -\frac{1}{2}$, but different y–intercepts).

Solving by the Substitution Method

Example 1   Solve for x and y by the substitution method

$$\begin{cases} 3x + 4y = 4..........(A) \\ x + 2y = 6..........(B) \end{cases}$$

Step 1: Solve for one of the variables (either x or y) in equation (A) or (B).
It is easier to solve for x in equation (B) since the coefficient of x is 1.

$$x + 2y = 6$$
$$x = 6 - 2y$$

Step 2: **Substitute** $(6 - 2y)$ for x in equation (A).   (The **parentheses** enclosing the 6 - 2y are important)
(That is replace x by $(6 - 2y)$ in the other equation)
Then, $3(6 - 2y) + 4y = 4$

Step 3: Solve for y.

$$3(6 - 2y) + 4y = 4$$
$$18 - 6y + 4y = 4$$
$$18 - 2y = 4$$
$$-2y = -14$$
$$y = \frac{-14}{-2}$$
$$y = 7$$

Step 4:   To find the value of x
substitute $y = 7$ in equation (B) and solve for x.

$$x + 2y = 6$$
$$x + 2(7) = 6$$
$$x + 14 = 6$$
$$x = -8$$

Now, $x = -8, y = 7$

Step 5**:** To check for $x= -8, y = 7$ in equations (A) and (B).
Substitute $x = -8$ and $y = 7$ in these equations.

Check # 1:   Checking in (A)
$$3(-8) + 4(7) \overset{?}{=} 4$$
$$-24 + 28 \overset{?}{=} 4$$
$$4 = 4 \quad \text{True (left-hand side of equation equals right-hand side of equation)}$$

Therefore, the ordered pair (-8,7) is a solution of equation (A)
Check # 2: Checking in (B)

$$-8 + 2(7) \overset{?}{=} 6$$
$$-8 + 14 \overset{?}{=} 6$$
$$6 = 6 \text{ True}$$

Therefore, the ordered pair (-8,7) is a solution of equation (B)
The solution of the system of equations is the ordered pair (-8,7)

Note.   Solving by the substitution method: In Step 1, it is easier to solve for the variable whose
coefficient is 1, **if** there is a variable with this coefficient.

Note also that if the problem does not specify which method (addition or substitution) to use,
then we can use either method. Experience will dictate which method to use. It must also be
remarked that there are relative merits for both methods.

Solving by the Equality Method

Example Solve for *x* and *y:*

$$\begin{cases} 2x + 5y = -11 & (1) \\ 3x - 4y = 18 & (2) \end{cases}$$

Step 1: Solve each equation for *y* or *x*.
We solve for *y*.

From equation (1) , $y = -\frac{2}{5}x - \frac{11}{5}$ (3)

From equation (2) , $y = \frac{3}{4}x - \frac{9}{2}$ (4)

Step 2: **Equate** the right-hand sides of equations (3) and (4) (since if $A = B$ and $A = C$ then $B = C$)

Then $-\frac{2}{5}x - \frac{11}{5} = \frac{3}{4}x - \frac{9}{2}$

Step 3: Solve for *x*:

$$-\frac{2(20)}{5}x - \frac{11(20)}{5} = \frac{3(20)}{4}x - \frac{9(20)}{2}$$
$$-8x - 44 = 15x - 90$$
$$46 = 23x$$
$$2 = x$$

Step 4: To find *y*, substitute $x = 2$ in equation (1):

Then $2(2) + 5y = -11$
$$4 + 5y = -11$$
$$5y = -15$$
$$y = -3$$

From Steps 3 and 4, the solution is the ordered pair (2, -3).

Lesson 57 Exercises

A   Solve the systems of equations for *x* and *y* by the elimination method:

1. $\begin{cases} 5x + 2y = 24 \\ 15x + 3y = 6 \end{cases}$ **2..** $\begin{cases} 7x + 3y = 15 \\ 3x - 2y = 13 \end{cases}$ **3.** $\begin{cases} 2x - 2y = 9 \\ x + 2y = 3 \end{cases}$ **4.** $\begin{cases} 3x + 5y = 6 \\ 6x + 5y = 12 \end{cases}$

Answers: **1.** $x = -4, y = 22$;   **2.** $x = 3, y = -2$;   **3.** $x = 4, y = -\frac{1}{2}$;   **4.** $x = 2, y = 0$

B Solve the systems of equations for *x* and *y* by the substitution method

1. $\begin{cases} y = 7 - 2x \\ x - 5y = 9 \end{cases}$ **2.** $\begin{cases} 4y + 3x = 4 \\ y + 3x = 4 \end{cases}$

Answers: **1.** $x = 4, \; y = -1$; **2.** $x = \frac{4}{3}, \; y = 0$

C Solve the systems for *x* and *y*:

1. $\begin{cases} 3x + 4y = 4 \\ 2x + y = 6 \end{cases}$ **2.** $\begin{cases} 2x + 5y = -11 \\ 3x - 4y = 18 \end{cases}$

Answers: **1**. (4, -2) ;   **2**. (2, -3)

Lesson 58
Graphical Method of Solving Systems of Equations

Solve for x and y graphically: $\begin{cases} y = x + 1 \\ y = -x + 5 \end{cases}$

Solution: (See Figure 1 below)

Step 1: Refer to Lesson 53B and review how to draw the graph of a straight line.

Step 2: Draw the graph of $y = x + 1$ in a rectangular coordinate system of axes (on graph paper).

Step 3: On the same system of axes as in Step 2, draw the graph of $y = -x + 5$.

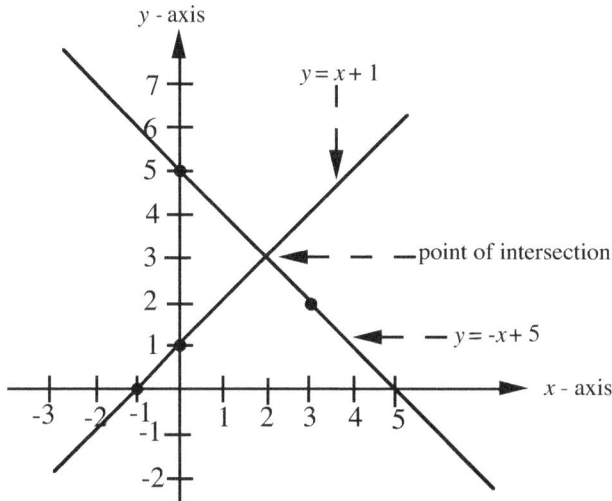

Figure 1

Step 4: From the graph, locate the point of intersection of the two lines drawn.
  Read the x- and y-coordinates of **the point of intersection** of the lines drawn.

Step 5: From the graph, the x-coordinate = 2, the y-coordinate = 3
  The solution is the ordered pair (2,3).

Note: The point of intersection is where the two lines meet. In drawing the lines, it is good practice to extend each line in both directions so that at least both the x- and y-axes are intercepted; and better, extend the lines to "fill" the graphing area.

Note also that : **1**. If the two lines are parallel (do not intersect), then there is no solution.
  2. If the two lines coincide, there are infinitely many solutions.

Lesson 58 Exercises

Solve by the graphical method

1. $\begin{cases} y = 3x - 5 \\ y = -x + 11 \end{cases}$

2... $\begin{cases} y = 7 - 2x \\ x - 5y = 9 \end{cases}$

Answer: 1. (4,7); 2. x=4. y = -1

CHAPTER 19
Applications of Solutions of Systems of Linear Equations

Lesson 59: **Coin Problems and Coin-Like Problems**
Lesson 60: **Distance-Time-Speed Problems: Travel by Water or by Air**

Lesson 59
Coin Problems and Coin-Like Problems

Example 1

Mary has 15 coins in dimes and nickels. The total value of the coins is $1.10. How many of each type of coin has she?

To solve the above problem, we can use either one variable (unknown) and one equation; or we can use two variables (unknowns) and two equations.
We will cover both methods.

Method 1:   Using two variables:

We need two equations. One of the equations relates or connects the number of each type of coin and the total number of coins. The other equation relates the sum of the values of the coins and the total value of the coins.

Now, back to the problem.
Let the number of dimes be d and let the number of nickels be n.

The value of d dimes is $10d$ cents. (Note that each dime is worth 10 cents)
The value of n nickels is $5n$ cents. (Each nickel is worth 5 cents)
(Examples: Value of 6 dimes is 6 x 10 = 60 cents. Value of 8 dimes is 8 x 10 = 80 cents).
        Also, $1.10 x 100 = 110 cents.

We form the two equations now:
$n + d = 15$................. (A)
$5n + 10d = 110$.................. (B)
Solve for n and d; by the elimination method or any method
We solve by the elimination method.
$-5(n + d = 15)$............(C)
$-5n - 5d = -75$...........(C)
$\underline{5n + 10d = 110}$...........(B)
Add (C) and (B): $5d = 35$
$d = 7$

To find n, substitute $d = 7$ in (A) and solve for n.
Then $n + 7 = 15$
$\underline{ -7 \quad -7}$
$n = 8$
$d = 7$

Check: 8 nickels are worth $8(5) = 40$ cents
7 dimes are worth $7(10) = 70$ cents
Total value = 110 cents.
Mary has 7 dimes and 8 nickels.

Method 2:   Using a single variable

> Let the numbers of dimes be d
> Then the number of nickels $= (15 - d)$
> Value of d dimes $= 10d$ cents
> Value of $(15 - d)$ nickels $= 5(15 - d)$ cents.

The equation: $10d + 5(15 - d) = 110$

Solve for d:
$$10d + 5(15 - d) = 110$$
$$10d + 75 - 5d = 110$$
$$5d + 75 = 110$$
$$\underline{ - 75 \quad - 75}$$
$$5d = 35$$
$$d = 7$$

Number of dimes $= 7$
Number of nickels $= (15 - 7) = 8$ nickels
Again, we obtain the same solution as in Method 1.

Note: In using a single variable, the equation needed is a relationship between the value of each type of coin and the total value of the coins; and also, the number of one type of coin is expressed in terms of the number of the other type of coin. **In some coin problems, we are not given the total number of coins, but rather we are given how the number of one type of coin is related to (connected with) the number of another type of coin.** Example: James has 6 more nickels than dimes.

Example 2   A student has 6 more nickels than quarters. If the student has $1.20 altogether, how many of each type of coin has the student?

Method 1:   Using one variable
> Let   the number of quarters be q.
> Then, the number of nickels is $(q + 6)$
> (6 more than q means $q + 6$)
> Value of the quarters $= 25q$ cents.
> Value of the nickels $= 5(q + 6)$

The equation: $25q + 5(q + 6) = 120$
Solve for q:
$$25q + 5q + 30 = 120$$
$$30q + 30 = 120$$
$$\underline{ - 30 \quad -30}$$
$$30q = 90$$
$$\frac{30q}{30} = \frac{90}{30}$$
$$q = 3$$

Number of quarters $= 3$
Number of nickels $= (3 + 6) = 9$ nickels
Check: 3 quarters are worth $3(25) = 75$ cents
   9 nickels are worth $5(9) = 45$ cents
   Total value $= 120$ cents.
The student has 3 quarters and 9 dimes.

Method 2:   **Using two variables**
Recall that we need two equations.

Let the number of quarters $= q$
Let the number of nickels $= n$

From the problem, "the number of nickels is 6 more that the number of quarters" translates to :

$n = q + 6$ . . . .(A)<------ This equation relates the number of coin types.

$5n + 25q = 120$. . . . . .(B)<------This equation relates the sum of the values
of the coins and the total value of the coins.

We solve for n and q by the substitution method.
Substitute for n from (A) in (B)

$5(q + 6) + 25q = 120$
$5q + 30 + 25q = 120$
$30q + 30 = 120$
$\underline{\qquad\qquad - 30 \quad\; -30}$
$30q = 90$
$q = 3$

To find n, substitute $q = 3$ in (A)
Then $n = 3 + 6 = 9$
$n = 9$

Again we obtain the same solution as by Method 1 (using one variable).

Coin-Like Problems

There are problems which do not use coins as the types of items whose total values are given, but which can be solved in an identical manner as if they were coin problems.

Example   Let us call the following type " Stamp Problem". In fact, we will reword
Example 1 (Lesson 59) and replace coins by stamps.

Mary bought 15 stamps, some of which were worth 10 cents each, and the others 5 cents each. If she spent altogether $1.10 on these stamps, how many 10-cent and 5-cent stamps did she buy?

Solution   We solve this problem the same way as we did for Example 1, above,)
Let the number of 5-cent stamps $= x$
Let the number of 10-cent stamps $= y$
Then $x + y = 15$.................(A)
$5x + 10y = 110$.............(B)
Answer: $x = 8$, $y = 7$
$\therefore$ the number of 5-cent stamps Mary bought is 8 and
the number of 10-cent stamps Mary bought is 7

Another Coin-Like Problem

Example   We can call this type " a ticket problem". Tickets to a concert are $5 for children and
$10 for adults. If 21 tickets are sold and the total receipts are worth $165,
how many children's tickets and how many adult's tickets were sold.
Let the number of children's tickets $= x$
Let the number adult's tickets $= y$.

Then, $x + y = 21$. ..(A)<---**This equation relates the number of each type of**
ticket to the total number of tickets.
$5x + 10y = 165$. . . . . .(B) <---**This equation relates the sum of the values of**
the tickets to the   total value of the tickets.

Solve for x and y:

Multiply (A) by -5 : $-5x - 5y = -105$. . . . .(C)
$\underline{5x + 10y = \ 165}$ . . . .(B)
Add (C) and (B) $5y = 60$
$y = 12$

To find x, substitute $y = 12$ in (A)

$x + 12 = \ 21$
$\underline{-12 \ = -12}$
$x \ = \ 9$

Check: 9 children's tickets are worth $9(5) = \$45$
12 adult's tickets are worth $12(10) = \underline{120}$
Total value $\$165$
Therefore 9 children's tickets and 12 adult's tickets were sold.

Some Expressions for Coin and Coin-Like Problems

Example 1   James has x dimes and y quarters. How much money has James?

Solution: 1 dime is worth 10 cents
x dimes are worth $10x$ cents

1 quarter is worth 25 cents.
$\therefore$ y quarters are worth $25y$ cents.

The total amount of money James has is $(10x + 25y)$ cents

Example 2   Maria has p ten-dollar bills and r twenty-dollar bills. How much money
has Maria?

Solution: Every ten-dollar bill is worth 10 dollars
$\therefore$ p ten -dollar bills are worth $10p$ dollars

Every twenty-dollar bill is worth 20 dollars
$\therefore$ r twenty-dollar bills are worth $20r$ dollars
The total amount of money Maria has is $(10p + 20r)$ dollars

Lesson 59 Exercises

A Betty has 15 coins, all in dimes and nickels, in her purse. The value of these coins is $1.05
How many of each type of coin has she?

Answer: 6 dimes and 9 nickels

B Lisa has four less dimes than nickels. The value of these coins is $1.40.
How many of each type of coin has she?

Answer: 12 nickels and 8 dimes

C 1. Peter has 9 coins in dimes and quarters. The total value of these coins is $1.35.
How many of each type of coin has he?

2. Nancy has two more dimes than nickels. If the total value of these coins is $1.10.
(*a*) How many of each type of coin has she?
(*b*) How many coins has she?

Answers: **1.** Ans: 6 dimes and 3 quarters ; **2.** (*a*) Ans: 8 dimes and 6 nickels ; (*b*) 14 coins

D 1. Tickets to a concert are $2 for children and $5 for adults. If 125 tickets are sold and the total
receipts are worth $505, how many children's tickets and how many adults' tickets were sold?

2. Maria bought 31 stamps, some of which were worth 29 cents each, and the others 40 cents each.
If she spent altogether $10.20 on these stamps, how many 29-cent and 40-cent stamps did she
buy?

3. The difference between two positive numbers is 92. The larger of the two numbers is 24 times the
smaller. Find these two numbers.

4. A ball thrown downwards from the top of a tower of height 240 feet. If the height, *s*, of this
tower is related to the time,*t*, in seconds the ball takes to strike the ground, according to the
equation $s = 32t + 16t^2$, how long does the ball take to strike the ground?

Answers:
1. # of children's tickets = 40; # of adult's tickets = 85.   **2** .# of 29-cent stamps = 20; # of 40-cent stamps = 11
3. The smaller number is 4, and the larger number is 96;   **4.** The ball takes 3 seconds to strike the ground.

Lesson 60

Distance-Time-Speed Problems: Travel by Water or by Air

The approach in solving problems involving travel on water (e.g., boat rowing or swimming) and air travel (e.g., aircraft or bird flight) is similar to the distance-time-speed problems covered already, except that we also have to consider the effect of the flow of water or air on the speed of the moving object. We introduce a new quantity, the **speed** of the **water** (river) **current** or the **wind current**. The equations derived below are based on **speed**. Also, see Lesson 43 techniques.

Boat rowing in a river (or any water body) downstream and upstream

Let S = the speed of boat (or ship or swimmer) in still (calm) water.

Let C = the speed of the river current.

Let S_N = the net speed of the boat.   These quantities are related as follows:

$$S_N = S \pm C \qquad \text{(The net speed is either the sum or the difference of } S \text{ and } C.)$$

where we use the **plus** sign if the boat travels downstream or with the river current, and the **minus** sign if the boat travels upstream or against the river current. **Note** that traveling with the river current implies traveling in the direction of the current.

Example 1 A boat traveled 40 miles downstream a river in 4 hours and returned upstream to the starting point in 8 hours.
Find (a) the speed of the boat in still water ; (b) the speed of the river current.

Solution We will write two equations in terms of two variables and solve for the variables.
We can rewrite this word problem as two simple statements:
1. A boat traveled 40 miles downstream a river in 4 hours.
2. The boat returned upstream, traveling 40 miles to the starting point, in 8 hours.

Step 1:   Let S = the speed of boat in still (calm) water.
Let C = the speed of the river current.
Let S_N = the net speed of the boat.

Traveling downstream: $S_N = \dfrac{40}{4}$ $(\text{speed} = \dfrac{d}{t})$

$S_N = S + C$ (We use the **plus** sign since the boat is traveling with the river current).

$\dfrac{40}{4} = S + C \qquad (1)$

Traveling upstream: $S_N = \dfrac{40}{8}$

$S_N = S - C$ We use the **minus** sign since the boat is traveling against the river current).

$\dfrac{40}{8} = S - C \qquad (2)$

Step 2: Solve equation (1) and (2) simultaneously for S and C .
From equation (1): $10 = S + C \qquad (3)$
From equation (2): $5 = S - C \qquad (4)$
Adding (3) and (4): $15 = 2S$

$\dfrac{15}{2} = S$

$7\frac{1}{2} = S$

To find C, substitute $S = 7\frac{1}{2}$ in equation (3) and solve for C

Then $10 = 7\frac{1}{2} + C$ and from which $C = 2\frac{1}{2}$

(a) The speed of the boat in still water is $7\frac{1}{2}$ mph ; and (b) the speed of the river current is $2\frac{1}{2}$ mph.

Aircraft (or bird) flight with wind, and against wind

The net speed is similar to the case of boat rowing downstream or upstream.
Let S = the speed of aircraft or bird in still air.
Let C = the speed of the wind current .
Let S_N = the net speed of the aircraft or bird. These quantities are related as follows:

$$S_N = S \pm C$$

where we use the **plus** sign if the aircraft travels **with** the wind current, and the **minus** sign if the aircraft travels **against** the wind current. **Note** that traveling with the wind current implies traveling in the direction of the current.

Example 2 A bird flew 40 miles with the wind current in 4 hours and flew back against the wind current to the starting point in 8 hours. Find (a) the speed of the bird in still air ; (b) the speed of the wind current.

Solution We will write two equations in terms of two variables and solve for the variables.
We can rewrite this word problem as two simple statements:
1. A bird flew 40 miles with the wind in 4 hours.
2. The bird flew back 40 miles against the wind to the starting point in 8 hours.

Step 1: Let S = the speed of bird in still (calm) air.
Let C = the speed of the wind current.
Let S_N = the net speed of the bird.

Flying with the wind: $S_N = \dfrac{40}{4}$ (speed = $\dfrac{d}{t}$)

$S_N = S + C$ (We use the **plus** sign since the bird is flying with the wind current).

$\dfrac{40}{4} = S + C$ $\qquad\qquad$ (1)

Flying against the wind: $S_N = \dfrac{40}{8}$ (speed = $\dfrac{d}{t}$)

$S_N = S - C$ We use the **minus** sign since the bird is flying against the wind current).

$\dfrac{40}{8} = S - C$ $\qquad\qquad$ (2)

Step 2: Solve equation (1) and (2) simultaneously for S and C .

From equation (1): $10 = S + C$ $\qquad$ (3)
From equation (2): $5 = S - C$ $\qquad$ (4)
Adding (3) and (4): $15 = 2S$

$\dfrac{15}{2} = S$

$7\tfrac{1}{2} = S$

To find C, substitute $S = 7\tfrac{1}{2}$ in equation (3) and solve for C.
Then $10 = 7\tfrac{1}{2} + C$ and

$C = 2\tfrac{1}{2}$

(a) The speed of the bird in still air is $7\tfrac{1}{2}$ mph ; and (b) the speed of the wind current is $2\tfrac{1}{2}$ mph.

Example 3 An aircraft can fly 1600 miles with the wind current in 4 hours, but against the wind current it can fly 1200 miles in 4 hours.
Find (a) the speed of the aircraft in still air ; (b) the speed of the wind current.

Solution

Step 1: Let S = the speed of aircraft in still (calm) air.
Let C = the speed of the wind current.
Let S_N = the net speed of the aircraft.

Flying with the wind: $S_N = \dfrac{1600}{4}$ (speed $= \dfrac{d}{t}$)

$S_N = S + C$ (We use the **plus** sign since the bird is flying with the wind current).

$\dfrac{1600}{4} = S + C$ (1)

Flying against the wind: $S_N = \dfrac{1200}{4}$ (speed $= \dfrac{d}{t}$)

$S_N = S + C$ (We use the **minus** sign since the bird is flying against the wind current).

$\dfrac{1200}{4} = S - C$ (2)

Step 2: Solve equation (1) and (2) simultaneously for S and C .

From equation (1): $400 = S + C$ (3)
From equation (2): $300 = S - C$ (4)
Adding (3) and (4): $700 = 2S$

$\dfrac{700}{2} = S$

$350 = S$

To find C, substitute $S = 350$ in equation (3) and solve for C

Then $400 = 350 + C$ and from which

$C = 50$

(a) The speed of the aircraft in still air is 350 mph ; and (b) the speed of the wind current is 50 mph.

Example 4 An aircraft flew 1200 miles with the wind in the same time as it flew 800 miles against the wind. If the speed of the aircraft in still air is 240 mph, find the speed of the wind.

Solution

Step 1: Let S = the speed of aircraft in still (calm) air.

Let C = the speed of the wind current.

Let S_N = the net speed of the aircraft.

Let t = the time taken to travel 1200 miles and also to travel 800 miles.

Also, $S = 240$.

Flying with the wind: $S_N = \dfrac{1200}{t}$

$$S_N = S + C \quad \text{(We use the \textbf{plus} sign since the aircraft is flying with the wind current).}$$

$$\frac{1200}{t} = 240 + C \qquad\qquad (1)$$

Flying against the wind: $S_N = \dfrac{800}{t}$

$$S_N = S - C. \text{We use the \textbf{minus} sign since the aircraft is flying against the wind current).}$$

$$\frac{800}{t} = 240 - C \qquad\qquad (2)$$

Step 2: Solve equation (1) and (2) simultaneously for C . (We are required to find C only and not t)

We can eliminate t by dividing equation (1) by equation (2):

$$\frac{1200}{t} \bullet \frac{t}{800} = \frac{240 + C}{240 - C}$$

$$\frac{3}{2} = \frac{240 + C}{240 - C} \qquad\qquad \text{(simplifying the left-hand side)}$$

$$2(240 + C) = 3(240 - C)$$

$$480 + 2C = 720 - 3C$$

$$5C = 240$$

$$C = 48$$

The speed of the wind current is 48 mph.

Note: We could have begun Step 2 by solving for t from each of equations (1) and (2) followed by equating the resulting expressions to each other.

A note about other units Example: 60 mph = 88 ft/sec (obtained from $\dfrac{60(5280)\,\text{ft}}{(60)(60)\,\text{sec}}$)

Lesson 60 Exercises

1. A boat traveled 60 miles downstream a river in 4 hours and returned upstream to the starting point in 6 hours. Find (a) the speed of the boat in still water ; (b) the speed of the river current.

2. A bird flew 60 miles with the wind current in 4 hours and flew back against the wind current to the starting point in 6 hours. Find (a) the speed of the bird in still air ; (b) the speed of the wind current.

3. An aircraft can fly 1600 miles with the wind current in 4 hours, but against the wind current it can fly 1200 miles in 4 hours. Find (a) the speed of the aircraft in still air ; (b) the speed of the wind current.

Answers: **1**. (a) $12\frac{1}{2}$ mph; (b) $2\frac{1}{2}$ mph; **2**. (a) $12\frac{1}{2}$ mph; (b) $2\frac{1}{2}$ mph; **3**. (a) 350 mph; (b) 50 mph.

Test # 11 -Student's Self-Test (Always, Test yourself before you are tested)

Attempt all questions on clean sheets of paper, **Do not write in the book** Show all necessary work.

---

1. Solve the systems of equations for x and y by the elimination method:

$$\begin{cases} 3x + 2y = 29 \\ 8x - 3y = 19 \end{cases}$$

2. Solve the systems of equations for x and y by the elimination method:

$$\begin{cases} 3x + 5y = 6 \\ 6x + 5y = 12 \end{cases}$$

---

3. Solve the systems of equations for x and y by the elimination method:

$$\begin{cases} 2x - 2y = 9 \\ x + 2y = 3 \end{cases}$$

4. Solve the systems of equations for x and y

$$\begin{cases} 3x - 4 = -4y \\ 2x + y = 6 \end{cases}$$

---

5. Solve the systems of equations for x and y by the substitution method:

$$\begin{cases} y = 7 - 2x \\ x - 5y = 9 \end{cases}$$

6. Solve the systems of equations for x and y by the substitution method

$$\begin{cases} 4y + 3x = 4 \\ y + 3x = 4 \end{cases}$$

---

7. Solve the systems for x and y:

$$\begin{cases} 3x + 4y = 4 \\ 2x + y = 6 \end{cases}$$

8. Solve the systems for x and y:

$$\begin{cases} 2x - 3y = 6 \\ 5x - 7y = 11 \end{cases}$$

---

9. Solve by the graphical method

$$\begin{cases} y = 3x - 5 \\ y = -x + 11 \end{cases}$$

10. Betty has 10 coins, all in dimes and nickels, in her purse. The value of these coins is 60 cents. How many of each type of coin has she?

---

11. Lisa has three more dimes than nickels. The value of these coins is $1.05. How many of each type of coin has she?

12. Peter has 17 coins in dimes and quarters. The total value of these coins is $2.60 How many of each type of coin has he?

---

13. Nancy has four more dimes than nickels. If the total value of these coins is $2.80.
    (a) How many of each type of coin has she?
    (b) How many coins has she?

14. Tickets to a concert are $3 for children and $7 for adults. If 80 tickets are sold and the total receipts are worth $480, how many children's tickets and how many adults' tickets were sold?

15. Maria bought 12 stamps, some of which were worth 35 cents each, and the others 20 cents each. If she spent altogether $3.45 on these stamps, how many 35-cent and 20-cent stamps did she buy?

16. The difference between two positive numbers is 92. The larger of the two numbers is 24 times the smaller. Find these two numbers.

17. If John gives $8 to James, they would have the same amount. If James gives John $8, John would have three times as much as James. How much does each have now?

18.

19.

20.

Answers: **1.** $x = 5$, $y = 7$; **2.** $x = 2$, $y = 0$; **3.** $x = 4$, $y = -\frac{1}{2}$; **4.** $x = 4$, $y = -2$;

5. $x = 4$, $y = -1$; **6.** $x = \frac{4}{3}$, $y = 0$; **7.** $x = 4$, $y = -2$; **8.** $x = -9$, $y = -8$;

9. $x = 4$, $y = 7$; **10.** Betty has 8 nickels and 2 dimes;

11. Lisa has 8 dimes and 5 nickels; **12.** Peter has 11 dimes and 6 quarters ;

13. Nancy has 20 dimes and 16 nickels. She has 36 coins ;

14. 60 adults' tickets and 20 children's tickets were sold;

15. The number of 35-cent stamps Maria bought is 7 and the number of 20-cent stamps she bought is 5.

16. The smaller number is 4 and the larger number is 96.

17. John has $40; James has $24.

CHAPTER 20A
Exponents and Radicals

Lesson 61: **Review of Exponents and Radicals**
Lesson 62: **Definitions and Simplification of Radicals**
Lesson 63: **Addition and Multiplication of Radicals**

Lesson 61

Review of Exponents and Radicals (see also Chapter 11)

Mnemonic Examples **for the Rules of Exponents and Radicals**

1. $x^2 x^3 = x^5$

2. $\dfrac{x^7}{x^4} = x^{7-4} = x^3$

3. $\left(x^4\right)^2 = x^8$

4. $(xy)^5 = x^5 y^5$

5. $\left(\dfrac{x}{y}\right)^6 = \dfrac{x^6}{y^6}$

6. $x^{-2} = \dfrac{1}{x^2}$

7. $x^0 = 1$

8. $\dfrac{x^4}{x^7} = x^{4-7} = x^{-3} = \dfrac{1}{x^3}$

9. $9^{1/2} = \sqrt{9}$

10. $\sqrt[3]{xy} = \sqrt[3]{x}\,\sqrt[3]{y}$

11. $8^{1/3} = \sqrt[3]{8} = 2$

12. $8^{2/3} = \left(\sqrt[3]{8}\right)^2 = \sqrt[3]{8^2} = 4$

13. $\left(\sqrt[4]{8}\right)^4 = \left(\sqrt[4]{8^4}\right) = 8$

14. $\sqrt[3]{\dfrac{x}{y}} = \dfrac{\sqrt[3]{x}}{\sqrt[3]{y}}$

Extra: Show that $x^{-2} = \dfrac{1}{x^2} \cdot \dfrac{x^7}{x^3} = x^{7-4} = x^3$

Note: $x^{-2} = \dfrac{x^{-2}}{1} \bullet \dfrac{x^2}{x^2} = \dfrac{x^{-2+2}}{x^2} = \dfrac{x^0}{x^2} = \dfrac{1}{x^2}$.

Example 1 Evaluate 7^{-2}

  Step 1: Express with positive exponents.

$$7^{-2} = \frac{1}{7^2}$$

  Step 2: Simplify. $\dfrac{1}{7^2} = \dfrac{1}{49}$

Example 2 Simplify, leaving answer with only positive exponents

$$\frac{x^{-3} y^2 z^6}{x^2 y^{-3} z}$$

Method 1 (Strictly, using the rules of exponents)

$$\frac{x^{-3}y^2z^6}{x^2y^{-3}z} = x^{-3-2}y^{2+3}z^{6-1}$$

$$= x^{-5}y^5z^5$$

$$= \frac{y^5z^5}{x^5}$$

Method 2 Taking powers across the division bar, and changing the signs of exponents, followed by cancellation

$$\frac{x^{-3}y^2z^6}{x^2y^{-3}z} = \frac{y^2y^3\cancel{z^6}\,z^5}{x^2x^3\cancel{z}}$$

(change x^{-3} to x^3 and write it in the denominator. Change y^{-3} to y^3 and write it in the numerator)
If the exponent is positive, leave the power where it is. The changes are for the powers with negative exponents).

$$= \frac{y^5z^5}{x^5}$$

 A note about cancellation in Method 2 above:
In order to apply cancellation (as done in arithmetic), the exponents of the powers involved in the cancellation must have the same sign; preferably, the exponents must be positive (or made positive by following the instructions outlined above) .

Example 3   Simplify $(4x)^0 - 4x^0$

Solution

$$(4x)^0 - 4x^0 = 4^0x^0 - 4x^0 \qquad\qquad \textbf{Note: } x^0 = 1,\ (4)^0 = 1$$

$$= (1)(1) - 4(1)$$

$$= 1 - 4$$
$$= -3$$

Example 4   Multiply $\left(\frac{-8a^2b}{5}\right)\left(\frac{9c}{3abc}\right)\left(\frac{-15}{27c}\right)$

Solution $\dfrac{-8(9)(-15)a^2bc}{5(3)(27)abc^2}$

$$= +\ \frac{8(9)(15)a^2bc}{5(3)(27)abc^2}$$

$$= \frac{8a}{3c} \qquad\qquad \text{(After canceling the common factors in the numerator and the denominator)}$$

Lesson 61 Exercises

A   Simplify the following:

1. $16^{1/2}$

2. $\sqrt[3]{x^6 y^{12}}$

3. $64^{1/3}$

4. $27^{2/3}$

5. $(\sqrt[4]{16})^3$

6. $\sqrt[3]{\dfrac{x^6}{y^{15}}}$

Answers: 1. 4 ; **2.** $x^2 y^4$; **3.** 4 ; **4.** 9; **5.** 8; **6.** $\dfrac{x^2}{y^5}$

B Simplify the following:

1. 3^{-3}

3. $\left(\dfrac{4}{9}\right)^{-3/2}$

4. $\dfrac{x^{-4} y^3 z^2}{x^4 y^{-2} z}$

5. $(6x)^0 - 6x^0$

2. $100^{-1/2}$

6. $\left(\dfrac{-4a^3 b}{3b}\right)\left(\dfrac{9c}{2bc}\right)\left(\dfrac{-12}{18c}\right)$

7. $81^{-1/4}$

Answers: 1. $\dfrac{1}{27}$; **2.** $\dfrac{1}{10}$; **3.** $\dfrac{27}{8}$; **4.** $\dfrac{y^5 z}{x^8}$; **5.** -5; **6.** $\dfrac{4a^3}{bc}$; **7.** $\dfrac{1}{3}$

Lesson 62
Definitions and Simplification of Radicals

Definitions of **Rational and Irrational Numbers**

Rational number : A rational number (a fraction) is a real number which **can** be written as the ratio of two integers. The word **rational** pertains to the word **ratio.**

Examples are (a) $\dfrac{2}{3}$; (b) $\dfrac{1}{5}$; (c) 4 (since $4 = \dfrac{4}{1}$)

(d) 0 (since $0 = \dfrac{0}{7} = \dfrac{0}{3}$ ... or $0 = \dfrac{0}{b}$, where b is an integer and $b \neq 0$)

(e) $\sqrt{4}$ (because $\sqrt{4} = 2 = \dfrac{2}{1}$)

Irrational number: An irrational number is a real number which **cannot** be written as the ratio of two integers. However, we can approximate irrational numbers as closely as we wish by rational numbers or decimals.

Examples of irrational numbers are $\sqrt{2}, \sqrt[3]{4}$, and π (pi). We can for example, approximate $\sqrt{2}$ by 1.414, and π by $\dfrac{22}{7}$ or 3.142. Except otherwise instructed, in simplifying radical expressions, we prefer to leave the irrational numbers in **radical** forms.

Radicals and Roots

Consider two real numbers r and A. We denote the **principal nth root** of A (n being a positive integer) by $\sqrt[n]{A}$. Also $\sqrt[n]{A}$ is called a **radical**. We define the principal nth root of A as follows:

$\sqrt[n]{A} = r$ if $r^n = A$ (i.e., the nth root of A = r if r^n = A) with the following qualifications:

1. Any root of zero is zero (i.e. $\sqrt[n]{0} = 0$)
2. If A is a positive number, then the principal nth root of A is the positive nth root of A.
3. If A is a negative number, then the principal nth root of A is the negative nth root of A.

Note from above that a **radical** is an expression used for indicating the root of a number.

We call n the index of the radical (n indicating the type of root being considered); A is called the radicand (the number of which a root is being taken). The radicand is thus the expression under the symbol " $\sqrt[n]{}$ " . Thus, the radical consists of the index n with root symbol " $\sqrt[n]{}$ " and the radicand.

If $n = 2$, we obtain $\sqrt[2]{A}$, the square root of A, We usually omit the "2" and write $\sqrt{A}$.

Therefore, $\sqrt[2]{A} = \sqrt{A}$. However, if explicit indication helps you to understand a problem, write the 2. When $n = 3$, we obtain $\sqrt[3]{A}$, the cube root A, but in this case, we have to write the "3". Similarly, for $n = 4$ and higher indices, we must always write the index. Some higher orders of the root of A are $\sqrt[4]{A}$,$\sqrt[5]{A}$, and $\sqrt[8]{A}$.

Some specific radicals are $\sqrt{2}, \sqrt[3]{4}$,$\sqrt{10}$, and $\sqrt{5}$.

Example 1 Name the parts of the radical $\sqrt{32}$

Solution   The index is understood to be 2, since, $\sqrt{32} = \sqrt[2]{32}$
The radicand is 32.

Example 2   Name the parts of the radical $\sqrt[3]{16}$

Solution   The index is 3
The radicand is 16

Square Roots

Example $\sqrt{9}$ or $\sqrt[2]{9}$

This radical is read as the square root of 9. The square root of 9 is 3 (because $3^2 = 9$).

The following definition is useful in finding the **square root** of a number:

The **principal square root** of a number (nonzero number) is one of the two equal positive factors of that number.

Thus, if we can "break up" a number into two equal positive factors, then, one of the positive factors is the square root.(Note that we exclude negative roots).

Square root of zero: $\sqrt{0} = 0$ (that is, the square root of zero is zero).
because $0^2 = 0$

Examples   (a) $\sqrt{9} = \sqrt{(3)(3)} = 3$;   (b) $\sqrt{64} = \sqrt{(8)(8)} = 8$ (one of (8)(8) is 8)

Cube Roots

Example $\sqrt[3]{8}$

The above radical consists of the index 3, the radical sign and the radicand 8.
This radical is read " the cube root of 8". The cube root of 8 is 2. (because $2^3 = 8$)

The following definition is also useful in finding the cube root of a number:

The **cube root** of a number is **one of the three equal** factors of that number. Here, we do not specify positive root, since the cube root may be positive or negative, depending on whether the given number is positive or negative.

Examples (a) $\sqrt[3]{8} = \sqrt[3]{(2)(2)(2)} = 2$ **(b)** $\sqrt[3]{-8} = \sqrt[3]{(-2)(-2)(-2)} = -2$

Check: $2^3 = 8$ Check: $(-2)^3 = -8$

Note: For **even roots** such as the square root of a , $(\sqrt{a})$, the fourth root of $a, (\sqrt[4]{a}\,)$, the sixth

root of $a, (\sqrt[6]{a}\,)$, $\sqrt[n]{a}$ implies the positive root.

For **odd roots** such as the cube root $(\sqrt[3]{a}\,)$, the fifth root $(\sqrt[5]{a}\,)$, or the seventh root $(\sqrt[7]{a}\,)$, the root may be positive or negative according to whether the given number is positive or negative.

More Examples

(a) $\sqrt[4]{16} = \mathbf{2}$ or $\sqrt[4]{16} = \sqrt[4]{(2)(2)(2)(2)} = \mathbf{2}$
because $2^4 = 16$

(b) $\sqrt[5]{32} = 2$ or $\sqrt[5]{32} = \sqrt[5]{(2)(2)(2)(2)(2)} = \mathbf{2}$
because $2^5 = 32$

(c) $\sqrt[5]{-32} = -2$ or $\sqrt[5]{-32} = \sqrt[5]{(-2)(-2)(-2)(-2)(-2)} = \mathbf{-2}$
because $(-2)^5 = -32$

Simplifying Radicals

Two approaches are considered: One approach depends on guessing correctly a perfect power which is a factor of the radicand. The other approach is the application of prime factorization, and the grouping of the factors into n equal factors: for example, for square root, we would like to have two equal **positive** factors. For cube root, we would like to have three equal factors.

Example Simplify the following: **(a)** $\sqrt{32}$; **(b)** $\sqrt{\dfrac{49}{64}}$; **(c)** $\sqrt[3]{80}$.

Solutions

(a) **Method 1**

$$\sqrt{32} = \sqrt{(16)(2)}$$
$$= \sqrt{16}\sqrt{2}$$
$$= 4\sqrt{2} \quad (\sqrt{16} \text{ is } \textbf{rational} \text{ but } \sqrt{2} \text{ is } \textbf{irrational})$$

Method 2 $\quad \sqrt{32} = \sqrt{(2)(2)(2)(2)(2)}$

$$= \sqrt{(4)(4)(2)} \quad (\textbf{Rule}: \sqrt[n]{ab} = \sqrt{a}\sqrt{b})$$
$$= 4\sqrt{2}$$

(b) $\quad \sqrt{\dfrac{49}{64}} = \dfrac{\sqrt{49}}{\sqrt{64}} = \dfrac{7}{8} \qquad \left(\textbf{Rule:}\ \sqrt[n]{\dfrac{a}{b}} = \dfrac{\sqrt[n]{a}}{\sqrt[n]{b}}\right)$

(c) $\quad \sqrt[3]{80} = \sqrt[3]{(2)(2)(2)(2)(5)}$

Scrapwork:

$$= \sqrt[3]{(2)(2)(2)}\ \sqrt[3]{10}$$

$$80 = 2 \times 2 \times 2 \times 2 \times 5$$

$$= \sqrt[3]{8}\ \sqrt[3]{10}$$

$$= 2\sqrt[3]{10} \quad \text{(We leave the product of the "extra" 2 and the 5 under the radical sign as 10)}$$

Simplify the following: **1.** $\sqrt{x^3 y^2}$; **2.** $\sqrt{27}$; **3.** $\sqrt{18x^3 y^4}$; **4.** $\sqrt{64x^7 y^{14} z^3}$; **5.** $\sqrt{\dfrac{8}{9}}$

Solutions

1. $\quad \sqrt{x^3 y^2} = \sqrt{x^2 y^2 x} = xy\sqrt{x}$

2. $\quad \sqrt{27} = \sqrt{(9)(3)} = \sqrt{9}\sqrt{3} = 3\sqrt{3}$

3. $\sqrt{18x^3 y^4} = \sqrt{(9)(2)x^2 xy^4} = \sqrt{9x^2 y^4 \cdot 2x} = 3xy^2\sqrt{2x}$

4. $\sqrt{64x^7 y^{14} z^3} = \sqrt{64x^6 y^{14} z^2 \cdot xz} = 8x^3 y^7 z\sqrt{xz};$

5. $\sqrt{\dfrac{8}{9}} = \dfrac{\sqrt{8}}{\sqrt{9}} = \dfrac{2\sqrt{2}}{3}$

Lesson 62 Exercises

A Simplify:

1. $\sqrt{18}$;   **2.** $\sqrt{50}$;   **3.** $\sqrt{24}$;   **4.** $\sqrt{72}$;   **5.** $\sqrt{288}$;   **6.** $\sqrt{48}$;   **7.** $\sqrt{\dfrac{81}{100}}$

Answers: **1.** $3\sqrt{2}$;   **2.** $5\sqrt{2}$;   **3.** $2\sqrt{6}$;   **4.** $6\sqrt{2}$;   **5.** $12\sqrt{2}$;   **6.** $4\sqrt{3}$;   **7.** $\dfrac{9}{10}$

B Simplify the following: **1.** $\sqrt{16x^6}$ **2.** $\sqrt{9x^8y^2}$ **3.** $\sqrt{27x^9y^4}$

4. $\sqrt{18x^3y^5}$;   **5.** $\sqrt{32x^8y}$;   **6.** $\sqrt{8x^4y^2z^3}$

Answers: **1.** $4x^3$;   **2.** $3x^4y$;   **3.** $3x^4y^2\sqrt{3x}$;   **4.** $3xy^2\sqrt{2xy}$;   **5.** $4x^4\sqrt{2y}$;   **6.** $2x^2yz\sqrt{2z}$

Lesson 63

304

Addition and Multiplication of Radicals

Addition of Radicals (Like Radicals or Similar Radicals)

Like radicals have the same index **and** the same radicand.

Examples: **1.** $4\sqrt{3}$ and $\sqrt{3}$ are like radicals

(index $= 2$, the square root index; radicand $= 3$).

2. $5\sqrt[3]{4}$ and $2\sqrt[3]{4}$ are like radicals

(index $= 3$, radicand $= 4$).

Unlike radicals: Unlike radicals have different indices **or** radicands or both.

Examples   **1.** $\sqrt{3}$ and $\sqrt{2}$ are unlike radicals. (They have different radicands.)

2. $\sqrt[3]{2}$ and $\sqrt[4]{2}$ are unlike radicals (They have different indices.)

Like radicals can be combined into a single radical (i.e., added or subtracted in much the same way as like terms of polynomials are added or subtracted).

Unlike radicals cannot be combined into a single radical (i.e., cannot be added). However, in some cases, by simplifying the given radical(s), like radicals may be obtained, and which then may be added.

To **add like radicals**, add the non-radical parts (coefficients) and keep the radical part.

Example 1   Add: $3\sqrt{11} + 5\sqrt{11}$ (same index = 2; same radicand =11)

Solution $3\sqrt{11} + 5\sqrt{11}$

$\qquad = (3 + 5) \sqrt{11}$

$\qquad = 8\sqrt{11}$

Note that $3\sqrt{11}$ means 3 times $\sqrt{11}$; but $\sqrt[3]{11}$ means the cube root of 11.

Example 2   Simplify : $3\sqrt{19} + 8\sqrt{10} + 6\sqrt{19} - 2\sqrt{10}$

Solution $3\sqrt{19} + 8\sqrt{10} + 6\sqrt{19} - 2\sqrt{10}$

$\qquad = 3\sqrt{19} + 6\sqrt{19} + 8\sqrt{10} - 2\sqrt{10}$ <-------- You may skip this step.

$\qquad = (3 + 6)\sqrt{19} + (8 - 2)\sqrt{10}$

$\qquad = 9\sqrt{19} + 6\sqrt{10}$

Example 3 Add : $\sqrt{50}   + \sqrt{72}$

Solution Step 1: Simplify the radicals first to see if there are any like radicals

$\qquad \sqrt{50} = \sqrt{(25)(2)} =$ or $\sqrt{50} = \sqrt{(5)(5)(2)}$

$\qquad \mathbf{\sqrt{50} = 5\sqrt{2}}$ $\qquad\qquad = 5\sqrt{2}$

$\qquad \sqrt{72} = \sqrt{(36)(2)}$ or $=\sqrt{(3)(3)(2)(2)(2)}$

$\qquad \mathbf{\sqrt{72} = 6\sqrt{2}}$ $\qquad\qquad = (3)(2)\sqrt{2} = 6\sqrt{2}$

Step 2: Now, we have like radicals and therefore, we can add.

$\qquad \sqrt{50} + \sqrt{72}$

$\qquad = 5\sqrt{2} + 6\sqrt{2}$

$\qquad = (5 + 6)\sqrt{2}$

$\qquad = 11\sqrt{2}$

In the future, Step 1 could be considered as scrapwork and show only Step 2.

Example 4   Simplify: $2\sqrt{75} - 4\sqrt{27}$

Solution

$$= 2\sqrt{75} - 4\sqrt{27}$$
$$= 2\sqrt{(25)(3)} - 4\sqrt{(9)(3)}$$
$$= 2(5)\sqrt{3} - 4(3)\sqrt{3}$$
$$= 10\sqrt{3} - 12\sqrt{3}$$
$$= (10 - 12)\sqrt{3}$$
$$= -2\sqrt{3}$$

Example 5 Simplify: $4x\sqrt{25y} - 6x\sqrt{16y}$.

Solution

$$4x\sqrt{25y} - 6x\sqrt{16y}$$
$$= 4x\,(5)\sqrt{y} - 6x(4)\sqrt{y} \qquad \text{Scrapwork: } \sqrt{25} = 5;\ \sqrt{16} = 4$$
$$= 20x\sqrt{y} - 24x\sqrt{y}$$
$$= (20x - 24x\,)\sqrt{y} \qquad \text{(Adding the coefficients of } \sqrt{y}.)$$
$$= -\,4x\sqrt{y}$$

Multiplication of Radicals

Compared to the conditions in the addition of radicals, the only condition here is that the radicals to be multiplied must have the **same index**. Thus, the radicands may be different or the same, but the index must be the same. Note however that radicals with different indices can always be changed to radicals with a common index by using fractional exponents (see p. 312 Examples 6 & 7, bottom of page) and then multiplying the resulting radicals (The common index will be the LCM of the different indices.)

Example 1   Multiply $\sqrt{3}$ and $\sqrt{5}$

Solution

$$(\sqrt{3})(\sqrt{5}$$
$$= \sqrt{(3)(5)}$$
$$= \sqrt{15}$$

Example 2 Multiply $7\sqrt{3}$ and $9\sqrt{5}$

Solution

Procedure: First multiply the non-radical parts (coefficients) and then multiply the radical parts and simplify if possible.

$$(7\sqrt{3}\,)(9\sqrt{5})$$
$$= (7)(9)\sqrt{(3)(5)}$$
$$= 63\sqrt{15}$$

Example 3   Multiply $4\sqrt{3}$ and $5\sqrt{3}$

Solution

$$(4\sqrt{3})(5\sqrt{3}) = (4)(5)\sqrt{(3)(3)}$$
$$= 20\sqrt{9}$$
$$= 20(3)$$
$$= 60$$

Example 4 Find the product of $2\sqrt{3}$ and $5\sqrt{2} - 4\sqrt{3}$

Solution $2\sqrt{3}\ (5\sqrt{2} - 4\sqrt{3})$<----------The parentheses are important.

$\quad = (2\sqrt{3})(5\sqrt{2})\ - (2\sqrt{3})(4\sqrt{3})$ <--------Application of the distributive rule.

$\quad = (2)(5)\sqrt{(3)(2)}\ - (2)(4)\sqrt{(3)(3)}$

$\quad = 10\sqrt{6}\ - 8\sqrt{9}$

$\quad = 10\sqrt{6}\ - 8(3)$

$\quad = 10\sqrt{6} - 24$

$\quad = -24 + 10\sqrt{6}$

Note: In Example 4, you may skip lines 2 and 3.

Example 5 Simplify: $\sqrt{8}\sqrt{24}$
Method 1

$\quad \sqrt{8}\sqrt{24}\ = \sqrt{(8)(24)}$

$\quad\quad\quad = \sqrt{(2)(2)(2)(3)(2)(2)(2)}$

$\quad\quad\quad = \sqrt{(2)(2)(2)\ (2)(2)(2)\ (3)}$

$\quad\quad\quad = \sqrt{(8)(8)3}$

$\quad\quad\quad = 8\sqrt{3}$

Method 2 $\sqrt{8}\sqrt{24}$
Step 1:   Simplify each radical first.

$\quad\quad\quad \sqrt{8}\sqrt{24}$

$\quad\quad = (2\sqrt{2})(2\sqrt{6})$

Step 2: Multiply the radicals.

$\quad\quad\quad = \ (2)(2)(\sqrt{2})(\sqrt{6})$

$\quad\quad\quad = 4\sqrt{12}$

$\quad\quad\quad = 4\sqrt{4}\sqrt{3}$

$\quad\quad\quad = 4(2)\sqrt{3}$

$\quad\quad\quad = 8\sqrt{3}$

Scrapwork:
1.$\sqrt{8}\ = 2\sqrt{2}$
2. $\sqrt{24} = 2\sqrt{6}$

Example 6 $(\sqrt{5})(\sqrt{5}) = \sqrt{25} = 5$

Example 7 $(\sqrt{6})(\sqrt{6}) = \sqrt{36} = 6$

Lesson 63 Exercises

A Add or subtract:

1. $2\sqrt{3} + 4\sqrt{3}$ **2.** $3\sqrt{5} + \sqrt{98} + \sqrt{20} + 8\sqrt{2}$;   **3.** $4\sqrt{48} - \sqrt{12} + \sqrt{75}$

4. $2x\sqrt{9y} - 5x\sqrt{4y}$

Answers: **1.** $6\sqrt{3}$;   **2.** $5\sqrt{5} + 15\sqrt{2}$;   **3.** $19\sqrt{3}$;   **4.** $-4x\sqrt{y}$

B Multiply and simplify:

1. $2\sqrt{5}$ and $3\sqrt{2}$;   **2.** $3\sqrt{5}$ and $4\sqrt{10}$;   **3.** $4\sqrt{2}$ and $5\sqrt{2}$

4. $4\sqrt{3}$ and $4\sqrt{3}$;   **5.** $2x\sqrt{3y}$ and $\sqrt{6y}$

Answers: **1.** $6\sqrt{10}$;   **2.** $60\sqrt{2}$;   **3.** 40 ;   **4.** 48 ;   **5.** $6xy\sqrt{2}$

CHAPTER 20B
More Radicals

Lesson 64: Division of Radicals; Rationalization of Denominators
Lesson 65: Fractional Exponents and Reduction of Indices

Lesson 64
Division of Radicals; Rationalization of Denominators,

Division of a Radical by a Rational Number

Example Simplify: $\dfrac{6 - \sqrt{12}}{2}$

Solution $\dfrac{6 - \sqrt{12}}{2}$

$= \dfrac{6}{2} - \dfrac{\sqrt{12}}{2}$

$= 3 - \dfrac{\overset{1}{\cancel{2}}\sqrt{3}}{\cancel{2}_1}$

Scrapwork: $\sqrt{12} = \sqrt{(2)(2)(3)} = = 2\sqrt{3}$

$= 3 - \sqrt{3}$

Rationalization of Denominators

To rationalize a denominator, we change a given "fraction" to an "equivalent fraction" so that there are no radicals in the denominator. The equivalent fraction may have radicals in the numerator (but not in the denominator).

Example 1 Rationalize the denominator: $\dfrac{3}{\sqrt{6}}$

We will multiply both the denominator and the numerator by a radical such that the radicand in the denominator becomes a perfect square.

We will multiply both the denominator and the numerator by $\sqrt{6}$

$\dfrac{3}{\sqrt{6}} = \dfrac{3}{\sqrt{6}} \dfrac{\sqrt{6}}{\sqrt{6}}$

$= \dfrac{3\sqrt{6}}{\sqrt{36}}$

$= \dfrac{3\sqrt{6}}{6}$

$= \dfrac{\overset{1}{\cancel{3}}\sqrt{6}}{\cancel{6}_2}$

$= \dfrac{\sqrt{6}}{2}$

A motivation for rationalizing denominators:

Note above that in practical applications, $\dfrac{\sqrt{6}}{2}$ is more convenient to use than $\dfrac{3}{\sqrt{6}}$; for example, it is

easier to divide the decimal approximation of $\sqrt{6}$ by 2 than to divide 3 by the decimal approximation of $\sqrt{6}$.

Example 2   Rationalize the denominator: $\sqrt{\dfrac{7}{3}}$ (one-term denominator)

Solution   We shall multiply both the denominator and the numerator by $\sqrt{3}$.

$$\sqrt{\frac{7}{3}} = \sqrt{\frac{7 \cdot 3}{3 \cdot 3}} \quad \text{or} \quad \frac{\sqrt{7}}{\sqrt{3}} \cdot \frac{\sqrt{3}}{\sqrt{3}}$$

$$= \frac{\sqrt{21}}{\sqrt{9}} \qquad \text{or} \quad \frac{\sqrt{21}}{\sqrt{9}}$$

$$= \frac{\sqrt{21}}{3} \qquad \text{or} \quad \frac{\sqrt{21}}{3}$$

Example 3   Rationalize the denominator : $\sqrt[3]{\dfrac{5}{4}}$

Solution

$$\sqrt[3]{\frac{5}{4}} = \sqrt[3]{\frac{5}{2 \cdot 2}}$$

$$= \sqrt[3]{\frac{5 \cdot 2}{2 \cdot 2 \cdot 2}} \quad \text{<-------One more 2 will make the denominator a perfect cube.}$$

$$= \frac{\sqrt[3]{10}}{\sqrt[3]{8}}$$

$$= \frac{\sqrt[3]{10}}{2} \; .$$

Example 4 Rationalize the denominator: $\sqrt[3]{\dfrac{8}{9}}$

Method 1 $\sqrt[3]{\dfrac{8}{9}} = \sqrt[3]{\dfrac{(2)(2)(2)}{(3)(3)}}$

$$= \sqrt[3]{\frac{(2)(2)(2)(3)}{(3)(3)(3)}} \quad \text{<--- one more 3 will make the denominator a perfect cube}$$

$$= \frac{2\sqrt[3]{3}}{3}$$

Method 2 $\sqrt[3]{\dfrac{8}{9}} = \dfrac{\sqrt[3]{8}}{\sqrt[3]{9}}$

$$= \frac{2}{\sqrt[3]{9}} \qquad\qquad (\sqrt[3]{8} = 2)$$

$$= \frac{2\sqrt[3]{3}}{\sqrt[3]{9} \, \sqrt[3]{3}}$$

$$= \frac{2\sqrt[3]{3}}{\sqrt[3]{27}}$$

$$= \frac{2\sqrt[3]{3}}{3}$$

Method 3   In the following approach, simplifying the numerator becomes more involved than Methods 1 & 2:

$$\sqrt[3]{\frac{8}{9}} = \frac{\sqrt[3]{8}}{\sqrt[3]{9}} \cdot \frac{\sqrt[3]{9}(\sqrt[3]{9})}{\sqrt[3]{9}(\sqrt[3]{9})} = \frac{\sqrt[3]{8 \cdot 9 \cdot 9}}{\sqrt[3]{9 \cdot 9 \cdot 9}} = \frac{\sqrt[3]{8 \cdot 9 \cdot 9}}{9}$$

$$= \frac{\sqrt[3]{2 \cdot 2 \cdot 2 \cdot 3 \cdot 3 \cdot 3 \cdot 3}}{9} = \frac{2 \cdot 3\sqrt[3]{3}}{9} = \frac{2\sqrt[3]{3}}{3}$$

Example 5 Rationalize the denominator: $\dfrac{\sqrt{5}}{\sqrt[3]{2}}$

Solution $\dfrac{\sqrt{5}}{\sqrt[3]{2}} = \dfrac{\sqrt{5}}{\sqrt[3]{2}} \cdot \dfrac{\sqrt[3]{2}(\sqrt[3]{2})}{\sqrt[3]{2}(\sqrt[3]{2})} = \dfrac{\sqrt{5} \cdot \sqrt[3]{4}}{2}$ or $\dfrac{\sqrt[6]{2000}}{2}$ (see also page 312 Examples 6 & 7, bottom of page)

Definition : The **conjugate** of a given binomial is another binomial that differs from the given binomial only in the sign of one of the terms. The conjugate of $a + b$ is $a - b$; and the conjugate of $a - b$ is $a + b$.

Example 6   Rationalize the denominator $\dfrac{5}{\sqrt{3} + \sqrt{2}}$ (two-term denominator).

Procedure: Multiply both the denominator and the denominator by the conjugate of $\sqrt{3} + \sqrt{2}$.

The conjugate of $\sqrt{3} + \sqrt{2}$ is $\sqrt{3} - \sqrt{2}.$ (You may also multiply by $-\sqrt{3} + \sqrt{2}$; try it later on.)

Step 1: $\dfrac{5}{\sqrt{3} + \sqrt{2}} = \dfrac{5 \ (\sqrt{3} - \sqrt{2})}{(\sqrt{3} + \sqrt{2})(\sqrt{3} - \sqrt{2})}$

Step 2: $= \dfrac{5(\sqrt{3} - \sqrt{2})}{(\sqrt{3})(\sqrt{3}) - (\sqrt{3})(\sqrt{2}) + (\sqrt{2})(\sqrt{3}) - (\sqrt{2})(\sqrt{2})}$

$= \dfrac{5(\sqrt{3} - \sqrt{2})}{\sqrt{9} + 0 - \sqrt{4}}.$ Note: $(-\sqrt{3})(\sqrt{2}) + (\sqrt{2})(\sqrt{3}) = 0$

$= \dfrac{5(\sqrt{3} - \sqrt{2})}{3 - 2}$

$= \dfrac{5(\sqrt{3} - \sqrt{2})}{1}$

$= 5(\sqrt{3} - \sqrt{2})$ or $5\sqrt{3} - 5\sqrt{2}$

Note above: As was suggested in the procedure, you could also have multiplied by $-\sqrt{3} + \sqrt{2}$ and obtained the same result. Therefore, it is not critical which of the terms differ in sign.

Example 7   Rationalize the denominator : $\dfrac{1}{\sqrt{8} - 2}$

Procedure: Multiply both the denominator and the numerator by the conjugate of $\sqrt{8} - 2$.

   The conjugate of $\sqrt{8} - 2$ is $\sqrt{8} + 2$ (You may also multiply by $-\sqrt{8} - 2$; try it later on.)

$$\frac{1}{\sqrt{8} - 2} = \frac{1}{(\sqrt{8} - 2)}\frac{(\sqrt{8} + 2)}{(\sqrt{8} + 2)}$$

$$= \frac{\sqrt{8} + 2}{(\sqrt{8})(\sqrt{8}) + 2\sqrt{8} - 2\sqrt{8} - 4} \qquad \text{<----After some practice, you may skip writing this step.}$$

$$= \frac{\sqrt{8} + 2}{\sqrt{64} + 0 - 4} \qquad\qquad (2\sqrt{8} - 2\sqrt{8} = 0)$$

$$= \frac{\sqrt{8} + 2}{8 - 4}$$

$$= \frac{2\sqrt{2} + 2}{4}$$

$$= \frac{2(\sqrt{2} + 1)}{4}$$

$$= \frac{\sqrt{2} + 1}{2} \quad \text{or} \quad \frac{1}{2} + \frac{\sqrt{2}}{2}$$

Note above that $\dfrac{1}{\sqrt{8} - 2} = \dfrac{1}{-2 + \sqrt{8}}$

* Perhaps, it would be better to say "a" conjugate instead of "the" conjugate since it does not matter which terms differ in sign, so far as the rationalization of denominators is concerned.

Lesson 64 Exercises

A **1.** Simplify: $\dfrac{8 + \sqrt{48}}{4}$ **2.** $\dfrac{12 - \sqrt{18}}{3}$

Answers: **1.** $2 + \sqrt{3}$;   **2.** $4 - \sqrt{2}$

B Rationalize the denominators:

1. $\dfrac{1}{\sqrt{3}}$;   **2.** $\dfrac{2}{\sqrt{5}}$;   **3.** $\dfrac{5}{\sqrt{2}}$;   **4.** $\dfrac{6}{\sqrt{5} - \sqrt{3}}$;   **5.** $\dfrac{\sqrt{3}}{4 + \sqrt{3}}$;

6. $\sqrt[3]{\dfrac{4}{9}}$;   **7.** $\sqrt[3]{\dfrac{8}{9}}$.   **8.** Is $\dfrac{\sqrt{3}}{\sqrt{2}} = \dfrac{3}{\sqrt{6}}$?

Answers: **1.** $\dfrac{\sqrt{3}}{3}$; **2.** $\dfrac{2\sqrt{5}}{5}$; **3.** $\dfrac{5\sqrt{2}}{2}$; **4.** $3\sqrt{5} + 3\sqrt{3}$; **5.** $\dfrac{-3 + 4\sqrt{3}}{13}$; **6.** $\dfrac{\sqrt[3]{12}}{3}$; **7.** $\dfrac{2\sqrt[3]{3}}{3}$; **8.** Yes

Lesson 65
Fractional Exponents and Reduction of Indices

Fractional Exponents

Algebraic operations involving radicals can sometimes be made easy by changing the radicals involved to their equivalent exponential forms, and then applying the laws of exponents.
The advice is that when a radical is easy to simplify as is, simplify accordingly; but if it is too complicated and not easy to simplify, consider changing to exponential form and then simplifying, followed by conversion back to radical form.

Interconversion between radicals and exponential expressions

Examples

1. $25^{1/2} = \sqrt[2]{25} = \sqrt{25} = 5$
2. $9^{1/2} = \sqrt{9} = 3$
3. $8^{1/3} = \sqrt[3]{8} = 2$
4. $8^{2/3} = (8^{1/3})^2 = (\sqrt[3]{8})^2 = 2^2 = 4$
5. $16^{3/4} = (16^{1/4})^3 = (\sqrt[4]{16})^3 = 2^3 = 8$

Rules: **1**. $\sqrt[n]{a} = a^{1/n}$

$\quad$ **2.** $\sqrt[n]{a^m} = a^{m/n}$

$\quad$ **3.** $\sqrt[n]{a^m} = (\sqrt[n]{a})^m$

Example 1 $\quad$ Simplify the following:

1. $a^{1/3} \cdot a^{2/3}$; $\quad$ **2.** $b^{2/3} \cdot b^{1/2}$; $\quad$ **3.** $4^{-1/2}$; $\quad$ **4.** $2^{2/5} \cdot 2^{1/4}$ $\quad$ **5.** $\sqrt[5]{x^2} \cdot \sqrt[5]{x^3}$

6. $\sqrt[5]{3^2} \cdot \sqrt[4]{3}$; $\quad$ **7.** $\sqrt{x} \cdot \sqrt[3]{x^2}$

Solutions

1. $a^{1/3} \cdot a^{2/3} = a^{1/3 + 2/3} = a^1 = a$

2. $b^{2/3} \cdot b^{1/2} = b^{2/3 + 1/2} = b^{7/6}$ $\quad\quad$ $(2/3 + 1/2 = 7/6)$

3. $4^{-1/2} = \dfrac{1}{4^{1/2}} = \dfrac{1}{\sqrt{4}} = \dfrac{1}{2}$

4. $2^{2/5} \cdot 2^{1/4} = 2^{2/5 + 1/4} = 2^{13/20}$

5. $\sqrt[5]{x^2} \cdot \sqrt[5]{x^3} = x^{2/5} x^{3/5} = x^{5/5} = x^1 = x$ $\quad$ or $\sqrt[5]{x^2} \cdot \sqrt[5]{x^3} = \sqrt[5]{x^2 \cdot x^3} = \sqrt[5]{x^5} = x$

6. $\sqrt[5]{3^2} \cdot \sqrt[4]{3} = 3^{2/5} \cdot 3^{1/4} = 3^{13/20} = \sqrt[20]{3^{13}}$ $\quad$ <---(Multiplying radicals with different indices)

7. $\sqrt{x} \cdot \sqrt[3]{x^2} = x^{1/2} \cdot x^{2/3} = x^{7/6} = \sqrt[6]{x^7} = x\sqrt[6]{x}$ $\quad$ <---(Multiplying radicals with different indices)

Example 2   Evaluate. $1000^{-1/3}$

Step 1: Express with positive exponents.

$$1000^{-1/3} = \frac{1}{1000^{1/3}}$$

Step 2:   Change the exponential form to radical form.

then, $\dfrac{1}{1000^{1/3}} = \dfrac{1}{\sqrt[3]{1000}} = \dfrac{1}{10}$

Example 3   Evaluate. $81^{-1/2}$

$$81^{-1/2} = \frac{1}{81^{1/2}}$$

$$= \frac{1}{\sqrt{81}}$$

$$= \frac{1}{9}$$

Example 4   Evaluate $\left(-\dfrac{8}{27}\right)^{-5/3}$

We will cover two methods.

Method 1

$$\left(-\frac{8}{27}\right)^{-5/3} = \left(\frac{-8}{27}\right)^{-5/3}$$ <--- (We can give the minus sign to either the "8"or the"27")

$$= \frac{1}{\left(\frac{-8}{27}\right)^{5/3}}$$

$$= \frac{1}{\left(\sqrt[3]{\frac{-8}{27}}\right)^{5}}$$

$$= \frac{1}{\dfrac{(\sqrt[3]{-8})^5}{(\sqrt[3]{27})^5}}$$

$$= \frac{1}{\dfrac{-32}{243}} \quad \text{or} \quad 1 \div \frac{-32}{243}$$

$$= -\frac{243}{32}$$

Scrapwork: 1. $\sqrt[3]{-8} = -2$

2. $(-2)^5 = \mathbf{-32}$

3. $\sqrt[3]{27} = 3$

4 . $3^5 = \mathbf{243}$

Method 2 (Much faster than Method 1)

$$\left(-\frac{8}{27}\right)^{-5/3} = \left(\frac{-8}{27}\right)^{-5/3}$$

$$= \left(\frac{27}{-8}\right)^{5/3} \quad \text{(Interchange the numerator and the denominator and change the sign of the exponent)}$$

$$= \frac{(\sqrt[3]{27})^5}{(\sqrt[3]{-8})^5} = \frac{(3)^5}{(-2)^5} = \frac{243}{-32}$$

$$= -\frac{243}{32}$$

Reduction of Indices

Example 1 Reduce the index: $\sqrt[4]{36}$

Step 1: Change the radical to exponential form (power form).

Then, $\sqrt[4]{36} = \sqrt[4]{6^2} = 6^{2/4}$

Step 2: Reduce the exponent $\frac{2}{4}$ to lowest terms.

then $\frac{2}{4} = \frac{1}{2}$

Step 3: $6^{2/4} = 6^{1/2}$

Step 4: Change back to the radical form.

then $6^{1/2} = \sqrt{6}$ **Note:** $6^{1/2} = \sqrt[2]{6} = \sqrt{6}$

Perfect powers

Examples: **1.** 9 is a perfect square.
            2. 8 is a perfect cube .
            3. $16y^2$ is a perfect square .
            4. $\frac{4}{9}$ is a perfect square .

What is meant by to simplify a radical?

1. It may mean remove all perfect powers from the radicand (page 302); or

2. It may mean rationalize the denominator (page 308)

3. It may mean reduce the index of the radical (page 314)

4. It may mean perform a combination of the above.

Lesson 65 Exercises

A   Simplify the following:

1. $x^{3/4} \cdot x^{1/4}$ **2.** $x^{2/3} \cdot x^{3/4}$ **3.** $3^{1/2} \cdot 3^{2/3}$

Answers: **1.** x ;   **2.** $x^{17/12}$;   **3.** $3^{7/6}$

B   Evaluate the following:

1. $100^{-1/2}$;   **2.** $25^{-1/2}$;   **3.** $\left(\frac{16}{81}\right)^{-1/4}$

Answers: **1.** $\frac{1}{10}$;   **2.** $\frac{1}{5}$;   **3.** $\frac{3}{2}$

C Reduce the index:

1. $\sqrt[4]{x^2}$;   **2.** $\sqrt[6]{x^8}$;   **3.** $\sqrt[4]{25}$;   **4.** $\sqrt[4]{4}$

Answers: 1 . $\sqrt{x}$;   **2.** $\sqrt[3]{x^4}$;   **3.** $\sqrt{5}$; **4.** $\sqrt{2}$.

CHAPTER 21

Lesson 66

The Right Triangle, the Pythagorean Theorem and Applications

A **right triangle** is a triangle which has one right angle. In a right triangle, the side opposite the right angle is known as the hypotenuse. Also, the hypotenuse is the longest side of a right triangle.

The Pythagorean theorem

The Pythagorean theorem states that in any right triangle, the square of the length of the hypotenuse is equal to the sum of the squares of the lengths of the other two sides.

Example 1 Find x:

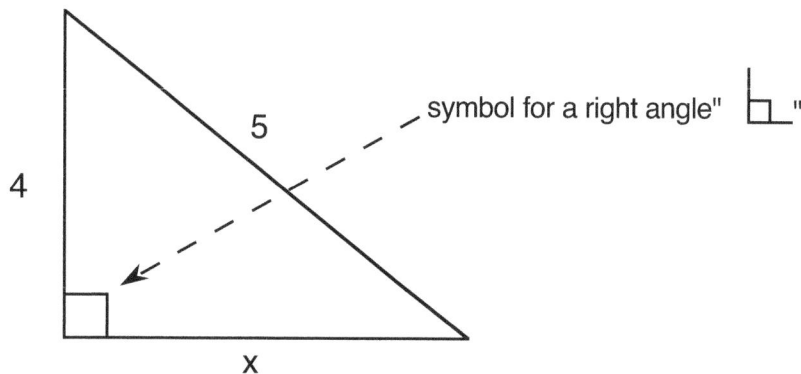

symbol for a right angle" ⌐ "

Step 1: The given triangle is a right triangle (by the right angle symbol shown.)
Pick the length of the hypotenuse first. We pick 5 first. (The length of the hypotenuse is 5)

Step 2: Apply the Pythagorean theorem.

$$5^2 = 4^2 + x^2 \longleftarrow \text{------------Note:}$$

Here, the unknown is not on one side of the equation by itself. Always, it is the hypotenuse term which is on one side of the equation by itself.

square of (length)
the hypotenuse

The squares of the (lengths)
other two sides

Continuing,

$$25 = 16 + x^2$$
$$\underline{-16 \quad -16}$$
$$9 = x^2$$
$$\sqrt{9} = x$$
$$3 = x \text{ or } x = 3$$

Example 2   Find h

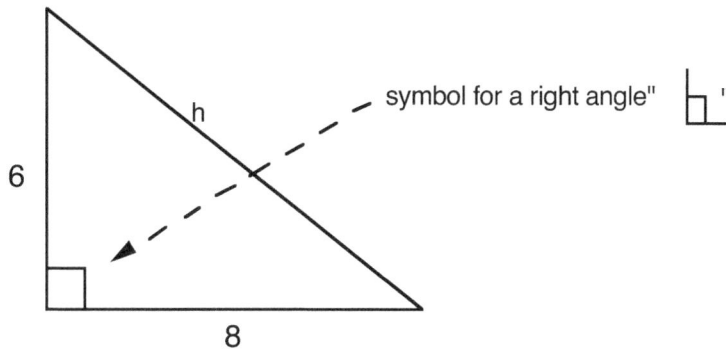

Solution: We will apply the Pythagorean theorem since the triangle is a right triangle.

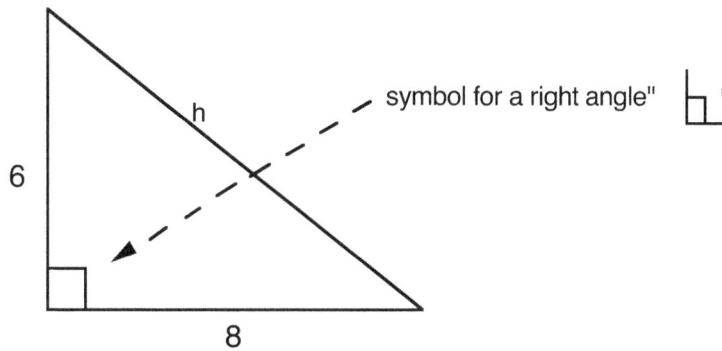

By the Pythagorean theorem,

$h^2 = 6^2 + 8^2$ (The length of the hypotenuse is h)
$h^2 = 36 + 64$
$h^2 = 100$
$h = \sqrt{100}$
$h = 10$

Example 3   Find x

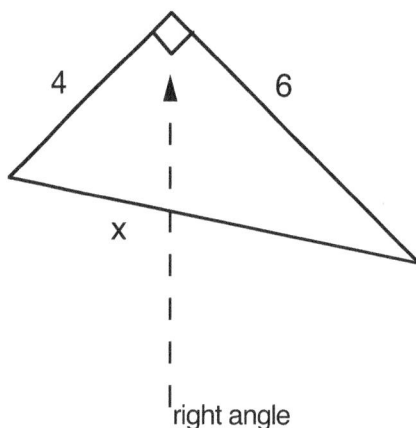

right angle

Step 1: Since the given triangle is a right triangle, we can apply the
          Pythagorean theorem. The length of the hypotenuse is x.

Step 2:   Applying the Pythagorean theorem,

$$x^2 = 4^2 + 6^2$$
$$x^2 = 16 + 36$$
$$x^2 = 52$$

$$x = \sqrt{52}$$ **Note:** 52 is not a perfect square but we

can simplify the radical $\sqrt{52}$.

$$x = \sqrt{(4)(13)}$$ (Remove any perfect square in 52)
$$x = \sqrt{4}\sqrt{13}$$
$$x = 2\sqrt{13}$$

Note: It is a good practice to always pick the hypotenuse first, whether or not it is known. Sometimes,
some students pick the unknown first and this may lead to the misapplication of the Pythagorean theorem.
  **Therefore, remember to always pick the hypotenuse (the side opposite the right angle)
first.** The Pythagorean theorem is **valid only for a right triangle** and therefore, we should **not** apply
it to any triangle.

Question: How do we know if the triangle is a right triangle?

Answer: The given information may state so, or the given information may
be used to deduce that the triangle is a right triangle.

The following symbolic form of the **Pythagorean theorem** will be useful in the future.
By convention, c is the length of the hypotenuse, a and b are the lengths of the other two sides.

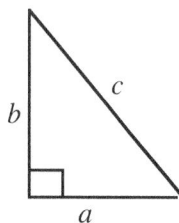

The Pythagorean theorem: $$c^2 = a^2 + b^2$$

Lesson 66 Exercises

A

1. Find c

2. Find b

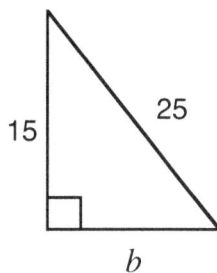

Answers: **1.** $c = 10$ **2.** $b = 20$

B In each of the following right triangles, find x:

1.

2.

3.

4.

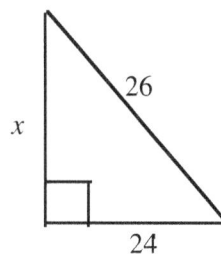

Answers: **1.** $x = 5$; **2.** $x = 17$;   **3.** $x = \sqrt{31}$;   **4.** $x = 10$

CHAPTER 22
Areas and Perimeters

Lesson 67: **Area and Perimeter of Triangles, Rectangles, and Trapezoids**

Lesson 68: **Area and Perimeter of a Circle; Arc Length and Sector of a Circle; Area and Perimeter of Composite Figures**

Lesson 67
Area and Perimeter of Triangles, Rectangles, and Trapezoids

Example 1 Find the area of figure (triangle) below.

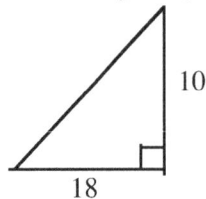

Solution : The figure is a right triangle with base 18 units and altitude 10 units.

$$\text{The area of a triangle} = \frac{1}{2} \text{ base} \times \text{altitude}$$

$$= \frac{1}{2}(18)(10)$$

$$= 90 \text{ square units}$$

Note: The altitude and the base always meet at right angles. In the above problem, it does not matter which of 18 or 10 we call the the base. We could have taken the base to be 10 and then, the altitude would have been 18. The altitude of the triangle is also known as the height of the triangle.

Example 2 In the figure below:

(a) Find the area of the figure.

(b) Find the perimeter of figure.

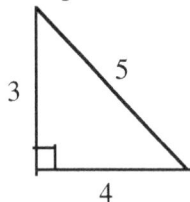

(a) $b = 4, h = 3$

$$\text{Area of the triangle} = \frac{1}{2} bh$$

$$= \frac{1}{2}(4)(3)$$

$$= 6 \text{ sq. units}$$

(b) Perimeter of the triangle = 3 + 4 + 5

$$= 12 \text{ units}$$

Example 3   In the figure below:
      (a) Find the area of the figure.
      (b) Find the perimeter of figure.

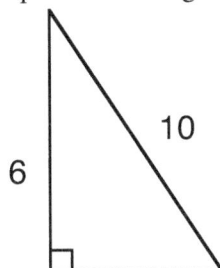

Solution
(a)

Let the base be b, and the altitude h.

Area of the triangle $= \frac{1}{2} bh$.

$h = 6$, but we are **not** given b; however

Step 1: We can find b by using the Pythagorean theorem since the triangle
      is a right triangle.

$$10^2 = 6^2 + b^2$$
$$100 = 36 + b^2$$
$$64 = b^2$$
$$8 = b$$
$$b = 8$$

Step 2: Now, $b = 8, h = 6$

Area of the triangle $= \frac{1}{2} bh$

$$= \frac{1}{2}(8)(6)$$
$$= 24 \text{ sq. units}$$

The area of the triangle is 24 sq. units

(b)   From figure below,
     Perimeter $= 6 + 8 + 10$
                $= 24 \text{ units}$

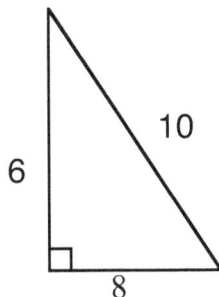

Note: Obtaining the same numerical answers in **(a)** and **(b)** is a coincidence; and therefore, the perimeter is
      not always numerically equal to the area.

Area and Perimeter of a Trapezoid

Example 4   Find the area of the figure shown below.

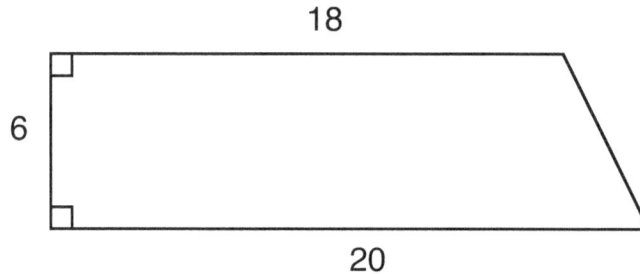

18

6

20

Method 1: The figure is a trapezoid.

The formula for the area, A, of a trapezoid is given

by $A = \frac{1}{2}(b_1 + b_2)h$, where b_1 and b_2 are the bases

(the bases are the parallel sides) and h is the altitude.

$b_1 = 20, b_2 = 18, h = 6$

$$A = \frac{1}{2}(20 + 18)(6)$$

$$= \frac{1}{2}(38)(6)$$

$$= 19(6)$$

$$= 114 \text{ square units}$$

The area of the figure is 114 square units.

Method 2: We break up the figure into a rectangle and a triangle.

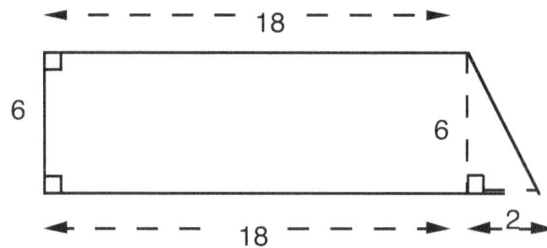

18

6

6

18

2

The area of the rectangle $= 6 \times 18$

$$= 108 \text{ sq. units}$$

The area of the triangle $= \frac{1}{2}(2)(6)$

$$= 6 \text{ sq. units}$$

Area of the figure = Area of the rectangle + Area of the triangle

$$= (108 + 6) \text{ sq. units}$$

$$= 114 \text{ sq. units}$$

Again, we obtain the same answer as in Method 1.

Example 5 Find the area of the figure below.

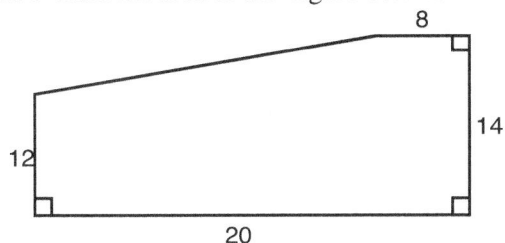

Solution   We break up the figure into a rectangle and a trapezoid.

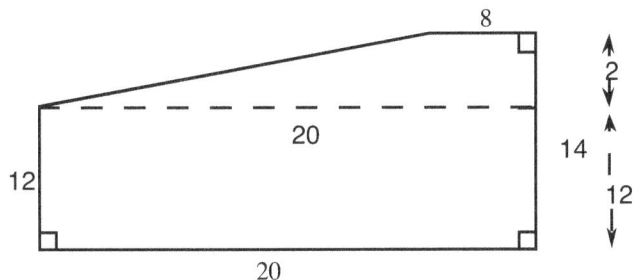

The area of the rectangle = 12 (20)

$\qquad\qquad\qquad\qquad$ =240 sq. units

The area of the trapezoid = $\frac{1}{2}(b_1 + b_2)h$

$\qquad\qquad\qquad\qquad = \frac{1}{2}(20 + 8)(2)$ $\qquad\qquad b_1= 20 , b_2 = 8, \ h = 2$

$\qquad\qquad\qquad\qquad = \frac{1}{2}(28)(2)$

$\qquad\qquad\qquad\qquad = 28$ sq. units

Total area = area of the rectangle + area of the trapezoid

$\qquad\qquad$ =(240 + 28) sq. units

$\qquad\qquad$ = 268 sq. units

Note in the above problem that the original figure given is **not** a trapezoid,
since it has five sides. By definition, a trapezoid is a quadrilateral (a four-sided figure) which has exactly a pair
of parallel sides.
Note also that in the above problems, given a complex figure for which we do not know any formulas, we
can break up the figure into simpler figures (such as a rectangle, a triangle, or a trapezoid), whose formulas we
know, and then apply the appropriate formulas accordingly to find the areas, which we can then add together.
Sometimes, we can extend the sides of the figures to form "known" figures and then, in this case, subtract
the extensions of the areas.

Lesson 67 Exercises

A Find (a) the area and (b) the perimeter of the figure below.

Answers: (a) 84 sq. units; (b) $(34 + 2\sqrt{13})$ units

B

1. Find (a) Area; (b) Perimeter of right triangle ABC.
2. Find (a) Area; (b) Perimeter of right triangle *DEF*.

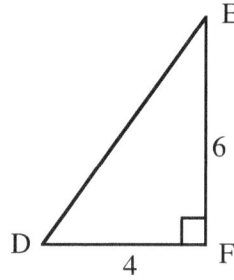

Answers: Triangle ABC:(a). 96 sq. units; (b). 48 units;   Triangle *DEF*: (a) 12 sq. units; (b) $10 + 2\sqrt{13}$

Lesson 68
Area and Perimeter of a Circle; Arc Length and Sector of a Circle; Area and Perimeter of Composite Figures

Area and Perimeter of a Circle

Area of a circle

The area, A , of a circle of radius , r, is given by $A = \pi r^2$.

π is approximately equal to 3.1416; in calculations, we will leave our answers in terms of π, unless instructed otherwise.

Note also that the radius of a circle $= \frac{1}{2}$ of the diameter of the circle.

Area of a semicircle (half-circle) $= \frac{1}{2}$ of the area of the given circle.

Area of a quarter-circle $= \frac{1}{4}$ of the area of the given circle.

Perimeter (or Circumference) of a Circle

The perimeter of a circle is the distance around the circle (i.e., the length of the circle).

The perimeter, P, of a circle of radius, r, is given by $P = 2\pi r$.

The perimeter of a semicircle $= \frac{1}{2}$ of the perimeter of the given circle.

 The perimeter of a quarter circle is $\frac{1}{4}$ of the perimeter of the given circle.

Example Find the area and the perimeter of a circle of radius 6 units.

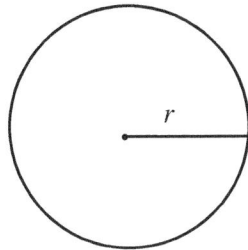

Finding the area
$r = 6$
Area of a circle $= \pi r^2$.
Area of the given circle $= \pi(6)^2$
$= \pi(36)$
$= 36\pi$ sq. units

Finding the perimeter
$r = 6$ units
Perimeter of a circle $= 2\pi r$
Perimeter of the given circle $= 2\pi(6)$
$= 12\pi$ units

Note: In the above problem, the area of a **semicircle** of radius, $r = 6$ is $\frac{1}{2}(36\pi)$ $= 18\pi$ sq. units;

and the perimeter of the semicircle of radius, $r = 6$, is $\frac{1}{2}(12\pi)$
$= 6\pi$ units.

Arc Length and Sector of a Circle

Arc length: Arc length is the length of a part of a circle between any two points on the circle. In a circle of radius r, the arc length s cut off by a central angle θ (θ in degrees) is given by

$s = \dfrac{\theta}{360}$ of the circumference of the circle.

$$\boxed{s = \dfrac{\theta}{360}(2\pi r) \text{ or } \dfrac{\theta \pi r}{180}}.$$

However, if the central angle is in radians,   (Note: $180° = \pi$ radians)

$$\boxed{s = r\theta}$$

Sector of a circle: A sector of a circle is the part of the interior of the circle bounded by two radii and the intercepted arc.

In a circle of radius r, the area A of a sector with central angle θ (θ in degrees) is given by

$A = \dfrac{\theta}{360}$ of the area of the circle.

$$\boxed{A = \dfrac{\theta}{360}(\pi r^2)}$$

However, if θ is in radians,

$$\boxed{A = \tfrac{1}{2}r^2\theta}$$

Example In the circle shown below, if the diameter is 6 and the central angle is 30°,
    (a) Find the circumference of the circle.
    (b) Find the arc length of CB
    (c) Find the area of sector COB.

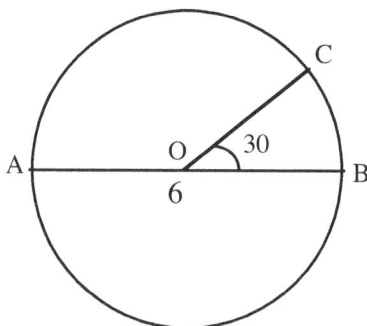

Figure 1

Solution

(a) Circumference of a circle of radius r is given by $2\pi r$.

Radius of a circle $= \dfrac{1}{2}$ of the diameter of the circle.

Therefore, $r = \dfrac{1}{2}(6) = 3$

Circumference of the given circle is $2\pi r. = 2\pi(3) = 6\pi$ units.

(b) arc length, s, of CB $= \dfrac{\theta}{360}$ of the circumference of the circle

$$s = \dfrac{30}{360} \cdot 2\pi(3) = \dfrac{\pi}{2} \text{ units}$$

OR if θ is in radians, $s = r\theta = 3 \cdot \dfrac{\pi}{6}$ (**Note**: $\theta = 30° = \dfrac{\pi}{6}$; $r = 3$.)

$$s = \dfrac{\pi}{2} \text{ units}.$$

(c) Area, A , of the shaded sector $= \dfrac{30°}{360°}$ of the area of the circle.

$$A = \dfrac{30°}{360°} (\pi r^2)$$
$$= \dfrac{30°}{360°} (\pi \cdot 3^2)$$
$$= \dfrac{30}{360} (9\pi)$$
$$= \dfrac{3\pi}{4} \text{ sq. units}$$

OR if θ is in radians, $A = \dfrac{1}{2} r^2 \theta$ (**Note**: $\theta = 30° = \dfrac{\pi}{6}$; $r = 3$.)

$$= \dfrac{1}{2} \cdot 3^2 \cdot \dfrac{\pi}{6}$$
$$= \dfrac{1}{2} \cdot \dfrac{9}{1} \cdot \dfrac{\pi}{6}$$
$$= \dfrac{3\pi}{4} \text{ sq. units}.$$

Areas and Perimeters of Composite Figures

Example 1   In the figure below:
(a) Find the area of the figure.
(b) Find the perimeter of figure.

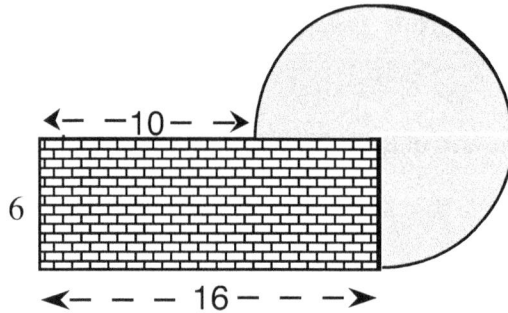

Solution

(a) Finding the area

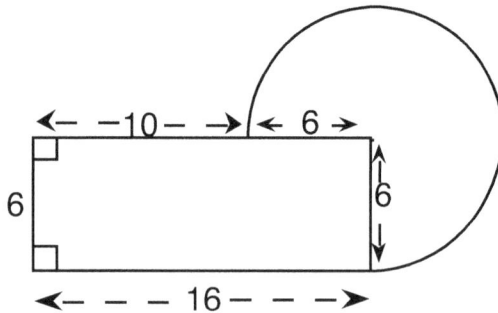

Radius, r, of the three-quarter circle $= 6$

Area of the three-quarter circle $= \frac{3}{4}\pi r^2$ ($\frac{3}{4}$ of the area of a circle of radius 6 units)

$$= \frac{3}{4}\pi(6)^2$$

$$= \frac{3}{4}\pi(36)^{9}$$

$$= 27\pi \text{ sq. units}$$

Area of the 16 by 6 rectangle $= 16 \times 6$ sq. units
$= 96$ sq. units
Area of the **whole figure** = area of rectangle + area of three-quarter circle
$= (96 + 27\pi)$ sq. units

(b) Finding the perimeter

Perimeter of the three-quarter circle $= \dfrac{3}{4} \, 2\pi r$

$$= \dfrac{3}{4} \, (2\pi)(6)$$

$$= 9\pi \text{ units}$$

Perimeter of the whole figure $= 6 + 16 + 9\pi + 10$

$$= (32 + 9\pi) \text{ units}$$

Begin here and go around counterclockwise

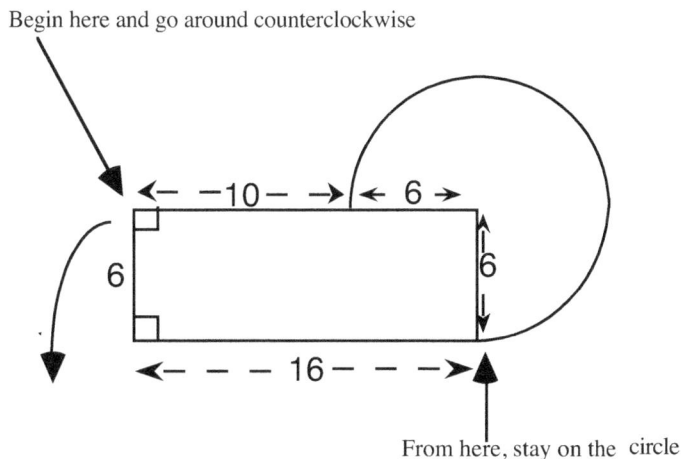

From here, stay on the circle

Note that in finding the perimeter we did **not** add the parts (6 and 6) of the rectangle which are also radii of the three-quarter circle, because the perimeter is the distance around the figure.

Lesson 68 Exercises

A

In the circle below, if the diameter of the circle is 18 units and the central angle has a measure $60°$,
(a) Find the circumference of the circle.
(b) Find the area of sector COB.
(c) Find the arc length of CB.

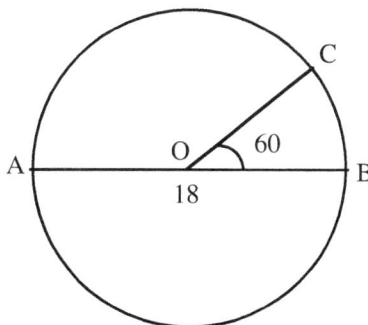

Answers (a) 18π units; (b) $\dfrac{27\pi}{2}$ sq. units; (c) 3π units

B

(a) Find the area of the figure.
(b) Find the perimeter of figure.

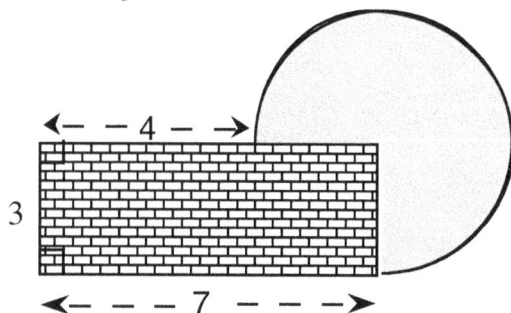

Answers: **(a)** $(21 + \dfrac{27\pi}{4})$ sq. units;   **(b)**. $(14 + \dfrac{9\pi}{2})$ units

Test # 12 -Student's Self-Test (**Always**, **Test yourself before you are tested**)

Attempt all questions on clean sheets of paper, **Do not write in the book** Show all necessary work.

1. Simplify: (a) $\sqrt{12}$; (b) $\sqrt{48}$; (c) $\sqrt{75}$.	Simplify: **2.** (a) $\sqrt{162}$; (b) $\sqrt{128}$; (c) $\sqrt{96}$; (d) $\sqrt{\dfrac{16}{81}}$.
3. Simplify: (a) $\sqrt{36x^8}$; (b) $\sqrt{4x^6y^4}$; (c) $\sqrt{8x^7y^2}$	**4.** Simplify the following: (a) $\sqrt{8x^5y^7}$; (b) $\sqrt{18x^{10}y}$; (c) $\sqrt{32x^6y^4z^5}$

5. Add or subtract:
(a) $5\sqrt{3} + 3\sqrt{3}$
 	
(b) $2\sqrt{6} + \sqrt{18} + \sqrt{24} - 2\sqrt{72}$
 	
(c) $\sqrt{12} + \sqrt{48} + \sqrt{75}$.
	6. Add or subtract:
	   (a) $4\sqrt{50} - \sqrt{24} + 5\sqrt{2}$
	   (b) $3x\sqrt{25y} - 2x\sqrt{16y}$

7. Multiply and simplify:

   (a) $4\sqrt{5}$ and $2\sqrt{6}$;   (b) $2\sqrt{5}$ and $3\sqrt{30}$;

   (c) $3\sqrt{2}$ and $6\sqrt{2}$

8. Multiply and simplify:

   (a) $5\sqrt{7}$ and $6\sqrt{7}$;   (b) $3x\sqrt{2y}$ and $\sqrt{3y}$.

9. Find c

10. Find b

11. Fund x:

12. Find x:

13. Find x:

14. Find x:

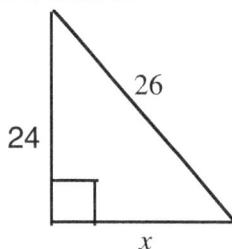

15. Find (a) Area; (b) Perimeter of right triangle ABC

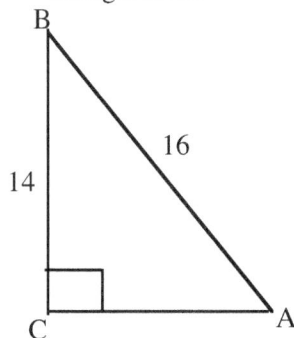

16. Find (a) Area; (b) Perimeter of right triangle DEF.

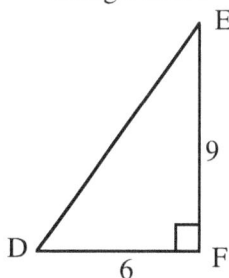

17. Find (a) the area and (b) the perimeter of the figure below.

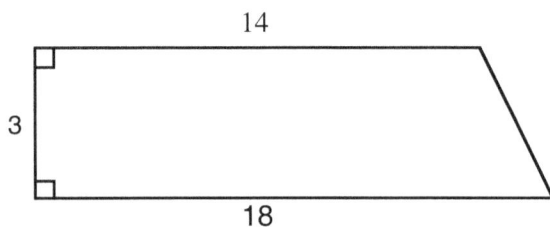

18. a) Find the area of the figure.
(b) Find the perimeter of figure.

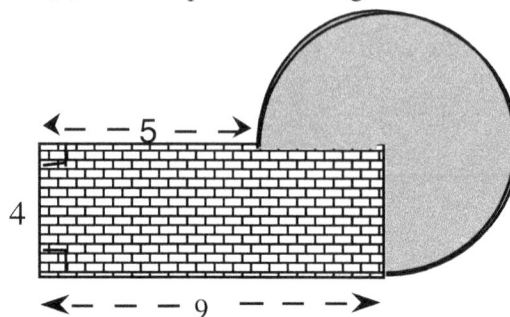

Answers: **1.** (a) $2\sqrt{3}$; (b) $4\sqrt{3}$; (c) $5\sqrt{3}$; **2.** (a) $9\sqrt{2}$; (b) $8\sqrt{2}$; (c) $4\sqrt{6}$; (d) $\frac{4}{9}$;

3. (a) $6x^4$; (b) $2x^3y^2$; **(c)** $2x^3y\sqrt{2x}$; **4.** (a) $2x^2y^3\sqrt{2xy}$; (b) $3x^5\sqrt{2y}$; (c) $4x^3y^2z^2\sqrt{2z}$;

5. (a) $8\sqrt{3}$; (b) $4\sqrt{6}-9\sqrt{2}$; (c) $11\sqrt{3}$; **6.** (a) $25\sqrt{2}-2\sqrt{6}$; (b) $7x\sqrt{y}$;

7. (a) $8\sqrt{30}$; (b) $30\sqrt{6}$; (c) 36; **8.** (a) 210 ; (b) $3xy\sqrt{6}$; **9.** $c=10$; **10** $b=20$;

11. $x=2\sqrt{13}$; **12.** $x=34$; **13.** $\sqrt{69}$; **14.** $x=10$; **15.** (a) $14\sqrt{15}$; (b) $(30+2\sqrt{15})$ units;

16. (a) 27 sq. units; (b) $(15+3\sqrt{13})$ units; **17.** (a) 48 sq. units; (b) 40 units ;

18. (a) $(36+12\pi)$ sq. units; (b) $(18+6\pi)$ units.

CHAPTER 23

ALGEBRAIC FRACTIONS

Lesson 69: Reduction, Multiplication, Division of Algebraic Fractions
Lesson 70: Addition of Fractions; Simplification of Complex Fractions

Lesson 69
Reduction, Multiplication, and Division of Algebraic Fractions

Reduction of Fraction to Lowest Terms

The most important (and perhaps the only) prerequisite in being able to reduce algebraic fractions to lowest terms is **knowing how to factor polynomials.** Therefore, go back and review the factoring of polynomials.

Example 1   Reduce to lowest terms: $\dfrac{x^2 - 5x}{x^2 - 4x - 5}$

Step 1   Factor both the numerator and the denominator

$$\dfrac{x^2 - 5x}{x^2 - 4x - 5}$$

$$= \dfrac{x(x - 5)}{(x + 1)(x - 5)}$$

Step 2:   Cancel (divide out) the common factor in the numerator and the denominator.

$$\dfrac{x\cancel{(x - 5)}}{(x + 1)(\cancel{x - 5})}$$

$$= \dfrac{x}{x + 1}$$

Example 2   Reduce to lowest terms

$$\dfrac{x^2 + 10x + 24}{3x + 12}$$

Step 1: Factor both the numerator and the denominator.

$$\dfrac{x^2 + 10x + 24}{3x + 12} = \dfrac{(x + 4)(x + 6)}{3(x + 4)}$$

Step 2: Cancel the common factors in the numerator and the denominator

$$\dfrac{x^2 + 10x + 24}{3x + 12} = \dfrac{\cancel{(x + 4)}(x + 6)}{3\cancel{(x + 4)}}$$

$$= \dfrac{x + 6}{3}$$

Example 3 Reduce to lowest terms $\dfrac{x^2 + 2x - 24}{12 - 3x}$

Solution

Step 1: Factor the numerator, factor the denominator:

Then, $\dfrac{x^2 + 2x - 24}{12 - 3x} = \dfrac{(x - 4)(x + 6)}{3(4 - x)}$

Step 2: $= \dfrac{(x - 4)(x + 6)}{-3(x - 4)}$ Note that $4 - x = -(-4 + x) = -(x - 4)$

Step 3: $= \dfrac{\cancel{(x - 4)}(x + 6)}{-3\cancel{(x - 4)}}$

Step 4: $= \dfrac{x + 6}{-3}$

Step 5: $= -\dfrac{x + 6}{3}$

A note about cancellation (dividing out):   When may we cancel?
We may cancel when the following two conditions have been satisfied simultaneously.

1. All quantities (factors) in the numerator are multiplying each other, and all the quantities (factors) in the denominator are multiplying each other.

2. There are common quantities (factors) in both the numerator and the denominator.

Also, we consider each expression (such as a binomial) within the parentheses as one quantity.

Example in which we **can** cancel: $\dfrac{2x(x + 2)(x - 1)(x + 3)}{x(x - 1)(x + 3)} = \dfrac{2\cancel{x}(x + 2)\cancel{(x - 1)}\cancel{(x + 3)}}{\cancel{x}\cancel{(x - 1)}\cancel{(x + 3)}} = 2(x + 2)$

Example in which we **cannot** cancel as is: $\dfrac{2x(x + 2) + (x - 1)(x + 3)}{x(x - 1)(x - 3)}$

because of the "+ " sign between the parentheses in the numerator.

Multiplication and Division of Fractions

As it was in the case of reduction of fractions to lowest terms, the most important (and perhaps the only) prerequisite in being able to multiply and divide algebraic fractions is **knowing how to factor polynomials.** Therefore, go back and review the factoring of polynomials.

Perform the indicated operations and simplify.

Example 1 $\dfrac{x}{x - 2} \cdot \dfrac{3x - 6}{6x}$

Solution: Factor and divide out (cancel) the common factors in the numerators and the denominators

$\dfrac{x}{x - 2} \cdot \dfrac{3x - 6}{6x} = \dfrac{x}{x - 2} \cdot \dfrac{3(x - 2)}{6x}$ Note: $3x - 6 = 3(x - 2)$

Consider ($x - 2$) as a single quantity

$= \dfrac{\overset{1}{\cancel{x}}}{\underset{1}{\cancel{x - 2}}} \cdot \dfrac{\overset{1}{\cancel{3(x - 2)}}}{\underset{2}{\cancel{6x}}^{1}}$

$= \dfrac{1}{2}$

Example 2 Divide: $\dfrac{3y^2}{y+2} \div \dfrac{6y}{y^2-4}$

$\dfrac{3y^2}{y+2} \div \dfrac{6y}{y^2-4} \quad = \dfrac{3y^2}{y+2} \cdot \dfrac{y^2-4}{6y}$ (Changing to multiplication and inverting the divisor)

$= \dfrac{\overset{1}{\cancel{3y^2}}\, y}{\cancel{y+2}} \cdot \dfrac{(y+2)(\,y-2)}{\underset{2}{\cancel{6y}}}$ Scrapwork: $y^2-4=(y+2)(y-2)$

$= \dfrac{y(y-2)}{2}$ or $\dfrac{y^2-2y}{2}$

Example 3 Simplify: $\dfrac{x^2-49}{x^4} \cdot \dfrac{x^2}{3x+21}$

Solution: Factor and cancel the common factors.

$\dfrac{x^2-49}{x^4} \cdot \dfrac{x^2}{3x+21} = \dfrac{(x+7)(x-7)}{x^4} \cdot \dfrac{x^2}{3(x+7)}$

$= \dfrac{\overset{1}{(\cancel{x+7})}(\,x-7)}{\underset{x^2}{\cancel{x^4}}} \cdot \dfrac{\cancel{x^2}^{\,1}}{\underset{1}{3\cancel{(x+7)}}}$

$= \dfrac{x-7}{3x^2}$

Lesson 69 Exercises

A Reduce to lowest terms:

1. $\dfrac{x+4}{x^2+7x+12}$;   **2.** $\dfrac{x^2-8x+15}{x^2-x-20}$ **3.** $\dfrac{6-x}{x^2-9x+18}$;   **4.** $\dfrac{2x^2-32}{x^2-11x+28}$

Answers: **1.** $\dfrac{1}{x+3}$; **2** $\dfrac{x-3}{x+4}$ **3.** $-\dfrac{1}{x-3}$ **4.** $\dfrac{2(x+4)}{x-7}$

B Reduce the following to lowest terms:

1. $\dfrac{2x-6}{x-3}$;   **2.** $\dfrac{x^2+x-6}{x-2}$;   **3.** $\dfrac{4-x}{x^2+x-20}$;   **4.** $\dfrac{x^2+x-30}{x^2-x-20}$

5. $\dfrac{x^2+4x-21}{x^2-49}$;   **6.** $\dfrac{x^2-81}{x^2-11x+18}$;   **7.** $\dfrac{x^2-8x+12}{4x^2-32x+48}$ **8.** $\dfrac{2x^2-32}{2x^2+2x-40}$

Answers: **1.** 2 ; **2.** $x+3$; **3.** $-\dfrac{1}{x+5}$; **4.** $\dfrac{x+6}{x+4}$; **5.** $\dfrac{x-3}{x-7}$; **6.** $\dfrac{x+9}{x-2}$; **7.** $\dfrac{1}{4}$; **8.** $\dfrac{x+4}{x+5}$

C Perform the indicated operations:

1. $\dfrac{x}{x-3} \cdot \dfrac{4x-12}{x+3}$;

2. $\dfrac{10x^2}{x-5} \div \dfrac{5x}{x^2-25}$;

3. $\dfrac{x^2-81}{x^6} \cdot \dfrac{x^4}{x^2+6x-27}$

Answers: **1.** $\dfrac{4x}{x+3}$; **2.** $2x^2+10x$; **3.** $\dfrac{x-9}{x^2(x-3)}$

D Simplify the following:

1. $\dfrac{6x^4}{7y^2} \cdot \dfrac{21y}{3x}$;

2. $\dfrac{x^2}{x+3} \div \dfrac{x}{2x+6}$;

3. $\dfrac{8-x}{x^2-64} \cdot \dfrac{3x+24}{x+3}$

4. $\dfrac{x^2-9}{4x^2-36} \cdot \dfrac{2x^2-50}{3x+15}$;

5. $\dfrac{8-x}{x^8} \div \dfrac{x^2-4x-32}{x^6+4x^5}$

Answers: **1.** $\dfrac{6x^3}{y}$; **2.** $2x$; **3.** $-\dfrac{3}{x+3}$; **4.** $\dfrac{x-5}{6}$; **5.** $-\dfrac{1}{x^3}$

Lesson 70

Addition of Fractions; Simplification of Complex Fractions

Addition and Subtraction of Fractions

Like Fractions (Fractions with the same denominator)

Example   Add: $\dfrac{2x}{x-3} + \dfrac{4}{x-3}$

Solution   Since the above fractions already have a common denominator, we add the numerators and keep the common denominator.

$$\dfrac{2x}{x-3} + \dfrac{4}{x-3}$$

$$= \dfrac{2x+4}{x-3}$$

Unlike Fractions (Fractions with different denominators)

In this case ,we have to change the fractions to (equivalent) like fractions , add the numerators and then reduce the fraction to lowest terms if possible.

Example 1   Add: $\dfrac{5}{x^2 + 7x + 12} + \dfrac{x+2}{x+3}$

Step 1:   Factor the denominators if factorable.

$$\dfrac{5}{(x+3)(x+4)} + \dfrac{x+2}{x+3}$$

Step 2: We will make the denominators the same by looking for missing factors. The factors of the denominator of the first fraction are ($x+3$) and ($x+4$), but the denominator of the second fraction has the factor $x+3$. To make the denominators of the two fractions identical, we multiply both the denominator and the numerator of the second fraction by $x+4$, since the denominator of the second fraction is "missing" $x+4$, compared to the denominator of the first fraction.

Note that any time we multiply the denominator by any quantity, we must also multiply the numerator by the same quantity (in order **not** to change the value of the original fraction.).

$$= \dfrac{5}{(x+3)(x+4)} + \dfrac{(x+2)(x+4)}{(x+3)(x+4)}$$

Step 3: Since the denominators are now the same, add the numerators. (We have also produced the LCD)

$$= \dfrac{5 + (x+2)(x+4)}{(x+3)(x+4)}$$ The LCD we produced (in Step 1) is $(x+3)(x+4)$

$$= \dfrac{5 + x^2 + 4x + 2x + 8}{(x+3)(x+4)}$$

$$= \dfrac{x^2 + 6x + 13}{(x+3)(x+4)}$$ (Simplifying the numerator)

Note that if the numerator were factorable, we would have factored it to determine if there are common factors in the numerator and the denominator, and in which case, we would have canceled the common factors. In this example, the numerator is not factorable. It is preferable to leave the denominator in the factored form.

Example 2 Add: $\dfrac{3}{x+2} + \dfrac{2}{x^2 - 4}$

Step 1: Factor the factorable denominator(s).

Then, $\dfrac{3}{x+2} + \dfrac{2}{x^2 - 4}$

$= \dfrac{3}{x+2} + \dfrac{2}{(x+2)(x-2)}$

Step 2: The denominator of the first fraction is "missing" $(x - 2)$ compared to the denominator of the second fraction. Therefore, multiply both denominator and numerator of the first fraction by $(x - 2)$; add the numerators of the resulting like fractions and reduce if possible.

$= \dfrac{3(x-2)}{(x+2)(x-2)} + \dfrac{2}{(x+2)(x-2)}$ \qquad (We have produced the LCD which is $(x+2)(x-2)$)

$= \dfrac{3(x-2) + 2}{(x+2)(x-2)}$

$= \dfrac{3x - 6 + 2}{(x+2)(x-2)}$

$= \dfrac{3x - 4}{(x+2)(x-2)}$

Example 3 Combine into a single fraction: $\dfrac{3x}{5y} + \dfrac{x}{2y}$

Solution

Step 1: Make the denominators identical by " looking for missing factors" and multiplying accordingly. (Multiply the first fraction by 2 and the second fraction by 5.)

$\dfrac{3x}{5y} + \dfrac{x}{2y}$

$= \dfrac{3x(2)}{5y(2)} + \dfrac{x(5)}{2y(5)}$

$= \dfrac{6x}{10y} + \dfrac{5x}{10y}$ <-------Like fractions (The LCD produced is 10y)

Step 2: Add the numerators since we have like fractions.

Then, $\dfrac{6x}{10y} + \dfrac{5x}{10y}$

$= \dfrac{11x}{10y}$

Example 4   Add: $\dfrac{7}{5} + \dfrac{3}{4n}$

Solution

Step 1:   Make the denominators the same by looking for missing factors and multiplying accordingly.
(Multiply the first fraction by 4n, and multiply the second fraction by 5)

$$\frac{7}{5} + \frac{3}{4n}$$

$$= \frac{7(4n)}{5(4n)} + \frac{3(5)}{4n(5)}$$

$$= \frac{28n}{20n} + \frac{15}{20n} \quad \text{<--------Like fractions (The LCD produced is } 20n)$$

Step 2:   Add the numerators.

Then, $\dfrac{28n}{20n} + \dfrac{15}{20n}$

$$= \frac{28n + 15}{20n}$$

Example 5   Combine into a single fraction: $\dfrac{4}{x} + \dfrac{1}{2x} + \dfrac{2}{5}$

$$\frac{4}{x} + \frac{1}{2x} + \frac{2}{5}$$

Step 1:   Multiply accordingly by 2, 5, and x.

$$= \frac{4(2)(5)}{x(2)(5)} + \frac{1(5)}{2x(5)} + \frac{2(x)(2)}{5(x)(2)} \qquad \text{(Making the denominators the same)}$$

$$= \frac{40}{10x} + \frac{5}{10x} + \frac{4x}{10x} \quad \text{<----Like fractions}$$

Step 2:   Add the numerators, since we have like fractions.

Then, $\dfrac{40}{10x} + \dfrac{5}{10x} + \dfrac{4x}{10x}$

$$= \frac{4x + 45}{10x}$$

Example 6. Add: $\dfrac{5}{x+4} - \dfrac{3x}{x-4} + \dfrac{2}{x^2-16}$

$$= \dfrac{5}{x+4} - \dfrac{3x}{x-4} + \dfrac{2}{(x+4)(x-4)}$$

$$= \dfrac{5(x-4)}{(x+4)(x-4)} - \dfrac{3x(x+4)}{(x-4)(x+4)} + \dfrac{2}{(x+4)(x-4)}$$

Now, since we have like fractions (all the denominators are the same), we add the numerators.

$$= \dfrac{5(x-4) - 3x\,(x+4) + 2}{(x+4)(x-4)}$$

$$= \dfrac{5x - 20 - 3x^2 - 12x + 2}{(x+4)(x-4)}$$

$$= \dfrac{-3x^2 - 7x - 18}{(x+4)(x-4)}$$

$$= -\dfrac{3x^2 + 7x + 18}{(x+4)(x-4)}$$

Example 7 Simplify: $\dfrac{3x+4}{x-3} - \dfrac{x-1}{x+1}$

Solution

$$= \dfrac{3x+4}{x-3} - \dfrac{x-1}{x+1}$$

$$= \dfrac{(x+1)(3x+4)}{(x+1)(x-3)} - \dfrac{(x-1)(x-3)}{(x+1)(x-3)} \quad \text{(Making the denominators the same and producing the LCD)}$$

$$= \dfrac{3x^2 + 4x + 3x + 4 - (x^2 - 3x - x + 3)}{(x+1)(x-3)} \quad \text{(The minus sign before the parentheses is important)}$$

$$= \dfrac{3x^2 + 7x + 4 - (x^2 - 4x + 3)}{(x+1)(x-3)} \quad \text{(The minus sign before the parentheses is important)}$$

$$= \dfrac{3x^2 + 7x + 4 - x^2 + 4x - 3}{(x+1)(x-3)}$$

$$= \dfrac{2x^2 + 11x + 1}{(x+1)(x-3)}$$

Example 8 Simplify : $3x + \dfrac{2x}{y}$

$$3x + \frac{2x}{y}$$

$$= \frac{3x(y)}{1y} \ + \ \frac{2x}{y}$$

$$= \frac{3xy}{y} \ + \ \frac{2x}{y}$$
$$= \frac{3xy + 2x}{y}$$

Example 9 Simplify: $x - 2 \ + \dfrac{3}{x - 4}$

$$= \frac{x - 2}{1} \ + \frac{3}{x - 4} \qquad \text{or} \qquad \frac{x}{1} \ - \frac{2}{1} \ + \frac{3}{x - 4}$$

$$= \frac{(x - 4)(x - 2)}{(x - 4)1} \ + \ \frac{3}{x - 4} \qquad \text{or} \qquad \frac{x(x - 4)}{1(x - 4)} \ - \frac{2(x - 4)}{1(x - 4)} \ + \frac{3}{x - 4}$$

$$= \frac{(x - 4)(x - 2) + 3}{x - 4} \qquad \text{or} \qquad \frac{x(x - 4) - 2(x - 4) + 3}{x - 4}$$

$$= \frac{x^2 - 2x - 4x + 8 + 3}{x - 4} \qquad \text{or} \qquad \frac{x^2 - 4x - 2x + 8 + 3}{x - 4}$$

$$= \frac{x^2 - 6x + 11}{x - 4} \qquad \text{or} \qquad \frac{x^2 - 6x + 11}{x - 4}$$

The above "or" means that you may follow the steps on the left or the steps on the right.

Example 10 Add: $x + 4 \ + \dfrac{3}{x - 2}$

$$= \frac{x}{1} \ + \frac{4}{1} \ + \frac{3}{x - 2}$$

$$= \frac{x(x - 2)}{1(x - 2)} \ + \frac{4(x - 2)}{1(x - 2)} \ + \frac{3}{x - 2}$$

$$= \frac{x(x - 2)}{(x - 2)} \ + \frac{4(x - 2)}{(x - 2)} \ + \frac{3}{x - 2}$$

(Now, we can add the numerators since we have like fractions, and write the denominator only once.)

$$= \frac{x(x - 2) + 4(x - 2) + 3}{x - 2}$$
$$= \frac{x^2 - 2x + 4x - 8 + 3}{x - 2}$$

$$= \frac{x^2 + 2x - 5}{x - 2}$$

Example 11   Add : $\dfrac{x^2 + 7x + 12}{x^2 + 8x + 15} + \dfrac{3}{x - 5}$

Solution

Step 1: Factor and reduce the first expression before proceeding to add. (Failure to reduce to lowest terms first, in this problem, will result in having to deal with higher powers of x.)

$$\dfrac{x^2 + 7x + 12}{x^2 + 8x + 15} + \dfrac{3}{x - 5}$$

$$= \dfrac{(x + 3)(x + 4)}{(x + 3)(x + 5)} + \dfrac{3}{x - 5}$$

$$= \dfrac{x + 4}{x + 5} + \dfrac{3}{x - 5}$$

Step 2: Now, proceed to add as before.

$$= \dfrac{(x + 4)(x - 5)}{(x + 5)(x - 5)} + \dfrac{3 \cdot (x + 5)}{(x - 5)(x + 5)}$$

$$= \dfrac{x^2 - 5x + 4x - 20 + 3x + 15}{(x + 5)(x - 5)}$$

$$= \dfrac{x^2 + 2x - 5}{(x + 5)(x - 5)}$$

Note above that sometimes a given fraction may **not** be in its lowest terms; and in this case, it is good practice to reduce to lowest terms before proceeding to add.

Simplification of Complex Fractions

Example 1   Simplify: $\dfrac{\dfrac{3x}{ax}}{\dfrac{5x}{bx}}$

Solution $\dfrac{\dfrac{3x}{ax}}{\dfrac{5x}{bx}} = \dfrac{3x}{ax} \div \dfrac{5x}{bx}$

$\qquad\qquad = \dfrac{3x}{ax} \cdot \dfrac{bx}{5x}$ (inverting the divisor, $\dfrac{5x}{bx}$).

$\qquad\qquad = \dfrac{3b}{5a}$ (canceling the common factors).

Example 2 Simplify:

$$\dfrac{\dfrac{y}{y^2 + 4\,y + 3}}{\dfrac{1}{y^2 - 1} + 1}$$

Solution **Method 1** (Adding the denominator first)

$$\dfrac{\dfrac{y}{y^2 + 4\,y + 3}}{\dfrac{1}{y^2 - 1} + 1} = \dfrac{y}{y^2 + 4\,y + 3} \div \dfrac{1}{y^2 - 1} + 1$$

Scrapwork

$$\dfrac{1}{(y + 1)(\ y - 1)} + \dfrac{1}{1}$$

$$= \dfrac{1 + (\ y + 1)(\ y - 1)}{(y + 1)(\ y - 1)}$$

$$= \dfrac{y}{y^2 + 4\,y + 3} \div \dfrac{y^2}{(y + 1)(\ y - 1)}$$

$$= \dfrac{1 + \ y^2 - 1}{(y + 1)(\ y - 1)}$$

$$= \dfrac{y}{y^2 + 4\,y + 3} \cdot \dfrac{(y + 1)(\ y - 1)}{y^2}$$

$$= \dfrac{y^2}{(y + 1)(\ y - 1)}$$

$$= \dfrac{1 \quad y}{1 \ (y + 1)(\ y + 3)} \cdot \dfrac{1 (y + 1)(\ y - 1)}{y^2 \ y}$$

$$= \dfrac{y - 1}{y(y + 3)}$$

Method 2: Multiply both the numerator and denominator of the complex fraction by the LCD of all the fractions in the numerator and the denominator. The LCD $= (y + 1)(y - 1)(y + 3)$

$$\dfrac{\dfrac{y}{y^2 + 4y + 3}}{\dfrac{1}{y^2 - 1} + 1}$$

$$= \dfrac{\dfrac{y(y + 1)(\ y - 1)(\ y + 3)}{(y + 1)(\ y + 3)}}{\dfrac{(y + 1)(\ y - 1)(y + 3)}{(y + 1)(\ y - 1)} + \dfrac{1(y + 1)(y - 1)(\ y + 3)}{1}}$$

$$= \dfrac{\dfrac{y(y + 1)(\ y - 1)(y + 3)}{(y + 1)(\ y + 3)}}{\dfrac{(y + 1)(\ y - 1)(y + 3)}{(y + 1)(y - 1)} + \dfrac{(y + 1)(\ y - 1)(\ y + 3)}{1}}$$

$$= \dfrac{y(y - 1)}{(y + 3) + (y^2 - 1)(y + 3)} \qquad \text{Note :} y^2 - 1 = (y + 1)(y - 1)$$

$$= \dfrac{y(y - 1)}{(y + 3)[1 + \ y^2 - 1]}$$

$$= \dfrac{y(y - 1)}{(y + 3)\ y^2}$$

$$= \dfrac{1 \ y(y - 1)}{(y + 3)\ y^2 \ y}$$

$$= \dfrac{(y - 1)}{y(y + 3)}$$

Lesson 70 Exercises

A Carry out the indicated operation and simplify:

1. $\dfrac{3}{x-4} + \dfrac{x+1}{x^2-16}$;

2. $\dfrac{2}{xy} - \dfrac{4}{y}$;

3. $\dfrac{3x}{x-4} + \dfrac{2}{x-4}$;

4. $x - 2 + \dfrac{2}{x-3}$

5. $\dfrac{5}{x^2-4} + \dfrac{2x}{x+2} - \dfrac{4}{x-3}$.

Answers: **1.** $\dfrac{4x+13}{(x-4)(x+4)}$; **2.** $\dfrac{-4x+2}{xy}$; **3.** $\dfrac{3x+2}{x-4}$; **4.** $\dfrac{x^2-5x+8}{x-3}$; **5.** $\dfrac{2x^3-14x^2+17x+1}{(x-3)(x-2)(x+2)}$;

B Carry out the indicated operation and simplify:

1. $\dfrac{x}{3} + \dfrac{4}{x+2}$

2. $\dfrac{2}{x-3} - \dfrac{x}{4}$

3. $\dfrac{x}{y} + \dfrac{2}{n}$

4. $\dfrac{2}{x} + \dfrac{3}{y}$;

5. $\dfrac{2}{x-5} + \dfrac{3}{x-5}$;

6. $\dfrac{x+1}{x+2} + \dfrac{x-3}{x-2}$;

7. $\dfrac{2x-1}{x+4} - \dfrac{x-5}{x-3}$

Answers: **1.** $\dfrac{x^2+2x+12}{3(x+2)}$; **2.** $\dfrac{-x^2+3x+8}{4(x-3)}$; **3.** $\dfrac{nx+2y}{ny}$; **4.** $\dfrac{3x+2y}{xy}$;

5. $\dfrac{5}{x-5}$; **6.** $\dfrac{2x^2-2x-8}{(x+2)(x-2)}$; **7.** $\dfrac{2x^2-6x+23}{(x-3)(x+4)}$

C Simplify the following:

1. $\dfrac{\dfrac{x}{x^2+7x+12}}{\dfrac{16}{x^2-16}+1}$;

2. $\dfrac{\dfrac{2ab}{cd}}{\dfrac{cd}{ab}}$;

3. $\dfrac{3-\dfrac{2}{x}}{9x}$;

4. $\dfrac{\dfrac{x}{y}+3}{\dfrac{y}{x}-3}$;

5. $\dfrac{\dfrac{5c^2y}{3bx}}{\dfrac{25cy}{6bx^2}}$

Answers: **1.** $\dfrac{x-4}{x(x+3)}$; **2.** $\dfrac{2a^2b^2}{c^2d^2}$; **3.** $\dfrac{3x-2}{9x^2}$; **4.** $\dfrac{x^2+3xy}{y(y-3x)}$; **5.** $\dfrac{2cx}{5}$

CHAPTER 24
Inequalities

Lesson 71: **Set Notation, Set Operations, Interval Notation**
Lesson 72: **Linear Inequalities; Compound Inequalities; Word Problems**

Lesson 71

Set Notation, Set Operations, Interval Notation

Sets, Set Notation

Set A set is a well-defined collection of objects or things. The objects or things are called the elements or members of the set. A set may contain many elements; it may contain only one element; or it may contain no element. We call the set which contains no element the empty set or the null set.

Representation of sets: We will cover the roster method and the set-builder notation.

Roster method: In the roster method, we list the elements of the set and enclose them by braces.
 Example: We can denote the set of numbers $2, 3, 4$, as $\{ 2, 3, 4\}$.
 We separate the elements by commas. Note that the order in which the elements are listed does not matter. We may **name** a set by a capital letter . For example, if we denote the set of numbers $2, 3, 4$ by A, then we write $A = \{ 2, 3, 4\}$.
 We read this as "A is the set whose elements are $2, 3, 4$.

Set-builder notation: In the set builder notation, we state the conditions which the elements of the set must satisfy.
Example: Let E be the set of all odd integers between 4 and 12. Then if we use x to represent an arbitrary element of this set, we write $E = \{x \mid x$ is an odd integer between 4 and 12$\}$.
This is read " E is the set of numbers (elements) x such that x is an odd integer between 4 and 12" . We read the vertical line " $\mid$" as "such that".
(Note that by the roster method, the set E in this example would be represented by
$E = \{5, 7, 9, 11\}$

We denote the empty set by $\{ \}$ or $\emptyset$.

We will use the **set-builder notation** in stating the solutions of inequalities.

Definition

An **inequality** is a mathematical statement that two expressions are not equal.

An inequality, like an equation, has three parts, namely the left-hand side, the inequality symbol which may be " $>, <, \geq, \leq$ or $\neq$", and the right-hand side.

Examples 1. $2x - 5 > 6x + 3$ read " $2x$ minus 5 **is greater than** $6x$ plus 3".

2. $2x + 3 \geq 6x + 7$ read " $2x$ plus 3 **is greater than or equal to** $6x$ plus 7".

3. $4x + 5 < 8x - 2$ read "$4x$ plus **five is less than** $8x$ minus 2".

4. $3x - 1 \leq 5x + 7$ read "$3x$ minus 1 **is less than or equal to** $5x$ plus 7".

5. $x > 6$ read " x is greater than 6"

Sense of an inequality (direction or order of an inequality)
 The sense of an inequality refers to whether the inequality symbol is the greater than symbol " $>$" or the less than symbol "$<$".

Set Notation, Interval Notation and Graphs

Set notation Interval notation

$\downarrow$ $\downarrow$

$\{x \mid x > a\} = (a, +\infty)$

$\{x \mid x \geq a\} = [a, +\infty)$

$\{x \mid x < b\} = (-\infty, b)$

$\{x \mid x \leq b\} = (-\infty, b]$

$\{x \mid a < x < b\} = (a, b)$

$\{x \mid a \leq x \leq b\} = [a, b]$

$\{x \mid a < x \leq b\} = (a, b]$

$\{x \mid a \leq x < b\} = [a, b)$

$\{x \mid x \text{ is a real number}\}$

$= (-\infty, +\infty)$

Graphs

Example 1 Draw the graph for $\{x \mid x > 2\} = (2, +\infty)$

Solution

Graph for $\{x \mid x > 2\} = (2, +\infty)$

Note: The hollow circle at 2 indicates that 2 is **not** part of the solution set.

Example 2 Draw the graph for $\{x \mid x \geq 2\} = [2, +\infty)$

Solution

Graph for $\{x \mid x \geq 2\} = [2, +\infty)$

Note: The solid circle at 2 indicates that 2 is part of the solution set.

Example 3

$$\{x \mid x \leq -8\} = (-\infty, -8]$$

Graph for $\{x \mid x \leq -8\} = (-\infty, -8]$

Set Operations Involving Inequalities

Union of two sets

The union of two sets A and B, written $A \cup B$, is the set containing all the elements that belong to either set A or set B (or belong to both sets).

Symbolically, $A \cup B = \{x \mid x \in A, \text{ or } x \in B\}$

Example 1   If $A = \{x \mid x < -3\}$ and $B = \{x \mid x > 4\}$ then

$$A \cup B = \{x \mid x < -3\} \cup \{x \mid x > 4\}$$
$$= \{x \mid x < -3 \text{ or } x > 4\}$$

Graph for $\{x \mid x < -3 \text{ or } x > 4\}$

Intersection of two sets

The intersection of two sets A and B, written $A \cap B$ is the set containing the elements that are common to A and B. Thus the elements of the intersection are the elements that belong to both sets simultaneously.

Symbolically, $A \cap B = \{x \mid x \in A \text{ and } x \in B\}$

Example 2 If $A = \{x \mid x > -3\}$ and $B = \{x \mid x < 2\}$ then
$$A \cap B = \{x \mid x > -3\} \cap \{x \mid x < 2\}$$
$$A \cap B = \{x \mid -3 < x < 2\}$$

Example 3 If $A = \{x \mid x < -3\}$ and $B = \{x \mid x > 4\}$ then

$$A \cap B = \{x \mid x < -3\} \cap \{x \mid x > 4\}$$

$A \cap B = \{ \ \}$, the empty set.
(Since a number cannot be less than -3 and greater than 4 at the same time.)

Compare the solutions to Examples 1 and 3, above, and note that even though they have identical graphs, Example 1 has a solution but Example 3 has no solution.

Lesson 71 Exercises

1. What is a set?
2. What is an inequality?
3.What is meant by the sense or direction or the order of an equality?
4. Draw graphs for the following:
a. $\{x \mid x > 3\}$; **b.** $\{x \mid x \geq 3\}$; **c.** $\{x \mid x < 3\}$; **d.** $\{x \mid x \leq 3\}$; **e.** $\{x \mid x < -5\}$

Answers:**4.**

(a)

$\{x \mid x > 3\}$

(b)

$\{x \mid x \geq 3\}$

(c)

$\{x \mid x < 3\}$ (d) $\{x \mid x \leq 3\}$

(e)

$\{x \mid x < -5\}$

Lesson 72

Solving Linear Inequalities; Compound Inequalities; Word Problems

Solving Linear Inequalities

The solution set of a linear inequality is the set of real numbers each of which when substituted for the variable makes the inequality true.

The techniques for solving linear inequalities are similar to the techniques for solving linear equations, except that when an inequality is divided or multiplied by a negative number, the sense of the inequality must be reversed as follows: The symbol " > " when reversed becomes " < "; the symbol " < " when reversed becomes " > "; the symbol " ≥ " when reversed becomes " ≤ "; and the symbol " ≤ " when reversed becomes " ≥ ".

Example 1 Solve and graph the solution set of the inequality,
$$3x - 4 > 0$$
Step 1: To undo the -4, add + 4 to both sides of the inequality.

$$
\begin{array}{r}
3x - 4 > 0 \\
+\,4\ +4 \\
\hline
3x > 4
\end{array}
$$

Step 2: Divide both sides of the inequality by 3.

$$\frac{3x}{3} > \frac{4}{3} \qquad x > \frac{4}{3}$$

The solution set is $\{x \mid x > \frac{4}{3}\}$.   (The solution set is all real numbers greater than $\frac{4}{3}$)

Graph for $x > \frac{4}{3}$

Note: The hollow circle at $\frac{4}{3}$ indicates that $\frac{4}{3}$ is **not** part of the solution set.

Example 2 Solve and graph the solution set of the inequality
$$3x - 4 \geq 0$$
Step 1: To undo the -4, add + 4 to both sides of the inequality.

$$
\begin{array}{r}
3x - 4 \geq\ 0 \\
+4\ \ +4 \\
\hline
3x \geq 4
\end{array}
$$

Step 2: Divide both sides of the inequality by 3.

$$\frac{3x}{3} \geq \frac{4}{3}$$

$$x \geq \frac{4}{3}$$

The solution set is $\{x \mid x \geq \frac{4}{3}\}$. (The solution set is all real numbers greater than or equal to $\frac{4}{3}$)

Graph for $x \geq \frac{4}{3}$ **Note**: The solid circle at $\frac{4}{3}$ indicates that $\frac{4}{3}$ is part of the solution set.

Example 3 $5x + 5 > 9x - 6$

Solution

Step 1: We undo the $9x$ by adding $-9x$ to both sides of the inequality.

$$5x + 5 > 9x - 6$$
$$\underline{-9x \qquad - 9x}$$
$$-4x + 5 > -6$$

Step 2: To undo the 5, add -5 to both sides of the inequality.

$$-4x + 5 > -6$$
$$\underline{\quad -5 \quad -5}$$
$$-4x > -11$$

Step 2: Divide both sides of the inequality by -4 and **reverse the sense** of the inequality.

$$\frac{-4x}{-4} < \frac{-11}{-4} \quad \text{(change this " > ' to that " < ")}$$

$$x < \frac{11}{4}$$

The solution set is $\{x \mid x < \frac{11}{4}\}$.   (The solution set is all real numbers less than $\frac{11}{4}$.)

Graph for $x < \frac{11}{4}$

Example 4   Solve for x: $2(2x + 1) \geq 5(x + 2)$

Solution

$$2(2x + 1) \geq 5(x + 2)$$
$$4x + 2 \geq 5x + 10$$
$$\underline{-5x \qquad -5x} \quad,$$
$$-x + 2 \geq 10$$
$$\underline{\quad -2 \quad -2}$$
$$-x \geq 8$$
$$x \leq -8 \qquad \text{(Multiplying or dividing both the left-hand side and the right-hand side}$$
by -1 and reversing the sense of the inequality or changing the signs of both
sides and reversing the sense of the inequality)

The solution set is $\{x \mid x \leq -8\}$.

Graph for $x \leq -8$

Compound Inequalities

A compound inequality is formed by connecting two inequalities with the words "**and**" or " **or**".

Example We will call the following type an "**AND**" problem. (More formally, a **conjunction** problem)

$$3x - 2 < 4 \text{ and } 2x + 10 > 4$$

Because of the connective "and", we want a solution set (a set of numbers, if any) which is common to both solution sets of the two inequalities (i.e., the intersection of the solution sets). Any solution set must satisfy both inequalities simultaneously.

Example Solve for x: $3x - 2 < 4$ and $2x + 10 > 4$

Procedure: Solve each inequality separately and find the intersection (common set of numbers) of tbe solutions.

$$
\begin{array}{ll}
3x - 2 < 4 & \text{and} \quad 2x + 10 > 4 \\
\underline{+2\ +2} & \underline{- 10\ -10} \\
3x < 6 & 2x > -6 \\
x < 2 \quad \text{and} & x > -3
\end{array}
$$

The solution is the intersection of $\{x \mid x < 2\}$ and $\{x \mid x > -3\}$ i.e., $\{x \mid x < 2\} \cap \{x \mid x > -3\}$
Graph the solutions to determine if the solutions intersect.
The solution is $\{x \mid -3 < x < 2\}$

Graph for $-3 < x < 2$

Another "**AND**" problem

Example 2 Solve for x: $\qquad -5 < 2x + 1 \le 7$

Solution: The above inequality is equivalent to the compound inequality

$$-5 < 2x+1 \text{ and } 2x+ 1 \le 7.$$

We can solve this compound inequality using the method used in Example 1 above, however, we will use a faster method called the "**Condensed**" or "**continued inequality**" method as follows:

$$
\begin{array}{l}
-5 < 2x + 1 \le 7 \\
\underline{-1 -1 \ -1} \\
-6 < 2x \le 6
\end{array}
\qquad \text{(Adding -1 to all three sides of the inequality)}
$$

$$
\frac{-6}{2} < \frac{2x}{2} \le \frac{6}{2}
\qquad \text{(Dividing each of all the three sides of the inequality by 2)}
$$

$$-3 < x \le 3$$

Graph for $-3 < x \le 3$ $\qquad$ (Note the hollow circle at -3 and the solid circle at 3)

Example 3 Solve for x: $-6x > 12$ and $x + 1 > 5$

Solution: For the first inequality, divide by -6 and reverse the sense of the inequality.
For the second inequality, add -1 to both sides of the inequality.

$$
\begin{array}{ll}
-6x > 12 & \text{and} \quad x + 1 > 5 \\
x < -2 \quad \text{and} & x > 4
\end{array}
$$

Graph the solution set for $x < -2$ and for $x > 4$ and determine if the two graphs intersect. If they intersect, there is a solution, but if they do not intersect, then there is no solution.

$x < -2$ $\qquad\qquad\qquad\qquad\qquad$ $x > 4$

From the graph, the two solution sets do not intersect. (i.e., there are no numbers which belong to both solutions).
Therefore, there is no solution or the solution set is the empty set.

An "OR " Compound Inequality Problem (More formally, a **disjunction** problem)

Example 4 Solve for x: $2x - 3 > 5$ or $3x + 1 < -8$

Because of the connective "or" we want the solution set (if any) to be the union of the solution sets of the two inequalities. The solution is the set of real numbers which satisfies either inequality or both.

$$2x - 3 > 5 \text{ or } 3x + 1 < -8$$
$$\underline{+\ 3 + 3} \qquad \underline{\quad -1 \ \ -1\quad}$$
$$2x > 8 \qquad\quad 3x < -9$$
$$x > 4 \quad \text{or} \quad x < -3$$

Same as $x < -3$ or $x > 4$

Graph

-9 -8 -7 -6 -5 -4 -3 -2 -1 0 1 2 3 4 5

Solution set is $\{x \mid x > 4\} \cup \{x \mid x < -3\} = \{x \mid x < -3 \text{ or } x > 4\}$.

Comparison of Example 3 and Example 4

Note that even though the two problems have similar graphs, Example 3 has no solution because it is an " **and**" problem but Example 4 has a solution because it is an " **or** " problem".

More on Intervals   (Finite Intervals)

Open interval The **open** interval (Figure) from a to b , written (a, b) or $a < x < b$, (where x is real number between a and b) is the set of all numbers between a and b but excluding the end-points a and b. (" open" means the endpoints are **excluded**)

Closed interval The **closed** interval from a to b , written $[a, b]$ or $a \leq x \leq b$ (where x is real number between a and b) is the set of all numbers between a and b and including the end-points a and b. (" closed " means the endpoints are **included**)

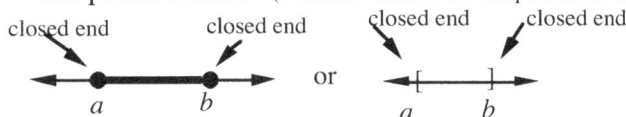

Half-closed and half-open Interval The half-closed and half-open interval from a to b is the set of all numbers between a and b, including either the endpoint a or the endpoint b but not both.

In Figure 1   the interval half-open on the left (or half-closed on the right) is the set of all numbers between a to b , including the endpoint b but excluding the endpoint a..

This interval is symbolized $(a, b]$ or $a < x \leq b$.

Figure 1

In Figure **2**, the interval half-open on the right (or half-closed on the left) is the set of all numbers between a to b , including the endpoint a but excluding the endpoint b.

This interval is symbolized $[a, b)$ or $a \leq x < b$.

Figure 2

Note: Within any interval, there are infinitely many rational numbers and infinitely many irrational numbers.

Applications (Word Problems Using Inequalities)

Example During the semester James scored 85, 70, 75 on the first three tests. He wants his class average for the term to be at least 80. There is one more test to take. What is the score he needs on the last test ?

Solution Let x be the last score.

Then, the total score for the four tests will be $85 + 70 + 75 + x$.

$$\text{Average score} = \frac{\text{Total score}}{\text{Number of tests}} = \frac{85 + 70 + 75 + x}{4}$$

Now, we want this average score to be at least 80 (i.e., 80 or more).

$$\frac{85 + 70 + 75 + x}{4} \geq 80 \qquad \text{Scrapwork}$$

$$\frac{230 + x}{4} \geq 80 \qquad 85+70+75=230$$

We solve for x.

$$\frac{4(230 + x)}{4} \geq 80(4)$$

$$230 + x \geq 320$$

$$\underline{-230 \qquad\quad -230}$$

$$x \geq 90$$

$\therefore$ James needs at least a score of 90 on the last test. (i.e., 90 or more).

Some words to help translate inequality word problems

1. " is at least " translates to " $\geq$ ".

2. " is no less than" translates to " $\geq$ ".

3 " is greater than or equal to " translates to " $\geq$"

4 " is more than" translates to " $>$ ".

5. " is greater than" translates to " $>$ ".

6. " is over " translates to " $>$ "

7. " is no more than" translates to " $\leq$ ".

8." is at most " translates to " $\leq$ ".

9. " is less than or equal to " translates to " $\leq$ ".

10 " is less than " translates to " $<$ ".

11. " is under " translates to " $<$ "

Also: x is between a and b translates to " $a < x < b$ " .

x is greater than a but less than b translates to " $a < x < b$ " .

Note the difference between **"more than"** and **"is more than "** as in the following:

Example: 3 **more than** twice a number **is more than** 5 less than the number

If the number is x, the translation is $2x + 3 > x - 5$. (**"Is more than"** translates to " $>$")

Some Reciprocal Relationships

Lesson 72 Exercises

A Solve and graph the solution set:

1. $3x - 6 > 0$;   2. $-4x + 1 > 13$;   3. $2x - 1 \leq 7x - 11$;   4. $2(x + 4) \geq 3(x - 2)$

5. $x + 4 > -5$;   6. $x + 4 < 6$;   7. $x + 4 \leq 6$;   8. $x + 4 \geq 6$;   9. $-4x > 12$

10. $-5x \leq -30$;   11. $\frac{-x}{4} \leq 9$;   12. $\frac{x}{4} - 6 > 14$;   13. $2(x - 4) + 3 \leq 17$

Answers; **1**. $\{x \mid x > 2\}$;   **2**. $\{x \mid x < -3\}$;   **3**. $\{x \mid x \geq 2\}$;   **4**. $\{x \mid x \leq 14\}$;   **5**. $\{x \mid x > -9\}$;   **6**. $\{x \mid x < 2\}$;   **7**. $\{x \mid x \leq 2\}$; **8**. $\{x \mid x \geq 2\}$; **9**. $\{x \mid x < -3\}$; **10**. $\{x \mid x \geq 6\}$; **11**. $\{x \mid x \geq -36\}$; **12**. $\{x \mid x > 80\}$; **13**. $\{x \mid x \leq 11\}$;

B Solve for: **1**. $2x - 3 < 1$ and $3x + 5 > 2$;   **2**. $-4 < 3x - 1 < 5$;   **3**. $2x - 3 < 1$ and $3x - 5 > 7$

4. $-1 \leq 2x + 5 < 3$;   **5**. $3x + 7 < 1$ and $2x - 2 < 5x + 7$

Answers: **1**. $\{x \mid -1 < x < 2\}$;   **2**. $\{x \mid -1 < x < 2\}$ **3**. No solution: $x < 2$ and $x > 4$ do not intersect.   **4**. $\{x \mid -3 \leq x < -1\}$;   **5**. $\{x \mid -3 < x < -2\}$

C Solve for x:

1. $3x - 2 > 4$ or $3x - 1 > -7$;   **2**. $3x + 4 > 1$ or $2x + 1 < -9$;   **3**. $2x - 3 < 1$ or $3x - 5 > 7$

4. $2x + 3 < 5$ or $x + 1 > 4$;   **5**. $3x - 2 > 4$ or $x - 2 < 6$;   **6**. $x - 1 < \frac{1}{2}$ or $x + 3 > 5$

Answers: **1**. $\{x \mid x > -2\}$;   **2**. $\{x \mid x < -5 \text{ or } x > -1\}$;   **3**. $\{x \mid x < 2 \text{ or } x > 4\}$; **4**. $\{x \mid x < 1 \text{ or } x > 3\}$;   **5**. All real numbers since $x > 2$ or $x < 8$; **6**. $\{x \mid x < \frac{3}{2} \text{ or } x > 2\}$;

D During the semester, Betty scored 75 and 78 on the first two tests. She wants her class average for the term to be at least 80. There is one more test to take. What is the score she needs on the last test ?

Answer: Betty needs at least a score of 87 on the last test

E Solve for x:

1. $\frac{x}{2} + 1 < x - 14$;   **2**. $3x - 2 < \frac{10 - x}{-2}$;   **3**. $\frac{x + 3}{2} \leq \frac{2 + x}{-2}$

Answers **1**. $x > 30$;   **2**. $x < -\frac{6}{5}$;   **3**. $x \leq -\frac{5}{2}$

CHAPTER 25
Factoring Polynomials

Lesson: **Greatest Common Factor**
Lesson: **Common Monomial Factoring**
Lesson: **Difference between two Squares**
Lesson: **Factoring by Grouping**
Lesson: **Monic Quadratic Trinomials**
Lesson: **Non-monic Quadratic Trinomials**
Lesson: **Factoring trinomials quadratic in form**
Lesson: **Factoring the Sum and Difference between two Cubes.**

Use Chapters 16A &16B

The above was designed to avoid duplication and to reduce the volume of this book, and also, such a design is good for the environment.

CHAPTER 26

Lesson: **Rectangular Coordinate System**
Lesson: **Drawing the Graph of a Straight Line**
Lesson: **Points on a Line; Slopes of lines; Intercepts**
Lesson: **Equations of Straight Lines**
Lesson: **Equations of Parallel and Perpendicular Lines**

Use Chapters 17A & 17B

The above was designed to avoid duplication and to reduce the volume of this book, and also, such a design is good for the environment.

CHAPTER 27
Exponents and Radicals

Lesson **Review of Exponents and Radicals**
Lesson **Definitions and Simplification of Radicals**
Lesson **Addition and Multiplication of Radicals**

Lesson **Division of Radicals; Rationalization of Denominators**
Lesson **Fractional Exponents and Reduction of Indices**

Use Chapters 20A & 20B

The above was designed to avoid duplication and to reduce the volume of this book, and also, such a design is good for the environment.

CHAPTER 28
Quadratic Equations

Lesson 73: **Solving by Factoring**
Lesson 74: **Solving by the Square Root Method**
Lesson 75: **Solving by Completing the Square**
Lesson 76: **Solving by Quadratic Formula**
Lesson 77: **Applications of the Quadratic Equation**

Standard form of the quadratic equation: $ax^2 + bx + c = 0$, where a, b, and c are constants and $a \neq 0$

Examples **1.** $x^2 - 3x - 28 = 0$ <--------in standard form
2. $x^2 - 2x = 0$ <--------- in standard form
3. $x^2 = 5x + 14$ <---------**not** in standard form
4. $5x^2 = 8x$ <---------**not** in standard form

We shall consider four methods:

Method 1: **By factoring** (For easily recognizable factors and for cases in which the constant term is missing, i.e., $c = 0$)

Method 2: **By the square root method** (For cases in which the x-term is missing i.e., $b = 0$)

Method 3: **By the quadratic formula** (This always works)

Method 4: **By completing the square** (If asked to use this method, otherwise choose from Methods 1, 2, and 3 above).

Note that Method 4 also always works. **Note also** that the quadratic formula can be derived from the quadratic equation **by completing the square** and solving for x.

Lesson 73
Solving Quadratic Equations by Factoring

Principle of zero products: If $ab = 0$ then either $a = 0$, or $b = 0$ (or both $= 0$).

Example 1 Solve by factoring
$$x^2 - 3x - 28 = 0$$

Step 1: Factor the quadratic trinomial.

$$(x + 4)(x - 7) = 0$$

Step 2: Set each factor equal to zero and solve each equation for x.

$$x + 4 = 0 \qquad \text{or} \qquad x - 7 = 0$$
$$\underline{-4 \ -4} \qquad\qquad \underline{+7 \ \ +7}$$
$$x = -4 \qquad\qquad\qquad x = 7$$

$$\therefore \ x = -4, \text{ or } x = 7$$

The solutions are -4 and 7.

Example 2   Solve by factoring.

$$3x^2 - 4x = 0$$

Solution

Step 1: Factor. $3x^2 - 4x = 0$

$$x(3x - 4) = 0 \qquad (\text{performing common monomial factoring}).$$

Step 2: Set each factor equal to zero and solve for x.   (Only each factor containing a variable is set to zero)

$$x = 0 \quad \text{or} \quad 3x - 4 = 0$$
$$\underline{+4 \quad +4}$$
$$\frac{3}{3}x = \frac{4}{3}$$

$$x = \frac{4}{3}$$

$$\therefore x = 0, \text{ or } x = \frac{4}{3} \qquad \textbf{Note} \text{ that 0 is also a solution }.$$

The solutions are 0 and $\frac{4}{3}$

Note: A common **wrong** approach in the above problem:

$$3x^2 - 4x = 0$$
$$3x^2 = 4x$$

$3x = 4$ (Canceling, that is dividing out an x on both sides of the equation ; such a cancellation excludes $x = 0$ as a solution.)

$x = \dfrac{4}{3}$ (Of course, you still obtain one of the solutions but lose the other solution)

Example 3   Solve for x by factoring.

$$\frac{x^2}{2} + \frac{10x}{3} + 2 = 0$$

Step 1: Undo the denominators by multiplying the equation by the LCM of 2 and 3 which is 6.

$$(6)\frac{x^2}{2} + (6)\frac{10x}{3} + 2(6) = 6(0)$$

$$\overset{3}{\cancel{(6)}}\frac{x^2}{\cancel{2}} + \frac{2\cancel{(6)}10x}{\cancel{3}} + 2(6) = 0$$
$$\qquad \quad 1 \qquad \quad 1$$

$$3x^2 + 20x + 12 = 0 \ ....(A)$$

Step 2:   Factor by the substitution method or otherwise

$$3(3x^2) + 20x(3) + 12(3) = 0$$
$$9x^2 + 20(3x) + 36 = 0$$
$$(3x)^2 + 20(3x) + 36 = 0 \ ....(B)$$

Step 3:   Let $3x = s$ in equation (B), and factor.

$$s^2 + 20s + 36 = 0$$
$$(s + 2)(s + 18) = 0..................................(C)$$

Step 4:   Replace s by $3x$ in equation (C)

$$(3x + 2)(3x + 18) = 0 \ ....(D)$$

Step 5: Set each factor equal to zero and solve for x.

$$3x + 2 = 0 \qquad\qquad 3x + 18 = 0$$
$$\underline{-2 \quad -2} \qquad\qquad \underline{-18 \quad -18}$$
$$\frac{3}{3}x = \frac{-2}{3} \qquad\qquad \frac{3}{3}x = \frac{-18}{3}$$
$$x = -\frac{2}{3} \qquad\qquad x = -6$$

The solutions are $-\frac{2}{3}$ and -6. or solution set $= \{-\frac{2}{3}, -6\}$

Note that in the above problem , since it is an equation, it was not necessary to factor completely the left-hand side of equation (D) of Step 4.

Factorability of a quadratic trinomial

A quadratic trinomial is factorable if the discriminant $b^2 - 4ac$ is a perfect square. (see also p. 234.)

Lesson 73 Exercises

A Solve by factoring: **1.** $x^2 - 11x + 18 = 0$; **2.** $x^2 - 5x - 36 = 0$;
3. $x^2 - 18x = 0$; **4.** $3x^2 = 24x$

Solutions 1. $\{2, \ 9\}$; **2.** $\{-4, 9\}$; **3.** $\{0,18\}$; **4.** $\{0,8\}$

B Solve by factoring: **1.** $x^2 - 4x - 21 = 0$; **2.** $x^2 + 4x - 45 = 0$; **3.** $x^2 - 25 = 0$

4. $9x^2 - 6x = 0$ **5.** $14t = 7t^2$ **6.** $x^2 = 11x - 18$

7. $2x^2 + 2x - 144 = 0$ **8.** $9x^2 - 36 = 0$ **9.** $ax^2 = -bx$

10. $x^2 - mx + nx - mn = 0$

Answers: **1.** $\{-3, 7\}$; **2.** $\{-9, 5\}$; **3.** $\{-5, 5\}$; **4.** $\{0, \frac{2}{3}\}$; **5.** $\{0, 2\}$; **6.** $\{2, 9\}$; **7.** $\{-9, 8\}$; **8.** $\{-2, 2\}$;

9. $\{0, -\frac{b}{a}\}$; **10.** $\{m, -n\}$

Lesson 74

Solving Quadratic Equations by the Square Root Method
(For cases in which $b = 0$, i.e., the x-term is missing)

Principle: If $x^2 = k$, then $x = \pm\sqrt{k}$ (i.e., $x = +\sqrt{k}$ or $x = -\sqrt{k}$)

Example 1 Solve by the square root method.
$$x^2 - 9 = 0$$

Solution
$$x^2 - 9 = 0$$
$$\underline{+9 \quad +9}$$
$$x^2 = 9$$
$$x = \pm\sqrt{9}$$
$$x = \pm 3 \quad (\text{i.e. } x = 3 \text{ or } x = -3)$$
The solutions are -3 and 3.

Example 2   Solve by the square root method.
$$9x^2 - 36 = 0$$

Solution
$$9x^2 - 36 = 0$$
$$\underline{+36 \quad +36}$$
$$9x^2 = 36$$
$$\frac{9}{9}x^2 = \frac{36}{9}$$
$$x^2 = 4$$
$$x = \pm\sqrt{4}$$
$$x = \pm 2$$
The solutions are -2 and 2.

Lesson 74 Exercises

A Solve by the square root method:

1. $x^2 - 16 = 0$;

2. $4x^2 - 20 = 0$;

3. $x^2 - 49 = 0$;

Read page 427-428 before attempting Problem 4:   **4.** $x^2 + 49 = 0$

Answers: **1.** $\{-4, 4\}$;   **2.** $\{-\sqrt{5}, \sqrt{5}\}$;   **3.** $\{-7, 7\}$; 4. $\{-7i, 7i\}$

B Solve the following for x:

1. $x^2 - 9 = 0$

2. $-8 + x^2 = 0$

3. $ax^2 - c = 0$

4. $x^2 + 9 = 0$

5. $s = ax^2$

6. $x^2 - \frac{1}{4} = 0$

Answers: **1.** $\{-3, 3\}$;   **2.** $\{-2\sqrt{2}, 2\sqrt{2}\}$;   **3.** $\left\{-\sqrt{\frac{c}{a}}, \sqrt{\frac{c}{a}}\right\}$;   **4.** $\{-3i, 3i\}$;

5 $\left\{-\sqrt{\frac{s}{a}}, \sqrt{\frac{s}{a}}\right\}$;   **6.** $\left\{-\frac{1}{2}, \frac{1}{2}\right\}$

Lesson 75

Solving Quadratic Equations by Completing the Square

Example 1   Solve by completing the square.

$$x^2 - 12x + 8 = 0$$

Step 1: Eliminate the constant term (the "8") from the left -hand side.

(Note that the 8 ends up on the right-hand side of the equation as -8).

$$
\begin{array}{r}
x^2 - 12x + 8 = 0 \\
-\ 8\quad -8 \\
\hline
x^2 - 12x = -8
\end{array}
$$

Step 2:   Add the square of half the coefficient of the x-term to both sides of the equation.

(i.e., add the square of $\dfrac{b}{2}$ to both sides of the equation)

$$x^2 - 12x + \left(\tfrac{-12}{2}\right)^2 = -8 + (-6)^2 \quad (\ b = -12,\ \tfrac{b}{2} = -6;\ \text{and the square of } \tfrac{b}{2} = (-6)^2$$

$$x^2 - 12x + (-6)^2 = -8 + (-6)^2$$

Step 3:   Complete the square on the left-hand side of the equation

$$(x - 6)^2 = -8 + 36$$
$$(x - 6)^2 = 28$$

Step 4:   " Take the square root " of both sides of the equation.

$$x - 6 = \pm\sqrt{28}$$

Step 5: Solve for x and simplify right-hand side.

$$
\begin{array}{r}
x -\ 6 = \pm\sqrt{28} \\
+\ 6\quad +6 \\
\hline
x = 6 \pm \sqrt{28}
\end{array}
$$

$$\boldsymbol{x = 6 \pm 2\sqrt{7}}$$

Scrapwork:

$$\sqrt{28} = \sqrt{4}\sqrt{7} = 2\sqrt{7}$$

Example 2   Solve by completing the square.

$$3x^2 - 9x - 2 = 0$$

Step 1: $3x^2 - 9x = 2$

Step 2:   Divide the equation by the coefficient of the x^2-term (we want this coefficient to be 1, for this method)

$$\frac{3x^2}{3} - \frac{9x}{3} = \frac{2}{3}$$

$$x^2 - 3x = \frac{2}{3}$$

Step 3:   Add the square of the coefficient of the x-term to both sides of the equation, and complete the square.

$$x^2 - 3x + \left(\frac{-3}{2}\right)^2 = \frac{2}{3} + \left(\frac{-3}{2}\right)^2 \qquad \textbf{Note:} \quad \frac{b}{2} = \frac{-3}{2} \;;\; \left(\frac{b}{2}\right)^2 = \left(\frac{-3}{2}\right)^2$$

$$\left(x - \frac{3}{2}\right)^2 = \frac{2}{3} + \frac{9}{4}$$

Multiply this out

$$\left(x - \frac{3}{2}\right)^2 = \frac{35}{12}$$

Scrapwork: $\frac{2}{3} + \frac{9}{4} = \frac{35}{12}$

$$x - \frac{3}{2} = \pm\sqrt{\frac{35}{12}}$$

Step 4: Solve for x and simplify right-hand side.

$$x - \frac{3}{2} = \pm\sqrt{\frac{35}{12}}$$
$$+\frac{3}{2} \qquad\qquad +\frac{3}{2}$$

$$x = +\frac{3}{2} \pm \sqrt{\frac{35}{12}}$$

$$x = \frac{3}{2} \pm \frac{\sqrt{105}}{6}$$

Scrapwork: $\sqrt{\frac{35}{12}} = \sqrt{\frac{35 \cdot 3}{12 \cdot 3}}$

Solution is $\left\{ \frac{3}{2} \pm \frac{\sqrt{105}}{6} \right\}$ or $\left\{ \frac{3}{2} + \frac{\sqrt{105}}{6} , \frac{3}{2} - \frac{\sqrt{105}}{6} \right\}$

Lesson 75 Exercises

A Solve by completing the square:

1. $x^2 + 12x + 10 = 0$;

2. $x^2 - 4x + 6 = 0$;

3. $x^2 - 3x - 8 = 0$

4. $x^2 - 12x - 35 = 0$

Answers: **1.** $\{ -6 + \sqrt{26}, -6 - \sqrt{26}\}$; **2.** $\{2 + i\sqrt{2},\ 2 - i\sqrt{2}\}$; **3.** $\{ \frac{3}{2} + \frac{\sqrt{41}}{2},\ \frac{3}{2} - \frac{\sqrt{41}}{2} \}$;

4. $\{6 + \sqrt{71}, 6 - \sqrt{71} \}$

B Solve by completing the square:

1. $x^2 - 11x + 18 = 0$; **2.** $2x^2 + 5x - 12 = 0$; **3.** $x^2 - 4x - 1 = 0$; **4.** $x^2 - 4x + 8 = 0$.

5. $4x^2 + 48x + 40 = 0$; **6.** $2x^2 - 7x - 12 = 0$; **7.** $\frac{1}{2}at^2 + bt + k = 0$; **8.** $ax^2 + bx + c = 0$

Solutions: **1.** $\{2, 9\}$; **2.** $\{-4, \frac{3}{2}\}$; **3.** $\{ 2 + \sqrt{5}, 2 - \sqrt{5}\}$; **4.** $\{2 + 2i, 2 - 2i\}$; **5.** $\{-6 + \sqrt{26}, -6 - \sqrt{26} \}$;

6. $\{ \frac{7}{4} + \frac{\sqrt{145}}{4}, \frac{7}{4} - \frac{\sqrt{145}}{4} \}$; **7.** $\{ -\frac{b}{a} \pm \frac{\sqrt{b^2 - 2ak}}{a} \}$; **8.** $x = -\frac{b}{2a} \pm \frac{\sqrt{b^2 - 4ac}}{2a}$

or $x = \frac{-b \pm \sqrt{b^2 - 4ac}}{2a}$

Lesson 76

Solving Quadratic Equations by the Quadratic Formula

Example 1   Solve for x by the quadratic formula

$$3x^2 - 6x - 2 = 0$$

Step 1: $a = 3,\ b = -6,\ c = -2$

Step 2: $$x = \frac{-b \pm \sqrt{b^2 - 4ac}}{2a}$$

Step 3: $$x = \frac{-(-6) \pm \sqrt{(-6)^2 - 4(3)(-2)}}{2(3)} \quad \text{(Substituting for } a, b, \text{ and } c)$$

$$= \frac{+6 \pm \sqrt{36 + 24}}{6}$$

$$= \frac{6 \pm \sqrt{60}}{6}$$

$$= \frac{6}{6} \pm \frac{\sqrt{60}}{6}$$

$$= 1 \pm \frac{2\sqrt{15}}{6}$$

$$x = 1 \pm \frac{\sqrt{15}}{3}$$

Solution set is $\left\{ 1 - \frac{\sqrt{15}}{3},\ 1 + \frac{\sqrt{15}}{3} \right\}$

Example 2   Solve by the quadratic formula:

$$x^2 - 16 = 8 - 2x$$

Step 1:   Place the equation in standard form (i.e. rewrite the equation so that the only term on the right-hand side is zero)

$$
\begin{array}{rl}
x^2 - 16 &= 8 - 2x \\
+2x & \quad +2x \\
\hline
x^2 + 2x - 16 &= 8 \\
-8 & \quad -8 \\
\hline
x^2 + 2x - 24 &= 0 \;\text{<-----------(Standard Form)}
\end{array}
$$

Step 2: $a = 1$, $b = 2$, $c = -24$

Step 3: $x = \dfrac{-b \pm \sqrt{b^2 - 4\,ac}}{2a}$ <----------The quadratic formula

Step 4 : $x = \dfrac{-2 \pm \sqrt{(2)^2 - 4(1)(-24)}}{2(1)}$ (Substituting for a, b, and c)

$x = \dfrac{-2 \pm \sqrt{4 + 96}}{2}$

$x = \dfrac{-2 \pm \sqrt{100}}{2}$

$x = \dfrac{-2 \pm 10}{2}$

$x = \dfrac{-2 + 10}{2}$, or $x = \dfrac{-2 - 10}{2}$

$x = \dfrac{8}{2}$, or $x = \dfrac{-12}{2}$

$x = 4$ or $x = -6$

The solutions are 4 and -6.

Solving by any method

Example Solve by any method: $x^2 = 6x$

$$x^2 = 6x$$

Step 1: Write the equation in standard form.

$$x^2 - 6x = 0$$

Step 2: We solve by factoring (since $c = 0$, the quadratic is easily factorable)

$$x(x - 6) = 0$$

Step 3: Set each factor equal to 0 and solve for x:

$$
\begin{array}{rl}
x = 0 \text{ , or} & x - 6 = 0 \\
& +6 \quad +6 \\
\hline
& x = +6
\end{array}
$$

$x = 0$, or $x = 6$

The solutions are 0 and 6; or the solution set is {0,6}.

Discriminant of the quadratic equation: nature of roots and graphs of the equation

The expression $b^2 - 4ac$ is called the **discriminant** of the quadratic equation. The value of the discriminant determines the nature of the roots (or solutions) of the quadratic equation:

1. If $b^2 - 4ac > 0$, the equation $ax^2 + bx + c = 0$ has two real and unequal roots. Also if $b^2 - 4ac$ is a perfect square, the roots are rational. Graphically, the curve of the corresponding quadratic function, $f(x) = ax^2 + bx + c$, crosses the x-axis at two different points.

2. If $b^2 - 4ac = 0$, the quadratic equation has real equal roots (double root or repeated root). We therefore have only one real solution. Graphically, the curve of the corresponding quadratic function $f(x) = ax^2 + bx + c$, touches the x-axis but does not cross it.

3. If $b^2 - 4ac < 0$, the equation has two non-real (complex) roots. Graphically, the curve of the corresponding quadratic function, $f(x) = ax^2 + bx + c$, does not touch or cross the x-axis The curve is either entirely above the x-axis or entirely below the x-axis.

Note also that a quadratic equation is factorable if $b^2 - 4ac$ is a perfect square. (see also p. 234.)

Lesson 76 Exercises

A Solve by the quadratic formula:
1. $x^2 - 11x + 18 = 0$; **2.** $2x^2 + 5x - 12 = 0$; **3.** $x^2 - 4x - 1 = 0$; **4.** $x^2 - 4x + 8 = 0$.
5. $x^2 + 12x + 10 = 0$; **6.** $x^2 + 6 - 4x = 0$ **7.** $x^2 = 3x + 8$

Solutions: **1.** $\{2, 9\}$; **2.** $\{-4, \frac{3}{2}\}$; **3.** $\{2 + \sqrt{5}, 2 - \sqrt{5}\}$; **4.** $\{2 + 2i, 2 - 2i\}$; **5.** $\{-6 \pm \sqrt{26}\}$

6. $\{2 \pm i\sqrt{2}\}$; **7.** $\{\frac{3 \pm \sqrt{41}}{2}\}$

B Solve by any applicable method:
1. $x^2 - 6x + 4$; **2.** $6x^2 + 11x - 10 = 0$; **3.** $2x^2 - 3x + 6 = 0$; **4.** $2x(x - 4) = 6$

Solutions: **1.** $\{3 + \sqrt{5}, 3 - \sqrt{5}\}$; **2.** $\{\frac{2}{3}, -\frac{5}{2}\}$; **3.** $\{\frac{3 + i\sqrt{39}}{4}, \frac{3 - i\sqrt{39}}{4}\}$ **4.** $\{2 + \sqrt{7}, 2 - \sqrt{7}\}$

Lesson 77

Applications of the Quadratic Equation

Example 1   The width of a rectangle is 3 units less than the length. If the area of the rectangle is 28 sq. units, determine the dimensions of this rectangle.

Solution
Let the length of the rectangle $= x$.
Then, the width of this rectangle $= (x - 3)$
Area of a rectangle is given by LW (Where L is the length and W is the width).
The required equation: $x(x - 3) = 28$.
$$x^2 - 3x - 28 = 0$$
$$(x - 7)(x + 4) = 0$$
$$x = 7 \text{ or } x = -4 \qquad \text{(Solving by factoring)}$$
We reject the negative value, -4, since the dimension of a rectangle cannot be negative.
When the length is 7 units, the width is $7 - 3 = 4$ units.

(**Check**: If the length is 7 and the width is 4, the area is $7(4) = 28$ sq. units; and also the width, 4, is 3 less than the length, 7. Therefore, the conditions in the original word problem have been satisfied.)
The length is 7 units and the width is 4 units.

Example 2 The perimeter of a rectangle is 34 units, and its area is 30 sq. units.
              Find the dimensions of this rectangle.
Preliminaries:
Perimeter of a rectangle $= 2L + 2W$ (where L is the length, and W is the width)
Area of a rectangle $= LW$

Solution

Let the dimensions be x and y, where the larger of x and y is the length and the smaller of x and y is the width.

$$2x + 2y = 34 \longleftarrow \text{-----------Perimeter equation} \qquad (1)$$
$$xy = 30 \longleftarrow \text{-----------Area equation} \qquad (2)$$

We will solve the above system of equations simultaneously by the substitution method.
Step 1 : Solve for y from equation (1).
        Then, $y = (17 - x)$ (3)   (obtained by subtracting $2x$ from both sides of the equation
                                                         followed by dividing the equation by 2)

Step 2: Substitute for y from (3) in (2).
      $x(17 - x) = 30$
      $17x - x^2 = 30$
      $17x - x^2 - 30 = 0$
      $x^2 - 17x + 30 = 0$ (Multiplying the equation by -1 and rearranging terms)

Step 3: Solve this quadratic equation by any method.
　　　　We solve by factoring since the factors are easily recognizable.
　　　　$(x - 2)(x - 15) = 0$
　　　　$x = 2$ or $x = 15$

Step 4: When $x = 2$,
　　　　$2y = 30$; and from which $y = 15$.
　　　　when $x = 15$,
　　　　$15y = 30$; and from which $y = 2$.
From Step 4, we obtain the dimensions 15 and 2.

The dimensions are 15 units and 2 units.　(The length is 15 units and the width is 2 units)

Lesson 77 Exercises

A The perimeter of a rectangle is 36 units, and its area is 77 sq. units.
　Find the dimensions of this rectangle.

Answers:　Dimensions are 11 and 7.

B 1.　A number is 4 more than another number. If their product is 165, find them.

　2. The sum of two numbers is 20, and their product is 96.　Find these numbers.

　3. The length of a rectangle is 5 units more than the width. If the area of this rectangle is 126 sq,
　　units, find the length and width of this rectangle.

Ans: **1.** 15 and 11;　and also -11 and -15;　**2.** 8 and 12;　**3.** Length = 14, width = 9.

CHAPTER 29
FUNCTIONS

Lesson 78: **Sets, Relations, Functions, Comparison of Relations and Functions**

Lesson 79: **Functional Notation; Defined Functions; Excluded Values, Domain and Range**

Lesson 79 Extra: **Algebra of Functions**

Lesson 78

Sets, Relations, Functions, Comparison of Relations and Functions

Ordered Pair

An **ordered pair** of numbers is an arrangement of two numbers in a specified order. In an x-y rectangular coordinate system of axes, the first element (or component) is the x-value and the second element is the y-value.

Example 1   (a) $(1, 2)$ <--- $(x = 1, y = 2)$
  (b) $(2, 3)$ <---- $(x = 2, y = 3)$
  (c) $(5, -1)$ <---- $(x = 5, y = -1)$

Note that each ordered pair represents a point in an x-y coordinate system of axes.

Set of numbers

A **set of numbers** is a well-defined collection of numbers. The numbers are called the elements or members of the set.

Example 2: If we denote the set of the numbers 2, 5 and 6 by A, then we may write $A = \{2, 5, 6\}$

Example 3: The set B of the ordered pairs $(1, 2), (2, 3),$ and $(5,-1)$ is given by
  $B = \{(1, 2), (2, 3), (5,-1)\}$.

Relation

A **relation** is a set of ordered pairs. (A collection of ordered pairs of numbers)

Example 4: The set $E = \{(2, 3), (2, 5), (4, 6)\}$ is a relation, <---There are three ordered pairs.

Example 5: The set $C = \{(6, 2), (7,4), (11, 5)\}$ is a relation.

Definition of a Function

A function may be defined in a number of ways, namely,

(a) in terms of ordered pairs; (b) in terms of a rule involving two variables;
(c) in terms of a rule for inputs and outputs; (d) in terms of correspondence of two sets; (e) as a graph

Definition 1:   In terms of ordered pairs
A **function** is a relation in which no two distinct ordered pairs have the same first component; or a function is a set of ordered pairs in which for any two different ordered pairs, the first elements are different. The set in Example 5, above, is a function but the set in Example 4 is not a function, because the first two ordered pairs have the same first element, namely 2.

The set of all the first elements of the ordered pairs is called the **domain** of the function; and the set of all the second elements is called the **range** of the function.

Example 6: In the function $C = \{(6, 2), (7, 4), (11, 5)\}$. The domain, $D = \{6, 7, 11\}$ (first elements)
  The range, $R = \{2, 4, 5\}$. (second elements)

Other definitions of a function

Definition 2: If x and y are two variables. then we say that y is a function of x if there is a rule which gives just one corresponding value of y for **each** value of x. The variable x is called the independent variable, and a variable y is called the dependent variable. The rule may be specified in the form of a set, in the form of a graph, in the form of a table, or in the form of an equation or formula. The **domain** of a function is the set of numbers that can be assigned to x (the independent variable). The **range** of a function is the set of all the corresponding numbers y (the dependent variable) associated by the function (rule) with the numbers, x, in the domain .

We symbolize that f is a function of x by $f(x)$, where x is called the independent variable, y is called the dependent variable. **Note**: $f(x)$s is read as f of x.

The following are examples of how the rules for functions may be specified:

(a) In the form of an equation or a formula: $y = 2x$.
(b) In the form of a set: $\{(2,3),(1,4),(7,5)\}$.
(c) In the form of a table for x and y: See Table 1.
(d) In the form of a graph. See Figure

Table 1:: $y = 2x$

$x =$	0	1	2	3	4
$y =$	0	2	4	6	8

Figure: Graph of $y = 2x$

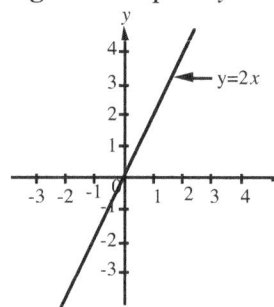

Definition 3 In terms of the correspondence of two sets (Fig. 1)

A function is a correspondence between a first set, say set A and a second set, say Set B such that **each** element of set A corresponds to exactly one element of set B. The set of all the elements of set A is a called the **domain** of the function, and the set of all the corresponding elements of set B is called the **range** of the function.

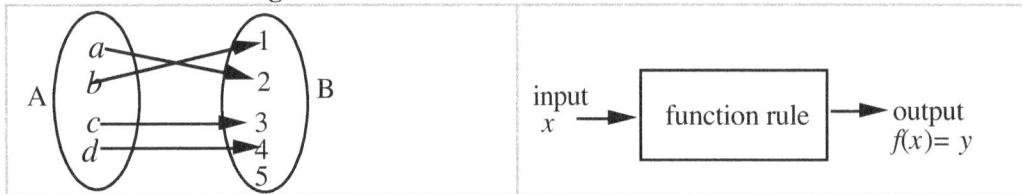

Fig 1   **Fig 2**

Definition 4   In terms of a rule for inputs and outputs (Fig. 2)

A function is a rule which assigns to each input number exactly one output number.The set of all input numbers that the rule is applicable to is called the **domain** of the function; and the set of all the corresponding output numbers is called the **range** of the function.

A variable representing an input number is called the independent variable, and a variable representing an output number is called the dependent variable.

Given a graph   (Vertical line test)

A given graph is that of a function if every possible vertical line drawn to intersect the graph intersects (cuts) the graph exactly once (i.e., at one point only).

Comparison of a Function and a Relation 371

Similarities: Each is a set of ordered pairs.

Differences: In a relation, two or more ordered pairs may have the same first component; but in a function, no two distinct ordered pairs may have the same first component.

Example: The set $D = \{(1,6),(3,4),(3,5),(4,6)\}$ is only a relation and **not** a function. because the second and third ordered pairs have the same first component, which is 3.

Example: The set $E = \{(1,2),(2,3),(4,5),(7,5)\}$ is a function (even though the second components of the third and fourth ordered pairs are the same).

A function is a relation, but a relation is not necessarily a function.

Determining if a given graph is a relation or a function

We will use the so-called **vertical line test.**

Procedure

Step 1: Draw as many vertical lines as possible (This can be done visually.) to intersect the graph.

Step 2: If any of the possible lines intersects (cuts) the graph at more than one point, then the given graph is not a function but a relation. However, if each of the possible vertical lines intersects the graph only once (at one point only), then the graph represents a function. Figure.. is a graph which is a relation but not a function.

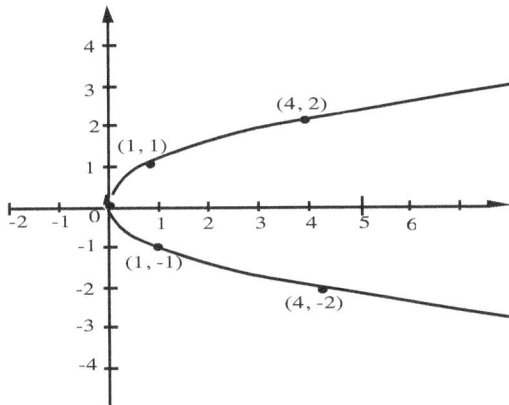

Figure: Graph of $y = \pm\sqrt{x}$ or $x = y^2$
This graph is a relation but not a function

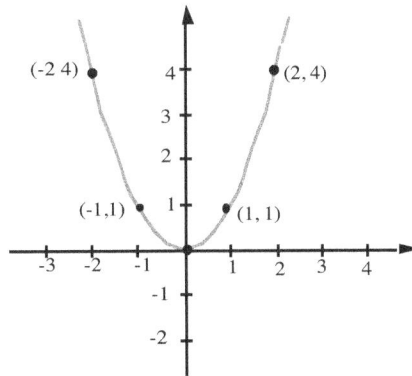

Figure: Graph of $y = x^2$.
This graph is that of a function.

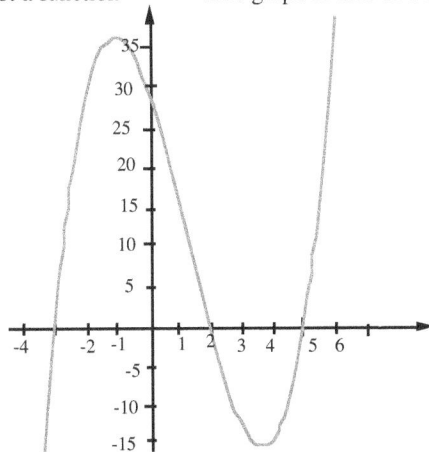

Figure: Graph of $y = (x - 5)(x - 2)(x + 3)$. This graph is that of a function.

Lesson 78 Exercises

Determine which of the following are graphs of functions.

Figure (a)

Figure (b)

Figure (c)

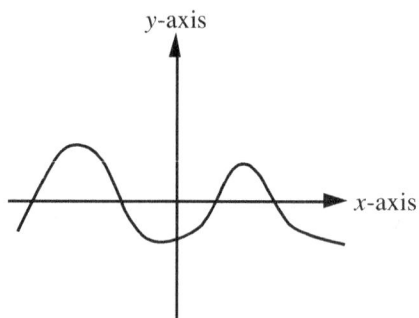

Figure (d)

Answers: (a) A function; (b) Not a function; (c) Not a function; (d) A function

Lesson 79

Functional Notation; Defined Functions; Excluded Values; Domain and Range

Functional Notation

Let a function $f(x)$ be specified by the rule $f(x) = x^2 + 3$.   (1)

In equation (1), $f(x)$ is read "f of x" or f is a function of x.

To evaluate a function for a particular value of x, we substitute that value of x in the rule that defines $f(x)$. Note that $f(x)$ does not mean f times x but that $f(x)$ is written as a symbol.

Example 1   Given that $f(x) = x^2 + 3$, find (a) $f(-1)$.; (b) $f(x_0 + h) - f(x_0)$

Solution: (a) $f(x) = x^2 + 3$

$$f(-1) = (-1)^2 + 3 \qquad \text{(replacing } x \text{ in the given equation by -1)}$$
$$= 1 + 3$$
$$f(-1) = 4$$

(b) (Replace x in the given equation by $(x_0 + h)$ and x_0, accordingly in $f(x) = x^2 + 3$

$$f(x_0 + h) \;-\; f(x_0) = [(x_0 + h)^2 + 3] - (x_0^2 + 3)$$
$$= [x_0^2 + 2hx_0 + h^2 + 3] - x_0^2 - 3$$
$$= x_0^2 + 2hx_0 + h^2 + 3 - x_0^2 - 3$$
$$= 2hx_0 + h^2$$

Example 2   Find $f(2)$, given that $f(x) = x + 7$

Solution

$$f(x) = x + 7$$
$$f(2) = 2 + 7$$
$$= 9$$

Example 3   If $f(x) = 2 - \dfrac{1}{x - 4}$, find $(a)\, f(-3)$;   (b) $f(x_0 + h)$;   (b) $f(-x)$.

Solution   (a) $f(x) = 2 - \dfrac{1}{x - 4}$

$$f(-3) = 2 - \frac{1}{(-3) - 4} \qquad \text{(replacing } x \text{ in the given equation by -3)}$$
$$= 2 - \frac{1}{-7}$$
$$= 2 + \frac{1}{7}$$
$$= 2\tfrac{1}{7}$$

(c) (Replace x in the given equation by $x_0 + h$

$$f(x) = 2 - \frac{1}{x-4} \quad \text{<- - - - given equation}$$

$$f(x_0 + h) = 2 - \frac{1}{(x_0 + h) - 4} \quad \text{<- - -replacing } x \text{ by } x_0 + h$$

$$= 2 - \frac{1}{x_0 + h - 4}$$

$$= \frac{2(x_0 + h - 4) - 1}{x_0 + h - 4}$$

$$= \frac{2x_0 + 2h - 8 - 1}{x_0 + h - 4}$$

$$= \frac{2x_0 + 2h - 9}{x_0 + h - 4}$$

(c) (Replace x in the given equation by $-x$)

$$f(x) = 2 - \frac{1}{x-4} \quad \text{<- - - - given equation}$$

$$f(-x) = 2 - \frac{1}{(-x) - 4} \quad \text{<- - - - -(replacing } x \text{ by } -x)$$

$$= 2 - \frac{1}{-x - 4}$$

$$= \frac{2(-x - 4) - 1}{-x - 4}$$

$$= \frac{-2x - 8 - 1}{-x - 4}$$

$$= \frac{-2x - 9}{-x - 4}$$

$$= \frac{-(2x + 9)}{-(x + 4)} \quad (factoring\ out\ -1)$$

$$= \frac{2x + 9}{x + 4}$$

Note above that in (b) the final result contains x. This is so, because we replaced x by $-x$. In the case of (a), we replaced x by the integer -3, and the final result was purely a numerical value. Furthermore, in Example 3, $f(-a) = \frac{2a + 9}{a + 4}$

Defined Real-Valued Function

Meaning of a defined function of x

A real-valued function $f(x)$ is said to be defined for a variable x if the following conditions are satisfied:

> **1.** The x and $f(x)$ must be real (i.e., x and $f(x)$ should not be the square root or an even root of a negative number). Thus, a value such as $\sqrt{-4}$ or $2i$ is not allowed.
>
> For example, in $f(x) = \sqrt{x-4}$, we have to make sure that $x \geq 4$, since otherwise, we obtain imaginary numbers.

> **2.** The value of x when substituted in the functional equation should yield specific real numbers. (i.e., the value of x when substituted in the functional equation should **not** make the denominator become zero.)

Condition (2) implies that the function should not become undefined when the value of x is substituted in the functional equation. When the function involves a denominator, we have to make sure that the denominator is not allowed to be zero. A particular example of this function occurs when the given function is the ratio of two polynomial functions. (We call such functions rational functions.)

In this book, it is agreed that a function is real-valued unless otherwise specified.

Examples of rational functions are: (a) $f(x) = \dfrac{x^2 + 4}{x - 1}$;   (b) $f(x) = \dfrac{1}{(x-3)(x+4)}$

Excluded Values, Domain and Range

The **excluded values** are (usually) the values which when substituted in the functional equation make the function either undefined or imaginary.

The **domain** (say , D) of a function $f(x)$ consists of those real values of the independent variable say, x, for which $f(x)$ is real and defined.

The **range** (say, R) of the function consists of the corresponding values which $f(x)$ assumes for x in the domain of the function .

Specification of Domain and Range

The domain and the range of a given function may be specified in several ways, namely in the form of a set, in the form of a table, in the form of an equation, in the form of an inequality, or in the form of a graph.

(In some of the examples that follow, we will use a graphing calculator or a computer grapher to generate graphs which will be useful in determining the range of some of the functions. Determining the range of some of the functions analytically may require advanced methods (which we do not cover in this book) , and therefore, we use graphing as an aid.)

Example 1: In the form of a set

Find the domain and range of the function specified by $\{(1, 2), (3, 2), (4, 4), (5, 5)\}$.

Solution: The domain consists of the first components of the ordered pairs.
       The domain, $D = \{1, 3, 4, 5\}$.
       The range consists of the second components of the ordered pairs.
       The range, $R = \{2, 4, 5\}$.

Example 2: In the form of a table of values

Find the domain and range of the function specified by the table of values below.

x	y
1	2
3	2
4	4
5	5

Solution: The domain consists of (the x-values) 1, 3, 4, and 5.

The range consists of the (y-values) 2, 4 and 5.

Note that this form is the set form written in a different format.

Example 3: In the form of an equation:

The domain consists of those real values of x for which the function is defined.
The corresponding values of $f(x)$ form the range of the function.

Example 4: In the form of a graph

Find the domain and range of the function specified by the graph below.

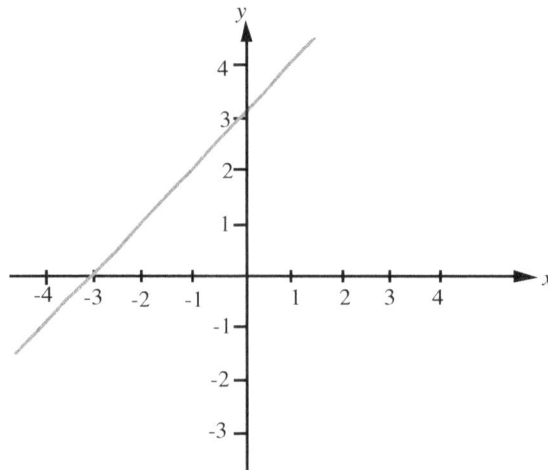

Figure: The graph of $y = x + 3$

Solution: From the graph, the function is defined for all real values of x.

The domain consists of all real x-values. The range consists of all real y-values.

Implicit and Explicit Specification of the Domain of a Function

The domain of a function may be specified either implicitly or explicitly.

Consider the function $f(x) = x^2$

As it stands, the domain is implicitly specified. The above function is that of a polynomial and as such the domain consists of all real numbers. Here, we assume the largest possible domain.

Now, consider $f(x) = x^2$ $0 \le x \le 5$

In this case, the inequality written to the right of the function specifies explicitly and restricts the domain. The domain of this function is such that x is between 0 and 5, including 0 and 5. If the inequality to the right had not been indicated, the domain would have consisted of all real x-values.

The restrictions on the domains are very important and useful in sketching the graphs of functions.

Other functions with explicitly specified domains are

$(a)\ \ y = \sin x$ $0 \le x \le 2\pi$

$(b)\ \ y = \cos x$ $-2\pi \le x \le 2\pi$

Determining the Excluded Values, Domain and Range of a Function 378

Case 1: Polynomial functions

Example 1 (a) For what values of x is the following function not defined? (b) what is the domain? (c) what is the range? $f(x) = x^2 - 3x + 1$

Solution: All polynomial functions are defined for all real values of the independent variable.

(a) Since the given function is a polynomial function, it is defined for all real values of x. We may note that the right-hand side of the equation does not involve denominators or square roots (or even roots) of the independent variable. There are **no** excluded values.

(b) Generally, determining the range of a polynomial function may require advanced methods. However, since the given function is a quadratic function , we may apply the range inequality

formula, $y \geq \dfrac{4ac - b^2}{4a}$

From the function, $a = 1, b = -3$, and $c = 1$.

Substituting these values, $y \geq \dfrac{4(1)(1) - (-3)^2}{4(1)} = -\dfrac{5}{4}$ (i.e., $y \geq -\dfrac{5}{4}$)

The range consists of all real numbers greater than or equal to $-\dfrac{5}{4}$.

(Note that any horizontal line drawn below the point $(\frac{3}{2}, -\frac{5}{4})$ will **not** intersect the curve, but any horizontal line through or above this point will intersect the curve)

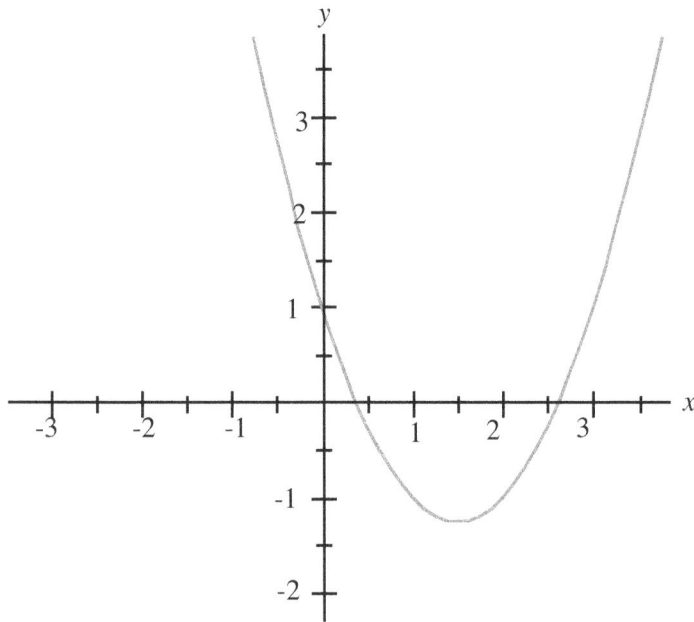

Figure: Graph of $f(x) = x^2 - 3x + 1$

Case 2: **Rational functions**

Note: A rational function is a function which is the ratio of two polynomial functions.

Example 2 (a) For what values of x is the following function not defined?
(b) what is the domain? (c) what is the range?

$$f(x) = \frac{3x - 2}{x - 1}$$

Solution   Step 1: Setting the denominator to zero,
$$x - 1 = 0$$
Step 2: Solving for x, $x = 1$.

(a) The function is not defined when $x = 1$. (The excluded value of x is 1.)

(b) The domain is all real values of x, except 1. same as $\{x \mid x \text{ is a real number and } x \neq 1 \}$

(c) The range (from graph) if found by being guided by the horizontal asymptote, $y = 3$.
The range is given by the set $\{y \mid y < 3 \text{ or } y > 3\}$ or simply $\{y \mid y \neq 3\}$
(Note that a horizontal line drawn through $(0, 3)$ will **not** intersect the curve)

Checking for $x = 1$, $f(1) = \frac{3(1) - 2}{1 - 1} = \frac{3 - 2}{0} = \frac{1}{0}$ which is undefined .(The right-hand side is division by zero).

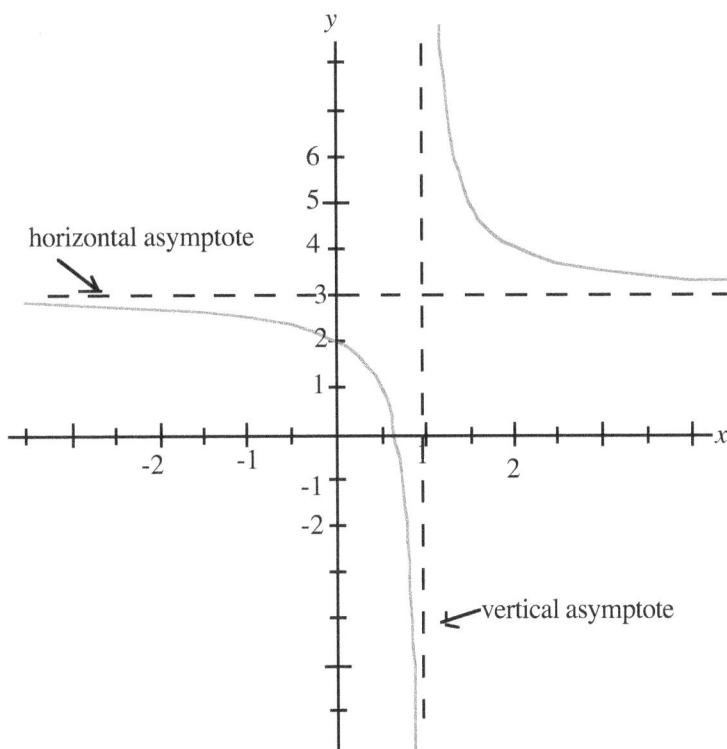

Figure: Graph of $f(x) = \frac{3x - 2}{x - 1}$

Example 3 (a) For what values of x is the given function not defined? (b) what is the domain? 380
(c) what is the range?

$$f(x) = \frac{2(x - 1)}{(x - 2)(x + 4)}$$

Solution Step 1: Setting the denominator to zero,
$$(x - 2)(x + 4) = 0$$

Step 2: Solving for x, $x = 2$, or $x = -4$.

(a) The function is not defined when $x = 2$ and -4. (The excluded values of x are 2 and -4.)

(b) The domain is all real values of x, except 2 and -4.
    Set-builder notation: $\{x \mid x$ is a real number and $x \neq -4$, $x \neq 2\}$

(c) The range (from graph) is all real y.
    Set-builder notation: $\{y \mid y$ is a real number$\}$
(Note that any horizontal line drawn through the y-axis will intersect the curve)

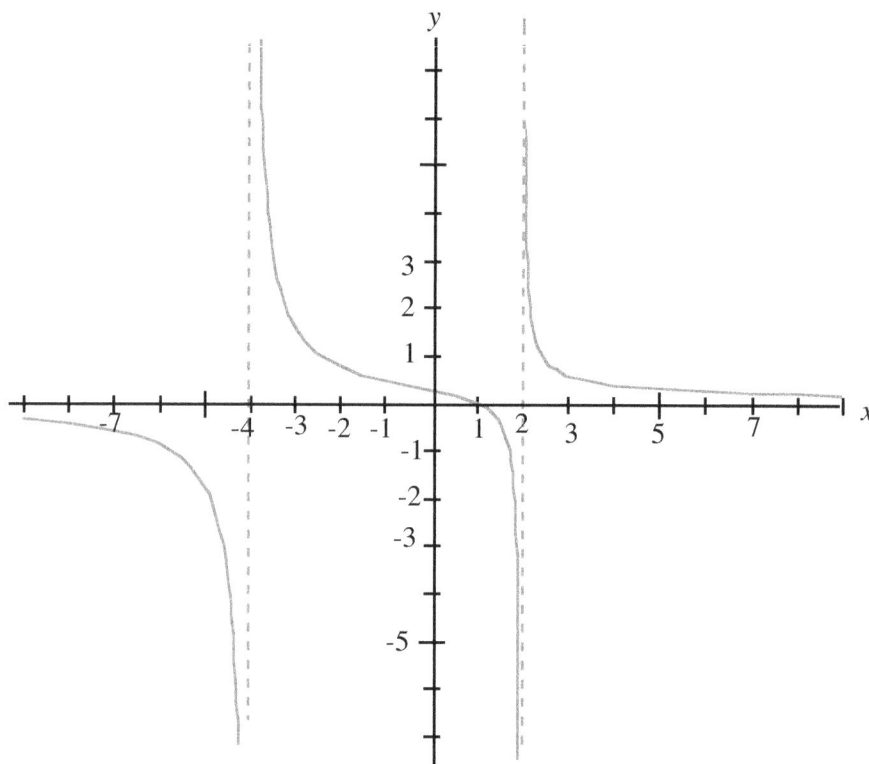

Figure: Graph of $f(x) = \dfrac{2(x - 1)}{(x - 2)(x + 4)}$

Example 4 (a) For what values of x is the given function not defined? (b) what is the domain?

(c) what is the range?

$$f(x) = \frac{1}{x}$$

Solution Setting the denominator to zero,
$$x = 0$$

(a) The function is not defined when $x = 0$. (The excluded value of x is 0.)

(b) The domain is all real values of x, except 0.
Set-builder notation: $\{x \mid x \text{ is a real number and } x \neq 0\}$

(c) The range (from graph) is given by the set $\{y \mid y < 0 \ \text{ or } \ y > 0\}$ or simply $\{y \mid y \neq 0 \}$
(the horizontal asymptote is $y = 0$)
(Note that any horizontal line drawn through the y-axis, except through the point $(0, 0)$, will intersect the curve.

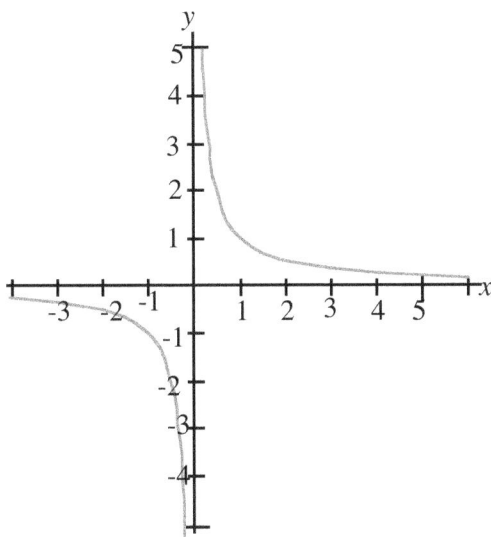

Figure: Graph of $f(x) = \frac{1}{x}$

Example 5 (a) For what values of x is the given function not defined? (b) what is the domain?   382
(c) what is the range?

$$f(x) = \frac{x^3 + x^2 + 2}{x^2 - 16}$$

Solution: Setting the denominator to zero and solving,

$$x^2 - 16 = 0$$
$$(x + 4)(x - 4) = 0$$
$$x = -4 \text{ or } x = 4$$

The function is not defined when $x = -4$ or 4. (The excluded values of x are -4 and 4.)

(a) The domain is all real x except -4 and 4.

   $\{x \mid x$ is a real number and $x \neq -4$, $x \neq 4 \}$

(b) The range (from graph) is all real values of y. same as $\{y \mid y$ is a real number$\}$

 (Note that any horizontal line drawn through the y-axis will intersect the curve)

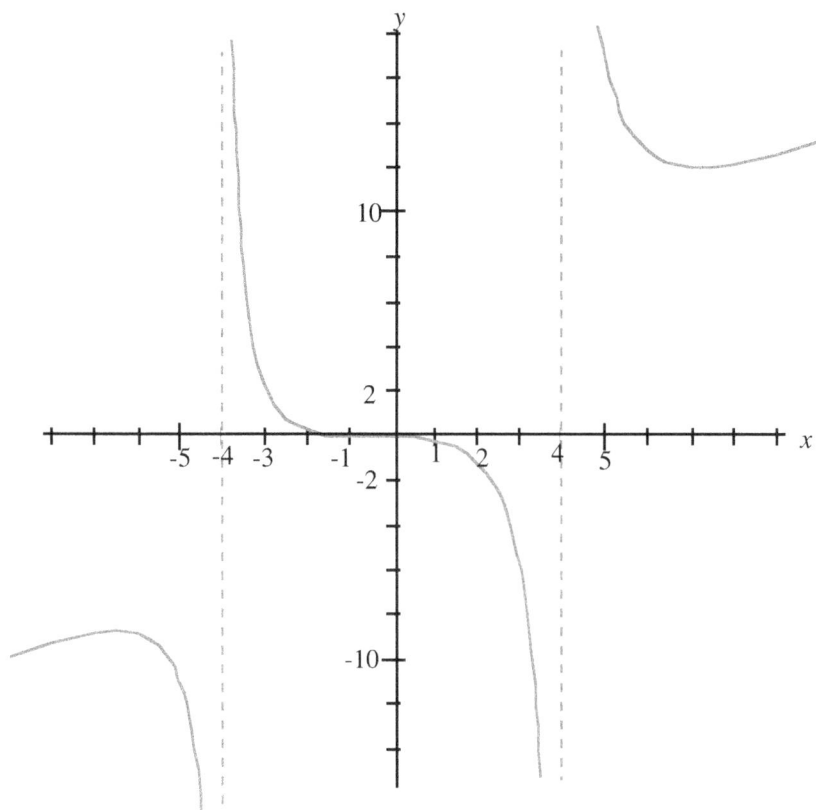

Figure: Graph of $\dfrac{x^3 + x^2 + 2}{x^2 - 16}$

Example 6 (a) For what values of x is the given function not defined? (b) what is the domain?

(c) what is the range?

$$f(x) = \frac{8}{x^2 - 4}$$

Solution Step 1: Setting the denominator to zero,

$$x^2 - 4 = 0$$
$$(x + 2)(x - 2) = 0.$$

Step 2: Solving for x, $x = 2$, or $x = -2$.

(a) The function is not defined when $x = 2$ and -2. (The excluded values of x are 2 and -2.)

(b) The domain is all real values of x except 2 and -2.

Same as $\{x \mid x \text{ is a real number and } x \neq -2, x \neq 2\}$

(c) The range (from graph) is given by the set $\{y \mid y \leq -2 \text{ or } y > 0\}$

(Note that any horizontal line drawn through or below $y = -2$ or above $y = 0$ will intersect the curve)

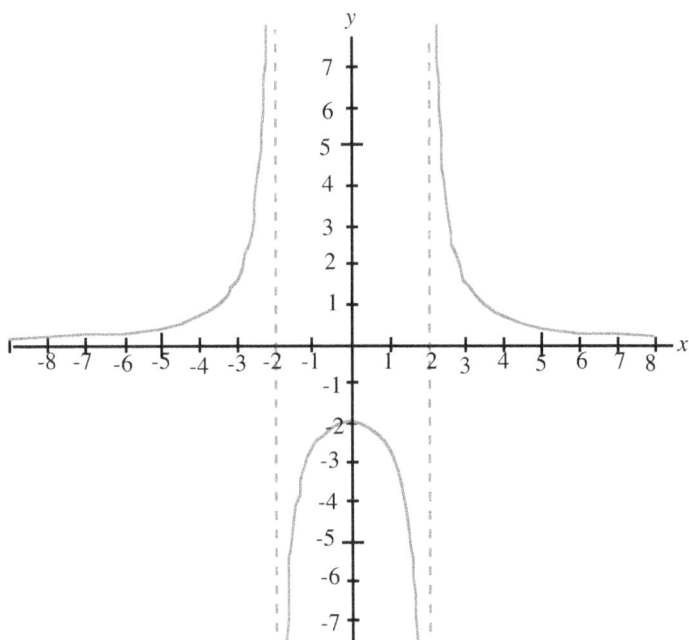

Figure: Graph of $f(x) = \dfrac{8}{x^2 - 4}$

Example 7 (a) For what values of x is the given function not defined? (b) what is the domain?   384

(c) what is the range?

$$f(x) = \frac{8}{x^2 + 4}$$

Solution   Step 1: Setting the denominator to zero, and solving, we obtain non-real values. Since we are dealing with real-valued functions, we conclude that

Step 2: (a) There are **no** excluded values.

$x^2 + 4$ is positive for all real values of x and never zero, since the square of any nonzero real number is always positive.

(b)   The function is defined for all real values of x.

(c) The range (from graph) is given by the set $\{y \mid 0 < y \le 2\}$

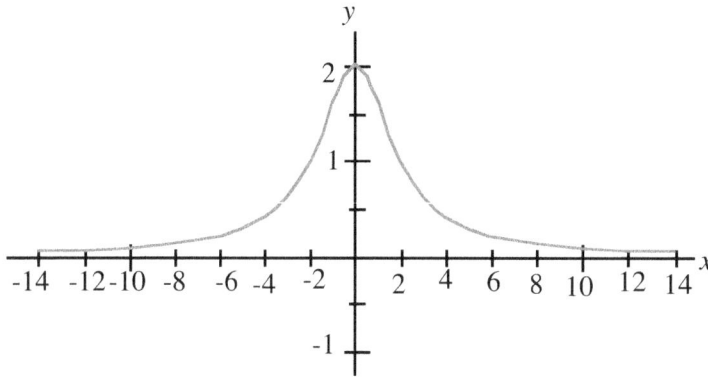

Figure: Graph of $f(x) = \dfrac{8}{x^2 + 4}$

Case 3: Functions containing even roots (e.g., square root) of polynomials

Example 8 (a) For what values of x is the function $f(x) = \sqrt{x - 5}$ not real?

(b) What is the domain?   What is the range?

Solution (a) For real roots, the radicand, $(x - 5)$ must be positive or zero.

Symbolically, $x - 5 \ge 0$

$x \ge 5$ (solving for x).

The function is not real when $x < 5$ (because the square root would be that of a negative number. The square root of a negative number is imaginary. Any x-value less than 5 is excluded.

Checking for some specific values of x:

1. When $x = 5$, $f(5) = \sqrt{x - 5}$ $= \sqrt{5 - 5} = 0$, which is real.   (i.e., for $x = 5$).

2. When $x = 6$, $f(6) = \sqrt{6 - 5}$ $= 1$, which is real.   (i.e., for $x > 5$).

3. When $x = 4$, $f(4) = \sqrt{4 - 5}$ $= \sqrt{-1}$, which is imaginary (non-real (i.e., for $x < 5$).

(b) The domain consists of all real values of x such that $x \ge 5$.

(c) The range for $y = \sqrt{x - 5}$ is from $y = 0$ to $y = \infty$.

(The range is obtained by considering $x \ge 5$ ($x = 5$ and $x > 5$) in $y = \sqrt{x - 5}$, or from graph.

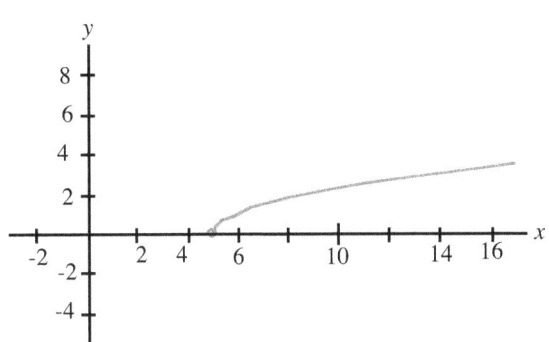

Figure 2: Graph of $f(x) = \sqrt{x - 5}$

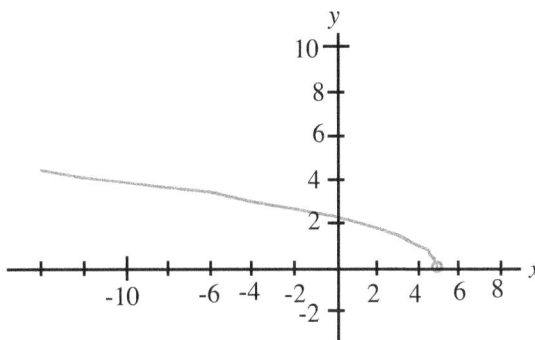

Figure 3: Graph of $f(x) = \sqrt{5 - x}$

Example 9 (a) For what values of x is $f(x) = \sqrt{5 - x}$) not real?

(b) What is the domain? What is the range?

Solution

(a) $5 - x \geq 0$ (for real values of $f(x)$, the radicand must be positive or zero).

$x \leq 5$ (solving for x)

The function is not real if $x > 5$. (Any x-value greater than 5 is excluded).

(b) The domain is all real x such that $x \leq 5$.

(c) The range is from $y = 0$ to $y = \infty$ (obtained by considering $x \leq 5$ ($x = 5$ and $x < 5$) in $y = \sqrt{5 - x}$., or (from graph)

Example 10 Find (a) the domain and (b) the range of $f(x) = \dfrac{\sqrt{x + 6}}{x - 1}$

Solution Here, two conditions must be satisfied.

Condition 1: $x + 6 \geq 0$ and from which $x \geq -6$ (For real root, the radicand must be positive or zero)

Condition 2: $x - 1 \neq 0$ and from which $x \neq 1$.

(a) The domain consists of all real numbers greater than or equal to -6, except 1, that is $x \geq -6$, $x \neq 1$.

(b) The range (from graph) is given by the set $\{y \mid y \text{ is a real number}\}$. Note that when $x = -6$, $y = 0$

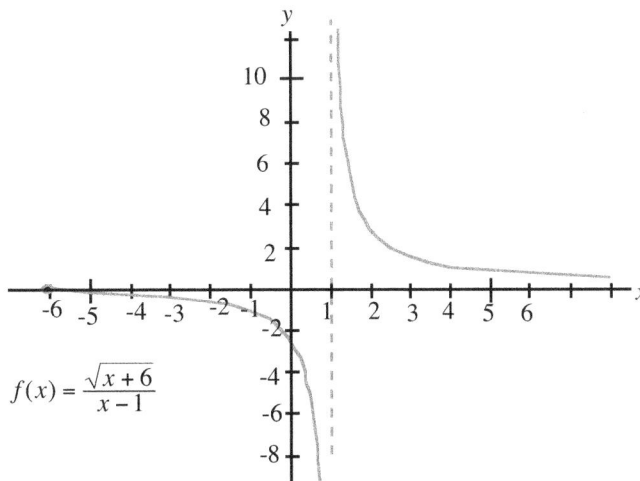

Fig.4: $f(x) = \dfrac{\sqrt{x + 6}}{x - 1}$

A

1. If $(x) = x^2 - 5x + 2$, find (a) $f(-2)$; (b) $f(-1)$, (c) $f(-x)$, (d) $f(a + h)$, (e) $\dfrac{f(a + h) - f(a)}{h}$;

2. If $(x) = 4 - \dfrac{1}{x - 2}$, find (a) $f(-3)$; (b) $f(-x)$; (c) $f(a)$.

3. If $(x) = x^3 - x^2 - x - 1$, find $f(-1)$.

Answers: **1.** (a) 16; (b) 8; (c) $x^2 + 5x + 2$; (d) $a^2 - 5a + 2ah - 5h + h^2 + 2$; (e) $2a + h - 5$

2. (a) $4\frac{1}{5}$; (b) $\dfrac{4x + 9}{x + 2}$; (c) $\dfrac{4a - 9}{a - 2}$; **(3)** -2

B

1. What is meant by the domain of a function?

2. What conditions do you look out for in determining the domain of a function?

3. What is meant by the range of a function?

Find the domain and range in each of the following:

4. $A = \{(2, 4), (3, 9), (4, 16), (5, 25)\}$; **5.** $B = (2, 1), (5, 1), (6, 1), (7, 3)$

6. $y = x + 2$; **7.** $y = x^2 + 3x + 7$; **8.** $y = -x - 4$; **9.** $y = x^3 + x^2 - 5$

Answers: **4.** Domain : $\{2, 3, 4, 5\}$;, range: $\{4, 9, 16, 25\}$

5. Domain : $\{2, 5, 6, 7\}$;, range: $\{1, 3\}$ Note: we do not repeat the "1".
6. Domain : All real values of x: range: all real values of y.
7. Domain : All real x; range: all real y such that $y \geq 4.75$.
8. Domain : All real values of x: range: all real values of y.
9. Domain : All real values of x.

C Determine the excluded values (if any) for the following:

1. $y = \dfrac{2}{x - 3}$;   **2.** $f(x) = \dfrac{2x - 4}{x^2 - 1}$; **3.** $f(x) = \dfrac{3x}{x^2 + 1}$; **4.** $f(x) = \dfrac{1}{x}$; **5.** $f(x) = \dfrac{(x - 1)}{(x - 1)(x + 2)}$

In Problems 6 & 7 below, for what values of x is the function not real?

6. $y = \sqrt{x - 3}$;   **7.** $y = \sqrt{4 - x}$ **8.** Find the domain of $f(x) = x - \dfrac{1}{x}$

Extra: Why is it important to note the excluded values in dealing with functions and solving equations?

Answers: **1.** 3; **2.** $\{1, 1\}$; **3.** None; **4.** 0; **5.** $\{-2, 1\}$; **6.** $x < 3$; **7.** $x > 4$; **8.** All real x except

D   Show and determine which of the following functions are one-to-one

1. The set $\{(4, 5), (3, 4), (3, 2)\}$; 2. The set $\{(2, 2), (4, 3), (5, 7)\}$; 3. $f(x) = 3x - 2$;

4. $f(x) = \sqrt{x + 3}$; 5. $f(x) = \sqrt{16 - x^2}$; 6. $f(x) = |x|$; 7. $f(x) = |x - 2|$; 8. $f(x) = x^3 + 2$.

Answers: **1.** Yes; **2.** Yes; **3.** Yes; **4. Yes**; **5.** No; **6. No**; **7.** Yes.

Lesson 79 Extra
Algebra of Functions

In the past, we covered the addition, subtraction, multiplication and division of algebraic expressions. Below, we similarly cover the same basic operations on functions.

Let f and g be functions of x such that x is in the domains of both f and g. Then

1. Sum: $(f + g)(x) = f(x) + g(x)$ (Domain: All real numbers common to the domains of f and g)

2. Difference: $(f - g)(x) = f(x) - g(x)$ Domain: All real numbers common to the domains f and g)

3. Product: $(f \bullet g)(x)$ or $(fg)(x) = f(x) \bullet g(x)$ or $f(x)g(x)$ (Domain: All real numbers common to the domains f and g

4. Quotient $\left(\dfrac{f}{g}\right)(x) = \dfrac{f(x)}{g(x)}$ (Domain: All real numbers common to the domains f and g and $g(x) \neq 0$.

Examples

If $f(x) = x^2 + 2$ and $g(x) = x - 2$; find (a) $(f + g)(x)$; (b) $(f - g)(x)$; (c) $(f \bullet g)(x)$;

(d) $\left(\dfrac{f}{g}\right)(x)$; (e) $(f + g)(3)$; (f) $\left(\dfrac{f}{g}\right)(3)$; (g) $\left(\dfrac{f}{g}\right)(2)$.

Solution

(a) $(f + g)(x) = f(x) + g(x)$
$$= (x^2 + 2) + (x - 2)$$
$$= x^2 + 2 + x - 2$$
$$= x^2 + x$$

(b) $(f - g)(x) = f(x) - g(x)$
$$= (x^2 + 2) - (x - 2)$$
$$= x^2 + 2 - x + 2$$
$$= x^2 - x + 4$$

(c) $(f \bullet g)(x) = f(x) \bullet g(x)$
$$= (x^2 + 2)(x - 2)$$
$$= x^3 - 2x^2 + 2x - 4$$

(d) $\left(\dfrac{f}{g}\right)(x) = \dfrac{f(x)}{g(x)}$
$$= \dfrac{x^2 + 2}{x - 2} \text{ or } x + 2 + \dfrac{6}{x - 2} \text{ (long division)}$$

(e) From (a), $(f + g)(x) = f(x) + g(x)$
$$= x^2 + x$$
$$(f + g)(3) = (3)^2 + 3 = 12$$

Method 2: $(f + g)(3) = f(3) + g(3)$
$$f(3) = (3)^2 + 2 = 11$$
$$g(3) = 3 - 2 = 1$$
$$(f + g)(3) = 11 + 1 = 12$$

(f) From (d), $\left(\dfrac{f}{g}\right)(x) = \dfrac{f(x)}{g(x)}$
$$= \dfrac{x^2 + 2}{x - 2}$$
$$\left(\dfrac{f}{g}\right)(3) = \dfrac{(3)^2 + 2}{(3) - 2}$$
$$= \dfrac{9 + 2}{3 - 2}$$
$$= 11.$$

Method 2: $f(3) = (3)^2 + 2 = 11$
$$g(3) = 3 - 2 = 1$$
$$\left(\dfrac{f}{g}\right)(3) = \dfrac{f(3)}{g(3)} = \dfrac{11}{1} = 11$$

(g) From (d), $\left(\dfrac{f}{g}\right)(x) = \dfrac{x^2 + 2}{x - 2}$
$$\left(\dfrac{f}{g}\right)(2) = \dfrac{(2)^2 + 2}{(2) - 2}$$
$$= \dfrac{6}{0} \text{ is undefined.}$$

Therefore, $\left(\dfrac{f}{g}\right)(2)$ is undefined. Note: $g(2) = 0$.

On the next page, we cover the domains for algebraic of functions.

Domains for Algebra of Functions

In the first step, we determine the common domain (intersection of domains) of the functions. In the second step, we check for the domain of the resulting function. (sum. difference, product or quotient). We add any excluded values to those from Step 1 and if there is an overlap in the domains, we determine the intersection of the domains from Step 1 and Step 2. Pay attention to radical functions such as $f(x) = \sqrt{x-2}$, determine the domain and find the intersection for the overall domain.

Example

Given $f(x) = \frac{3}{x}$ and $g(x) = \frac{3x-4}{x+2}$, find the domains of the following:

(a) $(f+g)(x)$; (b) $(f-g)(x)$; (c) $(f \bullet g)(x)$; (d) $\left(\frac{f}{g}\right)(x)$;

Solution:

Domain of $f : \{x \mid x$ is a real number and $x \neq 0\}$

Domain of $g : \{x \mid x$ is a real number and $x \neq -2\}$.

For **(a), (b), and (c):**

domain of $f+g$ = domain of $f-g$ = domain of $f \bullet g = \{x \mid x$ is a real number and $x \neq 0, x \neq -2\}$

(d) $\left(\frac{f}{g}\right)(x) = \frac{f(x)}{g(x)}$

$= \frac{3}{x} \div \frac{3x-4}{x+2}$ $(x \neq 0, x \neq -2)$

$= \frac{3}{x} \bullet \frac{x+2}{3x-4}$

$(x \neq \frac{4}{3}$. See $= \frac{3(x+2)}{x(3x-4)}$ below)

Note: $\frac{f(x)}{g(x)} = \frac{\frac{3}{x}}{\frac{3x-4}{x+2}}$

Setting $x(3x-4) = 0$

$x = 0$ or $3x-4 = 0$ and from which $x = \frac{4}{3}$

The excluded values are $-2, 0$ and $\frac{4}{3}$.

Domain of $\left(\frac{f}{g}\right)(x)$: $\{x \mid x$ is a real number and $x \neq 0, x \neq -2$ and $x \neq \frac{4}{3}\}$

Extra

Example: The intersection of the domains of $f(x) = \sqrt{x-2}$ and $g(x) = \frac{x-4}{x+3}$ is $\{x \mid x$ is a real number and $x \geq 2\}$ or $[2, \infty)$ (from $\{x \mid x \geq 2\} \cap \{x \mid x \neq -3\}$)

It seems the radical function is "King".

Lesson 79 Extra Exercises

If $f(x) = x^2 + 2$ and $g(x) = x - 2$; find (a) $(f+g)(x)$; (b) $(f-g)(x)$;(c) $(f \bullet g)(x)$; (d) $\left(\frac{f}{g}\right)(x)$;

(e) $(f+g)(3)$; (f) $\left(\frac{f}{g}\right)(3)$; (g) $\left(\frac{f}{g}\right)(2)$; (h). $(f-g)(4)$.

Ans: (a) $x^2 + x$; (b) $x^2 - x + 4$; (c) $= x^3 - 2x^2 + 2x - 4$; (d) $\frac{x^2+2}{x-2}$ or $x+2+\frac{6}{x-2}$;

(e) 12; (f) 11 (g) undefined; (h) 16.

CHAPTER 30

Sketching Parabolas

Lesson 80: **By General method**

Lesson 81: **By location of vertex, axis of symmetry and intercepts.**

The quadratic functions $y = f(x) = ax^2 + bx + c$ $(a \neq 0)$ and their graphs

We will cover two methods: a general method and the method of the location of the vertex, axis of symmetry and the x- and y-intercepts.

Lesson 80

General method of sketching the graph of the quadratic function

Example 1   Sketch the graph of $y = x^2 - 5$ for $x = 0, \pm 1, \pm 2, \pm 3$.

ordered pairs

Step 1: When $x = 0$, $y = (0)^2 - 5$
$$y = -5 \qquad \Big\} \rightarrow \quad (0,-5)$$

When $x = 1$, $y = (1)^2 - 5$
$$y = 1 - 5$$
$$y = -4 \qquad \Big\} \rightarrow (1,-4)$$

When $x = 2$, $y = (2)^2 - 5$
$$y = 4 - 5$$
$$y = -1 \qquad \Big\} \longrightarrow (2,-1)$$

When $x = 3$, $y = (3)^2 - 5$
$$y = 9 - 5$$
$$y = 4 \qquad \Big\} \rightarrow (3,4)$$

When $x = -1$, $y = (-1)^2 - 5$
$$y = 1 - 5$$
$$y = -4 \qquad \Big\} \longrightarrow (-1,-4)$$

When $x = -2$, $y = (-2)^2 - 5$
$$y = 4 - 5$$
$$y = -1 \qquad \Big\} \longrightarrow (-2,-1)$$

When $x = -3$, $y = (-3)^2 - 5$
$$y = 9 - 5 \qquad \Big\} \rightarrow (-3,4)$$
$$y = 4$$

Step 2: Plot the points (0,-5), (1,-4), (2,-1), (3,4), (-1,-4), (-2,-1) and (-3,4) on a rectangular coordinate system of axes (on graph paper) and connect the points by a smooth curve noting that the parabola is U-shaped and that since the coefficient of the x^2-term is positive, the parabola opens upwards (**Figure 1**).

Note above that the y-values are the same for $x = 1$ and $x = -1$; the same for $x = 2$ and $x = -2$ and the same for $x = 3$ and $x = -3$.

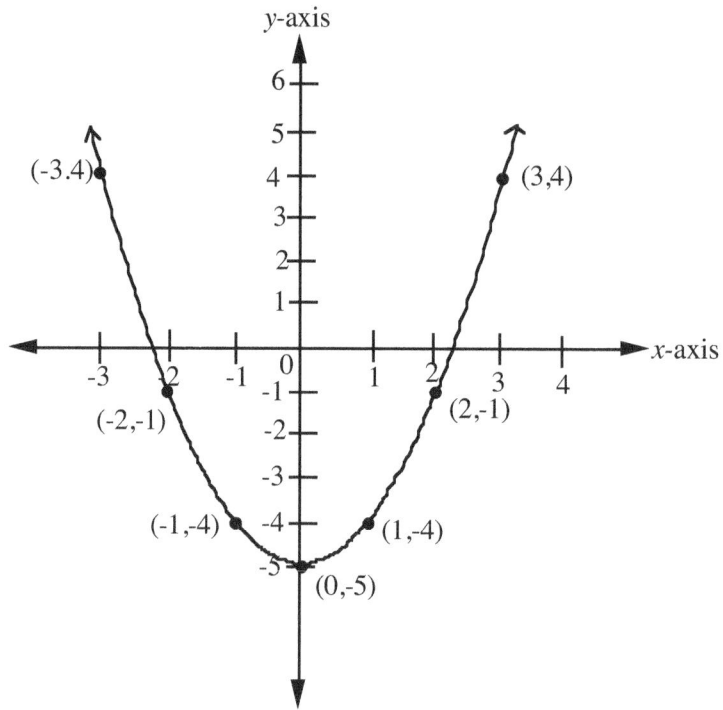

Figure 1:   Graph of $y = x^2 - 5$

Example 2 Sketch the graph of $y = -x^2 + 5$ for $x = 0, \pm 1, \pm2, \pm3$.

Step 1: See Example 1 above and perform similar calculations to obtain
(0,5), (1,4), (2,1), (3,-4), (-1,4), (-2,1), (-3,-4).

Step 2: Plot the points obtained from Step 1 on a rectangular coordinate system of axes (on graph paper) and connect the points by a smooth curve noting that the parabola is U-shaped and that since the coefficient of the x^2-term is negative (the coefficient of $-x^2$ is -1) , the parabola opens downwards (inverted U-shape). See Figure 2 below.

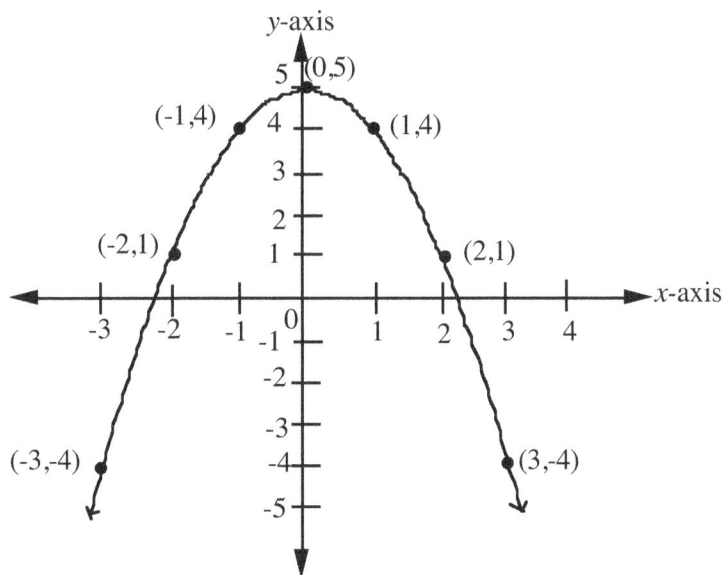

Figure 2: Graph of $y = -x^2 + 5$

Lesson 80 Exercises

Sketch the graphs of the following:
1. $y = x^2 - 4$ for $x = 0, \pm1, \pm2, \pm3$;   **2.** $y = -x^2 + 1$ for $x = 0, \pm1, \pm2, \pm3$.

1. Parabola opens upwards;   **2**. parabola opens downwards

Lesson 81

392

By location of vertex, axis of symmetry and intercepts

An alternative method of sketching the graphs of quadratic functions is by the location of the vertex, axis of symmetry and the x- and y-intercepts.

Example 1 Given the equation of the parabola $y = f(x) = x^2 - 6x + 8$, perform the following:
1. Find the x- and y-intercepts.
2. Find the axis of symmetry.
3. Find the coordinates of the vertex (or turning point)
4. Find the domain and range.
5. Sketch the graph of $y = f(x) = x^2 - 6x + 8$ (The graph of a quadratic function is called a parabola.)

Solution

1. To find the x-intercept, let $y = 0$ in $y = x^2 - 6x + 8$.

  Then, $0 = x^2 - 6x + 8$ or

  $x^2 - 6x + 8 = 0$ <-------This is a quadratic equation. Solve by any method
  $(x - 2)(x - 4) = 0$ <------We solve by factoring since the factors are easily recognizable.
  Solving, $x = 2$, or $x = 4$
  The x-intercepts are 2 and 4 . (The x-intercepts are at the points (2,0), and (4,0))

  To find the y-intercept, let $x = 0$ in $y = x^2 - 6x + 8$
  Then, $y = (0)^2 - 6(0) + 8$
  $y = 0 - 0 + 8$
  $y = 8$
  The y-intercept $= 8$.   (The y-intercept is at the point (0,8).)

2. $a = 1$ (the coefficient of the x^2-term), $b = -6$ (the coefficient of the x-term).

  Axis of symmetry is the line $x = -\dfrac{b}{2a}$ <------formula.
  $= -\dfrac{-6}{2(1)}$
  $= 3$

The axis of symmetry is the line $x = 3$. (see also p. 264)
The x-coordinate of the vertex $= 3$,

The y-coordinate of the vertex $= y = f(-\dfrac{b}{2a}) = f(3) = 3^2 - 6(3) + 8$
$= 9 - 18 + 8$
$y = -1$.

3. The vertex is at (3,-1). This point is the turning point on this curve. It is also a minimum point, since a is positive. This minimum point is also the lowest point on this curve.

4. The domain is all real x. (The given function is a polynomial.)
  The range is specified by using the y-coordinate of the vertex obtained in Step 2.
  The range is given by the inequality $y \geq -1$.
  The range consists of all real numbers greater than or equal to -1.

(Note: If we did not know the x-coordinate of the vertex, we could obtain the range by

Substituting $a = 1$. $b = -6$, $c = 8$, in $y \geq \dfrac{4ac - b^2}{4a}$ (Note the " $\geq$") to obtain

$y \geq \dfrac{4(1)(8) - (-6)^2}{4(1)}$ $\geq \dfrac{32 - 36}{4}$ $\geq \dfrac{-4}{4} \geq -1$;   That is $y \geq -1$)

5. To sketch the graph: On graph paper and using a rectangular coordinate system of axes,

Step 1: Locate the x- and y-intercepts from (**1**) above, by plotting the points (2,0), (4,0) and (0,8).

Step 2: Using a broken line draw the line $x = 3$, the axis of symmetry (from (**2**) above)

Step 3: Locate the vertex by plotting the point (3,-1). (from (**3**) above).

Step 4: Locate the point B using symmetry as follows: From the point A (the y-intercept), count horizontally say h units ($h = 3$ in this problem) to the axis of symmetry (the line $x = 3$). Next, continue and count horizontally h units (h = 3) from the axis of symmetry and stop at the point B (this is another point on the curve). You may also locate the point B by the intersection of the lines $y = 8$ and $x = 6$.

Step 5: Connect the above points by a smooth U-shaped curve (**Figure 1**), noting that the parabola opens upwards (is concave up), since a is positive ($a = 1$), and also noting that the graph is symmetric about the axis of symmetry (the line $x = 3$, in this example).

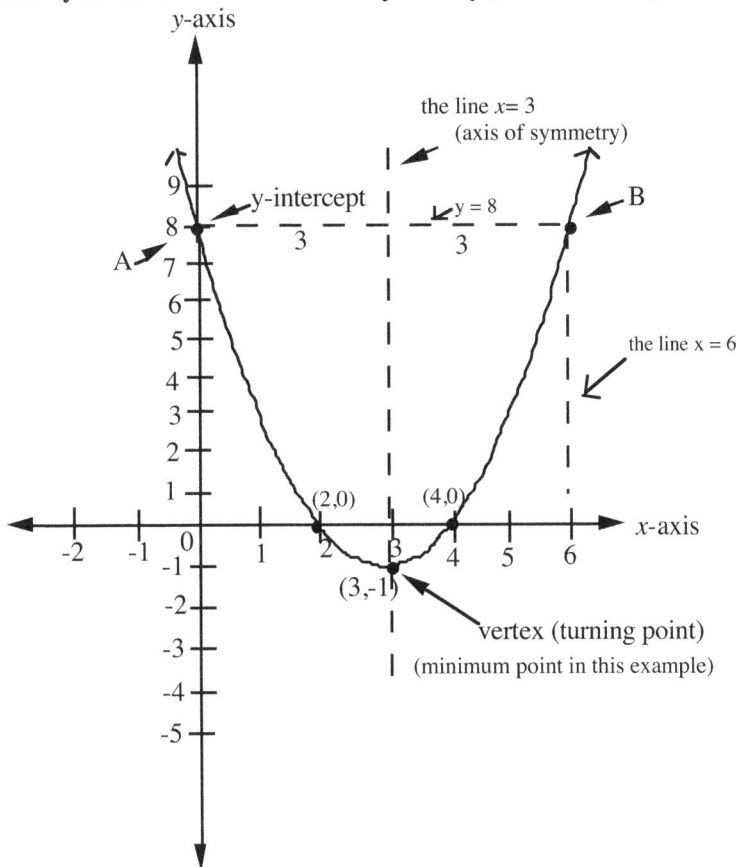

Figure 1: Graph of $y = x^2 - 6x + 8$

Note above that in sketching the graph, you may choose convenient x-values on either side of the axis of symmetry, calculate corresponding y-values to obtain additional points, especially, when there are no (real) x-intercepts.(This occurs when in solving the quadratic equation, we obtain solutions involving imaginary numbers.)**Again:** When in doubt as to the location of the curve, pick x-values, calculate y, plot the points and connect them.

Example 2 Given the equation of the parabola $y = f(x) = -x^2 + 6x - 8$, perform the following:

1. Find the x- and y-intercepts. **2.** Find the axis of symmetry.

3. Find the coordinates of the vertex (or turning point). **4.** Find the domain and range.

5. Sketch the graph of $y = f(x) = -x^2 + 6x - 8$

Solution

1. Using the procedure in Example 1,

The x-intercepts are 2 and 4.   (The x-intercepts are at the points (2,0), and (4,0))

The y-intercept = -8.   (The y-intercept is at (0,-8))

2. $a = -1$ (the coefficient of the x^2-term), $b = +6$ (the coefficient of the x-term)

Axis of symmetry is the line $x = -\dfrac{b}{2a} = -\dfrac{+6}{2(-1)} = 3$.

The axis of symmetry is the line $x = 3$. The x-coordinate of the vertex = 3,

and the y-coordinate of the vertex = $y = f(-\dfrac{b}{2a}) = f(3) = -3^2 + 6(3) - 8 = 1$

3. The vertex is at the point (3, 1). Since a is negative ($a = -1$), the vertex is a maximum point.

4. The domain is all real x. (The given function is a polynomial)

The range is specified by using the y-coordinate of the vertex obtained in Step 2.

The range is given by the inequality $y \leq 1$.   (Note the " $\leq$ ")

The range consists of all real numbers less than or equal to 1.

(Note: If we did not know the x-coordinate of the vertex, we could obtain the range by substituting. Substituting

$a = -1$. $b = +6$, $c = -8$ in $y \leq \dfrac{4ac - b^2}{4a}$ (Note the " $\leq$ ") to obtain $y \leq \dfrac{4(-1)(-8) - (+6)^2}{4(-1)}$; $y \leq 1$.)

5. To sketch the graph, plot the points from (**1**), (**2**), (**3**) above; similarly locate the point B (as was done in Example 1) and connect the points by a smooth U-shaped curve, noting that the parabola opens downwards (**Figure 2**), since a is negative ($a = -1$). Also note the symmetry as in Example 1.

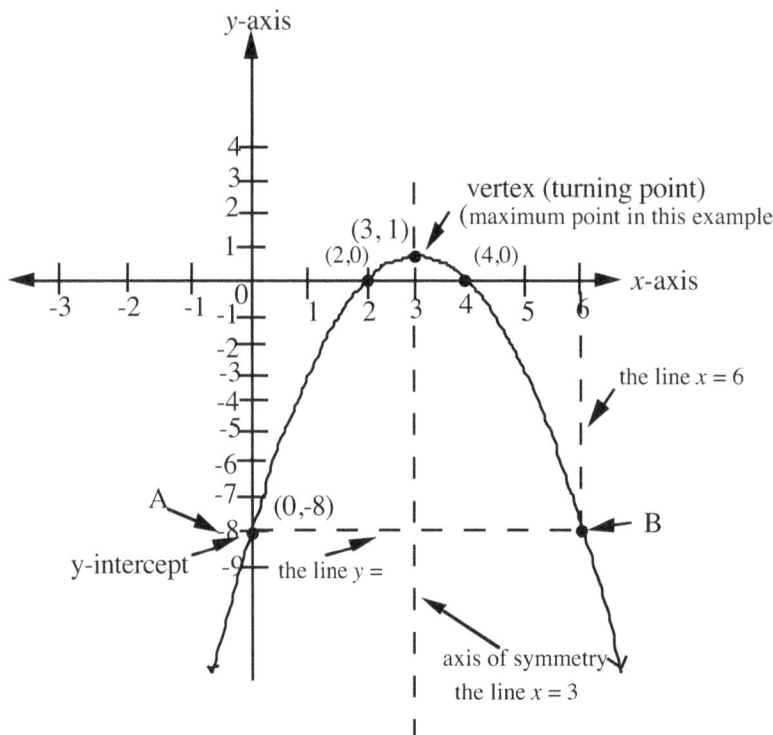

Figure 2: Graph of $y = -x^2 + 6x - 8$

EXTRA: Axis of Symmetry , Vertex and Minimum Value   395

Example 1

Given $f(x) = x^2 - 12x + 8$ (A)

Let $y = x^2 - 12x + 8$ (B)

$= x^2 - 12x + (-6)^2 - (-6)^2 + 8$

$= x^2 - 12x + (-6)^2 - 36 + 8$

$y = (x - 6)^2 - 28$ <---- knowing this

$y + 28 = (x - 6)^2$ <----- or this

Vertex is at $(6, -28)$

Axis of symmetry is the line $x = 6$

Minimum value is -28.

Alternatively

From (B), $a = 1$, $b = -12$

Axis of symmetry is given by $x = -\frac{b}{2a}$

$= -\frac{-12}{2(1)}$

$= \frac{12}{2}$

$x = 6$

$f(6) = 6^2 - 12(6) + 8$ (subst.. in (A)

$= 36 - 72 + 8$

$= -36 + 8$

$y = -28$

when $x = 6$. $y = -28$

Vertex is at $(6, -28)$

Axis of symmetry is the line $x = 6$

Minimum value is -28.

Example 2

Given $f(x) = x^2 - 6x + 8$ (A)

Let $y = x^2 - 6x + 8$ (B)

$= x^2 - 6x + (-3)^2 - (-3)^2 + 8$

$= x^2 - 6x + (-3)^2 - 9 + 8$

$y = (x - 3)^2 - 1$ <---- knowing this

$y + 1 = (x - 3)^2$ <----- or this

Vertex is at $(3, -1)$

Axis of symmetry is the line $x = 3$

Minimum value is -1.

Alternatively

From (B), $a = 1$, $b = -6$

Axis of symmetry is given by $x = -\frac{b}{2a}$

$= -\frac{-6}{2(1)}$

$= \frac{6}{2}$

$x = 3$

$f(3) = 3^2 - 6(3) + 8$ (subst.. in (A)

$= 9 - 18 + 8$

$= -9 + 8$

$y = -1$

when $x = 3$. $y = -1$

Vertex is at $(3, -1)$

Axis of symmetry is the line $x = 3$

Minimum value is -1.

Lesson 81 Exercises

Sketch the graphs of the following by the location of the vertex, axis of symmetry and the intercepts.:

1. $y = x^2 - 4$;   **2.** $y = -x^2 + 1$.   **3.** $y = x^2 - 6x + 8$

Answers: See above

CHAPTER 31
Inequalities
Lesson: Set Notation, Set Operations, Interval Notation
Lesson: Linear Inequalities, Compound Inequalities, Word Problems

Use Chapter 24

The above was designed to avoid duplication and to reduce the volume of this book, and also, such a design is good for the environment.

CHAPTER 32

Graphing Linear Inequalities in two Variables

Lesson 82: **Graphing and determining solution plane by point-testing**
Lesson 83: **Graphing and determining solution plane by the "Seesaw" view Method**
Lesson 84: **Graphing Special Cases of Linear Inequalities**

Geometrically, the solution set of a linear inequality in two variables is a section of the *x*-*y* plane. The plane may be open or closed. If the plane is closed, the symbol used would be " ≥ " or " ≤ "; but if plane is open, the symbol would be " > " or "<". The solution set for the closed half-plane includes the boundary line, but the solution set for the open half-plane does not include the boundary line. We will learn how to graph the solution sets of some inequalities.

Lesson 82

Graphing and determining solution plane by point-testing

General steps in graphing linear inequalities

Step 1: Replace the inequality symbol by the equality symbol " = ".

Step 2: Using convenient points, graph the equation (now a linear equation) from Step 1. To do this, choose convenient *x*-values and calculate corresponding y-values. Connect the points by a straight solid line if the inequality is the symbol " ≥ " or " ≤" ; but use broken (dashed or dotted) lines if the given inequality is the symbol " > " or " < ".

Note: You may also use any other specialized methods (for graphing straight lines) such as the "slope-intercept" method or the *x*- and *y*-intercepts method.

Step 3: We will determine which side of the straight line forms the solution. Choose any convenient point not on the boundary (of the straight line drawn) and substitute its *x*- and *y*- coordinates in the original inequality. If these values satisfy the inequality, then the chosen point is in the solution set of the inequality, and consequently, that side of the line (half-plane) where this point is located is the solution set. However, if the tested point does not satisfy the original inequality, then it is the other side of the line (the half-plane) which is the solution set We must note that the straight line drawn in Step 2 always divides the *x*-*y* plane (area or surface) into two halves.

Step 4: Shade the half-plane which is the solution set.

Example 1 Graph the solution set for the inequality
$$3x + 2y \geq 4$$

Solution

Step 1: Replace the inequality symbol by the equality symbol and solve for y.
$$3x + 2y = 4$$
$$y = -\tfrac{3}{2}x + 2$$

Step 2: Choose convenient *x*-values and calculate corresponding *y*-values
(To avoid fractional values of y, solve for *y* first, and choose *x*-values divisible by the denominator of the *x*-term)

For $x = 0, y = 2 \Leftrightarrow (0,\ 2)$

For $x = 2, y = -1 \Leftrightarrow (2, -1)$

Step 3: Plot the ordered pairs (0, 2) and (2, -1). See Figure .

Step 4: Connect the two points by a **solid straight line** since the inequality symbol is " $\geq$ "
    (That is, the half-plane is closed)

Step 5: Now, determine which half of the half-plane to shade (or is the solution set). Let us choose the origin $(0,0)$ to test the inequality. Substituting $x = 0, y = 0$ in the original inequality.

$$3(0) + 2(0) \overset{?}{\geq} 4$$

$$0 \overset{?}{\geq} 4 \qquad \textbf{No} \quad \text{(False inequality)}$$

Thus, the origin and the half-plane in which the origin is located **is not** in the solution set. Therefore, the solution set is **the other** half-plane. The solution contains, for example, the point $(3,4)$.

Step 5: Shade this half-plane (the solution half-plane) as in Figure 43.

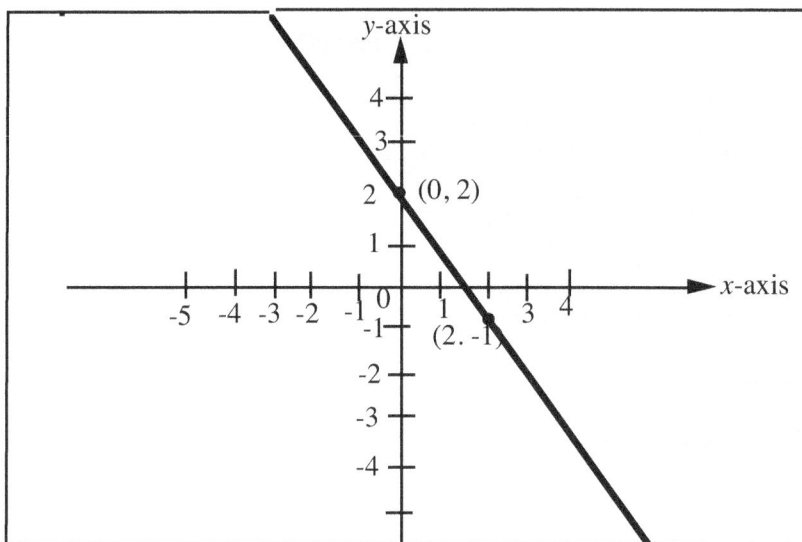

Figure Graph of $3x + 2y \geq 4$ (Note the solid boundary line)

The use of the origin (0 ,0) in determining the solution half-plane

Using the origin, $x = 0, y = 0$ makes the checking relatively easy and fast because the product of zero and a number is zero; and also when zero is added to a number, the number remains unchanged. However, if the boundary line passes through the origin $(0,0)$, then choose some other convenient point for testing. (Do not use $(0,0)$ for testing if there are inequalities such as $y \geq ax,\ y \leq ax,\ x \geq 0$ and $y \geq 0$.)

Example 2 Graph the solution set for the inequality: $3x + 2y > 4$

Solution

Steps 1, 2, 4, and 5 are the same as those in the last example (Example 1) except that in Step 3, we use **broken** (dashed) **lines** in connecting the two points , since the inequality symbol is " > ". and consequently, the solution half is open. See Figure 44.

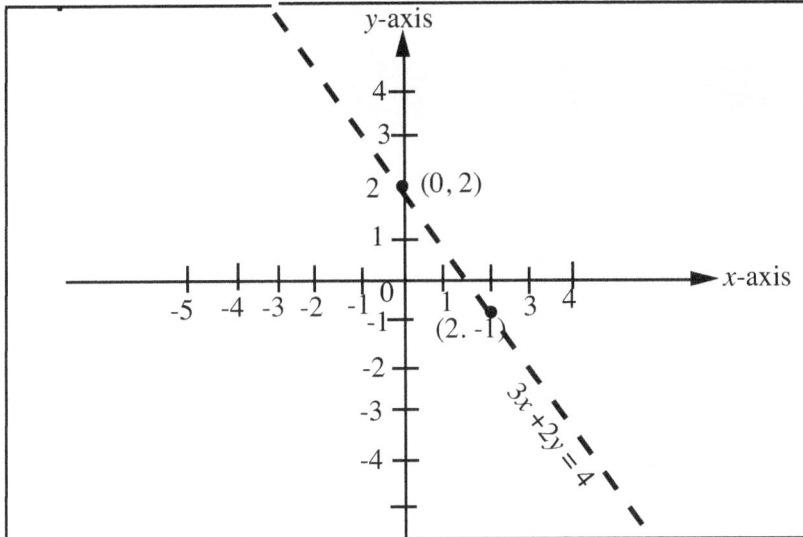

Figure   The graph of $3x + 2y > 4$ (Note the broken boundary line)

Example 3   Graph the solution set for the inequality
$$3x + 2y \leq 4$$

Solution

Steps $1, 2, 3,$ and $4,$ are the same as those in Example 1, except that in Step 5. we have

$$3(0) + 2(0) \overset{?}{\leq} 4$$

$$0 \overset{?}{\leq} 4 \qquad \textbf{Yes} \quad (\text{True inequality})$$

The true statement shows that the origin $(0, 0)$ **is in** the solution set. We therefore shade the solution half-plane containing $(0, 0)$.. See Figure

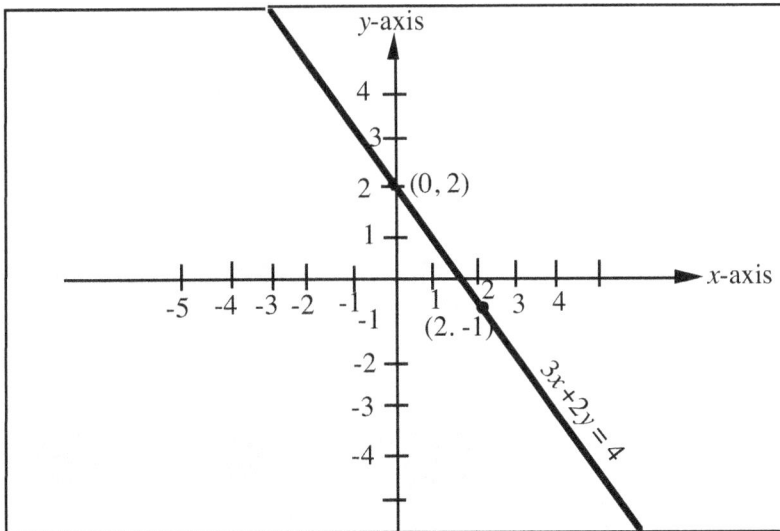

Figure: The graph of $3x + 2y \leq 4$

Example 4   From Example 3, the solution set for the inequality $3x + 2y < 4$ is

graphed in Figure 46. Note the broken line (compared to the solid line in Example 3).

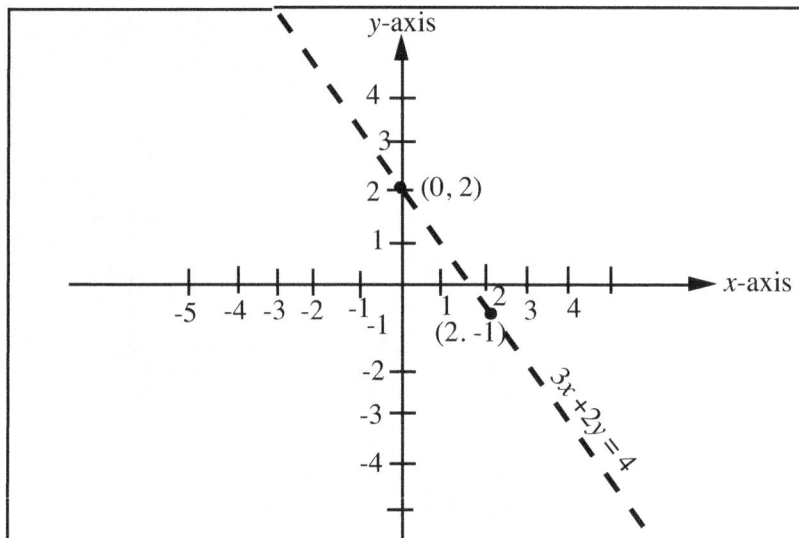

Figure: The graph of $3x + 2y < 4$

Lesson 82 Exercises

1. Graph the solution set for the inequality $3x + 2y \geq 4$

2. Graph the solution set for the inequality: $3x + 2y > 4$

3 Graph the solution set for the inequality $3x + 2y < 4$ is

4. Graph the solution set for the inequality $3x + 2y \leq 4$

Solution: See above

Lesson 83

Graphing: Determining Solution Plane by the "Seesaw" View Method
(Another method of determining the solution half-plane)

This method depends on two main principles. First, we solve the inequality for y. Secondly we should be able to point out which side of the boundary line is " above", and which side is "below" the boundary line. (When the inequality is solved for y, the inequality would be of the form , $y \geq ax + b$, $y \leq ax + b$, $y < ax + b$, or $y > ax + b$)

The symbols " > and ≥" will imply " above "; and the symbols " ≤ and < " will imply "below ". To determine the " above" and the " below " of the boundary line (the straight line drawn) we use the following mnemonic device:

Assume the line, ℓ, and the y-axis form a seesaw system, with the y-axis forming the pivot at where the axis meets the line. We will call the side where one sits or stands " the side above the line and we will call the other side " the side below the line ". In Figures (a), (b), (c), (d), we have shaded the half-planes above the lines for clarity.

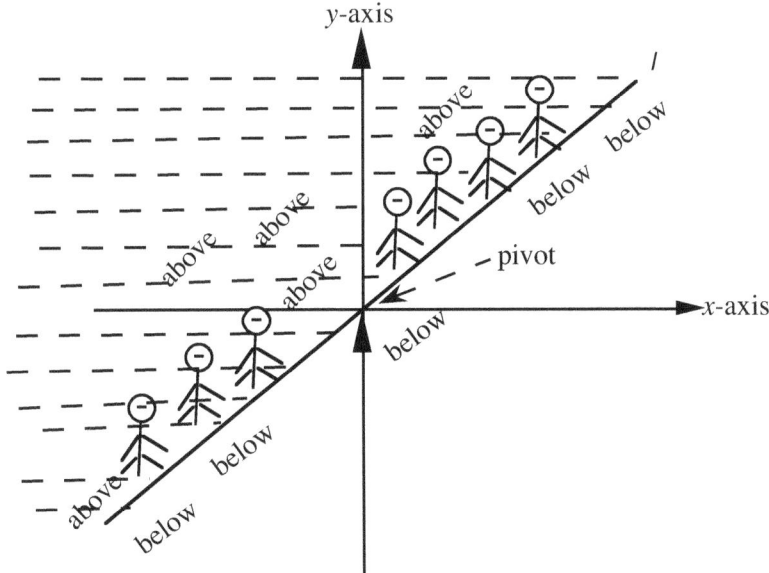

Figure: Seesaw-View Method

Note: The seesaw-view method is an original method devised by the author.

Figure: Seesaw-View Method

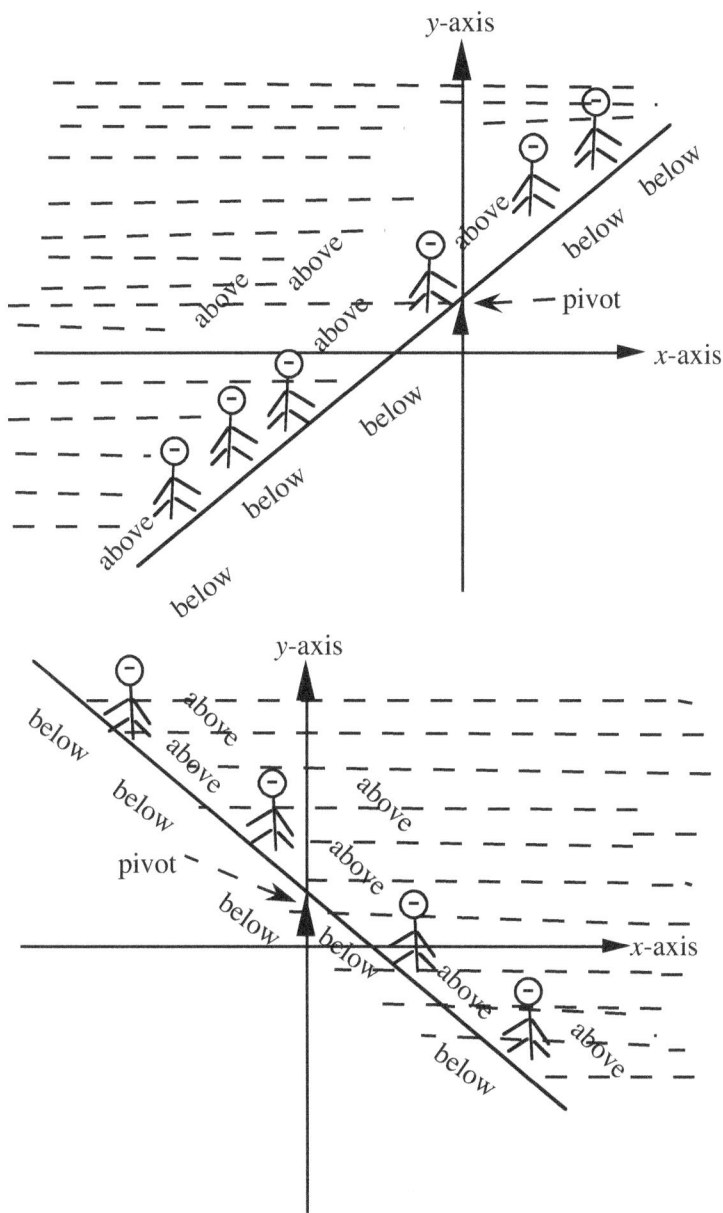

Figure: Seesaw-View Method

We will redo the previous examples using the " seesaw view" method, noting that::
the symbols " **>** and **≥** " (greater than symbols) will imply " **above the line**" and the symbols "**<** and **≤** (less than symbols) will imply "**below the line** ". The inequality must be placed in the form $y \geq$ (or $>, \leq, <$) $ax + b$. (That is, solve the inequality for y first before using this method.)

Example 5 (Example 1 redone)

Graph the solution set for the inequality

$$3x + 2y \geq 4$$

Solution (The solution is a closed half-plane including the boundary because of the " $\geq$ ".)

Step 1: Solve the inequality for y.

$$3x + 2y \geq 4$$

$$y \geq -\tfrac{3}{2}x + 2$$

Step 2: Replace the inequality symbol by the equality symbol and find two convenient ordered pairs to graph the straight line.

$$y = -\tfrac{3}{2}x + 2$$

For $x = 0, y = 2 \Leftrightarrow (0, \ 2)$

For $x = 2, y = -1 \Leftrightarrow (2, -1)$

Step 2: Plot the ordered pairs $(0, 2)$ and $(2, -1)$. See Fig. .

Step 3: Connect the two points by a **solid straight line** since the inequality symbol is " $\geq$ "
(That is, the half-plane is closed)

Step 4: Since the sense of the inequality with y solved for is " $>$ " , the **greater than** symbol,
we shade half-plane **above** the line . See Figure.
The solution set is a closed half-plane including .

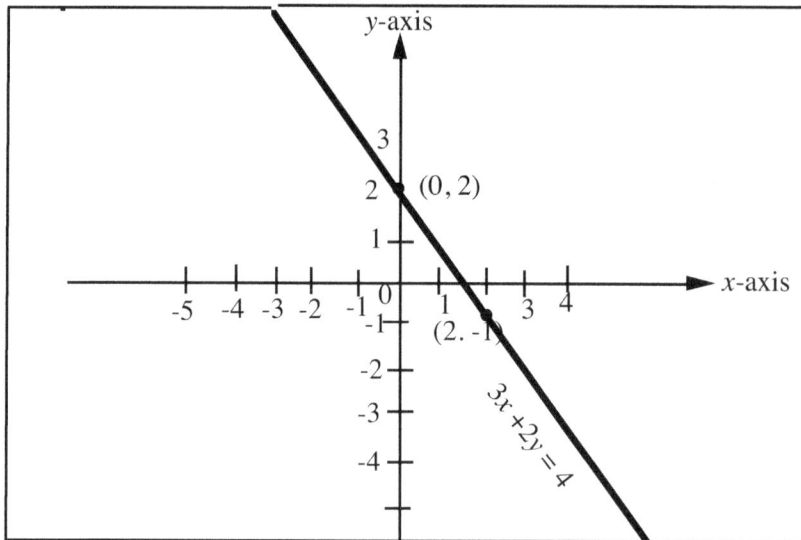

Figure: Graph of $3x + 2y \geq 4$

Example 6 (Example 3 redone)

Graph the solution set for the inequality

$$3x + 2y \le 4$$

Step 1: Solve the inequality e for y.

$$3x + 2y \le 4$$

$$y \le -\tfrac{3}{2}x + 2$$

Step 2: Replace the inequality symbol by the equality symbol and find two convenient ordered pairs to graph the straight line.

$y = -\tfrac{3}{2}x + 2$

For $x = 0, y = 2 \Leftrightarrow (0, \ 2)$

For $x = 2, y = -1 \Leftrightarrow (2, -1)$

Step 2: Plot the ordered pairs (0, 2) and (2, -1). See Figure..

Step 3: Connect the two points by a **solid straight line** since inequality symbol is " $\ge$ "
(That is, the half-plane is closed)

Step 4: Since the sense of the inequality with y solved for is" $<$ " , the **less than** symbol , we
shade half-plane **below** the line $y = -\tfrac{3}{2}x + 2$. See Figure
The solution set is a closed half-plane including the boundary.

We will now use the " seesaw view method " to graph more inequalities without explanation

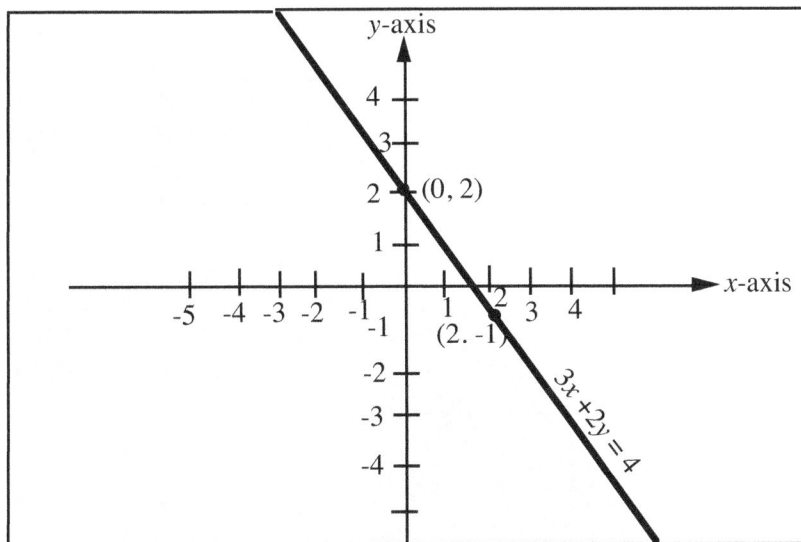

Figure Graph of $3x + 2y \le 4$

Example 7 Graph the solution set for the inequality $y + 2x \geq 6$.

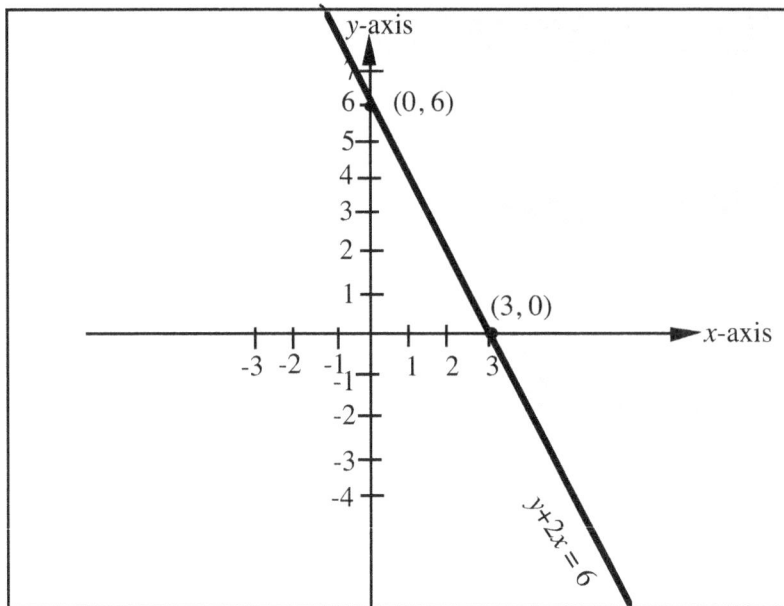

Figure : The graph of $y + 2x \geq 6$.

Example 8: Graph the solution set for the inequality $y + 2x \leq 6$

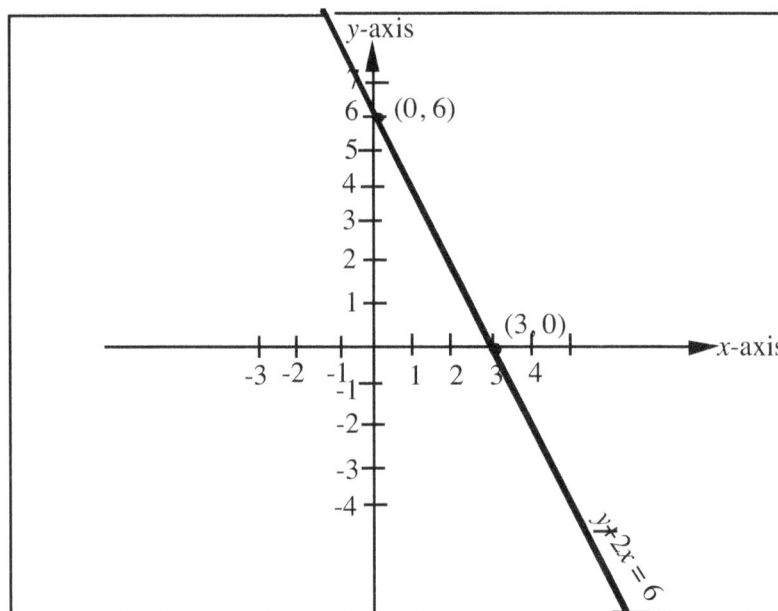

Figure The solution set for the inequality $y + 2x \leq 6$

Example 9 Graph the solution set for the inequality

$$4x - 2y \leq -5 \qquad \text{(A)}$$

Step 1: Solve the inequality for y.

$$4x - 2y \leq -5$$

$$-2y \leq -5 - 4x$$

$$y \geq \frac{-5}{-2} - \frac{4x}{-2} \qquad \text{(The sense of the inequality reverses because of division by a negative number)}$$

$$y \geq \frac{5}{2} + 2x$$

$$\boxed{y \geq 2x + \frac{5}{2}} \qquad \text{(B)}$$

Step 2: Replace the inequality symbol by the equality symbol and find two convenient ordered pairs to graph the line.

$y = 2x + \frac{5}{2}$

For $x = 0$, $y = \frac{5}{2}$, and we obtain the ordered pair $(0, \frac{5}{2})$

For $y = 0$, $x = -\frac{5}{4}$, and we obtain the ordered pair $(-\frac{5}{4}, 0)$

Step 2: Plot the ordered pairs $(0, \frac{5}{2})$ and $(-\frac{5}{4}, 0)$ See Fig..

Step 3: Connect the two points by a **solid straight line**   (boundary included).

Step 4: Since the sense of the inequality with y solved for is the " **greater than** "symbol inequality B) we shade half-plane **above** the line $y = 2x + \frac{5}{2}$ See Figure   . For a quick check, test with $(0, 0)$

Note: In Step 4, do not use implicitly inequality (A) but use inequality (B).

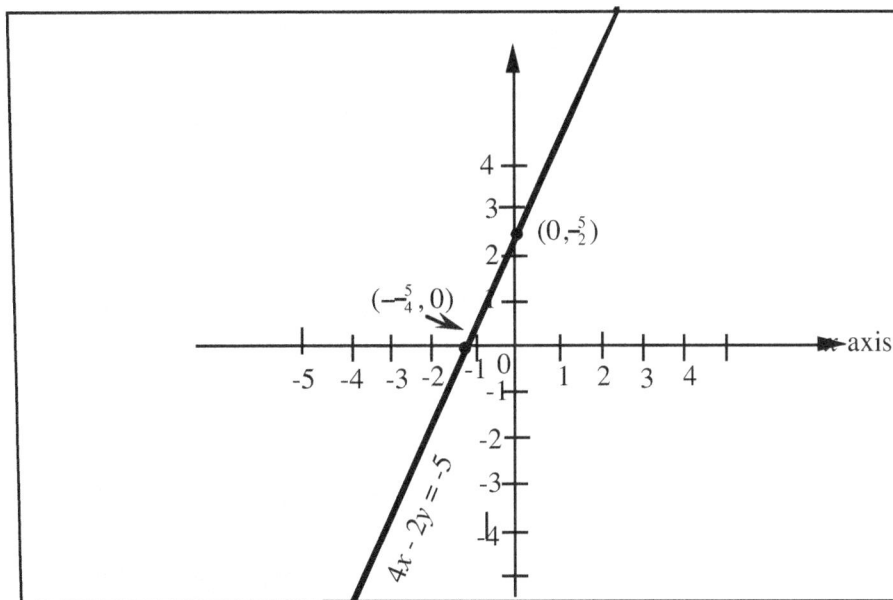

Figure : Graph for $4x - 2y \leq -5$

Lesson 83 Exercises

Use the "see-saw view" method to graph the following:

1. Graph the solution set for the inequality $3x + 2y \geq 4$

2. Graph the solution set for the inequality: $3x + 2y > 4$

3 Graph the solution set for the inequality $3x + 2y < 4$ is

4. Graph the solution set for the inequality $3x + 2y \leq 4$

5. Graph the solution set for the inequality $y + 2x \geq 6$.

6. Graph the solution set for the inequality $4x - 2y \leq -5$

Solution: See above

Lesson 84

Graphing Special Cases of Linear Inequalities and Distinction between a Point, a Set of Points, a Line, and a Half-plane

We must distinguish between, for example, the set of points $x \geq 4$ on a number line and the half-plane $x \geq 4$ just as we distinguish between the point $x = 4$ on a number line and the line $x = 4$ in an x-y plane.

The graph of the point $x = 4$:

The graph of the set of points $x \geq 4$:

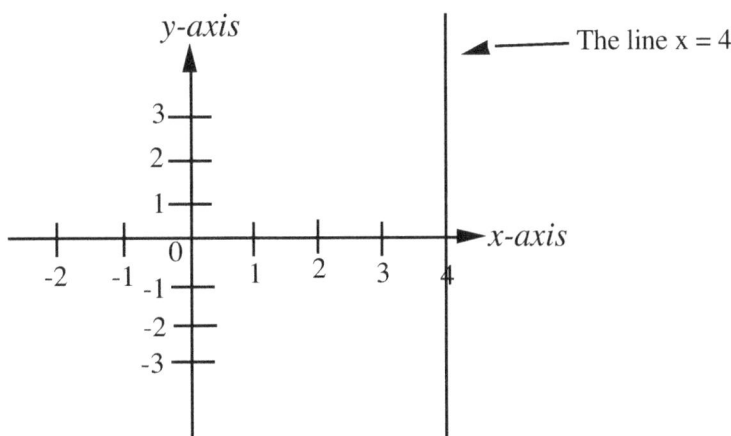

Figure : The graph of the line $x = 4$

Example 8 Graph the linear inequality $x \geq 4$.

Step 1: Replace " > " by " = ".
   Then $x = 4$

Step 2: Draw the vertical line $x = 4$ using a solid line (Fig. 50)
   To determine the half-plane which is the solution set, either choose a point and test or use the sense of the inequality.

   Point test: Substitute $x = 0$, $y = 0$, (the origin) in $x \geq 4$, then $0 \geq 4$, which is false. Therefore, the origin is not in the solution. The solution set is the other side of the line $x = 4$

   Sense test: Since the sense of the inequality symbol is " >", the solution half-plane is the half-plane to the right of the line $x = 4$ (noting that $x > 4$). The solution includes the boundary line.

Step 4: Shade the half-plane to the right of the line $x = 4$.

The line $x = 4$

y-axis

The graph of the half-plane $x \geq 4$:

x-axis

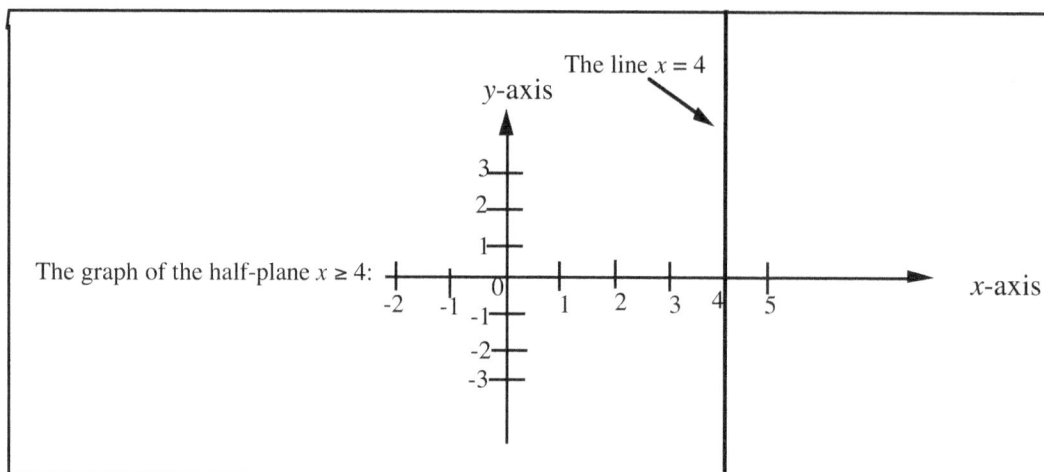

We present, below, similar special cases without the procedures:

(a) For the linear plane $x \leq 4$, the solution plane is the half-plane to the left of the line $x = 4$ (noting that $x < 4$), including the boundary. (Figure)

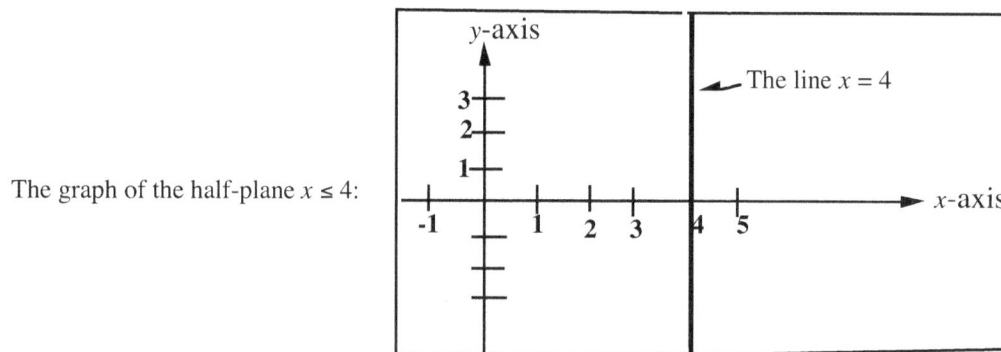

y-axis

The line $x = 4$

The graph of the half-plane $x \leq 4$:

x-axis

(b) For the linear plane $y > 2$, the solution plane (Figure ..) is the half-plane above (for the **greater** than symbol) the horizontal line $y = 2$, excluding the boundary; and we use a broken (dotted) line to draw the boundary, since the inequality symbol does **not include** the equality symbol.

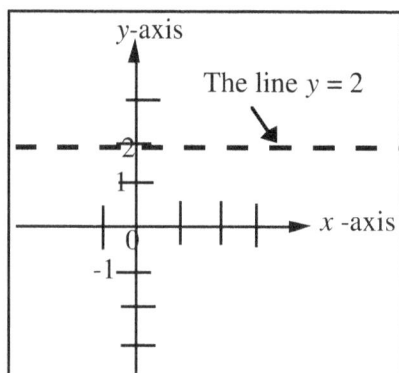

y-axis

The line $y = 2$

x -axis

Figure: half-lane $y > 2$,

(c) For the linear plane $y \leq -4$, the solution plane (Fig. 53) is the half-plane below (for the **less** 411 than symbol) the horizontal line $y = -4$, including the boundary and we use a solid line to draw the boundary, since the inequality symbol **includes** the equality symbol.

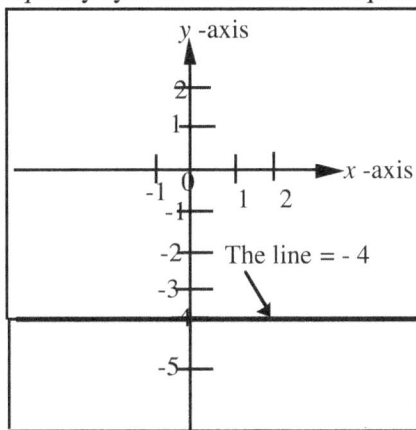

Figure: The graph of the linear plane $y \leq -4$

Lesson 84 Exercises

Questions 1-16: Graph the solution-plane of each of the following:

1. $y + 4x < 2$;   2. $x + 5y + 5 \geq 0$;   3. $-4y + x \leq 12$;   4. $-x + y \geq 0$;

5. $3x - 2y < 6$; 6. $y \geq |x|$; 7. $y \leq |x|$; 8. $y \leq -|x|$; 9. $y < x$ 10. $y \geq x$ 11. $y \geq 4$; 12. $y \geq -2$;

13. $y \leq 4$; 14. $x \leq -2$; 15. $|y| \geq 2$; 16. $|y| - 2 \leq -|x|$;

Questions 17-20: Draw the graphs for each of the following:

    17. The point $x = 3$;   18. The line $x = 3$;

    19. The set of points $x \geq 3$;   20. The half–plane $x \geq 3$.

Answers:

1. $y + 4x < 2$

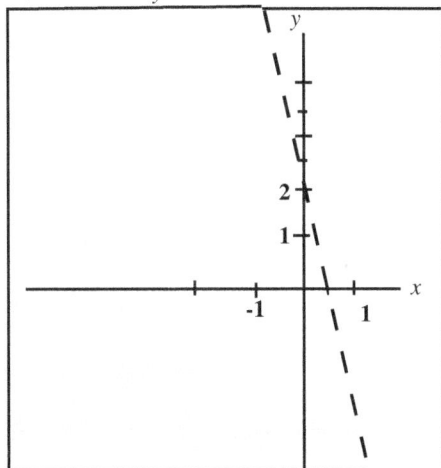

2. $x + 5y + 5 \geq 0$

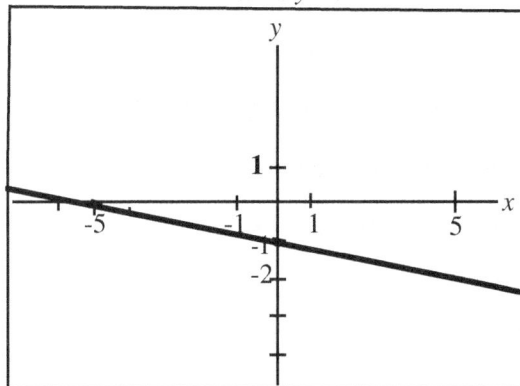

3.

$-4y + x \leq 12$

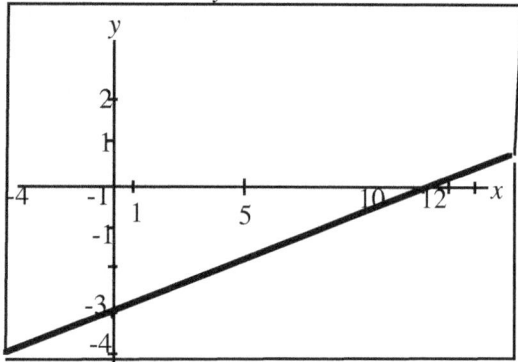

4.

$-x + y \geq 0$

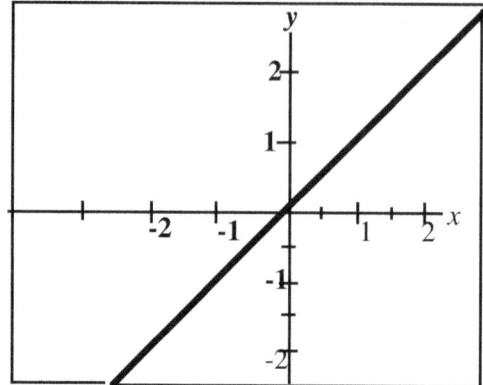

5.

$3x - 2y < 6$

6.

$y \geq |x|$

7. $y \leq |x|$

8. $y \leq -|x|$

9. $y < x$

10. $y \geq x$

11. $y \geq 4$

12. $y \geq -2$

13.

$y \le 4$

14.

$x \le -2$

15.

$|y| \ge 2$

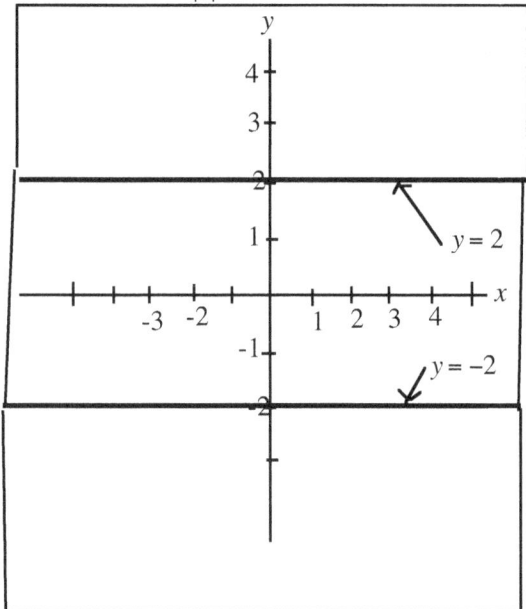

16.

$|y| - 2 \le -|x|$ or $|y| + |x| \le 2$

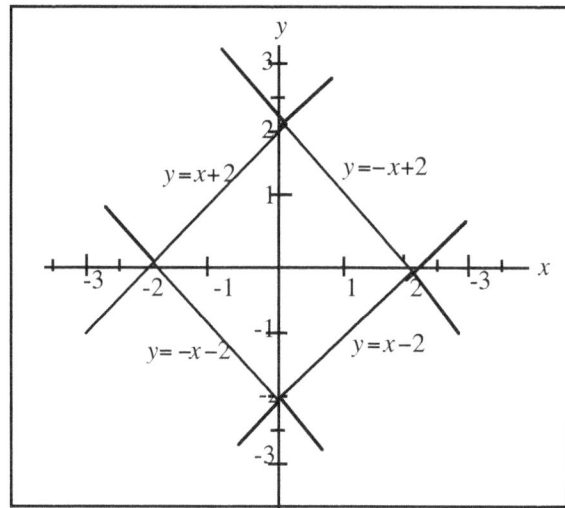

For solutions to Problems 17-20, see page 409-411 of the present Lesson and imitate.

CHAPTER 33

Lesson 85: **Absolute Value Equations**
Lesson 86: **Absolute Value Inequalities**

Lesson 85
Absolute Value Equations

Geometric definition: The absolute value of a real number is the distance between the number and zero on the number line.

Algebraic definition: The absolute value of x symbolized, $|x|$, is defined by a two-part rule as follows:

$$|x| = \begin{cases} x \text{ if } x \geq 0 \\ -x \text{ if } x < 0 \end{cases}$$

In words, the absolute value of a real number x is x if x is a positive number or zero; but it is $-x$ if x is a negative number. The above two-part rule in the definition requires us to consider two cases when solving absolute value equations. Understanding the above definition very well will save us from having to memorize the solution methods for absolute value equations.

Absolute value equations: Absolute value equations are equations in which the absolute value terms contain variables.

Example l Solve for x: $|x| = 2$

We will consider two methods.

Method 1 (Using the geometric definition)

$|x| = 2$ means the distance between x and 0 on the number line is 2.
Therefore $x = 2$ or -2 Check: $|2| = 2$; $|-2| = 2$.

Solution set is $\{-2, 2\}$

Method 2: If $x \geq 0$, then $|x| = x = 2$
If $x < 0$, then $|x| = -x = 2$, and from which $x = -2$.
Again the solution set is $\{-2, 2\}$

Graph for $\{-2, 2\}$

Example 2 Solve for x: $|x - 3| = 4$

Method 1 (Using the geometric definition)
$|x - 3| = 4$ means, on the number line, the distance between x and 3 is 4.

Step l: Locate the point 3 on the number line. (To obtain 3, set $x - 3$ to zero and solve for x.)

Step 2: From this point, count 4 units to the right and stop, the coordinate of this stopping point, 7, is one of the solutions. Similarly from the point 3, count 4 units to the left and stop, the coordinate of this stopping point is -1, and it is another solution. The solution set is $\{-1, 7\}$

Graph for $\{-1, 7\}$

Lesson 85: Absolute Value Equations

Method 2 (Using the algebraic definition)

Solution $|x - 3| = 4$ is equivalent to

$$x - 3 = 4 \text{ or } -(x - 3) = 4$$

Solve each equation for x:

$x - 3 = 4$ or	$-(x - 3) = 4$	(or $-x + 3 = 4$ and from which $x = -1$)
$x - 3 = 4$ or	$(x - 3) = -4$	(Multiplying each side of the second equation by -1)
$x - 3 = 4$ or	$x - 3 = -4$	

$$\underline{+3+ \quad 3} \qquad \underline{+3 \quad +3}$$

$x = 7$ or $\quad x = -1$ Check #1: $|7 - 3| = |4| = 4$;   Check #2: $|-1-3| = |-4| = 4$

The solution set is $\{-1, 7\}$

Example 3 Solve for x: $4 - |3x - 4| = -7$

Solution: $4 - |3x - 4| = -7$

$$\underline{-4 \qquad\qquad\qquad -4}$$

$$- |3x - 4| = -11$$

$$|3x - 4| = 11 \text{ (Multiplying both sides of the equation by -1)}$$

Now, we apply the algebraic definition of the absolute value:

Then, $3x - 4 = 11$ or $-(3x - 4) = 11$

$$\underline{+4 \quad +4} \qquad 3x - 4 = -11 \text{ (Multiplying both sides of the equation by -1)}$$

$$3x = 15 \qquad\qquad \underline{+4 \quad +4}$$

$$x = 5 \qquad\qquad 3x = -7 \quad \text{and } x = -\frac{7}{3}$$

Checking for $x = 5$:

$4 - |3(5) - 4| \overset{?}{=} -7$; $4 - |15 - 4| \overset{?}{=} -7$

$4 - |11| \overset{?}{=} -7$; $4 - 11 \overset{?}{=} -7$

$-7 = -7$ True (LHS = RHS)

Checking for $x = -\frac{7}{3}$: $4 - |3 \cdot (-\frac{7}{3}) - 4| \overset{?}{=} -7$; $4 - |-7 - 4| \overset{?}{=} -7$;

$4 - |-11| \overset{?}{=} -7$; $4 - 11 \overset{?}{=} -7$ (note: $|-11| = 11$)

$-7 = -7$ True

The solution set is $\{-\frac{7}{3}, 5\}$.

Example 4: Solve for x: $|x - 2| = |x + 6|$. (A) (See Appendix, p.526 for another approach)

(We apply the geometric definition which says that the absolute value of a real number is its distance from zero on the number line)

Applying the geometric definition, either the two expressions of the equation are the same number, or are opposites. For examples, $|3| = |3|$, $|-3| = |-3|$ (same number) or $|-3| = |3|$, $|3| = |-3|$ opposites.

| **Case 1:** If the two expressions represent the same number, then $x - 2 = x + 6$ $$x - 2 = x + 6$$ Solving, $\quad -2 = 6$, which is a contradiction. There is therefore no solution for **Case 1.** | **Case 2:** If the two expressions represent opposite numbers, then $x - 2 = -(x + 6)$ $$x - 2 = -x - 6 \text{, and from which}$$ $$x = -2$$ We check for $x = -2$. in (A) Then $|-2 - 2| \overset{?}{=} |-2 + 6|$; $|-4| \overset{?}{=} |4|$; $4 \overset{?}{=} 4$. Yes. $\boxed{\text{The solution is } -2.}$ |

Note above that sometimes, there may be solutions to both cases.

Lesson 85 Exercises

Solve for x: **1.** $|x - 5| = 2$; **2.** $|2x - 1| = 5$; **3.** $|2x + 1| = 4$; **4.** $3 - |2x - 1| = -8$

5. $|x + 4| = 3$; **6.** $|3x - 2| = 4$; **7.** $6 - |x - 3| = -2$; **8.** $|x - 10| = -5$

Answers: **1.** $\{3, 7\}$; **2** $\{-2,3\}$; **3.** $\{-\frac{5}{2}, \frac{3}{2}\}$; **4.** $\{-5,6\}$; **5.** $\{-1, -7\}$; **6.** $\{-\frac{2}{3}, 2\}$; **7.** $\{-5, 11\}$; **8.** No solution

Lesson 86
Absolute Value Inequalities

We shall consider two cases

Case **1**: $|ax + b| < c$. where c is positive. Example: $|x - 2| < 1$. (There is no solution if c is negative)

Case **2**: $|ax + b| > c$, where c may be positive or negative. Example: $|x - 2| > 1$.

Examples   **1**. $|x - 2| < 1$ means the distance from x to 2 is less than 1.

          2. $|x - 2| > 1$ means the distance from x to 2 is greater than 1.

Case 1: $|ax + b| < c$ is equivalent to

$$\boxed{-c < ax + b < c}, \text{ the " \textbf{and}" problem; or } -c < ax + b \text{ and } ax + b < c.$$

Therefore, we can solve this inequality using the condensed method.
Generally, the solution is a one-interval solution. Case 1 is a conjunction problem.

Case 2: $|ax + b| > c$ *is* equivalent to

       either $\boxed{ax + b > c}$ or $\boxed{-(ax + b) > c}$

Here, we solve an " or" problem. Generally, we have a two-interval solution.
Note that the form $-(ax + b) > c$ which is equivalent to $ax + b < -c$. (obtained by multiplying by -1 and reversing the sense) is easier to recall, since it follows immediately from the absolute value definition.
 Case 2 is a disjunction problem.

Example 1: Solve for x: $|x - 2| < 3$.
Solution $|x - 2| < 3$ is equivalent to
         $-3 < x - 2 < 3$
         $\underline{+2 \quad + 2 \quad +2}$
         $-1 < x \quad < 5$ (Solving for x by adding +2 to all three sides of the inequality)

Graph for $\{x \mid -1 < x < 5\}$

Example 2 Solve for x: $|x - 2| > 3$

Solution $|x - 2| > 3$ is equivalent to
     $(x - 2) > 3$ or $-(x - 2) > 3$
      $x - 2 > 3$ or $x - 2 < -3$ or $-x + 2 > 3$
      $\underline{+ 2 \quad +2}$ $\underline{+2 \quad +2}$ $\underline{-2 \quad - 2}$
       $x > 5$ or $x < -1$ $-x > 1$
                             $x < -1$

Solution set is $\{x \mid x < -1 \text{ or } x > 5\}$

Graph for $\{x \mid x < -1 \text{ or } x > 5\}$

Example 3   Solve for x: $|x + 3| \geq 4$ 418

Solution $|x + 3| \geq 4$ is equivalent to
$(x + 3) \geq 4$ or $-(x + 3) \geq 4$
$x + 3 \geq 4$ or $x + 3 \leq -4$
$\underline{ -3 \quad -3}$ $\underline{ -3 \quad -3}$
$x \geq 1$ or $x \leq -7$

Solution set is $\{x \mid x \leq -7 \text{ or } x \geq 1\}$

Graph for $\{x \mid x \leq -7 \text{ or } x \geq 1\}$

Lesson 86 Exercises

A   Solve for x: **1.** $|x - 4| < 2$;   **2.** $|x - 4| > 2$;   **3.** $|x + 5| < -4$

Answers: **1.** $\{x \mid 2 < x < 6\}$;   **2.** $\{x \mid x < 2 \text{ or } x > 6\}$;   **3.** No solution

B   Solve for: **1.** $|x - 5| < 2$;   **2.** $|x - 5| > 2$;   **3.** $|x - 5| \geq 2$;   **4.** $|x - 5| = 2$

5. $|2x + 3| \geq 3$

Answers: **1.** $\{x \mid 3 < x < 7\}$;   **2.** $\{x \mid x < 3 \text{ or } x > 7\}$;   **3.** $\{x \mid x \leq 3 \text{ or } x \geq 7\}$;   **4.** $\{3, 7\}$;
5. $\{x \mid x \leq -3 \text{ or } x \geq 0\}$

CHAPTER 34
ALGEBRAIC FRACTIONS
Lesson Reduction, Multiplication, Division of Algebraic Fractions
Lesson Addition of Fractions; Simplification of Complex Fractions

Use Chapter 23

The above was designed to avoid duplication and to reduce the volume of this book, and also, such a design is good for the environment.

CHAPTER 35
Lesson 87
Solving Fractional or Rational Equations

Example 1 Solve for x: $\dfrac{4}{x-1} = 5$

Solution

Step 1: Determine the excluded values. These are the x-values which make the denominators zero. Set the denominator to zero and solve for the variable.

$$x - 1 = 0$$
$$\underline{+1 \quad +1}$$
$$x = 1$$

The excluded value is 1.

Step 2: Multiply both LHS and RHS of the equation by $x - 1$

$$\frac{4}{x-1} = 5$$

$$\frac{\cancel{(x-1)}\,4}{\cancel{(x-1)}} = \frac{5(x-1)}{1}$$

$$4 = 5x - 5$$
$$\underline{+5 \qquad +5}$$
$$9 = 5x$$
$$\frac{9}{5} = x$$

Since $\dfrac{9}{5}$ is not an excluded value, it may be accepted as a solution; however, for good practice, we check it in the original equation.

Step 3:
$$\frac{4}{\frac{9}{5} - 1} \overset{?}{=} 5$$

$$\frac{4}{1\frac{4}{5} - 1} \overset{?}{=} 5$$

$$\frac{4}{\frac{4}{5}} \overset{?}{=} 5$$

$$\frac{4}{1} \times \frac{5}{4} \overset{?}{=} 5$$

$$5 = 5 \qquad \text{True} \quad (\text{RHS} = \text{LHS})$$

$\therefore$ the solution is $\dfrac{9}{5}$.

Example 2 Solve for x: $\quad \dfrac{4}{x-3} = \dfrac{3}{x-4}$

Step 1:   Determine the excluded values.
Set $x - 3 = 0$ and solve for x to obtain $x = 3$
Set $x - 4 = 0$ and solve for x to obtain $x = 4$
The excluded values are 3 and 4. Thus in solving the above equation, any value of x obtained may be checked for acceptance as a solution **except** 3 and 4.

Step 2: Multiply both sides of the equation by the LCM of the denominators.
(The LCD $= (x - 3)(x - 4)$

$$\dfrac{4}{(x-3)} \cdot \dfrac{(x-3)(x-4)}{1} = \dfrac{3}{(x-4)} \cdot \dfrac{(x-3)(x-4)}{1}$$

$$4(x-4) = 3(x-3)$$

$$4x - 16 = 3x - 9$$
$$\underline{-3x \qquad\quad -3x}$$
$$x - 16 = -9$$
$$\underline{+16 \quad +16}$$
$$x = 7$$

This value, 7, is not one of the excluded values (from Step 1) and so we may accept it as a solution but for good practice, we check it in the original equation.

$$\dfrac{4}{7-3} \overset{?}{=} \dfrac{3}{7-4}$$
$$\dfrac{4}{4} \overset{?}{=} \dfrac{3}{3} \quad \text{True} \qquad \text{(Also both the LHS and RHS of the equations are defined.)}$$
The solution is 7.

Example 3   Solve for y.

$$y + \dfrac{2y}{y-2} = \dfrac{4}{y-2}$$

Step 1: Set $y - 2 = 0$ and solve for y.
Then, $y - 2 = 0$, and $y = 2$

The excluded value is 2.
Step 2: Multiply every term of the equation by $y - 2$ (the LCD of the fractions).

$$y(y-2) + \dfrac{2y(y-2)}{y-2} = \dfrac{4(y-2)}{y-2}$$
$$y(y-2) + \dfrac{2y(y-2)}{y-2} = \dfrac{4(y-2)}{y-2}$$
$$y^2 - 2y + 2y = 4$$
$$y^2 = 4$$
$$y = \pm\sqrt{4}$$
$$y = \pm 2$$
i.e. $y = +2$, or $y = -2$

Since +2 is an excluded value (from Step 1), it is rejected; and only -2 may be accepted as a solution.

$$\text{Check: } -2 + \frac{2\,(-2)}{-2\,-2} \overset{?}{=} \frac{4}{-2\,-2}$$

$$-2 + \frac{(-4)}{-4} = \frac{4}{-4}$$

$$-2 + 1 = -1$$

$$-1 = -1 \quad \text{True}$$

Therefore, the solution is -2.

Note above that +2 is an extraneous solution.

Example 4   Solve for x.

$$\frac{3x}{x+5} - \frac{x+1}{x+3} = 2$$

Step 1. Find the excluded values.

Set $x + 5 = 0$ and solve for x to obtain $x = -5$.
-5 is an excluded value.

Similarly, set $x + 3 = 0$, and solve for x to obtain $x = -3$

The excluded values are -5 and -3. Note above that we could obtain the excluded values by inspection (mentally).

Step 2: Multiply each term of the equation by the LCM of the denominators (to undo the denominators). The LCM is $(x + 5)(x + 3)$.

$$(x + 5)(x + 3)\,\frac{3x}{x+5} - (x+5)(x+3)\frac{(x+1)}{x+3} = 2(x+5)\,(x+3)$$

$$\cancel{(x+5)}(x+3) \cdot \frac{3x}{\cancel{x+5}} - (x+5)(\cancel{x+3})\frac{(x+1)}{\cancel{x+3}} = 2(x+5)\,(x+3)$$

Step 3:   Simplify and solve for x.

$$(x+3)(3x) - (x+5)(x+1) = 2(x+5)(x+3)$$

$$3x^2 + 9x - (x^2 + 6x + 5) = 2(x^2 + 8x + 15)$$

$$3x^2 + 9x - x^2 - 6x - 5 = 2x^2 + 16x + 30 \qquad \text{(simplifying the left-hand side)}$$

$$2x^2 + 3x - 5 = 2x^2 + 16x + 30$$

$$2x^2 - 2x^2 + 3x - 16x - 5 - 30 = 0 \qquad \text{(eliminating the right-hand-side)}$$

$$-13x - 35 = 0$$

$$\frac{-35x}{-13} = \frac{+35}{-13}$$

$$x = -\frac{35}{13}$$

After checking (not shown), the solution is $-\frac{35}{13}$.

Note in Example **4** above, that in Step 3, the equation contained x^2-terms; but when we proceeded to solve the equation, the x^2-terms dropped out. Sometimes, this happens; in some cases these terms do not drop out, as in Example 3 above.

Note: 423

The excluded values for $\dfrac{8}{x^2 - 4} = 1$ are -2 and +2 (obtained by setting $x^2 - 4 = 0$ and solving for x);

however, note that for $\dfrac{8}{x^2 + 4} = 1$, $x^2 + 4$ is positive for all real values of x and never zero, since

the square of any nonzero real number is always positive. Therefore, there are no real excluded

values for $\dfrac{8}{x^2 + 4} = 1$. In fact the solution to $\dfrac{8}{x^2 + 4} = 1$ is the set $\{-2, 2\}$.

Lesson 87 Exercises

A Solve for x:

1. $\dfrac{9}{x + 4} = 1$; **2.** $\dfrac{2}{x - 3} = \dfrac{5}{x - 4}$; **3.** $\dfrac{x^2}{x + 2} = \dfrac{4}{x + 2}$ **4.** $\dfrac{7}{x - 5} + x = \dfrac{3}{x - 5}$

Answers: **1.** $\{5\}$; **2.** $\{\dfrac{7}{3}\}$ **3.** $\{2\}$ Note: $x \neq$ -2;   **4.** $\{1,4\}$

B Solve for x and state the excluded values:

1. $\dfrac{4}{x} + \dfrac{1}{2} = \dfrac{2}{5}$; **2.** $\dfrac{2}{3} - \dfrac{5}{x} = \dfrac{3}{4}$; **3.** $\dfrac{2x}{2x - 5} = 2$; **4.** $\dfrac{1}{x - 3} + \dfrac{1}{x + 3} = \dfrac{9}{x^2 - 9}$;   **5.** $\dfrac{1}{x - 2} = \dfrac{x^2}{x - 2} + 1$

Answers: **1.** -40 $(x \neq 0)$ **2.** -60 $(x \neq 0)$; **3.** 5 $(x \neq \dfrac{5}{2})$; **4.** $\dfrac{9}{2}$ $(x \neq$ -3 or 3); **5.** -$\dfrac{1}{2} \pm \dfrac{\sqrt{13}}{2}$ $(x \neq 2)$

CHAPTER 36
Lesson 88
Solving Radical Equations

We define **a radical equation** as an equation in which one of the variables occurs under one or more radical signs.

Examples   **1.** $\sqrt{x} = 6$

2. $\sqrt{x - 3} = 4$

3. $\sqrt{2x + 4} = \sqrt{x + 2}$

Principle of powers: If $a = b$

Then, $a^n = b^n$. However, this principle does not always produce equivalent equations, and therefore, if we use this principle to solve an equation, we must check the solutions in the original equation.

Procedure   The technique here involves the elimination of the radical signs by raising the radical expressions to integral (integer) powers: both sides of an equation are to be raised to the same integral power. We will then solve the resulting non-radical equation for the variable. If there are more than one radical, we may sometimes have to "undo" the radical signs in a number of steps. It will also be necessary that we test any solutions in the original equation for **extraneous solutions**. Before raising each side of an equation to an integral power, the radical whose radical sign we want to eliminate must be on one side of the equation by itself.

Example 1   Solve for x: $\sqrt{x} = 6$.

Solution

Step 1:   Square both sides of the equation to undo the square root symbol (i.e., raise both sides of the equation to the second power).
Then, $(\sqrt{x})^2 = (6)^2$
$x = 36$.
Step 2: Check the solution in the original equation.
$\sqrt{36} \stackrel{?}{=} 6$
$6 = 6$ True (left-hand side equals right-hand side, i.e., LHS = RHS).
$\therefore$ the solution is 36.

The above problem was so simple that we could have asked: If the square root of a number is 6, what is the number? Of course, the number is 36.

Example 2 Solve for x: $\quad \sqrt{x-3} = 4$

Solution

Step 1: Square both sides of the equation to undo the square root symbol.

Then, $(\sqrt{x-3})^2 = (4)^2$

$$x - 3 = 16$$

Step 2: Solve for x.

$$x - 3 = 16$$
$$\underline{+3 \quad +3}$$
$$x = 19$$

Step 3: Check the solution 19 in the original equation.

$$\sqrt{19-3} \overset{?}{=} 4$$
$$\sqrt{16} \overset{?}{=} 4$$
$$4 = 4 \quad \text{True} \quad \text{(LHS= RHS)}$$

∴ the solution is 19.

Note above that squaring the square root of a number eliminates the square root symbol.

Example 3 Solve for x:

$$\sqrt{2x+4} = \sqrt{x+2}$$

Solution

Step 1: Square both sides of the equation.

$$(\sqrt{2x+4})^2 = (\sqrt{x+2})^2$$

$$2x + 4 = x + 2$$

Step 2: Solve for x.

$$2x + 4 = x + 2$$
$$\underline{-x \qquad -x}$$
$$x + 4 = 2$$
$$\underline{\quad -4 \quad -4}$$
$$x = -2$$

Step 3: Check for $x = -2$ in the original equation

$$\sqrt{2(-2)+4} \overset{?}{=} \sqrt{(-2)+2}$$
$$\sqrt{-4+4} \overset{?}{=} \sqrt{0}$$
$$\sqrt{0} \overset{?}{=} \sqrt{0}$$
$$0 = 0 \qquad \text{True} \quad \text{(RHS = LHS)}$$

∴ the solution is -2.

Example 4   Solve for x: $\qquad$ $\sqrt{2x - 5} = x - 4$ $\qquad\qquad$

Solution

Step 1: $\qquad$ Square both sides of the equation.

$$(\sqrt{2x - 5})^2 = (x - 4)^2$$
$$2x - 5 = x^2 - 8x + 16$$
$$0 = x^2 - 10x + 21$$

or $x^2 - 10x + 21 = 0$

Step 2: $\qquad$ Solve the quadratic equation by any method.

We will solve by factoring, since the factors are easily recognizable.

$$x^2 - 10x + 21 = 0$$
$$(x - 3)(x - 7) = 0$$
$$x - 3 = 0 \text{ or } x - 7 = 0$$

$x = 3$ or $x = 7$ (Setting each factor equal to zero and solving for x)

Step 3: Checking for $x = 3$:

$$\sqrt{2(3) - 5} \overset{?}{=} 3 - 4$$
$$\sqrt{6 - 5} \overset{?}{=} -1$$
$$\sqrt{1} \overset{?}{=} -1$$
$$1 = -1 \text{ False }\quad (\text{LHS } \mathbf{not} \text{ equal to RHS})$$

Therefore 3 is not a solution. It is an extraneous root.

Checking for $x = 7$:

$$\sqrt{2(7) - 5} \overset{?}{=} 7 - 4$$
$$\sqrt{14 - 5} \overset{?}{=} 3$$
$$\sqrt{9} \overset{?}{=} 3$$
$$3 = 3 \quad \text{True }\quad (\text{RHS = LHS})$$

Therefore, 7 is a solution.

Since 3 is an extraneous root, the only solution is 7.

Lesson 88 Exercises

Solve for x:

1. $\sqrt{x} = 7$; $\qquad$ **2.** $\sqrt{x - 4} = 5$ $\quad$ **3.** $\sqrt{3x + 5} = \sqrt{x + 9}$

Answers: 1. $x = 49$; $\quad$ **2.** $x = 29$; $\qquad$ **3.** $x = 2$;

Solve and check: $\qquad$ **1.** $\sqrt{x - 5} + 6 = 2$; $\qquad$ **2.** $2\sqrt{x} = 6$; $\qquad$ **3.** $\sqrt{x^2 - 8x} = 3$;

4. $3\sqrt{x + 2} = 4\sqrt{x - 5}$; $\qquad\qquad$ **5.** $\sqrt{x + 5} - \sqrt{x} = 3$; $\qquad$ **6.** $\sqrt[3]{x + 1} = 2$

Answers: **1.** No solution (21 is extraneous) ; $\quad$ **2.** $\{\,9\,\}$; $\quad$ **3.** $\{\,-1, 9\,\}$; $\qquad$ **4.** $\{\,14\,\}$;

5. No solution. ($\frac{4}{9}$ is extraneous) ; $\quad$ **6.** $\{7\}$

CHAPTER 37
Complex Numbers

Lesson 89: **Definition, Powers of i, Square Root of Negative Numbers**
Lesson 90: **Addition and Subtraction of Complex Numbers**
Lesson 91: **Multiplication and Division of Complex Numbers**

Lesson 89
Definition, Powers of i, Square Root of Negative Numbers

Sometimes, in attempting to solve certain polynomial equations, we arrive at situations in which we have to find the square roots of negative numbers.

Example: If $x^2 = -1$, then $x = \pm \sqrt{-1}$

Since, for the set of real numbers, there is no provision for the square root of a negative number, we introduce a number "i" which we call the imaginary unit, with the following definition:

Definition: $i = \sqrt{-1}$ or $i^2 = -1$ (We will use both forms of the definition; memorize them)

For example, $(\sqrt{-1})(\sqrt{-1}) = (i)(i) = i^2 = -1$

Powers of i   (Cyclical property of i or i^2) : All powers of i are either equal to ± 1 or $\pm i$.
Examples

(a) $i^2 = -1$

(b) $i^3 = (i^2)i$
$\quad = -1(i)$
$\quad = -i$

(c) $i^4 = (i^2)(i^2)$
$\quad = (-1)(-1)$
$\quad = 1$

(d) $i^{10} = (i^4)(i^4)(i^2)$
$\quad = (1)(1)(-1)$
$\quad = -1$

Note above: Even powers of i are equal to ± 1. Odd powers of i are equal to $\pm i$.

Note that the imaginary unit "i" is only a tool in mathematics, and that it is not more imaginary (in the literal sense) than the real number 3. The introduction of the imaginary unit allows us to find the square roots of negative numbers.

Example   Find the square root: (a) -4 ;   (b) -25 ;   (c) Simplify: $\sqrt{-15}$.

Solution   (a) $\sqrt{-4} = (\sqrt{-1})(\sqrt{4})$
$\quad = i(2)$

Therefore $\sqrt{-4} = 2i$

(b) $\sqrt{-25} = \sqrt{-1}(\sqrt{25})$
$\quad = i(5)$
$\sqrt{-25} = 5i$

(c) $\sqrt{-15} = (\sqrt{-1})(\sqrt{15})$
$\quad = i\sqrt{15}$ or $\sqrt{15}\, i$

Note: In (c) we prefer the first form of the answer; because, sometimes, if we are not careful in writing "i ", the "i " may look as if it is under the radical sign . However, if no radical is involved we will leave the answers as in (a) and (b) above. In some old textbooks, you may find "i" written after the radical.

Generally, $\sqrt{-b} = (\sqrt{-1})(\sqrt{b})$ $(b \geq 0)$

           $= i\sqrt{b}$

The introduction of the imaginary unit helps us to expand the real number system to a more general system called the complex number system.

If we denote a complex number by z, then $z = a + bi$, where a and b are real numbers; a is called the real part and b is called the imaginary part. If $b = 0$, we have a pure real number (e.g., $z = 3 + 0i = 3$). If $a = 0$, we have a pure imaginary number ($z = 0 + 2i = 2i$). Thus, the product bi is called a pure imaginary number.

Distinction Between Square Roots of Negative Numbers and Roots of Equations

Roots of numbers (Principal Roots)**:** $\sqrt{-b} = i(\sqrt{b})$ $(b \geq 0)$

Example (a) $\sqrt{-4} = (\sqrt{-1})(\sqrt{4})$

              $= 2i.$

     (b) $\sqrt{-25} = \sqrt{-1}(\sqrt{25})$

             $= 5i.$

Roots of equations: Here, we must note that we have more than one root. (two roots.)

Example 1 Consider the solution to the equation $x^2 = -16$.

Solution $x^2 = -16$

                $x = \pm\sqrt{-16}$

                   $= \pm 4i$, which means $x = +4i$ or $x = -4i$.

Example 2   Solve for x:: $x^2 = -4$.

Solution

      $x = \pm\sqrt{-4}$

      $x = \pm 2i$ (or $x = +2i$ or $x = -2i$).

Lesson 89 Exercises

Find the square root or simplify:

1. $\sqrt{-9}$;   **2** $\sqrt{-49}$;   **3.** $\sqrt{-36}$;   **4.** $\sqrt{-18}$;   **5.** $\sqrt{\frac{4}{-9}}$;   **6.** i^{40};   **7.** i^{93} ;   **8.** i^{100}

9. $\sqrt{-28}$;   **10.** $\sqrt{-32}$

Answers: **1.** $3i$;   **2.** $7i$;   **3.** $6i$;   **4.** $3i\sqrt{2}$;   **5.** $\frac{2}{3}i$;   **6.** 1 ;   **7.** i ;   **8.** 1 ;   **9.** $2i\sqrt{7}$; **10.** $4i\sqrt{2}$

Lesson 90
Addition and Subtraction of Complex Numbers

In adding complex numbers, we will add the real parts, and then add the imaginary parts (in much the same way as we add like terms in polynomial addition).

Perform the indicated operations, leaving the answers in the form $a + bi$.

Example 1   Simplify: $(-3 + 5i) + (-2 + 7i)$

Step 1: Remove the parentheses.

$(-3 + 5i) + (-2 + 7i)$

$= -3 + 5i - 2 + 7i$

Step 2: Add the real parts, and add the imaginary parts.

$$-3 + 5i - 2 + 7i$$
$$= -5 + 12i \ .$$

Scrapwork:
For the real parts: -3 - 2 = - 5
For the imaginary parts: 5+ 7 = 12

Example 2   Simplify: $(6 - 5i) - (-3 + 2i)$

Solution   Remove the parentheses and add.

$(6 - 5i) - (-3 + 2i)$
$= 6 - 5i + 3 - 2i$

$= 9 - 7i$

Example 3 Simplify: $(5 + 2i) + (-3 - 6i)$

Solution

Remove the parentheses and add.
$(5 + 2i) + (-3 - 6i)$
$= 5 + 2i - 3 - 6i$
$= 5 - 3 + 2i - 6i$ <-------you may skip this step.
$= 2 - 4i$

or adding vertically:

$$\begin{array}{r} 5 + 2i \\ \underline{-3 - 6i} \\ 2 - 4i \end{array}$$ (Adding).

Example 4   Simplify : $(3 + 4i\} - (2 - 5i)$ (subtraction)

Solution

Remove the parentheses and add.

$(3 + 4i) - (2 - 5i)$
$= 3 + 4i - 2 + 5i$
$= 3 - 2 + 4i + 5i$ <-------you may skip this step.
$= 1 + 9i$

Example 5 Simplify : $(6 - 3i) + (2 + 3i)$

Solution

$(6 - 3i) + (2 + 3i)$
$= 6 - 3i + 2 + 3i$
$= 8 + 0i$
$= 8$

Lesson 90 Exercises

Simplify: **1.** $4 + 3i - 6 + 5i$;   **2.** $(7 - 8i) + (2 + 9i)$ **3.** $(2 - 3i) -(7 - 4i)$;

4. Subtract $(1 - i)$ from $4 - 2i$;   **5.** $4 + 2i - 3(4 - 6i) + i$;   **6.** $- i^5 + i^2$;   **7.** $4 - i + i^2$

Answers: **1.** $-2 + 8i$;   **2.** $9 + i$;   **3.** $-5 + i$;   **4.** $3 - i$;   **5.** $-8 + 21i$;   **6.** $-i - 1$;   **7.** $3 - i$

Lesson 91
Multiplication and Division of Complex Numbers

Multiplication of Complex Numbers

The approach here is multiply, replace i^2 by -1, (or higher powers of i by ± 1 or $\pm i$) add the real parts and add the imaginary parts.

Example 1 Multiply   -4 - 2i and -5 + i

Solution

Step 1:   Multiply as you multiply binomials
$$(-4 - 2i)(-5 + i)$$
$$= -4(-5 + i) + (-2i)(-5 + i) \quad \text{<--- You may skip this step.}$$
$$= 20 - 4i + 10i - 2i^2$$

Step 2:   (Replace i^2 by -1, since i^2 = -1 by definition)
$$= 20 - 4i + 10i - 2(-1)$$
$$= 20 + 6i + 2$$

Step 3:   (Add the like terms: add the real parts; and add the imaginary parts)
$$= 22 + 6i$$

Example 2   Multiply   3 + 2i and 4 + 5i
Solution
Procedure: Multiply, replace i^2 by -1, and simplify.

$$(3 + 2i)(4 + 5i)$$
$$= 12 + 15i + 8i + 10i^2$$
$$= 12 + 23i + 10(-1)$$
$$= 12 + 23i - 10$$
$$= 2 + 23i \quad (\text{Adding})$$

Example 3   Simplify : 4(6 - 3i)

Solution

$$4(6 - 3i)$$
$$= 24 - 12i$$

Example 4   Simplify: (4 + 2i)(3 - 6i)

Solution   The above implies multiplication.
Multiply, replace i^2 by -1, and add the like terms.
$$(4 + 2i)(3 - 6i)$$
$$= 12 - 24i + 6i - 12i^2$$
$$= 12 - 24i + 6i - 12(-1)$$
$$= 12 - 24i + 6i + 12$$
$$= 24 - 18i \qquad (\text{Note: } 12 + 12 = 24 \text{ and } -24i + 6i = -18i)$$

Example 5 Simplify: $(7 + 2i)(7 - 2i)$

Solution

$(7 + 2i)(7 - 2i)$
$= 49 - 14i + 14i - 4i^2$
$= 49 - 4(-1)$
$= 49 + 4$
$= 53$

Multiplication of the square roots of negative numbers

Example 1 Find the product of $(\sqrt{-8})$ and $(\sqrt{-2})$

Solution

Step 1: Change to complex number forms.

$\sqrt{-8} = \sqrt{-1}(\sqrt{8})$
$= i\sqrt{8}$

Similarly , $\sqrt{-2} = i\sqrt{2}$

Step 2: Multiply the complex forms now.

$(\sqrt{-8})(\sqrt{-2}) = i\sqrt{8} \, i\sqrt{2}$
$= i^2\sqrt{16}$ $(\sqrt{16} = 4)$
$= i^2 \, (4)$
$= (-1)(4)$ $(i^2 = -1)$
$= -4$

In the above problem (Example 1), you may skip Step 1 and show only Step 2.

Example 2 Find the product $(\sqrt{-49})(\sqrt{-25})$

Solution

Step 1: Change to complex number forms.

Then, $\sqrt{-49} = i\sqrt{49} = i(7) = 7i$

                                <---You may do Step 1 mentally and show only Step 2.

$\sqrt{-25} = i\sqrt{25} = i(5) = 5i$

Step 2: Multiply the complex forms now.

Then, $(\sqrt{-49})(\sqrt{-25}) = (7i)(5i)$
$= (7)(5)i^2$
$= 35(-1)$ $(i^2 = -1)$
$= -35$

We must note in the last two examples that it was necessary first to express the square roots of the negative numbers in terms of i before proceeding to multiply. Failure to do this may result in error such as the following:

Wrong procedure---> $(\sqrt{-8}(\sqrt{-2}) = \sqrt{(-2)(-8)}$
$= \sqrt{16}$
$= 4,$ **which is a wrong answer**.

Generally, it is true that $(\sqrt{a})(\sqrt{b}) = \sqrt{ab}$ if a and b are positive but it is not true if a and b are negative.

Complex conjugates

Definition : The **conjugate** of a given binomial is another binomial that differs from the given
binomial only in the sign of one of the terms. The conjugate of $a + b$ is $a - b$; and the
conjugate of $a - b$ is $a + b$.

The **conjugate** of $a + bi$ is $a - bi$. The conjugate of $a - bi$ is $a + bi$. A complex number and its
conjugate differ only in the sign of the imaginary part. To find the conjugate of a complex number,
change the sign of the imaginary part and keep the sign of the real part unchanged.

Examples:

The conjugate of $2 + 3i$ is $2 - 3i$.   The conjugate of $4 - 2i$ is $4 + 2i$.   The conjugate of $7 - 5i$ is
$7 + 5$.   The conjugate of $-3 + 6i$ is $-3 - 6i$. The conjugate of $-2 - 3i$ is $-2 + 3i$.
 The conjugate of $4i$ is $- 4i$. The conjugate of 7 is 7

The product of a complex number and its conjugate is a real number.

The product of $-2 - 3i$ and $-2 + 3i$. $= 4 - 6i + 6i - 9i^2 = 4 - 9(-1) = 4 + 9 = $ **13**, a real number.

The product of $a + bi$ and $a - bi = a^2 + b^2$

We can use the conjugate of a complex number to rationalize the denominator of a fraction or
to divide by a complex number.

Division of Complex Numbers

Example 1   Simplify : $\dfrac{2 + 3i}{4 - 5i}$ or divide $2 + 3i$ by $4 - 5i$

To simplify the above complex number is meant we are to write it in the form $z = a + bi$.
The operation here is similar to that of the rationalization of denominators of radical expressions.
To simplify the above expression, we **multiply both the denominator and the numerator**
 by the conjugate of the denominator.

Solution   The conjugate of $4 - 5i$ is **$4 + 5i$.**   (See also the note below)

Multiply both the denominator and the numerator by $4 + 5i$.

Then, we obtain:

$$\dfrac{2 + 3i}{4 - 5i}$$

$$= \dfrac{(2 + 3i)}{(4 - 5i)} \cdot \dfrac{(4 + 5i)}{(4 + 5i)}$$

$$= \dfrac{8 + 10i + 12i + 15i^2}{16 + 20i - 20i - 25i^2}$$

$$= \dfrac{8 + 22i + 15(-1)}{16 + 0 - 25(-1)}$$

$$= \dfrac{8 + 22i - 15}{16 + 25}$$

$$= \dfrac{8 + 22i - 15}{41}$$

$$= \dfrac{-7 + 22i}{41}$$

$$= -\dfrac{7}{41} + \dfrac{22}{41}i$$

We may observe above that the **denominator** does **not** contain the imaginary unit "i", even
 though the numerator contains the imaginary unit.

Note above that we could have multiplied by $-4 - 5i$. (Try it.) It is therefore not critical in this
 problem which terms should differ in sign, (so far as the rationalization is concerned)
 provided that either the real parts differ in sign or the imaginary parts differ in sign, but **not** both.
 We may produce a minus sign in the denominator which we can take care of as usual.

Example 2   Simplify: $\dfrac{5 + 4i}{-3 + 2i}$

Solution Multiply both the denominator and the numerator by -3 - 2i   (The conjugate of -3 + 2i).

$$\dfrac{5 + 4i}{-3 + 2i}$$

$$= \dfrac{(5 + 4i)}{(-3 + 2i)} \cdot \dfrac{\textbf{(-3 - 2i)}}{\textbf{(-3 - 2i)}}$$

$$= \dfrac{-15 - 10i - 12i - 8i^2}{9 + 6i - 6i - 4i^2}$$

$$= \dfrac{-15 - 10i - 12i - 8(-1)}{9 + 0 - 4(-1)}$$

$$= \dfrac{-15 - 22i + 8}{9 + 4}$$

$$= \dfrac{-7 - 22i}{13}$$

$$= -\dfrac{7}{13} - \dfrac{22}{13}i \qquad (-15 + 8 = -7)$$

Example 3 Simplify: $\dfrac{4 - 2i}{i}$

Solution $\dfrac{4 - 2i}{i}$

$$= \dfrac{(4 - 2i)}{i} \dfrac{\textbf{(-i)}}{\textbf{(-i)}} \quad \text{(The conjugate of } i \text{ is } -i \text{ ; but you could also use } i)$$

$$= \dfrac{(4 - 2i)(-i)}{(i)(-i)}$$

$$= \dfrac{-4i + 2i^2}{-i^2}$$

$$= \dfrac{-4i + 2(-1)}{-(-1)} \qquad (i^2 = -1)$$

$$= \dfrac{-4i - 2}{+1}$$

$$= -2 - 4i$$

Lesson 91 Exercises

A 1. Multiply: $4 + 3i$ and $2 + 5i$;   **2.** Multiply $2 - 5i$ and $-2 + 6i$; **3.** Simplify: $(6 - 7i)(2 + 3i)$;
4. Simplify $(1 - i)(1 + i)$;   **5.** $(2 + 3i)^2$; **6.** $(4i^2)(-8i)(i^2)$;   **7.** $(5 + 3i)(5 - 3i)$

Answers: **1.** $-7 + 26i$; **2.** $26 + 22i$; **3.** $33 + 4i$;   **4.** 2 ; **5.** $-5 + 12i$; **6.** $-32i$;   **7.** 34.

B Divide: **1.** $\dfrac{8 - 6i}{2}$; **2.** $\dfrac{6 + 8i}{2i}$;   **3.** $\dfrac{\sqrt{16}}{\sqrt{-16}}$;   **4.** $\dfrac{3}{4 - \sqrt{-9}}$; **5.** $\dfrac{4 - 2i}{3 + 5i}$

Answers: **1.** $4 - 3i$; **2.** $4 - 3i$; **3.** $-i$; **4.** $\dfrac{12}{25} + \dfrac{9}{25}i$; **5.** $\dfrac{1}{17} - \dfrac{13}{17}i$

C Simplify: **1.** $\dfrac{2 + 3i}{5 - 2i}$;   **2.** $\dfrac{4}{3 - 4i}$;   **3.** $\dfrac{5 + 2i}{-i}$ **4.** Divide $4 + 3i$ by $-2 + 3i$;

Simplify: **5.** $\dfrac{2 - 6i}{4 + 3i}$;   **6.** $\dfrac{i - 2}{2 - i}$;   **7.** $(3 - 2i)(4 + 2i)(i^2)$;   **8.** $(6 - 4i)(6 + 4i)$

Answers: 1. $\dfrac{4}{29} + \dfrac{19}{29}i$; **2.** $\dfrac{12}{25} + \dfrac{16}{25}i$; **3.** $-2 + 5i$; **4.** $\dfrac{1}{13} - \dfrac{18}{13}i$; **5.** $-\dfrac{2}{5} - \dfrac{6}{5}i$; **6.** -1 ; **7.** $-16 + 2i$; **8.** 52

CHAPTER 38
Logarithms

Lesson 92 Definitions, Basic Properties and Applications
Lesson 93 Evaluation of Logarithms; Common and Natural Logarithms

Lesson 92
Definition, Basic Properties and Applications

Definition: The logarithm of a number to a base is the power (or exponent) to which the base must be raised in order to produce that number. A logarithm is therefore an exponent.

For any positive numbers x and b $(b \neq 1)$, and y any real number,

$$\boxed{\log_b x = y \text{ if and only if } b^y = x}$$

(Read " log of x to the base b is y if and only if b to the power y is x.
or log to the base b of x is y if and only if b to the power y is x.)

Note: $\log_b x = y$ is equivalent to $b^y = x$.

Examples: **1.** $\log_{10} 100 = 2$ (because $10^2 = 100$)

       2. $\log_2 16 = 4$ (because $2^4 = 16$)

       3. $\log_{10} \dfrac{1}{100} = -2$ (because $10^{-2} = \dfrac{1}{10^2} = \dfrac{1}{100}$)

If you memorize and understand the definition of the logarithm very well, you will , in some evaluations be able to find the log mentally. For example, if asked to evaluate $\log_2 16$, ask yourself this question: To what power should 2 (the base) be raised in order to produce 16? ; or what is the exponent that must be placed on 2 in order to produce 16?.

Basic Properties of Logarithms

There are three basic rules which are properties of logarithms. Each rule is associated with a fundamental rule of exponents.

Property 1   For any positive numbers M, N, and b $(b \neq 1)$,

$$\log_b (MN) = \log_b M + \log_b N$$

In words, the log of a product equals the sum of the logs of the factors.

Examples   1. $\log_{10} [3(5)] = \log_{10} 3 + \log_{10} 5$.

        2. $\log_{10} [423(785)] = \log_{10} 423 + \log_{10} 785$.

Property 2 $\log_b \dfrac{M}{N} = \log_b M - \log_b N$ $(b \neq 1)$

In words, the log of a fraction equals the log of the numerator minus the log of the denominator.

Example: $\log_5 \dfrac{42}{23} = \log_5 42 - \log_5 23$

Property 3 For any positive numbers M and b and any real number p.

$$\log_b M^p = p \log_b M \qquad (b \neq 1)$$

In words, the log of the pth power of a number equals p times the log of the number.

Example: $\log_2 8^3 = 3 \log_2 8$

Some other useful properties of logarithms

Property 4 $\qquad \log_b b = 1 \qquad$ (equivalently, $b^1 = b$)

Property 5 $\qquad \log_b 1 = 0 \qquad$ (equivalently, $b^0 = 1$)

Property 6 $\qquad b^{\log_b M} = M$

Property 7 $\qquad$ If $M = N$ then

$\qquad\qquad \log_b M = \log_b N \qquad$ (We may call this step " taking logs of both sides of the equation")

Property 8 $\qquad$ If $\log_b M = \log_b N$, then

$\qquad\qquad\qquad M = N \qquad$ (Taking antilogs or " undoing the logs " of both sides of the equation")

A very useful property of equality of exponents when the bases are equal:

$$\boxed{\text{If } b^x = b^y \text{ then } x = y}$$

Applications of the above properties:

Express $\log C$ in terms of the logs of t, 5, and 7; and $\log S$ in terms of the logs of A, π, R and 5.

$$1. \quad C = \frac{t^2 \sqrt[3]{7}}{5} ; \qquad 2. \quad S = \sqrt{\frac{5A}{\pi R}}$$

Solution

1. $\log c = \log t^2 + \log (7)^{1/3} - \log$ $\qquad$ (Applying properties **7, 1, & 2** and changing radical to exponential form)

$\qquad = 2 \log t + \frac{1}{3} \log 7 - \log 5 \qquad$ (Applying property **3**)

2. $\qquad S = \sqrt{\frac{5A}{\pi R}}$

$\qquad\qquad S = \left(\frac{5A}{\pi R}\right)^{1/2} \qquad\qquad$ (Changing the radical to exponential form)

$\qquad\quad \log S = \log \left(\frac{5A}{\pi R}\right)^{1/2} \qquad\qquad$ (Taking logs)

$\qquad\quad \log S = \frac{1}{2} \left(\log \frac{5A}{\pi R}\right) \qquad\qquad$ (Applying property **3**)

$\qquad\qquad = \frac{1}{2} \left(\log 5 + \log A - \log \pi - \log R\right) \qquad$ (Applying properties **1, & 2**)

Note that $\dfrac{\log_b M}{\log_b N}$ is **not** equal to $\log_b \dfrac{M}{N}$ $\qquad$ ($\log_b \dfrac{M}{N} = \log_b M - \log_b N$)

In the first expression you find the logs first before dividing; in the second expression, you divide first before finding the log.

Antilogarithms

Definition : If $y = \log_a x$, then the **antilogarithm** (antilog) of y to the base a is x.

The antilog of a given number is, therefore, the number whose logarithm to a given base is the given number. Note that " taking logs" and "taking antilogs" are inverse operations of each other. Each operation reverses

the action of the other. Example: $\text{antilog}_2 5 = 2^5 = 32$, since $\log_2 32 = 5$.

Lesson 92 Exercises

Find the following:: **1.** $\log_{10} 100$; **2.** $\log_2 16$; **3.** $\log_{10} \dfrac{1}{100}$

4-5: Express $\log C$ in terms of the logs of t, 5, and 7; and $\log S$ in terms of the logs of A, π, R and 5.

4. $C = \dfrac{t^2 \sqrt[3]{7}}{5}$; **5.** $S = \sqrt{\dfrac{5A}{\pi R}}$

Answers: 1. 2; **2.** 4; **3.** -2; **4.** $\log c = 2\log t + \dfrac{1}{3}\log 7 - \log 5$;

5. $\log S = \dfrac{1}{2}(\log 5 + \log A - \log \pi - \log R)$

Lesson 93

Evaluation of Logarithms; Common and Natural Logs; Writing as a Single log term

Example 1 Evaluate $\log_2 16$.　　　　　　　　　　(1)

Method 1

Step 1: Use the equivalent definition of logarithms to change (1) to exponential form.

$$\text{Let } \log_2 16 = y$$
$$\text{Then } 2^y = 16$$
$$2^y = 2^4 \qquad\qquad (16 = 2^4)$$

Step 2: Since bases on both sides of the equation are equal, the exponents must be equal.

$$\therefore y = 4$$
$$\text{and } \log_2 16 = 4$$

Method 2

$$\begin{aligned}
\log_2 16 &= \log_2 2^4 \\
&= 4 \log_2 2 \qquad\qquad (\log_2 2 = 1)\\
&= 4(1) \\
&= 4
\end{aligned}$$

Again, we obtain the same result as by Method 1.

Method 3　Ask yourself this question and answer it : To what power should　2　(the base) be raised in order to produce 16? ; or what is the exponent that must be placed on 2 in order to produce 16?. Of course the answer is 4.

Example 2　　Evaluate $\log_3 9^{1/4}$

Method 1

Step 1: Let $\log_3 9^{1/4} = y$.

$$\text{Then, } 3^y = 9^{1/4}$$

Step 2: Express the right-hand side as a power of 3.

$$3^y = (3^2)^{1/4} \qquad\qquad (9 = 3^2)$$
$$3^y = 3^{2/4}$$
$$3^y = 3^{1/2}$$

Step 3: Since the bases are equal, equate the exponents.

$$\text{then, } y = \frac{1}{2}$$
$$\therefore \log_3 9^{1/4} = \frac{1}{2}$$

Method 2

$$\begin{aligned}
\log_3 9^{1/4} &= \log_3 (3^2)^{1/4} \\
&= \log_3 3^{2/4} \\
&= \log_3 3^{1/2} \\
&= \frac{1}{2} \log_3 3 \qquad\qquad (\log_3 3 = 1)\\
&= \frac{1}{2}
\end{aligned}$$

Note: In evaluations and proofs involving logs, whenever you do not know how to proceed, it is good practice to begin by saying " let the given log = say, y, and then apply the equivalent exponential definition and continue.

Common Logarithms

Logarithms to the base 10 are called common logarithms. The base 10 is sometimes omitted. Thus, $\log_{10} x$ may be written as $\log x$. In changing a logarithm to an exponential forms, it helps to indicate the base 10. The common log of a number is the power to which 10 must be raised to produce that number. Thus, the common log of a power of 10 is the exponent on 10.

Examples 1. $\log_{10} 100 = 2$ (because $10^2 = 100$)

2. $\log_{10} .01 = \log_{10} \dfrac{1}{100} = \log_{10} \dfrac{1}{10^2} = \log_{10} 10^{-2} = -2$ (because $10^{-2} = .01$)

3. $\log 2 = .3010$ (because $10^{.3010} = 2$)

More evaluations

Example 1: If $\log 2 = .3010$; $\log 3 = .4771$; $\log 7 = .8451$, evaluate the following:

     (a) $\log \dfrac{9}{4}$;   (b) $\log 24$;   (c) $\log 7^{1/2.}$

Solution

Procedure: Express each log in terms of $\log 2$, $\log 3$, $\log 7$ or $\log 10$, and simplify.

(a) $\log \dfrac{9}{4} = \log 9 - \log 4$ $(\log \dfrac{M}{N} = \log M - \log N)$

     $= \log 3^2 - \log 2^2$
     $= 2\log 3 - 2\log 2$ $(\log M^p = p\log M)$
     $= 2(.4771) - 2(.3010)$ (We are given that $\log 3 = .4771$; and $\log 2 = .3010$)
     $= .9542 - .6020$
     $= .3522$

(b) $\log 24 = \log 8(3)$
     $= \log 8 + \log 3$
     $= \log 2^3 + \log 3$
     $= 3 \log 2 + \log 3$
     $= 3(.3010) + .4771$
     $= .9030 + .4771$
     $= 1.3801$

(c) $\log 7^{1/2} = \dfrac{1}{2}\log 7$

     $= \dfrac{1}{2}(.8451)$

     $= .4226$

Example 2 Evaluate (a) $\log .00036$; (b) $\log 15$ using the given values in Example 1.
Solution

(a) $\log .00036 = \log (36 \times 10^{-5})$
     $= \log (4 \times 9 \times 10^{-5})$
     $= \log (2^2 \times 3^2 \times 10^{-5})$
     $= \log 2^2 + \log 3^2 + \log 10^{-5}$
     $= 2\log 2 + 2\log 3 - 5\log 10$
     $= 2(.3010) + 2(.4771) - 5(1)$ $(\log 2 = .3010; \log 3 = .4771; \log 10 = \log_{10} 10 = 1)$
     $= .6020 + .9542 - 5$
     $= -3.4438$

(b) $\log 15 = \log (3 \times 5)$

$\qquad = \log (3 \times \dfrac{10}{2})$

$\qquad = \log 3 + \log \dfrac{10}{2}$ <--------------You may skip this step.

$\qquad = \log 3 + \log 10 - \log 2$

$\qquad = .4771 + 1 - .3010 \qquad\qquad\qquad\qquad (\log 10 = \log_{10} 10 = 1)$

$\qquad = 1.1761$

Example 3   Given that $f(x) = \log_2 x$

$\qquad\qquad$ Find (a) $f(4)$;   (b) $f(\dfrac{1}{16})$.

Solution

$\qquad\qquad$ (a) $f(4) = \log_2 4 \qquad\qquad (x = 4)$

Method 1 $\qquad$ Let $y = \log_2 4$

$\qquad\qquad$ Then, $2^y = 4$

$\qquad\qquad\qquad 2^y = 2^2$ and

$\qquad\qquad\qquad y = 2 \qquad\qquad\qquad$ (Equating exponents)

Method 2 $\qquad \log_2 4$

$\qquad\qquad = \log_2 2^2$

$\qquad\qquad = 2 \log_2 2$

$\qquad\qquad = 2(1) \qquad\qquad\qquad (\log_2 2 = 1)$

Method 3 $\qquad$ Ask yourself this question and answer it : To what power should 2 (the base) be raised in order to produce 4 ? ; or what is the exponent that must be placed on 2 in order to produce 4 ?. Of course, the answer is 2.

(b) $\qquad\qquad f(\dfrac{1}{16}) = \log_2 (\dfrac{1}{16})$

Method 1

$\qquad \log_2 (\dfrac{1}{16}) = \log_2 1 - \log_2 16 \qquad\qquad (\log \dfrac{M}{N} = \log M - \log N)$

$\qquad\qquad\qquad = 0 - \log_2 2^4 \qquad\qquad\qquad (\log_2 1 = 0; \qquad \log_2 2 = 1)$

$\qquad\qquad\qquad = - 4 \log_2 2$

$\qquad\qquad\qquad = - 4(1)$

$\qquad\qquad\qquad = - 4$

Method 2

$\qquad\qquad \log_2 (\dfrac{1}{16}) = \log_2 (\dfrac{1}{2^4})$

$\qquad\qquad\qquad = \log_2 2^{-4}$

$\qquad\qquad\qquad = -4 \log_2 2 \qquad (\log M^p = p \log M)$

$\qquad\qquad\qquad = - 4(1)$

$\qquad\qquad\qquad = - 4$

Writing as a single log term

Example   Write as a single log term: $\log 4x + 2\log(x+1) - \log(x^2-1)$

$$\log 4x + 2\log(x+1) - \log(x^2-1)$$
$$= \log 4x + \log(x+1)^2 - \log(x^2-1)$$
$$= \log \frac{4x(x+1)^2}{x^2-1}$$
$$= \log \frac{4x(x+1)(x+1)}{(x+1)(x-1)} \qquad \text{(Applying } \log\frac{MN}{T} = \log M + \log N - \log T \text{ backwards)}$$
$$= \log \frac{4x(x+1)(\cancel{x+1})}{(\cancel{x+1})(x-1)}$$
$$= \log \frac{4x(x+1)}{x-1} \text{ or } \log \frac{4x^2+4x}{x-1}$$

Natural Logarithms

Logarithms to the base e (e ≈ 2.72) are called natural logarithms (or Napierian logarithms).

Example The natural log of x is written: $\ln x$ or $\log_e x$.

Conversions: $\ln x = \log_e x \approx 2.3\log_{10} x$ **or** $\log x = \log_{10} x \approx .43\log_e x$

Lesson 93 Exercises

A 1. $\log_3 81$;   2. $\log_6 6$;   3. $\log_2 64$;   4. $\log_3 27^2$;   5. $\log_5 \frac{1}{5}$; 6. $\log_2 32$;

  7.. $\log_3 81$;   8. $\log_3 243$; 9. $\log_{10} 1000$;   10. $\log_3 \frac{1}{27}$;   11. $\log .001$; 12. $\log_{10} 1$

Answers: **1**. 4 ;   **2**. 1 ;   **3**. 6 ;   **4**. 6 ;   **5**. -1 ; **6**. 5 ; **7**. 4 ; **8**. 5 ; **9**. 3 ; **10**. -3 ; **11**. -3 ; **12**. 0

B   If $\log 2 = .3010$; $\log 3 = .4771$; $\log 7 = .8451$, evaluate the following:

  1. $\log 16$;   **2.** $\log 20$;   **3.** $\log \frac{21}{2}$;   **4.** $\log .0072$;   **5.** $\log 50$;   **6.** $\log\sqrt{3}$

  7. $\log 5^{.01}$;   **8.** $\log \frac{1}{600}$;   **9.** $\log \frac{1}{2^4}$

Answers: **1**. 1.204 ;   **2**. 1.3010 ;   **3**. 1.0212 ;   **4**. - 2.1428 ;   **5**. 1.699 ;   **6**. 0.2386 ;
      7. 0.007 ;   **8**. -2.7781 ;   **9**. -1.204

C **a:** Given that $f(x) = \log_3 x$,   find **1.** $f(9)$;   **2.** $f(81)$; **3.** $f(\frac{1}{9})$;

   b: If $f(x) = \log_4 x$,   find **1.** $f(256)$;   **2.** $f(\frac{1}{64})$

Answers: **a 1**. 2 ;   **2**. 4 ;   **3**. - 2; **b 1.** 4 ; **2.** -3

D Write as a single log term: **1.** $\log(5x+10) - \log(x+2) + \log x$;

  2. $3\log(x^2+7x+12) - 3\log(x+3)$;   **3.** $\log_4 x - \log_4 y + \log_4 z$;   **4.** $2\log_4 5 + 3\log_4 2$;

  5. $\log(x^2-25) - \log(x+5)$;

Answers: **1**. $\log 5x$; **2**. $\log(x+4)^3$; **3**. $\log_4 \frac{xz}{y}$; **4**. $\log_4 200$; **5**. $\log(x-5)$

CHAPTER 39

Lesson 94: Exponential Equations
Lesson 95: Logarithmic Equations
Lesson 94

Exponential Equations

An **exponential equation** is an equation in which a variable occurs as an exponent.

Example 1   Solve the exponential equation:

$$15^x = 32$$

Step 1: Take logs of both sides of the equation.

$$\log 15^x = \log 32$$

$$x \log 15 = \log 32 \qquad (\log M^P = p \log M)$$

$$x = \frac{\log 32}{\log 15}$$

$$x = \frac{1.5051}{1.1761} \qquad \text{(From tables or calculator, } \log 32 = 1.5051; \log 15 = 1.1761)$$

$$x = 1.28$$

Example 2   Solve for x: $4^{x+1} = 16$

Method 1

$$4^{x+1} = (2^2)^{x+1} = 2^4$$

$$2^{2(x+1)} = 2^4$$

$2(x + 1) = 4$ (Equating exponents, since bases are equal)

$$2x + 2 = 4$$

$2x = 2$ and from which $x = 1$

Method 2:

$$4^{x+1} = 4^2$$

$x + 1 = 2$ (Equating exponents, since bases are equal)

$x = 1$

Method 3   Use the method in Example above.

Example 3   Solve for x:

$$7^{x^2 + x} = 34$$

Step 1: Take logs of both sides of the equation:

$$\log 7^{(x^2 + x)} = \log 34$$

$$(x^2 + x)\log 7 = \log 34$$

$$x^2 + x = \frac{\log 34}{\log 7}$$

$$x^2 + x = \frac{1.5315}{.8451} \qquad (\log 34 = 1.5315 ; \log 7 = .8451)$$

$$x^2 + x = 1.8122$$

$$x^2 + x - 1.8122 = 0$$

Step 2:   Solve for x by the quadratic formula.

$$x = \frac{-1 \pm \sqrt{1 + 7.2488}}{2} \qquad (a = 1, b = 1, c = -1.8122)$$

$$x = \frac{-1 \pm \sqrt{8.2488}}{2}$$

$$x = \frac{-1 \pm 2.8271}{2}$$

$$x = \frac{-1 + 2.8271}{2} \quad \text{or } x = \frac{-1 - 2.8271}{2}$$

$x = -1.9360$ or $.9360$.   The solution set is $\{-1.9360, .9360\}$

Lesson 94 Exercises

Solve for x: **1.** $12^x = 42$;   **2.** $3^{x+1} = 243$;   **3.** $2^{x^2+1} = 1024$;   **4.** $5^{x^2+x} = 42$

Answers: **1.** 1.5041 ;   **2.** 4 ;   **3.** Solution set is $\{-3, 3\}$;   **4.** Solution set is $\{-2.1038, 1.1038\}$

Lesson 95
Logarithmic Equations

A logarithmic equation is an equation which contains the logarithm of an expression involving the variable. We must check the solutions of logarithmic equations, since the logarithmic function is not defined for negative numbers.

Example 1   Solve for x: $\log(x + 9) - \log x = 1$

Step 1: Write the logarithmic term as a single logarithmic term.

$$\text{Then we obtain } \log_{10}\left(\frac{x+9}{x}\right) = 1 \qquad\qquad (\log A - \log B = \log \frac{A}{B})$$

Step 2: $10^1 = \dfrac{x+9}{x}$ (Applying the equivalent exponential definition)

$10 = \dfrac{x+9}{x}$

$10x = x + 9$

$x = 1$ (Solving for x)

Checking: Substitute $x = 1$ in the original equation.

$\log(1 + 9) - \log 1 \overset{?}{=} 1$

$\log 10 - \log 1 \overset{?}{=} 1$

$1 - 0 = 1$ $(\log_{10} 10 = 1)$

$1 = 1$ True

The solution is 1.

Example 2 Solve for x: $\log x + \log(x - 21) = 2$

Step 1: $\log x + \log(x - 21) = 2$

$\log[x(x - 21)] = 2$ (Writing as a single term)

$\log_{10}[x(x - 21)] = 2$ (Indicating the understood base so as to help transition to exponential form)

$10^2 = x(x - 21)$ (Applying the equivalent exponential definition)

$100 = x^2 - 21x$

Step 2:   Solve the quadratic equation by any method.

We solve by factoring, since the factors are easily recognizable.

$x^2 - 21x - 100 = 0$

$(x + 4)(x - 25) = 0$

$x = -4 \text{ or } x = 25$

Step 3: Since for $x = -4$, $\log(-4) + \log(-4 - 21)$ is not defined (The log of negative number is not defined) we reject the solution -4, and check for only 25.

Checking for $x = 25$: $\log 25 + \log(25 - 21) \overset{?}{=} 2$

$\log 25 + \log 4 \overset{?}{=} 2$

$\log 25(4) \overset{?}{=} 2$

$\log_{10} 100 \overset{?}{=} 2$ $(\log_{10} 100 = 2)$

$2 = 2$ True   (LHS = RHS)

The solution is 25.

Example 3   Solve for x: $\log \dfrac{3}{x} - 2 = -\log 6x^2$ $(x \neq 0)$

Solution

Step 1: Collect the log terms on the left-hand side.

$$\log \dfrac{3}{x} + \log 6x^2 = 2$$

Step 2: Write the log terms as a single log term.

$$\log \left[\dfrac{3}{x} \cdot 6x^2\right] = 2$$

Step 3: $\log \dfrac{18x^2}{x} = 2$

$\log_{10} 18x = 2$ (simplifying and indicating the base 10)

$10^2 = 18x$ (using the equivalent exponential definition)

$100 = 18x$

$\dfrac{100}{18} = x$, and $x = \dfrac{50}{9}$

After checking (not shown) the solution is $\dfrac{50}{9}$.

Example 4   Solve for x: $\log_x 2 = \dfrac{1}{3}$

Solution

Step 1: Apply the equivalent exponential definition.

If $\log_x 2 = \dfrac{1}{3}$

then $x^{1/3} = 2$

Step 2: $(x^{1/3})^3 = (2)^3$

(Cubing both sides of the equation)

$x = 2^3$

$x = 8$

Checking :

Step 1: $\log_8 2 \stackrel{?}{=} \dfrac{1}{3}$

Let $\log_8 2 = k$

Then, $8^k = 2$

(Cubing $(2^3)^k = 2$

$2^{3k} = 2^1$

$3k = 1$

$k = \dfrac{1}{3}$

Step 2: $\log_8 2 = \dfrac{1}{3}$

$\dfrac{1}{3} = \dfrac{1}{3}$ True

$\therefore$ the solution is 8.

Equivalent check:

$8^{1/3} \stackrel{?}{=} 2$

$\sqrt[3]{8} \stackrel{?}{=} 2$

$2 = 2$ True

Example 5 Solve for x: $\log(x + 2) + \log(x - 5) = \log 8$

Step 1: $\log(x + 2) + \log(x - 5) = \log 8$

$\log(x + 2)(x - 5) = \log 8$ $(\log MN = \log M + \log N)$

$(x + 2)(x - 5) = 8$ (If $\log M = \log N$, then $M = N$, ,undoing the logs.)

$x^2 - 3x - 10 - 8 = 0$

Step 2: $x^2 - 3x - 18 = 0$ (Solve this quadratic equation)

$(x + 3)(x - 6) = 0$

$x + 3 = 0$ or $x - 6 = 0$

$x = -3$ or $x = 6$

Step 3: We check the values of $x = -3$ and $x = 6$ in the original logarithmic equation.

Checking for $x = -3$: $\log(-3 + 2) + \log(-3 - 5) \stackrel{?}{=} \log 8$, is not defined , since the logarithm of a negative number is not defined, and we reject the value -3 as a solution.

Checking for $x = 6$: $\log(6 + 2) + \log(6 - 5) \stackrel{?}{=} \log 8$

$\log 8 + \log 1 \stackrel{?}{=} \log 8$

$\log 8(1) \stackrel{?}{=} \log 8$ $(\log MN = \log M + \log N)$ **or** $\log 8 + 0 \stackrel{?}{=} \log 8$ $(\log 1 = 0)$

$\log 8 \stackrel{?}{=} \log 8$ True   (LHS = RHS) $\log 8 \stackrel{?}{=} \log 8$

Since $x = 6$ satisfies the original logarithmic equation, the solution is 6.

General discussion about solving exponential and logarithmic equations   447

1.   Given the exponential equation, we (take logs) go through the logarithmic forms to
     simple algebraic equations (e.g., linear or quadratic equations).
Thus, **exponential** equation---->**logarithmic** equation (and then solve) or ---->**algebraic** equation and then solve.
     However, sometimes, it may not be necessary to use logarithms. Example: To solve for x, given $2^{x+3} = 8$,
     we may either do the following : $2^{x+3} = 2^3$; $x + 3 = 3$ (equating the exponents) and from which $x = 0$ or
     we may do the following: $(x + 3) \log 2 = \log 2^3$

$$x + 3 = \frac{3 \log 2}{\log 2}$$

$$x + 3 = 3 \text{ and from which } x = 0.$$

2.   Given the logarithmic equation, we change to the exponential equivalent form, and then to
     some basic algebraic equations.
Thus, **logarithmic** equation ---->**exponential** equation (and then solve) or---->**algebraic** equation and then solve.

Changing from one base to another base (Change of base formula)

Example   Change $\log_b M$ to base c.
Step 1: Let $\log_b M = y$.
          Then, $M = b^y$ (Equivalent exponential definition)
Step 2: Now, take logs of both sides of the equation and introduce the new base, c.
          Then, $\log_c M = \log_c b^y$
Step 3: Solve for y:
              $\log_c M = y \log_c b$

$$\frac{\log_c M}{\log_c b} = y \qquad \text{(Solving for } y)$$

Therefore, $\log_b M = \dfrac{\log_c M}{\log_c b}$ <-----This is known as the change of base formula.

(Here we changed from base b to base c.)

Lesson 95 Exercises

A Solve for x:

1. $\log (x + 4) - \log x = 1$;   **2.** $\log_2 x + \log_2 (x - 7) = 3$;   **3.** $\log (x + 26) = \log (x - 2) + \log 5$

Answers: 1. $\{\frac{4}{9}\}$;   **2.** $\{8\}$;   **3.** 9

B Solve for x: **1.** $7^x = 3$;   **2.** $2^x = 7$;   **3.** $5^{x-2} = 16$;   **4.** $\log x = 4$;

5. $\log (3x + 4) = 2$;   **6.** $\log_2 (x - 7) + \log_2 x = 3$;   **7.** $\log_3 (x - 2) = 4$;

8. Solve for t: $4^{2t} = \dfrac{9}{4}$

Answers: **1.** 0.5645 ;   **2.** 2.8076 ;   **3.** 3.7226 ;   **4.** 10,000 ;   **5.** 32 ; **6.** 8 ; **7.** 83 ;   **8.** 0.2925

CHAPTER 40
Basic Review For Geometry

Lesson 96: Points, Lines, Angles, Congruency, Complimentary and Supplementary Angles; Parallel and Perpendicular Lines

Lesson 97: Statements in Geometry, Axioms, Postulates, Parallel Lines and Transversal; Corresponding Angles, Triangles, Sum of the Measures of the angles of a triangle

Lesson 96
Points, Lines, Angles, Congruency, Complimentary and Supplementary Angles; Parallel and Perpendicular Lines

The undefined terms in geometry are a point, a line, and a plane.
By using these terms we can define other geometric terms.

Point: We use a point to indicate a position. On a piece of paper, we represent a point by a dot. A point has no size. (The "thinner" the dot, the better it represents a point; the tip of a well-sharpened pencil will be an acceptable size). We name a point by a capital letter written near the point (See the points P, A and B below).

P•   •   B
         A   •

Figure 1

Line: We may consider a line as an infinite set of points. A line may be a straight line or a curved line. A straight line may be extended indefinitely in both directions.
We may name a line by using a single lower case letter or two capital letters representing two points on the line, with a double-headed arrow over the capital letters (see Figure 2 below)

The line passing through the points A and B may be named as $\overleftrightarrow{AB}$ or   l

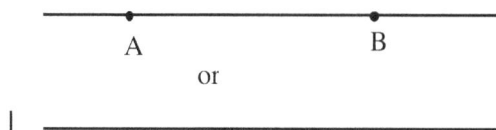

A   B

or

l

Figure 2

Line segment: A line segment or segment is a part of a line between two points on the line, including the two points (end points). We denote a line segment by placing a bar over the two capital letters which represent the two end points (see Figure 3 below).

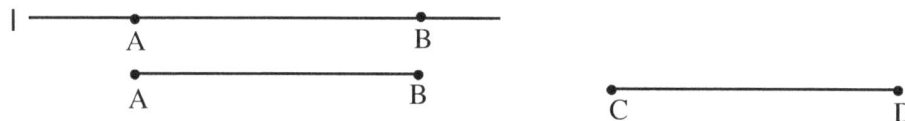

l
A   B

A   B   C   D

Line segment AB is represented by $\overline{AB}$ Line segment CD is represented by $\overline{CD}$

Figure 3

Ray: A ray is a part of a line to one side of a point on the line. This point is the endpoint of the ray. The ray begins at the end point and extends indefinitely in one direction. The arrow drawn begins at the end point . We name a ray by using two capital letters: the first letter represents the end point and the other letter represents any other point on the ray (Fig.4).

ray PQ is represented by $\overrightarrow{PQ}$ ray AB is represented by $\overrightarrow{AB}$

Figure 4

Measurement of a line segment : The **length** of a line segment is the distance between its end points.

The length of line segment $\overline{AB}$ is denoted by AB (i.e. without a bar over AB)

Figure 5

Angle: An angle (symbol: $\angle$) is the union of two rays having the same end point, the endpoint being called the vertex of the angle. An angle is formed when two line segments (or lines) meet at a common point (Figure 6).

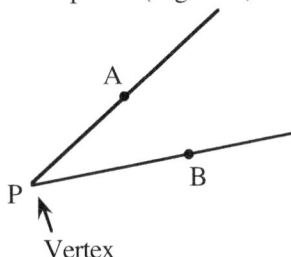

Figure 6

Naming of angles We may name an angle by using capital letters or lowercase letters as follows:

  1. If the angle is the only angle at the vertex, then we can use either a single capital letter or three capital letters. However, if there are two or more angles at the vertex, then we must use three capital letters. In using three capital letters, the middle letter is the letter at the vertex.

  2. We may use a lower case letter , a Greek letter, or a number .The letter or the number is placed inside the angle (as shown in Figure 7).

Examples: $\angle$ 2 and $\angle$ DAB denote the same angle
        $\angle$ 1 and $\angle$ CAD denote the same angle.

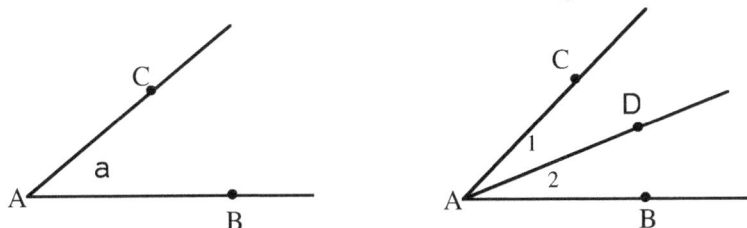

Figure 7

Congruent Figures: Congruent figures have the same shape and size.
Congruent figures can be made to coincide.
The symbol for "congruent" or "congruent to" is $\cong$.

Congruency of Line Segments: As shown in Figure 8, if we can place $\overline{AB}$ and $\overline{CD}$ so that
they coincide, then, we say that $\overline{AB}$ and $\overline{CD}$ are congruent,
and we write $\overline{AB} \cong \overline{CD}$ (read: line segment $\overline{AB}$ is
congruent to line segment $\overline{CD}$)

Congruent line segments have the same length.

$$\overline{AB} \cong \overline{CD} \text{ if and only if } AB = CD$$

(that is, $\overline{AB}$ is congruent to $\overline{CD}$ if the length of $\overline{AB}$ is equal to the length of $\overline{CD}$, and conversely,
the length of $\overline{AB}$ is equal to the length of $\overline{CD}$, if $\overline{AB}$ is congruent to $\overline{CD}$).

Indication of congruency of line segments: We may use the same number of slashes or
strokes (as shown in Figure 8) to indicate that two line segments are congruent.

Example (a) $\overline{AB} \cong \overline{CD}$ (b) $\overline{EF} \cong \overline{GH}$

Figure 8

Congruency of angles: As shown in Figure 9, if $\angle P$ can be placed on $\angle Q$ so that the angles
coincide, then we say that $\angle P$ is congruent to $\angle Q$ and we write

$$\angle P \cong \angle Q$$

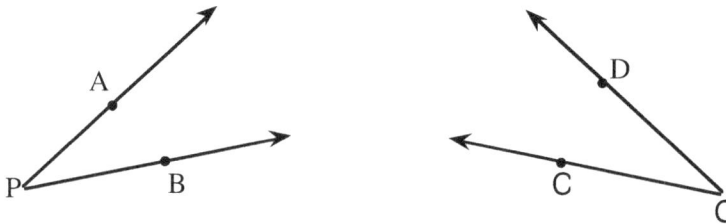

Figure 9

Note that congruent angles have equal measures.

Indication of congruency of angles: We can use arcs or a combination of arcs and slashes to indicate congruency (as shown in Figure 10)

Examples $\angle\, 1 \cong \angle\, 2$ (the angle marked 1 is congruent to the angle marked 2)
$\angle\, 3 \cong \angle\, 4$ (the angle marked 3 is congruent to the angle marked 4)
$\angle\, 5 \cong \angle\, 6$ (the angle marked 5 is congruent to angle marked 6)

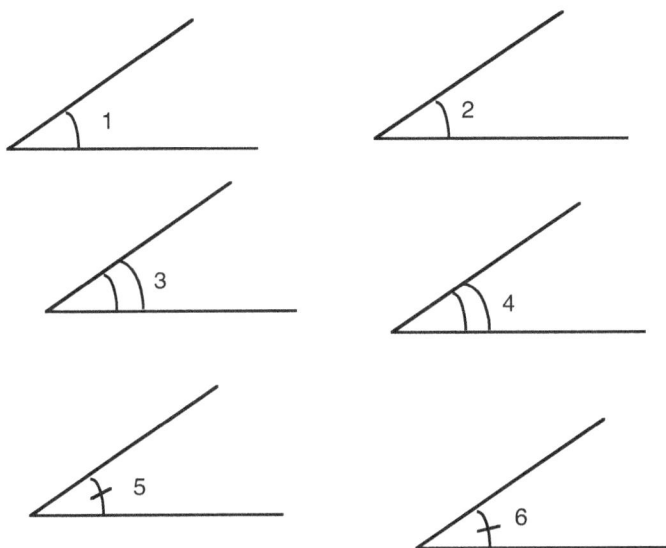

Figure 10

Units of Measuring angles : The main unit of measuring angles in geometry is the degree. There are smaller units of the degree, namely the minute, and the second.

The degree and the smaller units are as follows: 1 degree = 60 minutes
1 minute = 60 seconds

Classification of angles by the measures of the angles

Straight angle: An angle whose measure is 180°

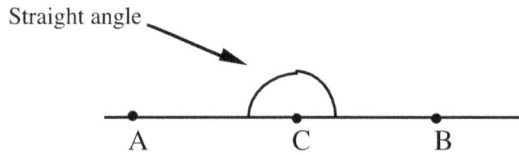

Right angle: A right angle is an angle whose measure is 90°.
In Figure 11, below, ∠ 5 is a right angle.

Figure 11

Acute angle: An angle whose measure is between 0° and 90°.
In Figure 12, below, ∠ 6 is an acute angle.

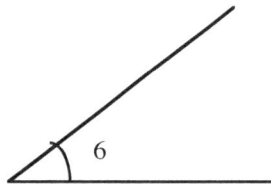

Figure 12

Obtuse angle: An angle whose measure is between 90° and 180°.
In Figure 13, below, ∠ 8 is an obtuse angle.

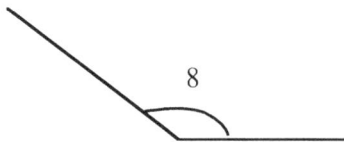

Figure 13

Reflex angle: An angle whose measure is between 180° and 360°.
In Figure 14, below, ∠ 7 is reflex angle.

Figure 14

Pairs of Angles

Complementary angles : If the sum of the measures of two angles is 90°, then the two angles are complementary.
The complement of a 40° angle is a 50° angle
In Figure 15, below, $\angle$ 1 and $\angle$ 2 are complementary angles.(i.e., m $\angle$ 1 + m $\angle$ 2 = 90°)

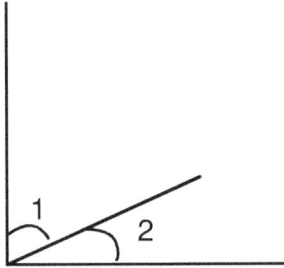

Figure 15

Supplementary angles: If the sum of the measures of two angles is 180°, then the two angles are supplementary.
The supplement of a 70° angle is a 110° angle.

In Figure 16, $\angle$ 3 and $\angle$ 4 are supplementary angles. (i.e., m $\angle$ 3 + m $\angle$ 4 = 180°)

Figure 16

Example 1 In the figure below , find x

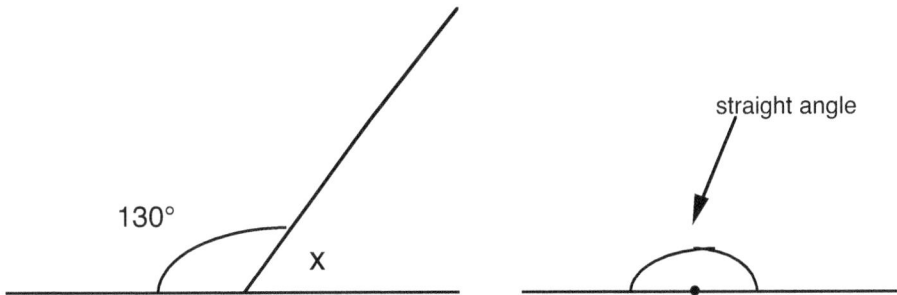

130°   straight angle

X

Solution: $x + 130° = 180°$ Reason : A straight angle has a measure of 180°
$x = 50°$

Vertical angles: As shown in Fig.17, below, when two lines intersect (cross each other) a pair of opposite angles such as ∠ 1 and ∠ 3 are called vertical angles. Similarly, ∠ 2 and ∠ 4 are vertical angles.

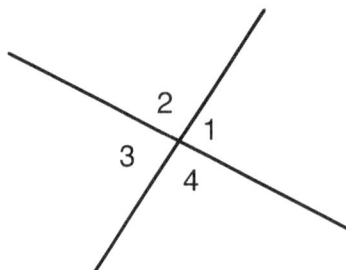

Figure 17

Theorem: When two lines intersect, the vertical angles are congruent.
In Figure 17, ∠ 1 and ∠ 3 are congruent.
∠ 2 and ∠ 4 are congruent

Adjacent angles: Two angles which are in the same plane and which have a common vertex and a common side . In Figure 17, ∠ 1 and ∠ 2 are adjacent angles .
∠ 2 and ∠ 3 are adjacent angles; ∠ 3 and ∠ 4 are adjacent angles.

Perpendicular lines (symbol for " perpendicular to": ⊥): Perpendicular lines are two lines which meet at right angles.
In Figure 18,. line l_1 is perpendicular to line l_2 and we write
$$l_1 \perp l_2$$

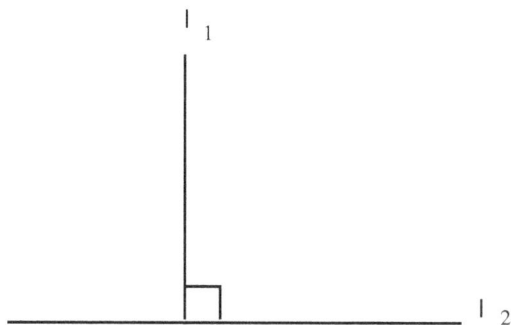

Figure 18

Example 2   In the figure below, find x

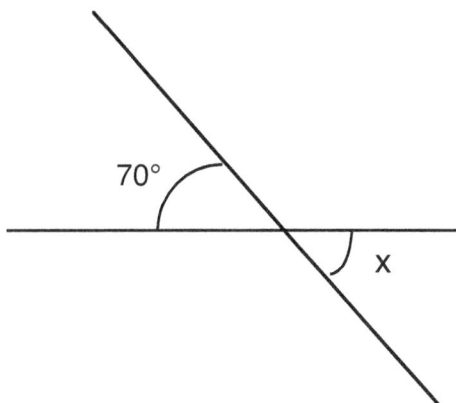

Solution: $x = 70^\circ$ **Reason:** Vertically opposite angles are congruent, and therefore have equal measures

Lesson 96 Exercises

1. In figure (a) below , find x

(a)

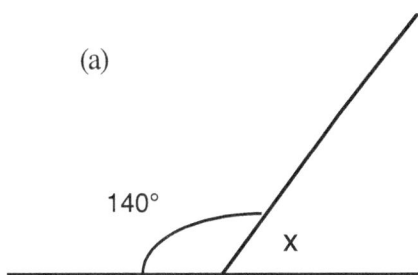

2. In figure (b) below , find x

(b)

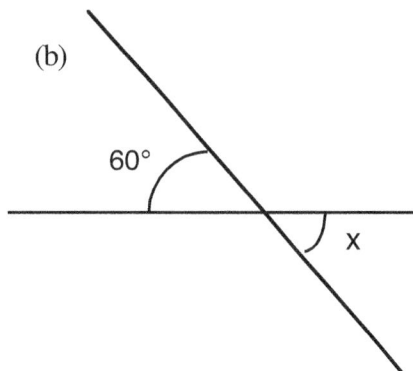

Answers: See above

Lesson 97

Statements in Geometry, Axioms, Parallel Lines and Transversals; Corresponding Angles, Triangles, Angles of a Triangle

Axioms and postulates (assumptions)

Axioms and postulates are statements which we accept as true, without any proof, in order to deduce other statements.

Axioms are applicable to mathematics in general.
 Example: The statement "If equals are added to equals, the sums are equal " is applicable to both algebra and geometry.

Postulates (in mathematics) are applicable to specific branches of mathematics.
Example: The statement "One and only one straight line can be drawn between any two points" is applicable only in geometry.

Theorems : A theorem is a statement that is proved by deduction.

Some Geometric Postulates

In geometric proofs and in geometric calculations, the following are some of the things that we can do if the need arises.
 (Try to avoid applying more than one postulate in one step, except when one of the postulates is the extension of a straight line; i.e., avoid applying two or more postulates in one step simultaneously, except when a line extension is one of the postulates.)

1. A geometric figure may be moved without changing its size or shape. (motion postulate)

2.   Between any two points, one and only one straight line can be drawn.
   (Two points determine a straight line)

3. A straight line segment may be extended indefinitely in both directions.

4. Two straight lines may intersect in one and only one point.

5. The shortest distance between two points is the straight line segment connecting these points.

6. A straight line segment has one and only one midpoint.

7.   An angle has one and only one bisector. (We may bisect an angle if the need arises)

8. At a given point on a line, one and only one perpendicular can be drawn to the line.

9. From a given point not on a given line, one and only one perpendicular can be drawn to the line.

10. One and only one circle can be drawn with a given point as center and a given line segment as a radius.

11. Through a given point not on a line, one and only one parallel line can be drawn parallel to the given line.

Transversal

If two lines are cut by a third line, the third line is called a transversal. Angles are formed at where the transversal intersects these lines. In Figure 19, angles 1 and 5 are called corresponding angles; angles 4 and 8 are corresponding angles; angles 2 and 6 are corresponding angles; angles 1, 2, 7, and 8 are called exterior angles; angles 3,4,5, and 6 are called interior angles.

Corresponding Angles

In Figure 19, below:

1. ∠ 1 and ∠ 5 are corresponding angles.

    (these angles are **above** lines ℓ_1 and ℓ_2 **and to (or on) the right** of the transversal).

2. ∠ 4 and ∠ 8 are corresponding angles**.**

    (these angles are **below** lines ℓ_1 and ℓ_2 **and to (or on) the right** of the transversal).

3. ∠ 2 and ∠ 6 are corresponding angles.

    (these angles are **above** ℓ_1 and ℓ_2 **and to (or on) the left** of the transversal).

4. ∠ 3 and ∠ 7 are corresponding angles.

    (These angles are **below** lines ℓ_1 and ℓ_2 **and to the left** of the transversal).

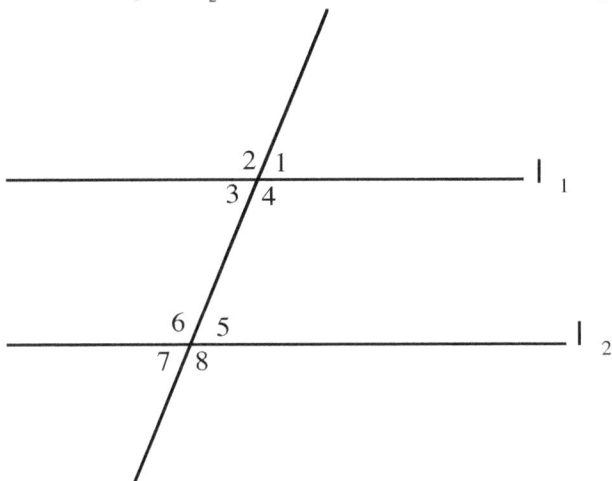

Figure 19

Even though the words " above", "below", "to (on) the right", "to (on) the left" were used to identify corresponding angles, if a given figure does not have the orientation of Figure 19, we can rotate the page to obtain the above orientation and then apply the above descriptions to identify the corresponding angles accordingly. Note that correspondence, as described above, requires that for any two angles to correspond, each angle must satisfy the same two conditions simultaneously; for example, the simultaneous conditions for ∠ 1 and ∠ 5 are "above the two lines" and to the right
(or on the right) of the transversal."

Alternate interior angles (Z- or N-angles)

∠ 4 and ∠ 6 are alternate interior angles (the lines ℓ_1 and ℓ_2 together with the transversal which form the angles are in the shape of the letter Z or N)

∠ 3 and ∠ 5 are alternate interior angles.

Lines , and l_3 h

Note above that, even though there are corresponding angles and alternate interior angles, we **cannot** say that
 any of these angles are congruent, since we do not know if the two lines are parallel. In the next section ,we
will be able say which angles are congruent, since the two lines are parallel.

Parallel Lines: Two lines l_1 and l_2 in the same plane are parallel if they never meet no

matter how far they may be extended. If line l_1 is parallel to l_2 we write $l_1 \parallel l_2$ (see Fig. 20)

Parallel lines, the transversal, and the pairs of angles formed

 In Figure 20, below, line l_1 is parallel to line l_2 .

 Theorem: If two parallel lines are cut by a transversal , then:

    1. the corresponding angles (F-angles) are congruent (see Figure 20).

         $\angle 1 \cong \angle 5$

         $\angle 4 \cong \angle 8$

         $\angle 2 \cong \angle 6$

         $\angle 3 \cong \angle 7$

    2. the alternate interior angles (Z- or N-angles) are congruent (see Figure 20).

         $\angle 4 \cong \angle 6$

         $\angle 3 \cong \angle 5$

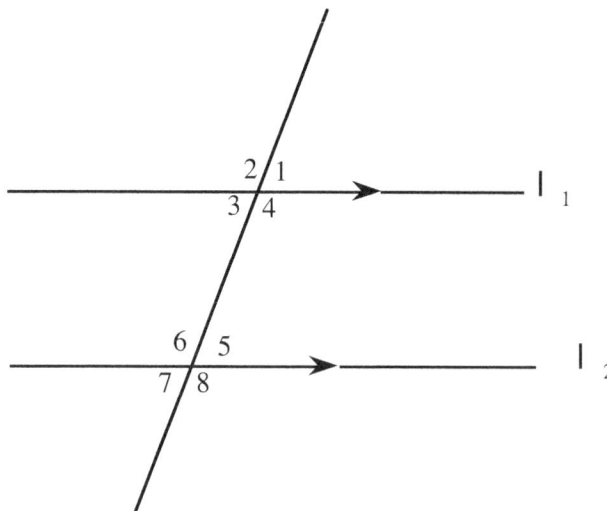

Figure 20

Note also that the vertical angles are congruent (whether or not there are parallel lines).

Theorem: If two lines cut by a transversal have congruent alternate interior angles, then the lines
         are parallel. (We can use this theorem to prove that two lines are parallel).

Theorem: Any pair of interior angles on the same side of the transversal are supplementary.

        m $\angle 4 +$ m$\angle 5 = 180°$ (Figure 20)

        m$\angle 3$ + m $\angle 6 = 180°$

Theorem: The exterior angles on the same side of the transversal are supplementary.

       m$\angle 1 +$ m $\angle 8 = 180°$ (Figure 20)

       m $\angle 2 +$ m $\angle 7 = 180°$

Example 3   In the figure below, find x given that line ℓ_1 is parallel to line ℓ_2

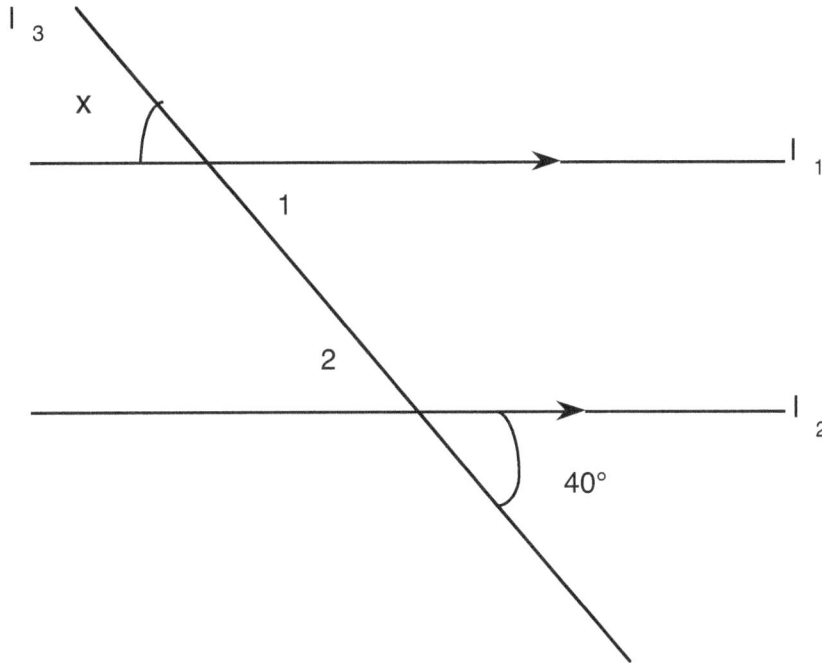

Solution:

$m \angle 2 = 40°$ (Vertical   angles are congruent)
$m \angle 2 = x$ (corresponding angles are congruent;

parallel lines ℓ_1, ℓ_2,   cut by transversal ℓ_3)

$x = 40°$

Triangles

 A triangle is a closed figure bounded by three straight line segments.
Triangles may be classified by angles: By this classification, a triangle may be acute, obtuse, or right.

Acute triangle: A triangle in which all the angles are acute.

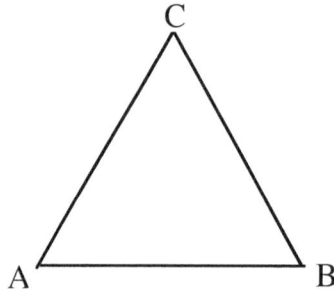

Obtuse triangle: A triangle in which there is an obtuse angle. **Note** that a triangle cannot have
more than one obtuse angle.

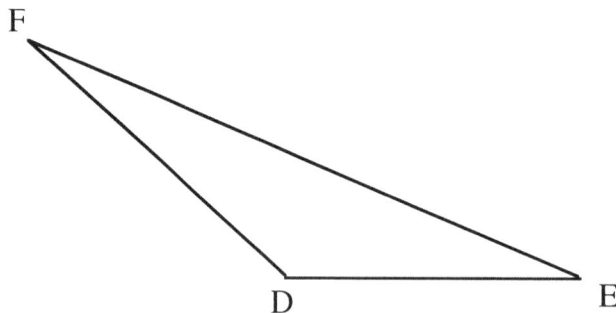

Right triangle (or right-angled triangle): A triangle which has one right angle (90°). **Note** that a
triangle cannot have more than one right angle.

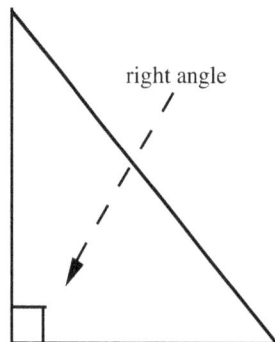

right angle

We can also classify triangles by sides. By this classification, a triangle may be equilateral,
 isosceles, or scalene.

Equilateral triangle: A triangle in which all three sides are congruent.
(The lengths of all three sides are equal)

All the interior angles of an equilateral triangle are congruent.
(Each angle has a measure of 60°).

Isosceles triangle: A triangle in which two sides are congruent. In an isosceles triangle, the angles opposite to the congruent sides are congruent.

Useful relationships concerning the relative measures of the sides and angles of a triangle

1. In any triangle, the **sum** of the **lengths** of any **two sides** is larger than the length of the **third** side.

2. In any triangle, the **longest side** is opposite to the **largest angle**, and the **shortest side** is opposite to the **smallest angle**.

3. In an **acute triangle**, the square of the length of **longest side is smaller than** the sum of the squares of the lengths of the other two sides.

4. In an **obtuse triangle**, the **square** of the length of the **longest side is larger than** the sum of the **squares** of the lengths of the other two sides.

5. In a **right triangle**, the **square** of the length of the **longest side** (the hypotenuse) **is equal** to the sum of the **squares** of the lengths of the other two sides. (**The Pythagorean theorem**)

Theorem: The sum of the measures of the interior angles of a triangle is 180°.
In the figure below, $m \angle A + m \angle B + m \angle C = 180°$

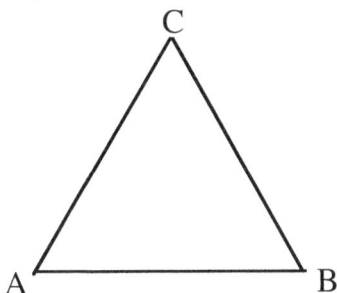

Theorem: If one side of a triangle is extended, the measure of the exterior angle so formed equals the sum of the measures of its remote interior angles.
In the figure below, $m \angle A + m \angle C = m \angle CBE$
($\angle A$ and $\angle C$ are the remote interior angles. $\angle CBE$ is the exterior angle)

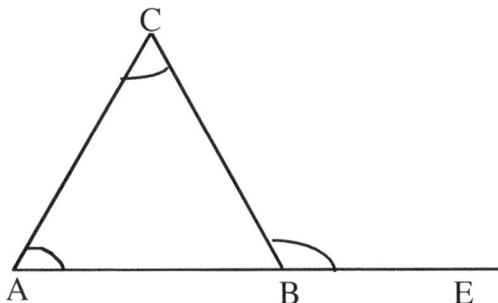

Example 1 In the figure below, find x.

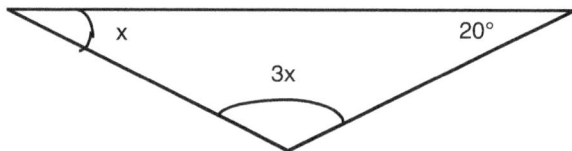

Solution: $x + 3x + 20° = 180°$ (The sum of the measures of the angles of a
$4x + 20° = 180°$ triangle is 180º)
$4x = 160°$
$x = 40º$

Example 2 In the right triangle shown below, find x

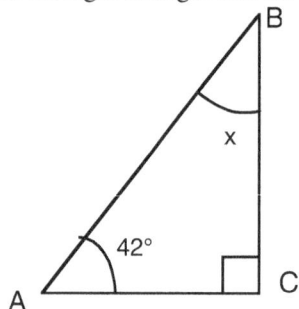

Solution: m $\angle$ C = 90º (A right angle has a measure of 90º)
m $\angle$ A + m $\angle$C + m $\angle$ B = 180º (The sum of the measures of the angles of a triangle is 180º)
$42°+ 90° + x = 180°$
$132º + x = 180°$
$x = 180° - 132°$
$x = 48º$

Example 3 Find x and y in the figure below.

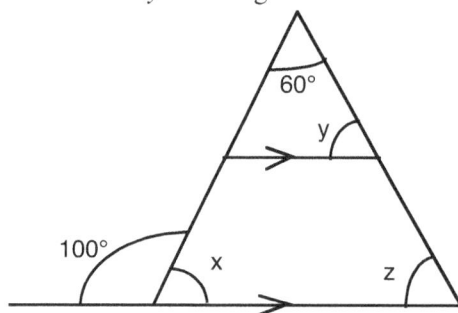

(a) $x + 100° = 180°$ (Supplementary angles)
$\underline{- 100° \quad - 100°}$
$x = 80°$

(b) Step 1: $x + 60° + z = 180°$
$80° + 60° + z = 180°$ (The sum of the measures of the angles of a triangle is 180˚)
$140° + z = 180°$
$\underline{- 140° \qquad - 140°}$
$z = 40°$

Step 2: $y = z = 40°$ (Corresponding angles are congruent.)
$y = 40$

Lesson 97 Exercises

1. Find x;   **2.** Find m $\angle$ A

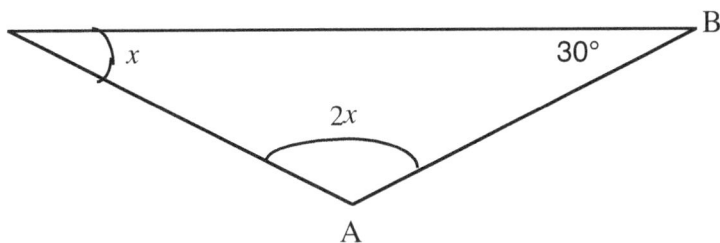

Answers: **1.** $x = 50$;   **2.** m $\angle$ A = 100

CHAPTER 41
Areas and Perimeters

Lesson **Area and Perimeter of Triangles; Rectangles, and Trapezoids**

Lesson **Area and Perimeter of a Circle; Arc Length and Sector of a Circle**
Area and Perimeter of Composite Figures

Use Chapter 22

The above was designed to avoid duplication and to reduce the volume of this book, and also, such a design is good for the environment.

CHAPTER 42
CONGRUENT TRIANGLES

Lesson 98: **Properties of Congruent Triangles**; **Theorems and Proofs**
Lesson 99: **Applications of Congruency Theorems; Isosceles Triangles**
Before proceeding, review page 448- page 461.

Lesson 98

Properties of Congruent Triangles; Theorems and Proofs

Congruent Figures: Congruent figures have the same shape and the same size.
Congruent figures can be made to coincide.
The symbol for "congruent" or "congruent to" is "≅ ".

Congruent Triangles: Congruent triangles have the same shape and the same size.

Identifying corresponding sides and angles of two congruent triangles

1.   Corresponding sides are opposite congruent angles.

How to identify corresponding sides in two congruent triangles say Δ #1 and Δ #2:

Step 1: Pick a side in Δ #1, note the angle this side is opposite to.

Step 2: Go to Δ #2 and locate the angle which is congruent to the angle noted in Δ #1. The side opposite to
this located angle in Δ #2 is the corresponding side. Note of course that if the sides in Δ #1 and Δ #2 are given
or known to have the same length, then these sides correspond and they are also congruent.

2.   Corresponding angles are opposite congruent sides.

How to identify corresponding angles in two congruent triangles say Δ #1 and Δ #2:

Step 1: Pick an angle in Δ #1, note the side this angle is opposite to.

Step 2: Go to Δ #2 and locate the side which is congruent to the side noted in Δ #1. The angle opposite to this
located side in Δ #2 is the corresponding angle. Note of course that if the angles in Δ 's #1 and #2 are
given or known have the same measure, then these angles correspond and they are also congruent.

Properties of congruent triangles

1. The corresponding angles are congruent (or the measures of corresponding angles are equal).

2. The corresponding sides are congruent (or the lengths of corresponding sides are equal).
Also, the ratio of each pair of corresponding sides equals 1.

Conditions or Criteria for Congruency

The conditions or criteria for proving that two triangles are congruent are embodied in the
theorems that follow.

Before proceeding to the theorems, review the axioms on pages 176 and 456.

Ordered naming of triangles in stating the congruency of triangles and identification of corresponding parts from the congruency statement

It is good communication to follow a certain order in stating that two triangles are congruent. The logical order is to match the letters of the vertices in stating that two triangles are congruent.

Example Stating that $\triangle ABC \cong \triangle DEF$ (read " triangle ABC is congruent to triangle DEF) implies that the vertex A (of $\triangle ABC$) corresponds to the vertex D (of $\triangle DEF$); the vertex B (of $\triangle ABC$) corresponds to the vertex E (of $\triangle DEF$); and the vertex C (of $\triangle ABC$) corresponds to the vertex F (of $\triangle DEF$).

By matching the vertices, the sides $\overline{AB}$ and $\overline{DE}$ naturally correspond, $\overline{BC}$ and $\overline{EF}$ correspond; and $\overline{AC}$ and $\overline{DF}$ correspond.

$$\triangle ABC \cong \triangle DEF$$

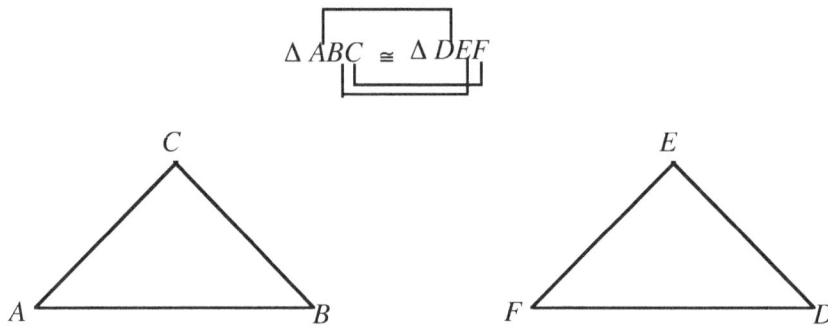

Figure 1

Sometimes, some authors may not follow the above-mentioned order strictly and merely state that some triangles are congruent, and in which case, additional information must be provided either in the form of further statements or in an accompanying diagram to indicate the corresponding parts.

Note for example that the triangle in Figure 2, below, may be named as either $\triangle ABC$, $\triangle BCA$, $\triangle CAB$, $\triangle ACB$, $\triangle CBA$ or $\triangle BAC$.

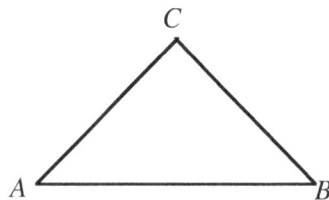

Figure 2

Theorem 1: If two sides and the included angle of one triangle are congruent to the corresponding two sides and the included angle of another triangle, the two triangles are congruent. Abbreviated **SAS** (Side-Angle-Side).

In the following example, we will apply this theorem to prove that two triangles given are congruent.

Example 1   In Figure 1 below, prove that $\triangle ABC \cong \triangle DEF$

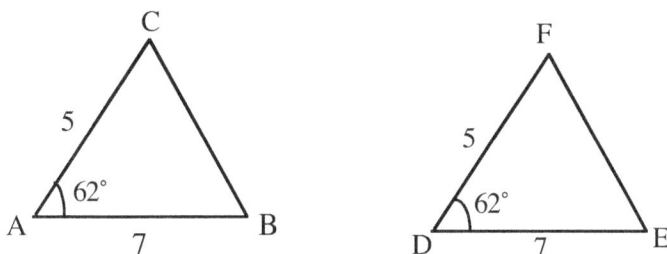

Figure 1

Given: **1.** In $\triangle$ ABC , AB = 7, AC $\cong$ 5 , m$\angle$ A $\cong$ 62°.
     2. In $\triangle$ DEF, DE =7, DF = 5, m$\angle$ D = 62°.

To prove: That $\triangle$ ABC $\cong$ $\triangle$ DEF

Proof: Statements   Reasons
     1. AB = 7   **1.** Given
     2. DE= 7   **2.** Given
     3. AB=DE   **3.** If two quantities are equal to the same quantity, they are equal to each other

     4. $\overline{AB} \cong \overline{DE}$ **4.** If two line segments have equal length, they are congruent .
     5. AC = 5   **5** Given
     6. DF= 5   **6.** Given
     7. AC = DF   **7.** If two quantities are equal to the same quantity, they are equal to each other.

     8. $\overline{AC} \cong \overline{DF}$ **8.** If two line segments have equal length, they are congruent.
     9.   m$\angle$ A = 62°   **9.** Given
     10. m$\angle$ D = 62°   **10.** Given
     11. m$\angle$ A = m $\angle$ D   **11.**   If two quantities are equal to the same quantity, they are equal to each other.
     12. $\angle$ A $\cong$ $\angle$ D   **13.** If two angles have the same measure, they are congruent.
     13. $\angle$ A and $\angle$ D are included angles.
     14. $\triangle$ **ABC** $\cong$ $\triangle$ **DEF**   **14.** SAS
               *Q.E.D

Some of the steps in the above proof could be skipped or combined with other steps. The thoroughness of the proof was to show the student how not to take too much for granted and also to show the need to give reasons for statements made in a geometric proof.

   * **Q.E.D** (or QED) is a Latin abbreviation for **quod erat demonstrandum,** which literally means "which had to be shown or proved". Written at the end of a proof, it signifies the **end of the proof**. Some authors signify the end of a proof by writing "and the proof is complete", "The proof is complete", or "Therefore, the proof is complete".

*Theorem 2: If two angles and the included side of one triangle are congruent to the corresponding two angles and the included side of another triangle, then the two triangles are congruent. Abbreviated **ASA** (Angle-Side-Angle).

Another form of Theorem 2 : If a side and both angles adjacent to that side of one triangle are congruent to the corresponding side and the angles adjacent to that side of another triangle, then the two triangles are congruent.

In the following example, we will apply this theorem to prove that the given triangles are congruent.

Example 2   In Figure 2 below, prove that $\triangle ABC \cong \triangle DEF$

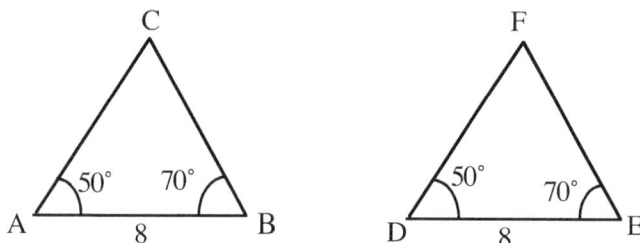

Figure 2

Given: **1.** In $\triangle ABC$, $AB = 8$, $m\angle A = 50°$, $m\angle B = 70°$.
　　　2. In $\triangle DEF$, $DE = 8$, $m\angle D = 50°$, $m\angle E = 70°$.

To prove: That $\triangle ABC \cong DEF$

Proof:

Statements	Reasons
1. AB = 8	**1.** Given
2. DE = 8	**2.** Given
3. AB = DE	**3.** If two quantities are equal to the same quantity, they are equal to each other
4. $\overline{AB} \cong \overline{DE}$	**4.** If two line segments have equal length, they are congruent
5. $\overline{AB}$ and $\overline{DE}$ are included sides	
6. $m\angle A = 50°$	**6.** Given
7. $m\angle D = 50°$	**7.** Given
8. $m\angle A = m\angle D$	**8.** If two quantities are equal to the same quantity, they are equal to each other.
9. $\angle A \cong \angle D$	**9.** If two angles have the same measure, they are congruent.
10. $m\angle B = 70°$	**10.** Given
11. $m\angle E = 70°$	**11.** Given
12. $m\angle B = m\angle E$	**12.** If two quantities are equal to the same quantity, they are equal to each other.
13. $\angle B \cong \angle E$	**13.** If two angles have the same measure, they are congruent.
14. $\triangle ABC \cong \triangle DEF$	**14.** ASA

Q.E.D.

*Note that another theorem abbreviated **AAS** or **SAA** is sometimes used (as a shortcut in proving two triangles congruent), since if two pairs of corresponding angles are congruent, the third pair of angles are also congruent, recalling that the sum of the measures of the angles of a triangle is 180°.
　In fact, we can easily prove **AAS or SAA** using **ASA.**

Theorem 3: If the three sides of one triangle are congruent to the corresponding three sides of 469 another triangle, then the two triangles are congruent. Abbreviated **SSS** (Side-Side-Side).

Example 3   In Figure 3 below, prove that △ ABC ≅ △ DEF

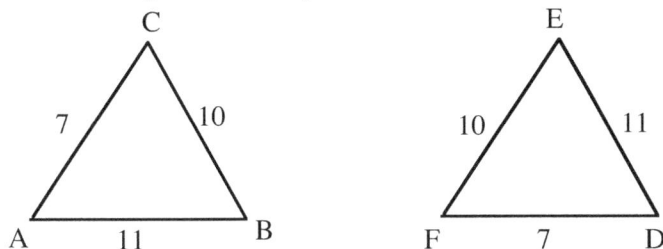

Figure 3

Given: **1**. In △ ABC, AB = 11, BC = 10, CA = 7.
       2. In △ DEF, DE = 11, EF = 10, FD = 7.

To prove: That △ ABC ≅   △ DEF

Proof: Statements   Reasons
       1. AB = 11   **1.** Given
       2. DE = 11   **2.** Given
       3. AB = DE   **3.** If two quantities are equal to the same quantity, they are equal to each other.

       4. $\overline{AB}$ ≅ $\overline{DE}$ **4.** If two line segments have equal length, they are congruent .
       5. BC = 10   **5.** Given
       6. EF = 10   **6.** Given
       7. BC = EF   **7.** If two quantities are equal to the same quantity, they are equal to each other.

       8. $\overline{BC}$ ≅ $\overline{EF}$ **8.** If two line segments have equal length, they are congruent.
       9. CA = 7   **9** Given
       10. FD = 7   **10.** Given
       11. CA = FD   **11.** If two quantities are equal to the same quantity, they are equal to each other.

       12. $\overline{CA}$ ≅ $\overline{FD}$ **12.** If two line segments have equal length, they are congruent.
       13. △ ABC ≅ △ DEF   **13.** SSS
                       Q.E.D.

Lesson 98 Exercises

Determine which of the following triangles are congruent and prove why they are congruent.

Fig. 1

Fig. 2

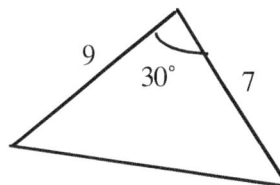

Fig.3

Answer: Fig1 and Fig 3 are congruent.

Lesson 99

More Applications of Congruency Theorems; Isosceles Triangles; Mid-Point Theorem

In proving that two triangles are congruent, it will be sufficient for us to show that any of the congruency theorems is true.

Example 1   In the figure below: (a)   Prove that $\triangle BAC \cong \triangle DEC$
                               (b)   Find x and y

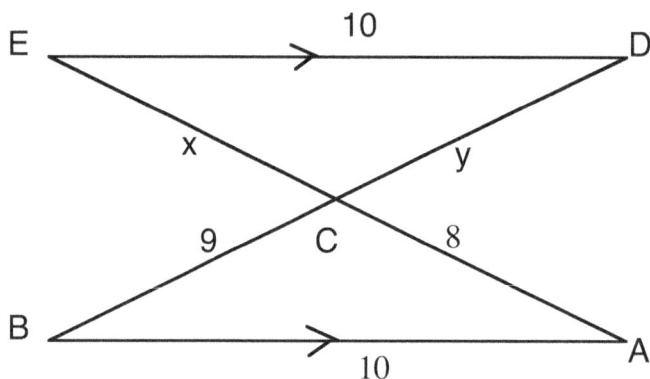

Solution (a): Proving that $\triangle BAC \cong \triangle DEC$

Given: In $\triangle$'s BAC and DEC , ED = 10, BA = 10; $\overline{ED}$ is parallel to $\overline{BA}$.
To prove: That $\triangle BAC \cong \triangle DEC$

Proof: Statements   Reasons
       1. ED = 10   **1.** Given
       2. BA = 10   **2.** Given
       3. ED = BA   **3.** If two quantities are equal to the same quantity, they are equal to each other.
       4. $\overline{ED} \cong \overline{BA}$ **4.** If two line segments have equal length, they are congruent.
       5. $\angle E \cong \angle A$ **5.** Alt. int. angles, parallel lines $\overline{ED}$, $\overline{BA}$ cut by transversal $\overline{EA}$
       6. $\angle D \cong \angle B$ **6.** Alt. int. angles, parallel lines $\overline{ED}$, $\overline{BA}$ cut by transversal $\overline{BD}$
       7. $\triangle BAC \cong \triangle DEC$ **7.** ASA
                           Q.E.D

Solution (b): Finding x and y:

$\overline{EC} \cong \overline{AC}$ (Corresponding sides of congruent $\triangle$'s are congruent)

EC = AC = 8   ($\overline{EC}$ is opposite $\angle D$; $\overline{AC}$ is opposite $\angle B$; and $\angle D \cong \angle B$)
$x = 8$

  Similarly, $y = DC = BC = 9$

  $\therefore x = 8, y = 9$

Example 2   Given the parallelogram in the figure below, prove that

(a) $\overline{AE} \cong \overline{EC}$;   (b) $\overline{BE} \cong \overline{ED}$

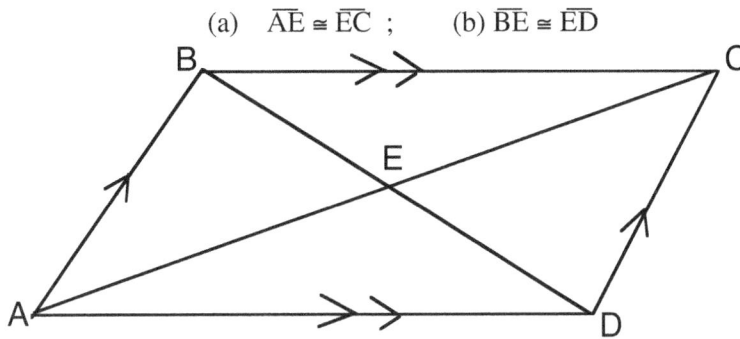

Given: ABCD is a parallelogram.

To prove: (a) That $\overline{AE} \cong \overline{EC}$; (b)   That $\overline{BE} \cong \overline{ED}$

Proof:

Statements	Reasons
1. ABCD is a parallelogram	1.Given
2. In Δ's BCE and DAE,	
$\overline{BC} \cong \overline{DA}$	**2. Opposite** sides of a parallelogram are congruent.
3. ∠ BCE ≅ ∠ DAE	**3. Alternate** interior angles are congruent.
	‖ lines $\overline{BC}, \overline{AD}$, cut by transversal $\overline{AC}$.
4. ∠ EBC ≅ ∠ EDA	**4.** Alternate interior angles are congruent.
	‖ lines $\overline{BC}, \overline{AD}$ cut by transversal $\overline{BD}$.
5. Δ BCE ≅ Δ DAE	**5.** ASA.
(a) **6.** ∴ $\overline{AE} \cong \overline{EC}$	**6.** Corresponding sides of congruent Δ's are congruent,
	noting that $\overline{AE}$ is the same as $\overline{EA}$
(b) Similarly, $\overline{BE} \cong \overline{ED}$	Noting that $\overline{ED}$ is the same as $\overline{DE}$
	Q.E.D

Alternative proof:

Statements	Reasons
1. ABCD is a parallelogram	1.Given
2. In Δ's BEA and DEC	
$\overline{AB} \cong \overline{CD}$	2. Opposite sides of a parallelogram are congruent.
3. ∠ ABE ≅ ∠ CDE	3. Alternate interior angles are congruent.
4. ∠ EAB ≅ ∠ ECD	4. Alternate interior angles are congruent.
5 Δ BEA ≅ Δ DEC	5. ASA
(a) **6.** ∴ $\overline{AE} \cong \overline{EC}$	6. Corresponding sides of congruent Δ's are congruent,
	noting that $\overline{AE}$ is the same as $\overline{EA}$
(b) Similarly, $\overline{BE} \cong \overline{ED}$	Noting that $\overline{ED}$ is the same as $\overline{DE}$
	Q.E.D

Isosceles Triangles

Definition: An isosceles triangle is a triangle in which two sides are congruent. In an isosceles triangle, the angles opposite the congruent sides are congruent.

Theorem 1: If two sides of a triangle are congruent, the angles opposite to these sides are congruent. (Another form : The base angles of an isosceles triangle are congruent.)

If in $\triangle$ ABC (Figure 1 below), $\overline{AB} \cong \overline{CB}$, then $\angle A \cong \angle C$ (The base is $\overline{AC}$)

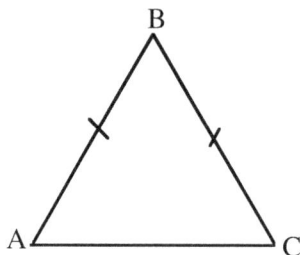

Figure 1

Theorem 2: The bisector of the vertex angle of an isosceles triangle bisects the base.

If in $\triangle$ ABC (Figure 2 below), $\overline{AB} \cong \overline{CB}$, and $\overline{BD}$ bisects $\angle$ B, then $\overline{AD} \cong \overline{CD}$. (The base is $\overline{AC}$)

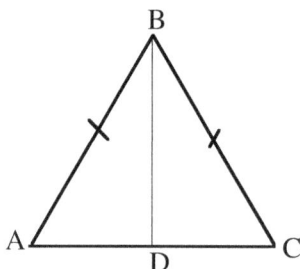

Figure 2

Theorem 3: The bisector of the vertex angle of an isosceles triangle is perpendicular to the base.

If in $\triangle$ ABC (Figure 3 below), $\overline{AB} \cong \overline{CB}$, and $\overline{BD}$ bisects $\angle$ B, then $\overline{BD} \perp \overline{AC}$. (The base is $\overline{AC}$)

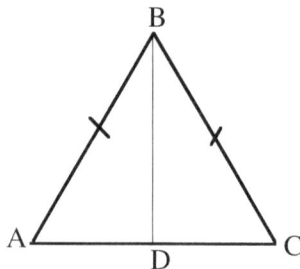

Figure 3

In trigonometry, we will apply a combination of Theorems 2 and 3, above, to find the dimensions of the 30°-60°-90° triangle.(See page 492.)
Note: An equilateral triangle is also isosceles.

Mid-Point Theorem

If a line is drawn from the midpoint of one side of a triangle and the line is parallel to a second side, then the line bisects the third side.

Given: $\triangle ABE$ in which $\overline{BC} \cong \overline{AC}$, $\overline{CD}$ is parallel to $\overline{BE}$.

Required: To prove that $\overline{AD} \cong \overline{ED}$

Construction: Through E, draw $\overline{EF}$ parallel to BA Reason : a line can be drawn parallel to another line

Plan: We first prove that $\overline{AC} \cong \overline{EG}$, followed by proving that $\triangle's ACD$ and EGD are congruent

Proof:

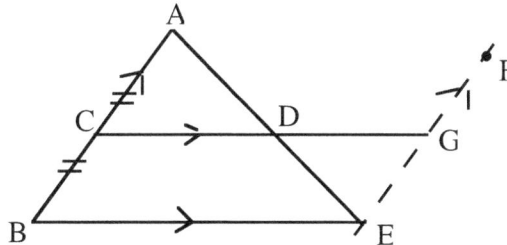

Statements	Reason
1. $\overline{BC} \parallel \overline{EG}$	1. by construction. $\overline{EF}$ is drawn parallel to BA
2. $\overline{CG} \parallel \overline{BE}$	2. Given
3. $\therefore$ quad $BEGC$ is a parallelogram	3. Definition If both pairs of the opposite sides of a quadrilateral are parallel, then the quadrilateral is a parallelogram
4. $\overline{BC} \cong \overline{EG}$	4. Opposite sides of a parallelogram are congruent
5. $\overline{BC} \cong \overline{AC}$	5. Given
6. $\overline{AC} \cong \overline{EG}$	6. Transitive property
7. In $\triangle's ACD$ and EGD, $\angle CAD \cong \angle GED$	7. Alt. int. angles; parallel lines $\overline{BA}$ and $\overline{EF}$ cut by transversal $\overline{AE}$
8. $\angle ACD \cong \angle EGD$	8. Alt. int. angles; parallel lines $\overline{BA}$ and $\overline{EF}$ cut by $\overline{CG}$
9. $\overline{AC} \cong \overline{EG}$	9. Proved in (6) above
10. $\triangle ACD \cong \triangle EGD$	10. ASA (Two angles and the included side)
11. $\overline{AD} \cong \overline{ED}$	11. Sides opposite congruent $\angle's$ are congruent

Q.E.D

Lesson 99 Exercises

In the figure below: (a) Prove that $\triangle BAC \cong \triangle DEC$; (b) Find x and y

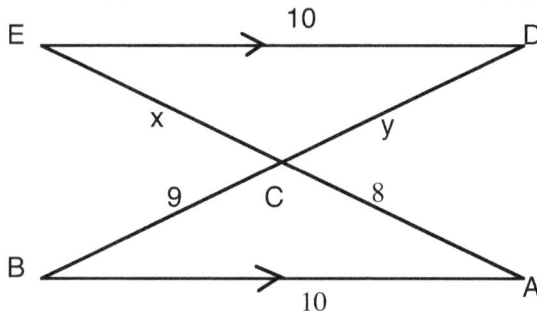

Answers: $x = 8$, $y = 9$

CHAPTER 43
Similar Triangles

Lesson 100: Properties of Similar Triangles; Theorems and Proofs
Lesson 101: Comparison of Congruency and Similarity of Triangles and Applications of Similarity Theorems

Lesson 100
Properties of Similar Triangles; Theorems and Proofs

Preliminaries

Ratio: The ratio of two quantities a and b (written a:b) is the fraction $\frac{a}{b}$ (i.e., is the first quantity divided by the second quantity). The quantities a and b are called the terms of the ratio.

But the ratio b: a is $\frac{b}{a}$

The ratio 3:4 is $\frac{3}{4}$; but the ratio 4:3 is $\frac{4}{3}$

Note above that order is important.

Proportion : A proportion is a statement that two ratios are equal.

The proportion a is to b as c is to d (symbolically a:b::c:d) is the equality

$$\frac{a}{b} = \frac{c}{d} \tag{1}$$

In the above proportion, a and d are called the extremes and b and c are called the means of the proportion.

Proportion Principles

1. We can invert both ratios of the proportion: From (1) above, we would obtain after the inversion:

$$\frac{b}{a} = \frac{d}{c}$$

2. We can interchange the extremes of a proportion: From (1) above we would obtain:

$$\frac{d}{b} = \frac{c}{a}$$

3. We can also interchange the means to obtain:

$$\frac{a}{c} = \frac{b}{d}$$

Proportion principles can be used to obtain desired ratios. For example, if we have $\frac{a}{b}$ as the ratio

of one side of a proportion, and we desire the ratio $\frac{b}{a}$, then proportion principle #1 above

(inversion of both sides of a proportion) can be used to obtain this ratio, $\frac{b}{a}$, from $\frac{a}{b}$.

Example: If $\frac{a}{b} = \frac{2}{3}$, find $\frac{b}{a}$.

Solution   By inverting both sides of the equation, we obtain $\frac{b}{a} = \frac{3}{2}$

Identifying corresponding sides and angles of two similar triangles

1. Corresponding sides are opposite congruent angles.
2. Congruent angles are opposite corresponding sides.

To identify corresponding sides in two similar triangles say, Δ #1 and Δ #2:

Case 1: Given or knowing the lengths of two sides and the included angle in each triangle (incl'd angles being congr.).
   The side with the smaller length in Δ #1 corresponds to the side with the smaller length in Δ #2.
   The side with the greater length in Δ #1 corresponds to the side with the greater length in Δ #2.
   The third side (length known or unknown) in Δ #1 corresponds to the third side of Δ #2.
Case 2: Given the lengths of all three sides of each triangle.
   The side with the smallest length in Δ #1 corresponds to the side with smallest length in Δ #2.
   The side with the greatest length Δ #1 corresponds to the side with the greatest length in Δ #2.
   The side with intermediate length in Δ #1 corresponds to the side with intermediate length in Δ #2,
Case 3: Given the measures of two pairs of congruent angles in the two triangles.
   Step 1: Pick a side in Δ #1, and note the angle this side is opposite to.
   Step 2: Go to Δ #2 and locate the angle which is congruent to the angle noted in Δ #1. The side opposite to this
        located angle in Δ #2 is the corresponding side.

To identify corresponding angles in two similar triangles say Δ's #1 and #2:

Given or knowing the corresponding sides
Step 1: Pick an angle in Δ #1, note the side this angle is opposite to.
Step 2: Go to Δ #2 and locate the side which corresponds to the side noted in Δ #1. The angle opposite to this
        located side in Δ #2 is the corresponding angle. Note of course that if the angles in Δ #1 and Δ #2 have
        the same measure, these angles correspond and they are also congruent.

Note: The above identification guidelines assume that the two triangles are given to be or have been proved to be
similar. Do not therefore apply them until you know that the two triangles are similar. However, a modified form
of this guide can be helpful in picking corresponding sides, even if the triangles are not similar.
        See also Examples 1 & 2, pages 478 & 480 respectively: the "Plan "parts.

Example Identify the corresponding lengths and corresponding angles in the Δ's given below

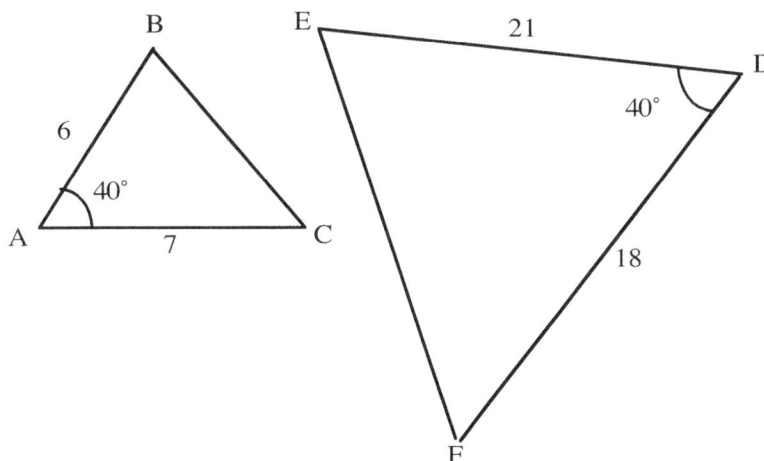

Solution: Smaller length of one triangle corresponds to the smaller length of the other triangle;
the larger length of one triangle corresponds to the larger length of the other triangle. Therefore
6 corresponds to 18; 7 corresponds to 21. ∠ D corresponds to ∠ A; ∠ C to ∠ E; and ∠ B to ∠ F.

Definition: Two triangles are similar ($\sim$) if and only if the corresponding angles are congruent and the ratio of corresponding sides is the same for each pair of corresponding sides.

Corresponding angles being congruent implies that corresponding angles have equal measures. The statement that "the ratio of corresponding sides is the same for each pair of corresponding sides" implies that the corresponding sides are in proportion.

Properties of Similar Triangles

1. Each pair of corresponding angles are congruent (i.e. have equal measure).
2. Corresponding sides are in proportion (that is, the ratio of the lengths of any pair of corresponding sides is equal to the ratio of the lengths of any other pair of corresponding sides).

Ordered naming of triangles in stating the similarity of triangles and identification of corresponding parts from the similarity statement

As it is in the case of the congruency of triangles, it is also good communication to follow a certain order in stating that two triangles are similar. The logical order is to match the letters of the vertices in stating that two triangles are similar.

Example   Stating that $\triangle ABC \sim \triangle DEF$ (read "triangle *ABC* is similar to triangle *DEF*"), implies that the vertex *A* (of $\triangle ABC$) corresponds to the vertex *D* (of $\triangle DEF$); the vertex *B* (of $\triangle ABC$) corresponds to the vertex *E* (of $\triangle DEF$); and the vertex *C* (of $\triangle ABC$) corresponds to the vertex *F* (of $\triangle DEF$).

By matching the vertices, the sides $\overline{AB}$ and $\overline{DE}$ naturally correspond, $\overline{BC}$ and $\overline{EF}$ correspond; and $\overline{AC}$ and $\overline{DF}$ correspond.

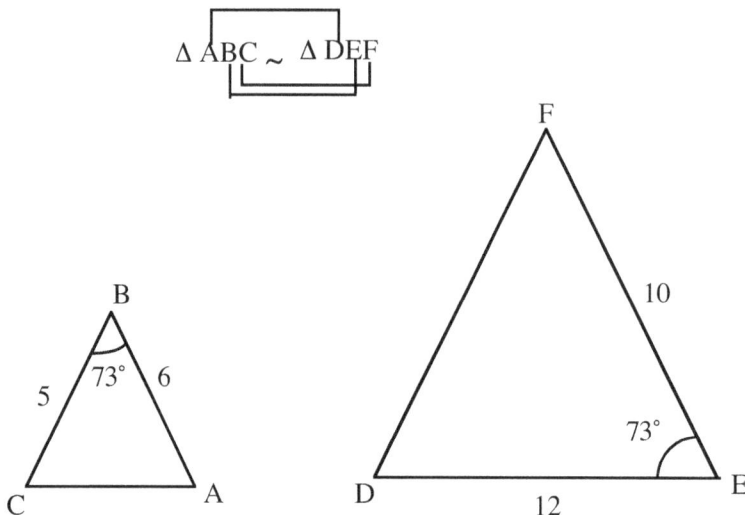

Theorems and Proofs

Conditions or criteria for proving similarity of triangles

* The conditions or criteria for proving that two triangles are similar are embodied in the following theorems:

Theorem 1: If two pairs of corresponding sides of two triangles are in proportion **and** included angles (the angles between the sides involved) are congruent, then the two triangles are similar. Abbreviated **SAS.**

In the figure below, Δ ABC ~ Δ DEF (read "triangle ABC is similar to triangle DEF"), since the conditions in Theorem 1 have been satisfied. In the next example we will prove the similarity of these two triangles.

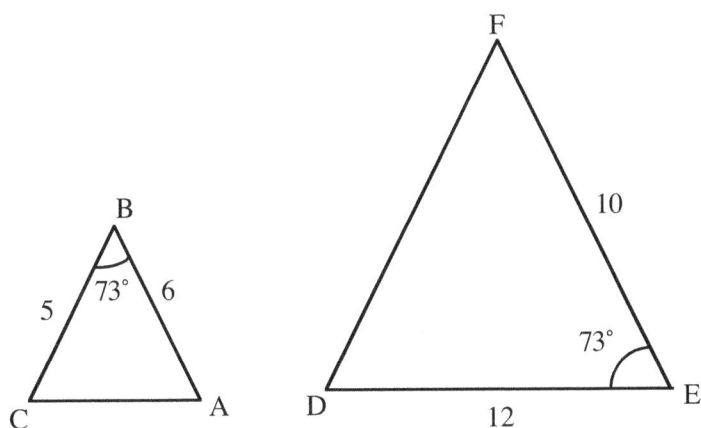

* In the Appendix, all the theorems on similar triangles covered have been stated in terms of the lengths of the sides, and the measures of the angles. These forms of the theorems may be useful when dealing with the lengths of the sides of the triangles.

Example 1 In the triangles shown (**Figure 1**), prove that $\triangle ABC \sim \triangle DEF$

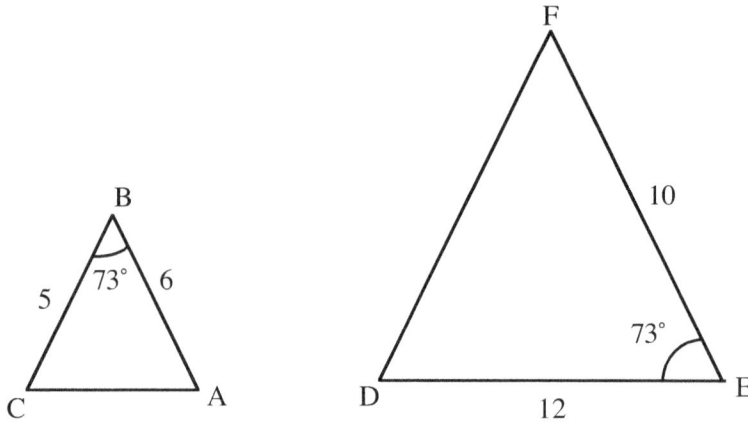

Figure 1

Given: 1. In $\triangle ABC$, $m\angle B = 73°$, $AB = 6$, $BC = 5$.

      2. In $\triangle DEF$, $m\angle E = 73°$, $DE = 12$, $EF = 10$.

To prove: That $\triangle ABC \sim \triangle DEF$

Plan: Arrange the dimensions of each triangle in increasing order (or decreasing order), i.e. smaller length first, etc.
For $\triangle ABC$, we have 5,6; for $\triangle DEF$, we have 10 ,12.
Divide the smaller length of $\triangle ABC$ by the smaller length of $\triangle DEF$, and note the line segments involved in the ratio. Next, divide the larger length of $\triangle ABC$ by the larger length of $\triangle DEF$ and again note the sides involved.. If we obtain the same ratio (same quotient), the sides of the triangles are in proportion, otherwise, they are not in proportion. We will also check to see if the given angles have the same measure (or are congruent) **and** are also included (i.e., the angles are also between the sides involved in the ratios).

Proof:

Statements	Reasons
1. The ratio, $5{:}10 = \dfrac{5}{10} = \dfrac{1}{2} = \dfrac{BC}{EF}$	**1.** The ratio of two quantities is the first quantity divided by the second quantity
2. The ratio, $6{:}12 = \dfrac{6}{12} = \dfrac{1}{2} = \dfrac{AB}{DE}$	**2.** The ratio of two quantities is the first quantity divided by the second quantity.
3. $\dfrac{AB}{DE} = \dfrac{BC}{EF}$	**3.** If quantities are equal to the same quantity, they are equal to each other.
4. AB, DE, BC and EF are in proportion.	**4.** A proportion is the equality of two ratios.
5. $m\angle B = 73°$	**5.** Given
6. $m\angle E = 73°$	**6.** Given
7. $m\angle B = m\angle E$	**7.** If quantities are equal to the same quantity, they are equal to each other.
8. $\angle B \cong \angle E$	**8.** If two angles have the same measure, they are congruent.
9. $\angle B$ and $\angle E$ are included angles.	
10. $\therefore \ \triangle ABC \sim \triangle DEF$	**10.** SAS

Q.E.D.

Pick the unknown side from one triangle and then pick the corresponding side in the other triangle and form a ratio. Go back to the first triangle and pick a known side whose corresponding side in the second triangle is known, and form another ratio. Equate the two ratios and solve for the unknown.

You may also pick corresponding sides and set up a proportion as "a is to b as c is to d" and translate to $\frac{a}{b} = \frac{c}{d}$ (where $a, b, c,$ and d are the lengths of the corresponding sides).

Example 1b If the two triangles below are similar, find x.

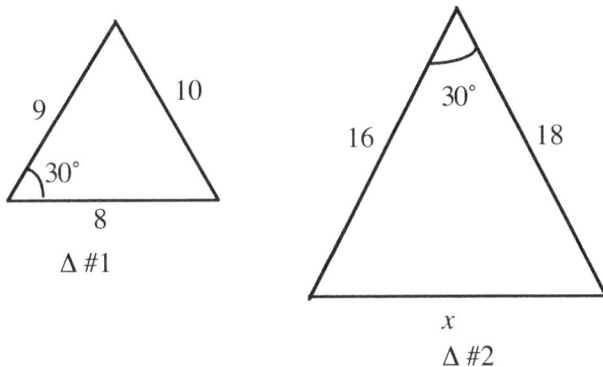

Solution
Step 1: Pick the x in say, Δ #2; note that x is opposite the angle whose measure is $30°$

Step 2 : Go to the other triangle say, Δ #1; pick the side opposite the angle whose measure is $30°$. The length of this corresponding side is 10.

Step 3: Form a ratio using these corresponding sides x and 10 to obtain $\frac{x}{10}$.

Step 4: Go back to Δ #2, pick a known side whose corresponding side in Δ #1 is also known. In this problem, we can pick either the "18" or the "16". We pick 16; and pick its corresponding side in Δ #1 which is 8.

Step 5: Form another ratio using 16 and 8 to obtain $\frac{16}{8}$.

Step 6 : Equate the ratios from Step 3 and Step 5, and solve for x.
$$\frac{x}{10} = \frac{16}{8}$$
$$8x = 10(16)$$
$$x = 20$$

Theorem 2: If all three pairs of corresponding sides of two triangles are in proportion (i.e., the ratio of corresponding sides is the same for each pair of corresponding sides) , then the two triangles are similar. Abbreviated **SSS** (Side-Side-Side: but note that implied are the **ratios of sides,** and **not** equality of sides)

In **Figure 2** below, $\triangle$ ABC ~ $\triangle$ FDE (read "triangle ABC is similar to triangle FDE", since the conditions in Theorem 2 have been satisfied. We will prove this similarity in the following example.

Example 2   In Figure 2 below, prove that $\triangle$ ABC ~ $\triangle$ FDE

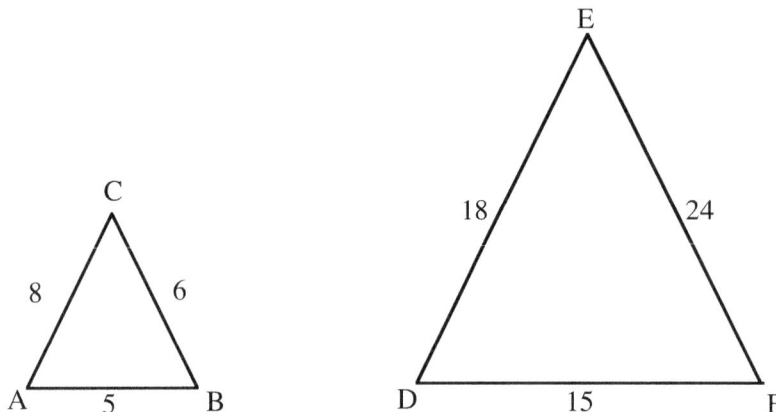

Figure 2

Given: **1**. In $\triangle$ ABC, AB $= 5$, BC $= 6$, CA $= 8$.
                2. In $\triangle$ FDE, DF $= 15$, FE $= 24$, ED $= 18$.

To prove: That $\triangle$ ABC ~ $\triangle$ FDE
Plan: Arrange the dimensions of each triangle in increasing order (or decreasing order) ,i.e. smallest length first, etc.
For $\triangle$ ABC, we have 5, 6, 8; for $\triangle$ FDE, we have 15 ,18, 24.
Divide the smallest length of $\triangle$ ABC by the smallest length of $\triangle$ FDE, and note the line segments involved in the ratio. Next, divide the larger length of $\triangle$ ABC by the larger length of $\triangle$ FDE and note the sides; and similarly divide the largest length of $\triangle$ ABC by the largest length of $\triangle$ FDE, and again note the sides.. If we obtain the same ratio (same quotient) for all three cases, the three pairs of sides of the triangles are in proportion, and the triangles are similar; otherwise, the sides are not in proportion, and the triangles are not similar.

Proof: Statements   Reasons
1. The ratio, $5\!:\!15 = \dfrac{5}{15} = \dfrac{1}{3} = \dfrac{AB}{DF}$ **1.** The ratio of two quantities is the first quantity divided by the second quantity.

2. The ratio, $6\!:\!18 = \dfrac{6}{18} = \dfrac{1}{3} = \dfrac{BC}{DE}$ **2.**The ratio of two quantities is the first quantity divided by the second quantity.

3. The ratio, $8\!:\!24 = \dfrac{8}{24} = \dfrac{1}{3} = \dfrac{AC}{FE}$ **3.** The ratio of two quantities is the first quantity divided by the second quantity.

4. $\dfrac{AB}{DF} = \dfrac{BC}{DE} = \dfrac{AC}{FE}$ **4.** If quantities are equal to the same quantity, they are equal to each other.

                Also all the three pairs of corresponding sides are in proportion.

5. $\triangle$ ABC~ $\triangle$ FDE   **5.** SSS
                Q.E.D

Theorem 3: If two angles of one triangle are congruent to two corresponding angles of another   4 8 1
triangle, then the two triangles are similar. Abbreviated **AA** (Angle-Angle).

Note that if two pairs of angles are congruent, the third pair are congruent, and therefore, AA implies
AAA.

In **Figure 3** below, △ ABC ~ △ DEF (read "triangle ABC is similar to triangle DEF"), since the
conditions in Theorem 3 have been satisfied. In the following example, will prove that the two triangles
given are similar .

Example 3   In Figure 3 below, prove that △ ABC ~ △ DEF.

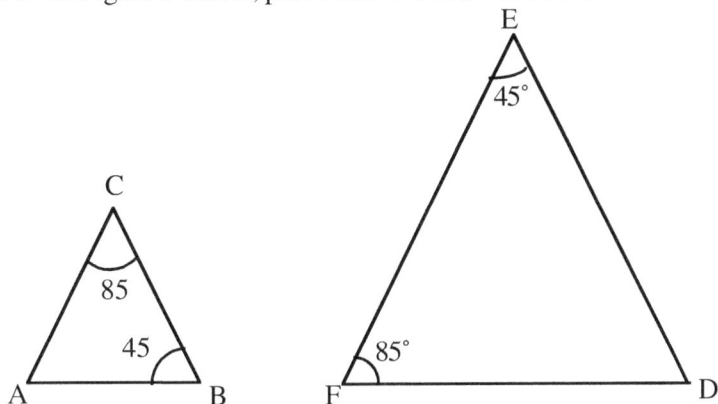

Figure 3

Given: **1**. In △ ABC, m∠ B = 45°, m∠ C = 85°
        2. In △ DEF, m∠ E = 45° , m∠ F = 85°

To prove: That △ ABC ~ △ DEF

Proof: Statements   Reasons

    1. m∠ B = 45°   **1.** Given
    2. m∠ E = 45°,   **2.** Given
    3. m∠ B = m∠ E   **3.** If quantities are equal to the same quantity, they are equal to each other.
    4. ∠ B ≅ ∠ E   **4.** If two angles have the same measure, they are congruent.
    5. m∠ C = 85°   **5.** Given
    6. m∠ F = 85°   **6.** Given
    7. m∠ C = m∠ F   **7.** If quantities are equal to the same quantity, they are equal to each other.
    8. ∠ C ≅ ∠ F   **8.** If two angles have the same measure, they are congruent.
    9. △ ABC ~ △ DEF   **9.** AA

                Q.E.D

A Determine which of the triangles in Fig.1, Fig.2 and Fig 3 are similar; and state the reason.

Fig.1

Fig. 2

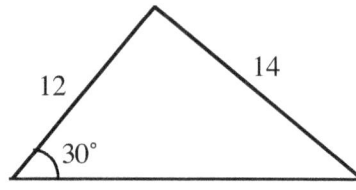
Fig. 3

Answers: **1.** Fig 1 and Fig.2 by SAS

B In the figures below, (a) Show that the Δ's are similar.
(b) Find x.

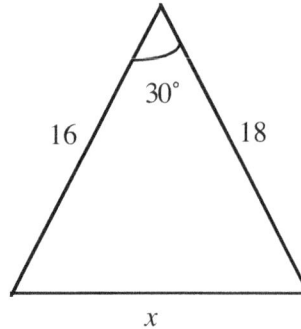

Solution: (a) See lesson and imitate; (b) 20.

C Determine if the two triangles in each problem with the give dimensions are similar:

1. Triangle #1 of sides 12, 16, 24 ; Triangle #2 of sides 8, 6, 12

2. Triangle #1 of sides 4, 6, 8; Triangle #2 of sides 8, 18, 12.

Answers: **1.** Similar;   **2.** Not similar

D

Determine if the two triangles in each of the following problems are similar according to the data given.

1. Δ #1 of sides 3 and 4; included angle 75°;   Δ #2 of sides 8 and 6, included angle 25°.

2. Δ #1 of sides 3, and 4; included angle 65°;   Δ #2 of sides 8, and 6, included angle 65°

Answers: **1.** Not similar; **2.** Similar

Lesson 101

Comparison of Congruency and Similarity of Triangles; Applications of Similarity Theorems

Comparison of congruency and similarity of triangles

1. Corresponding angles: For **both** congruency and similarity, the **corresponding angles** are **congruent.**
2. Corresponding sides: For congruency we have **equality of sides** (equality of lengths)
   For similarity, we have **equality of ratios of sides** (equality of ratio of lengths)
3. For congruency, each ratio of corresponding sides = 1. (Ratio of similitude = 1.)
   For similarity, each ratio of corresponding sides could be 1 or another number.
   (Ratio of similitude = 1, or another number)

Note 1: If two triangles are congruent, they are also similar; but if two triangles are similar, they are not necessarily congruent.

Note 2: There is no need for **ASA** for similar triangles (the "**S**" in **ASA** is redundant), since **ASA** implies **AA** which is sufficient for two triangles to be similar.

Example 4   In Figure 4 below, $\overline{DE}$ is parallel to $\overline{BC}$.
(a) Prove that $\triangle ADE \sim \triangle ABC$. (b) If BC = 12, DE = 4, AD = 6, Find AB.

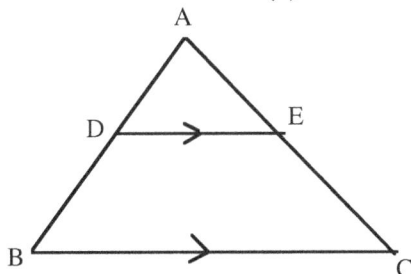

Figure 4

(a)   **Given** : $\triangle ABC$ in which DE || BC.
To prove: That $\triangle ADE \sim \triangle ABC$.

Proof: Statements   Reason

    1. $\angle ADE \cong \angle ABC$ 1. Corresponding angles, || lines $\overline{DE}$, $\overline{BC}$ cut by transversal $\overline{AB}$.

    2. $\angle AED \cong \angle ACB$ 2. Corresponding angles, || lines $\overline{DE}$, $\overline{BC}$ cut by transversal $\overline{AC}$.

    3. $\triangle ADE \sim \triangle ABC$ 3. AA

(b)
BC = 12 ,DE = 4, AD = 6.
    Let AB = x
Since $\triangle ADE \sim \triangle ABC$, the corresponding sides are in proportion:

Therefore, $\dfrac{AD}{x} = \dfrac{DE}{BC}$ (AD corresponds to AB. They are opposite congruent angles)
Substituting: AD = 6, DE = 4, BC =12.

$$\frac{6}{x} = \frac{4}{12}$$
$$4x = 72$$
$$x = 18$$
$$\therefore AB = 18$$

Lesson 101: Comparison of Congruency and Similarity of Triangles

Example 5 A tree cast a 18 ft. shadow at noon when a nearby upright pole of length 7 ft. cast a shadow of length 3 ft. Find the height of the tree assuming that both the tree and pole meet the ground at right angles.

Solution

We will assume that the measure of the angle between the sun's ray and the ground is the same for both the tree and the nearby pole.

Step 1: The tree, its shadow and the sun's ray form a right triangle. Similarly, the pole, its shadow together with the sun's ray form another right triangle. The two triangles formed are similar by AA. See Figure 1 below.

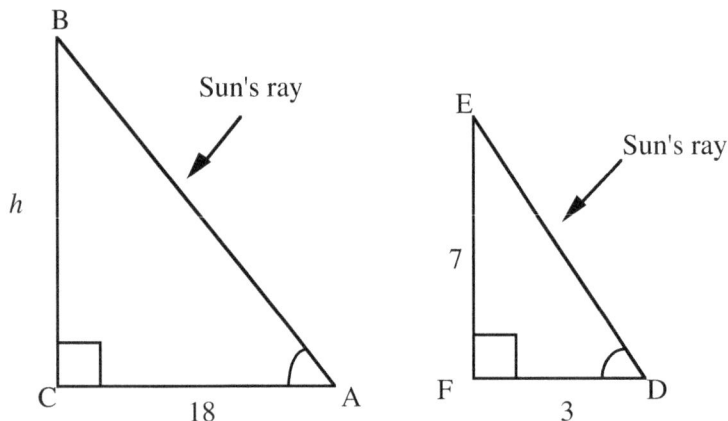

Figure 1

Step 2:

Since $\triangle ABC \sim \triangle DEF$,

$$\frac{h}{7} = \frac{18}{3}$$ ($\overline{BC}$ and $\overline{EF}$ are corresponding sides; $\overline{CA}$ and $\overline{FD}$ are corresponding sides)

and $h = \frac{18(7)}{3}$

$h = 42$

The height of the tree is 42 ft.

Extra If you study calculus in the future, you will apply similarity of triangles when covering problems on related rates and when finding volumes of solids of known cross-sections.

Lesson 101 Exercises

A tree cast a 18 ft. shadow at noon when a nearby upright pole of length 7 ft. cast a shadow of length 3 ft. Find the height of the tree assuming that both the tree and pole meet the ground at right angles.

Answer: 42 ft.

CHAPTER 44

Lesson 102: **Radian-Degree Conversions**
Lesson 103: **Right Triangle Trigonometry and Applications**
Lesson 104: **Special Δ's: The 30°-60°-90° and 45°-45°-90° triangles**

Angle: In **geometry**, we define an angle as being formed by two rays (or line segments) meeting at a common point called the vertex of the angle. The main unit for measuring angles in geometry is the **degree.** In **trigonometry**, we define an **angle** as being formed by rotating a line segment (or ray) about an end point, from an initial position to a final or terminal position. We call the initial position, the initial side and the terminal position the terminal side. The angle formed is measured by the amount of rotation. If the rotation is in a counterclockwise direction, then the measure of the angle formed is positive. If the rotation is in a clockwise direction, the measure of the angle formed is negative. In trigonometry, we measure angles in degrees as well as in **radians**.

Lesson 102
Radian-Degree Conversions

Converting from Degrees to Radians
Converting from one unit to another unit is a direct proportion problem, and as such any of the methods of solving direct proportion problems are applicable.

Example 1   Convert 240° to radians.

Solution $360° = 2\pi$ radians, or $\boxed{180° = \pi \text{ radians}}$

Method 1   Let x radians $= 240°$............(1)
π radians $= 180°$ ....(2)

Divide equation (1) by equation (2)

Then $\dfrac{x \text{ radians}}{\pi \text{ radians}} = \dfrac{240°}{180°}$, from which

$$x = \frac{\pi(240)}{180}$$

$$x = \frac{\pi(\cancel{240°})^{4}}{\cancel{3}\cancel{180°}}$$

$$x = \frac{4\pi}{3} ; \qquad \therefore 240° = \frac{4\pi}{3} \text{ radians.}$$

Method 2 Using the units label method

Step 1: $\dfrac{240 \text{ degrees}}{1} \times \dfrac{?}{?}$

Step 1: $\dfrac{240 \text{ degrees}}{1} \times \dfrac{\pi \text{ radians}}{180 \text{ degree}}$

$= \dfrac{240}{1} \times \dfrac{\pi \text{ radians}}{180}$

$= \dfrac{4\pi}{3}$ radians

Method 3   We will use simple proportion to solve the problem.
         Let x radians correspond to 240°.
         Then x radians is to 240° as π radians is to 180°.

Step 1:   " x radians is to 240° as π radians is to 180°" translates to:

$$\frac{x \text{ radians}}{240°} = \frac{\pi \text{ radians}}{180°}$$

Step 2:   Solve for x:

$$x = \frac{\pi(240)}{180}$$

$$x = \frac{\pi(240°)^4}{3^{180°}} = \frac{4\pi}{3} \quad ; \quad \text{Therefore} \quad 240° = \frac{4\pi}{3} \text{ radians.}$$

Converting from Radians to Degrees

Example 2   Convert $\dfrac{7\pi}{5}$ radians to degrees.

Method 1   Let $x° = \dfrac{7\pi}{5}$ radians..................................(1)

         $180° = \pi$ radians..(2)

Divide equation (1) by equation (2): $\dfrac{x°}{180°} = \dfrac{\frac{7\pi}{5} \text{ radians}}{\pi \text{ radians}}$

$$\frac{x°}{180°} = \frac{7\pi}{5} \frac{1}{\pi}$$

$$x = \frac{7\pi}{5} \frac{1(180)}{\pi}$$

Solving for x, $x = \dfrac{7\cancel{\pi}}{\cancel{5}}^1 \dfrac{1(\cancel{180})^{36}}{\cancel{\pi}}$

$$x = 252$$

$$\therefore \frac{7\pi}{5} \text{ radians} = 252°.$$

Method 2

   Step 1: $\dfrac{\frac{7\pi}{5} \text{ radians}}{1} \times \dfrac{?}{?}$

   Step 1: $\dfrac{\frac{7\pi}{5} \text{ radians}}{1} \times \dfrac{180 \text{ degrees}}{\pi \text{ radians}}$

         $= \dfrac{7}{5} \times \dfrac{180 \text{ degrees}}{1}$

         $= 252 \text{ degrees}$

   $\dfrac{7\pi}{5}$ radians $= 252°$

Method 3

Let $x°$ correspond to $\dfrac{7\pi}{5}$ radians.

Then, "$x°$ is to $\dfrac{7\pi}{5}$ radians as 180° is to π radians." translates to:

$$\frac{x°}{\dfrac{7\pi}{5} \text{ radians}} = \frac{180°}{\pi \text{ radians}}$$

Solving for x,

$$x = \frac{180}{\pi}\frac{7\pi}{5}$$
$$x = 252°$$
$$\therefore \frac{7\pi}{5} \text{ radians} = 252°.$$

We could have solved Examples 1 and 2 using proportion, knowing that π radians = 180°.

In Example 1: If 180° = π

Then, 240 °= $\dfrac{\pi \text{ radians}}{1} \cdot \dfrac{\overset{4}{\cancel{240}}}{\underset{3}{\cancel{180}}}$

$\therefore 240° = \dfrac{4\pi}{3}$ radians.

Lesson 102 Exercises

1. Convert 270° to radians

2. Convert $\dfrac{6\pi}{5}$ radians to degrees

Answers: **1**. $\dfrac{3\pi}{2}$ radians ; **2.** 216°

Lesson 103

Right Triangle Trigonometry and Applications

Trigonometric Ratios (Trigonometric Functions)
sine (**sin**); cosine (**cos**); tangent (**tan**); cosecant (**csc**); secant (**sec**); and cotangent (**cot**).

Definitions: The following definitions are applicable to **right** triangles only.

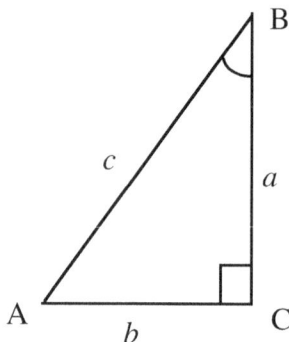

1. $\mathbf{sin\ A} = \dfrac{\text{opposite side}}{\text{hypotenuse}} = \dfrac{a}{c}$ (sin A is abbreviation for the sine of A)

2. $\mathbf{cos\ A} = \dfrac{\text{adjacent side}}{\text{hypotenuse}} = \dfrac{b}{c}$ (cos A is abbreviation for the cosine of A)

3. $\mathbf{tan\ A} = \dfrac{\text{opposite side}}{\text{adjacent side}} = \dfrac{a}{b}$ (tan A is abbreviation for the tangent of A)

4. $\mathbf{csc\ A} = \dfrac{1}{\sin A} = \dfrac{\text{hypotenuse}}{\text{opposite side}} = \dfrac{c}{a}$ (csc A is abbreviation for the cosecant of A)

5. $\mathbf{sec\ A} = \dfrac{1}{\cos A} = \dfrac{\text{hypotenuse}}{\text{adjacent side}} = \dfrac{c}{b}$ (sec A is abbreviation for the secant of A)

6. $\mathbf{cot\ A} = \dfrac{1}{\tan A} = \dfrac{\text{adjacent side}}{\text{opposite side}} = \dfrac{b}{a}$ (cot A is abbreviation for the cotangent of A)

Note above that $\mathbf{sin\ B} = \dfrac{b}{c}$; $\mathbf{cos\ B} = \dfrac{a}{c}$; $\mathbf{tan\ B} = \dfrac{b}{a}$

Note also that for $\angle$ B, the opposite side is $\overline{AC}$, and the adjacent side is $\overline{BC}$. For $\angle$ A,

the opposite side is $\overline{BC}$ and the adjacent side is $\overline{AC}$
Some mnemonic devices will be discussed after Example 1

Example 1 Solve the right triangle ABC for the unknown parts, given
that $b = 4$, m $\angle$ B = 30°

Solution

(a) m $\angle$ A $= 180° - (30° + 90°)$ (m$\angle$ C = 90°; sum of the measures of the $\angle$'s of a Δ is 180)

$$= 180° - 120°$$

$$\text{m} \angle \text{A} = 60°$$

(b) To find c:

c is the hypotenuse, $b = 4$ is opposite $\angle$ B.

$$\sin 30 = \frac{4}{c} \qquad \qquad (\text{m}\angle \text{ B} = 30°, b = 4)$$

$$c \sin 30° = 4$$

$$c = \frac{4}{\sin 30°} = \frac{4}{\frac{1}{2}} = 8 \qquad \qquad (\sin 30° = \frac{1}{2})$$

now $c = 8$, $b = 4$

Apply the Pythagorean theorem (We may apply this theorem since Δ ABC is a right triangle)

$$8^2 = 4^{2} + a^2$$

$$64 = 16 + a^2$$

$$48 = a^2$$

$$a = \sqrt{48} = \sqrt{16}\sqrt{3} = 4\sqrt{3}$$

Note: In (b), we could have found a first using the tangent relationship

$$\tan \text{A} = \frac{a}{4} \qquad \left(\frac{\textbf{O}\text{pposite}}{\textbf{A}\text{djacent side}} \text{ "\textbf{TOA}" } \texttt{<-------- } \text{mnemonic device} \right)$$

$$a = 4 \tan \text{A} \qquad\qquad\qquad\qquad \textbf{Also}, \cos 30° = \frac{a}{c}$$

$$a = 4 \tan 60° \qquad\qquad\qquad\qquad \frac{\sqrt{3}}{2} = \frac{a}{8} \text{ and } a = \frac{8(\sqrt{3})}{2} = 4\sqrt{3}$$

$$a = 4\sqrt{3}$$

Note: The determination of which trigonometric relationship to use depends on what we are given and what we want to find. A good practice is to scribble down the following three relationships:

$$\sin \theta = \frac{\text{opposite side}}{\text{hypotenuse}} \qquad \left(\frac{\textbf{O}\text{pposite}}{\textbf{H}\text{ypotenuse}} : \text{"\textbf{SOH}" } \texttt{<-------- } \text{mnemonic device} \right)$$

$$\cos \theta = \frac{\text{adjacent. side}}{\text{hypotenuse.}} \qquad \left(\frac{\textbf{A}\text{djacent}}{\textbf{H}\text{ypotenuse}} : \text{"\textbf{CAH}" } \texttt{<-------- } \text{mnemonic device} \right)$$

$$\tan \theta = \frac{\text{opposite side}}{\text{adjacent side}} \qquad \left(\frac{\textbf{O}\text{pposite}}{\textbf{A}\text{djacent}} : \text{"\textbf{TOA}" } \texttt{<-------- } \text{mnemonic device} \right)$$

and use the relationship in which you know all but one quantity. Sometimes, you may use two relationships simultaneously to find two unknowns.

Applications of Right Angle Trigonometry
490

Definition: For an observer sighting (looking) at something above the observer, **the angle of elevation** is the angle between the horizontal line (x-axis) from the observer's eye and the line of sight to the object, the angle being measured from the horizontal to the line of sight.

Example   Find the height reached by a kite, if the length of the kite is 400 ft, and the angle of elevation of the kite is 60°.

Step 1: Sketch a diagram for the system under consideration.
Assume that the hand holding on to the kite and the eye coincide at one point.

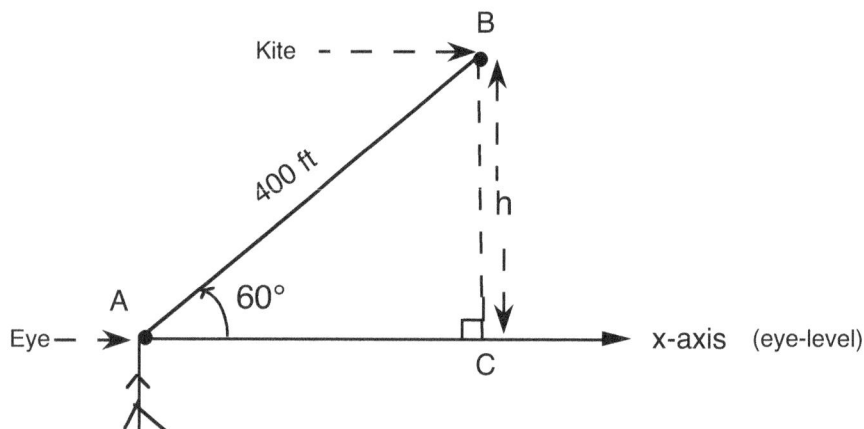

AB = Length of kite = 400 ft.
BC = h = height of kite (vertical distance from eye-level to the kite)

Step 2: $\triangle$ ABC is a right triangle, and therefore we may apply the trigonometric relationships.

$$\sin 60° = \frac{h}{400} \qquad \left(\text{SOH} : \frac{\textbf{O}\text{pposite}}{\textbf{H}\text{ypotenuse}}\right)$$

$$h = 400 \sin 60°$$

$$= 400 \left(\frac{\sqrt{3}}{2}\right) \text{ft.}$$

$$= 200 \sqrt{3} \text{ ft.} \qquad\qquad (\sqrt{3} \approx 1.73)$$

$$\approx 346.41 \text{ ft.}$$

The height reached is $200 \sqrt{3}$ ft. ($\approx$346.41 ft.)

Angle of depression

Definition:
For an observer sighting (looking) at something below the observer, the angle of depression is the angle between the horizontal line (x-axis) from the observer's eye and the line of sight to the object, the angle being measured from the horizontal to the line of sight.

In the figure below: $\angle \alpha$ is the angle of elevation of B from A.

$\qquad\qquad$ $\angle \beta$ is the angle of depression of A from B.

The angle of elevation is congruent to the angle of depression.

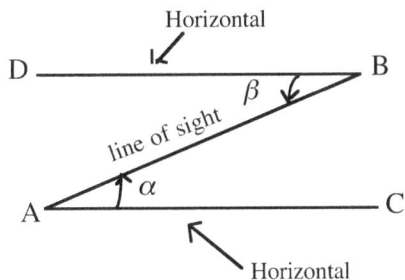

Lesson 103 Exercises

A Solve the right triangle ABC for the unknown parts, given that $b = 6$, and m $\angle$ A $= 60°$

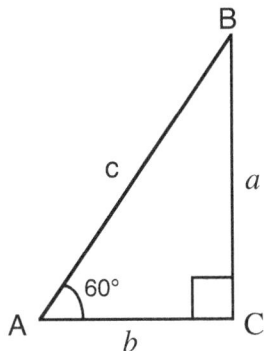

Answers: m $\angle$ B $= 30°$; $a = 6\sqrt{3}$; $c = 12$

B Find the height reached by a kite, if the length of the kite is 600 ft, and the angle of elevation of the kite is 30°.

Answer: 300ft.

Lesson 104

Special triangles: The 30º-60º-90º triangle and the 45º-45º-90º triangle

The trigonometric functions for 30º, 45º, 60º, and 90º occur very frequently in mathematics that students are usually required to memorize them.

The 45º-45º-90º triangle

By sketching an isosceles right triangle with measures of 45º, 45º, and 90º (Figure 1) we will be able, using the basic definitions, write down the trigonometric functional value for 45º.

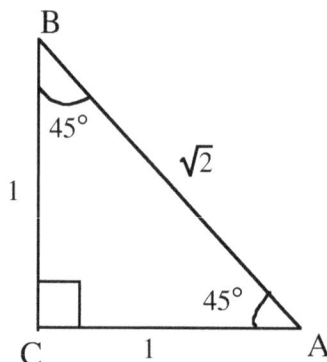

Figure 1

In an isosceles triangle, the sides opposite the congruent angles have the same length. If we let the length of the sides opposite the congruent angles be 1 unit each, we can calculate the length of the hypotenuse using the Pythagorean theorem.

$$AB^2 = AC^2 + BC^2$$
$$AB^2 = 1^2 + 1^2$$
$$AB^2 = 2$$
$$AB = \sqrt{2}.$$

Example $\sin 45º = \dfrac{\text{opposite side}}{\text{hypotenuse}} = \dfrac{1}{\sqrt{2}} = \dfrac{1}{\sqrt{2}} \cdot \dfrac{\sqrt{2}}{\sqrt{2}} = \dfrac{\sqrt{2}}{2}$ (Rationalizing the denominator)

$\cos 45º = \dfrac{\text{adjacent side}}{\text{hypotenuse}} = \dfrac{1}{\sqrt{2}} = \dfrac{1}{\sqrt{2}} \cdot \dfrac{\sqrt{2}}{\sqrt{2}} = \dfrac{\sqrt{2}}{2}$

$\tan 45º = \dfrac{\text{opposite side}}{\text{adjacent side}} \ \dfrac{1}{1} = 1.$

The 30º-60º-90º triangle

For the 30º-60º-90º triangle (Figure 2), the dimensions are obtained by considering an equilateral triangle whose sides are each 2 units long (see page 472, Theorems 2 and 3). From geometry, the length of the hypotenuse is twice the length of the shortest side. The third side is calculated by the Pythagorean theorem.

Thus if $AB = 2$, then $AC = 1$ and $2^2 = 1^2 + BC^2$ and from which $BC = \sqrt{3}$. Usually, after having gone through the derivation, using geometric considerations and the Pythagorean theorem, students become familiar with the dimensions $1, 2, \sqrt{3}$. The problem of recall that remains then is how to place these dimensions on the 30º-60º-90º triangle, if you do not want to draw the equilateral t riangle, but draw only $\triangle ABC$. The following mnemonic device will be helpful:

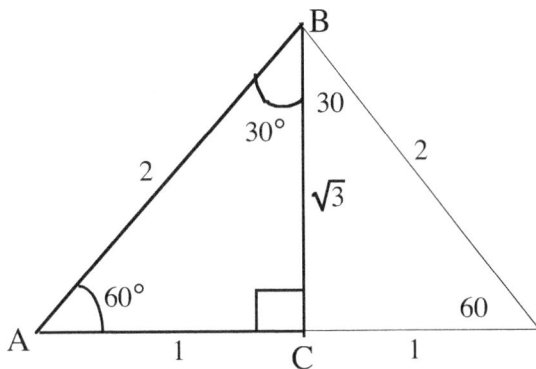

Figure 2

Mnemonic device: If you can remember the numbers $1, 2, \sqrt{3}$ as being the lengths of the sides of this triangle, and also remember that in any triangle, the longest side is opposite the largest angle, and that the shortest side is opposite the smallest angle, you should be able to place these dimensions accordingly on the triangle. Thus, 1 is opposite the 30º angle, 2 is opposite the 90º angle (the right angle) , and $\sqrt{3}$ is opposite the 60º angle. ($\sqrt{3} \approx 1.73$)

By applying the mnemonic devices, SOH , CAH, TOA (see p.489) we can write down some functional values.

Examples

$\sin 30º = \dfrac{\text{opposite side}}{\text{hypotenuse}} = \dfrac{1}{2}$

$\cos 30º = \dfrac{\text{adjacent side}}{\text{hypotenuse}} = \dfrac{\sqrt{3}}{2}$

$\sin 60º = \dfrac{\text{opposite side}}{\text{hypotenuse}} = \dfrac{\sqrt{3}}{2}.$

Lesson 104 Exercises

In Figure 2, above, find the following:

1. $\tan 30°$; **2.** $\tan 60°$; **3.** $\csc 30°$; **4,.** $\csc 60°$; **5.** $\sec 30°$; **6.** $\sec 60°$.

CHAPTER 45

Trigonometric Functional Value of any Angle

Lesson 105: **Introduction: Reference Angle**
Lesson 106: **Given the Measure of the Angle**
Lesson 107: **Given the Coordinates of a point on the terminal side of the Angle**
Lesson 108: **Given one Functional Value and the Quadrant of the Angle**

Lesson 105

Introduction: Reference Angle

Angle in standard position

Definition: An angle is in standard position in an *x-y* rectangular coordinate system if it's vertex is at the origin and its initial side is along positive *x* -axis.

Reference Angle

Definition: If an angle, θ, is standard position, then the reference angle is the positive acute angle between the terminal side of θ and the *x*- axis (the horizontal axis).

Example 1   Find reference angle for 230º.
Solution
Step 1:   Draw (or sketch) θ = 230º in standard position.
                 To draw the arc, begin from the positive *x*-axis and draw the arc counterclockwise.

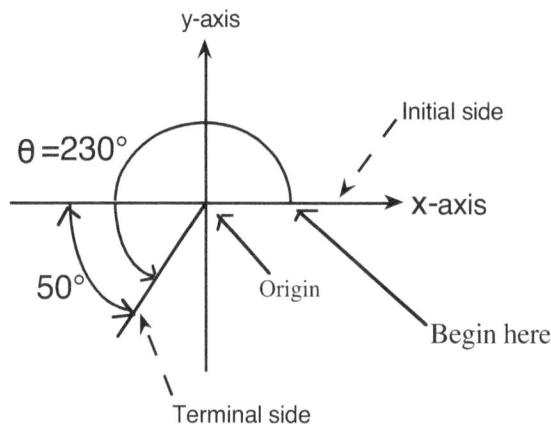

Step 2:   Determine the reference angle by difference. The reference angle for 230º is 230 - 180
           i.e. 50º
           The reference angle for 230º is 50º .

Note: The reference angle is always between the terminal side and the *x*-axis.

Example 2   Find the reference angle for 460°
Solution
Step 1:   Draw (sketch) θ = 460° in standard position.

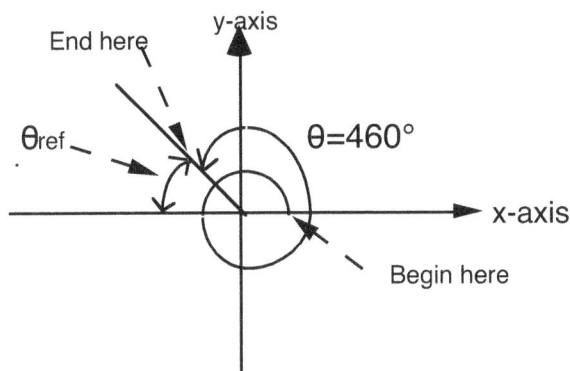

Step 2:   The reference angle is the difference between 180° and 100°.

i.e.,   180° - 100° = 80°

Note that 360° will bring you back to the initial side, leaving 460 - 360 = 100 to consider,

Lesson 105 Exercises

Find the reference angles for the following: **1.** 250° ; **2.** 295°; **3.** 560°; **4.** 75°;   **5.** 360°

Answers. **1.** 70° ;   **2.** 65° ;   **3.** 20°;   **4.**   75° ;   **5.** 0°

Lesson 106   496
Trigonometric Functional Value (Given the measure of the angle)

Example   (a) Find the value of sin 210° without using a calculator: use trigonometric tables.
                   (b) Verify your answer using a calculator.
Solution (a)
Step 1:   Draw (or sketch) the given angle in standard position.

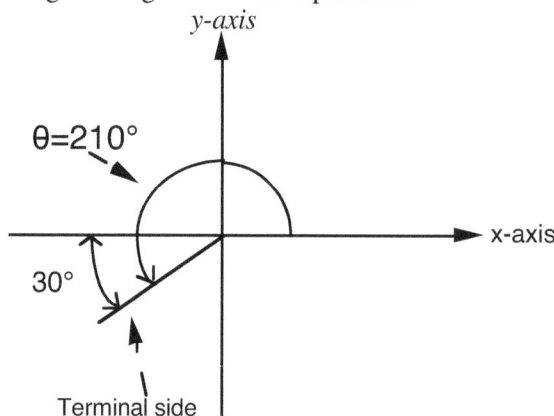

Figure 1

Step 2:   Find the reference angle.
           The reference angle is 210° - 180° = 30°
Step 3:   From tables or from memory, find sin 30°. This value will be positive, and we must then
 determine the sign of sin 210° according to the quadrant in which the terminal side of this angle lies.
(See Figure 2 below) The terminal side of 210° is in the 3rd quadrant where the sine function is negative.

$$\therefore \ \sin 210 = -\sin 30 = -\tfrac{1}{2}$$

Signs of Trigonometric Functions According to Quadrants

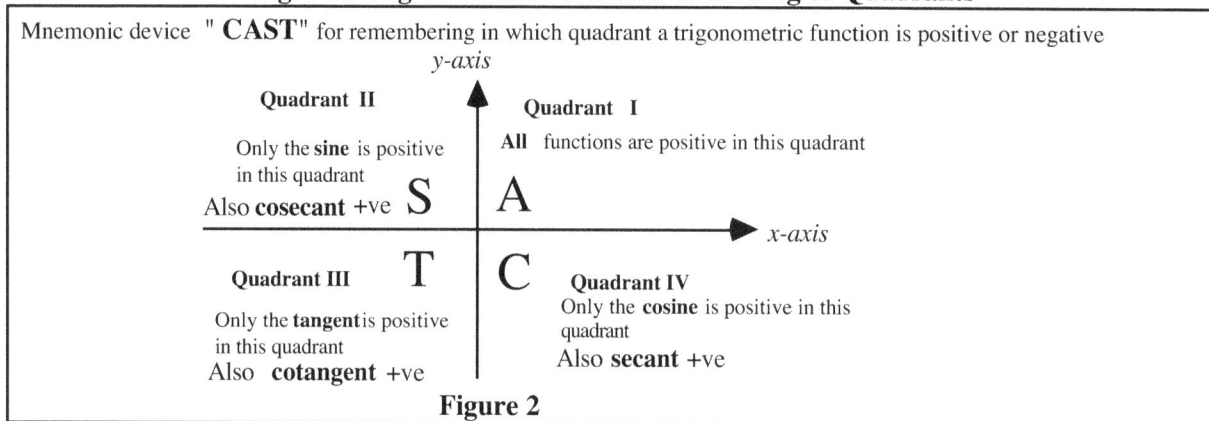

Figure 2

In the above device: The **A** in "**CAST**" means **A**ll functions are positive in this quadrant; **S** for
 only sine is positive in this quadrant; and similarly **T** for tangent; and **C** for cosine.
Note: The order of the above"CAST" is counterclockwise.There are other mnemonic devices
        such as "**A**ll **S**tudents **T**ake **C**alculus".
(b) With the calculator in degree mode, press 210; then press "the sin" key and read -0.5

Note above: The sign of a trigonometric function and its reciprocal are the same.

Lesson 106 Exercises

In the following problems, first use tables; and then verify your answers using a calculator.

Find the values of the following: **1**. sin 250° ; **2.** cos 250°;   **3.** sin 460°; **4.** sin 75°

Answers: **1**. - 0.94 ; **2.**   -0.34;   **3.** 0.98;   **4**. 0.96

Lesson 107
Finding Trigonometric Functional Value
(Given the coordinates of a Point on the Terminal Side of Angle)

Example Find the six trigonometric values of θ , (if θ is an angle in standard position) given that the terminal side of θ passes through the point (-3, 2√3).

Step 1: Draw (or sketch) θ with terminal side passing through the point (-3, 2√3)
(x-coordinate = -3; y-coordinate = 2√3)

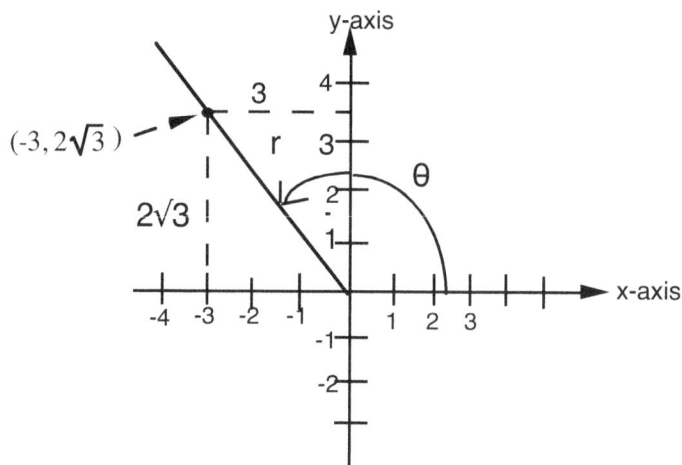

Step 2: Find r using the Pythagorean Theorem.
$$r^2 = (-3)^2 + (2\sqrt{3})^2 \quad (r^2 = x^2 + y^2)$$
$$r^2 = 9 + 4\sqrt{9}$$
$$r^2 = 9 + 12 \qquad \text{Note: } 4\sqrt{9} = 4(3) = 12)$$
$$r^2 = 21$$
$$r = \sqrt{21}$$

Step 3: Now, $r = \sqrt{21}$, $x = -3$, $y = 2\sqrt{3}$

(a) $\cos \theta = \dfrac{x}{r}$

$\qquad = \dfrac{-3}{\sqrt{21}}$

$\qquad = -\dfrac{3}{\sqrt{21}} \cdot \dfrac{\sqrt{21}}{\sqrt{21}}$ (Rationalizing the denominator)

$\therefore \cos \theta = -\dfrac{3\sqrt{21}}{21}$

(b) $\sin \theta = \dfrac{y}{r} \qquad (y = 2\sqrt{3},\ r = \sqrt{21})$

$\qquad = \dfrac{2\sqrt{3}}{\sqrt{21}}$

$$= \frac{2\sqrt{3}}{\sqrt{21}} \cdot \frac{\sqrt{21}}{\sqrt{21}} \quad \text{(Rationalizing the denominator)}$$

$$= \frac{2\sqrt{63}}{21}$$

$$= \frac{2\sqrt{9}\sqrt{7}}{21}$$

$$= \frac{2\cancel{3}\sqrt{7}}{\cancel{21}_{7}} \quad (\sqrt{9} = 3)$$

$$\sin\theta = \frac{2\sqrt{7}}{7}$$

(c) $\tan\theta = \dfrac{y}{x}$ $\quad\quad (y = 2\sqrt{3}, x = -3)$

$$= \frac{2\sqrt{3}}{-3}$$

$$\therefore \tan\theta = -\frac{2\sqrt{3}}{3}$$

(d) $\sec\theta = \dfrac{1}{\cos\theta}$

$$= \frac{1}{\dfrac{x}{r}} = \frac{r}{x} = \frac{\sqrt{21}}{-3}$$

$$\therefore \sec\theta = -\frac{\sqrt{21}}{3}$$

(e) $\csc\theta = \dfrac{1}{\sin\theta}$

$$= \frac{r}{y}$$

$$= \frac{\sqrt{21}}{2\sqrt{3}}$$

$$= \frac{\sqrt{21}}{2\sqrt{3}} \cdot \frac{\sqrt{3}}{\sqrt{3}} \quad\quad \text{(Rationalizing the denominator)} \quad = \frac{\sqrt{63}}{2\sqrt{9}}$$

$$= \frac{\sqrt{63}}{2(3)} \quad\quad (\sqrt{9} = 3)$$

$$= \frac{\sqrt{9}\sqrt{7}}{6}$$

$$= \frac{3\sqrt{7}}{6}$$

$$\therefore \csc\theta = \frac{\sqrt{7}}{2}$$

(f) $\cot\theta = \dfrac{1}{\tan\theta}$

$$= \frac{1}{\dfrac{y}{x}}$$

$$= \frac{x}{y}$$

$$= \frac{-3}{2\sqrt{3}} \qquad (x = -3, y = 2\sqrt{3})$$

$$= -\frac{3}{2\sqrt{3}} \cdot \frac{\sqrt{3}}{\sqrt{3}} \qquad \text{(Rationalizing the denominator).}$$

$$= -\frac{3\sqrt{3}}{2\sqrt{9}}$$

$$= -\frac{3\sqrt{3}}{2(3)} \qquad (\sqrt{9} = 3)$$

$$\therefore \cot \theta = -\frac{\sqrt{3}}{2}$$

Note in the last two examples that:

1. In Lesson 106,, we had to determine the sign of the function according to the quadrant (in which the terminal side of the angle lies) if we obtain the values of the acute angles from tables or from memory.

2. In Lesson 107, we did not have to determine the sign of the function. The signs were obtained solely from the signs of the x- and y-coordinates; but note also that r is always positive.

Lesson 107 Exercises

Find the six trigonometric values of θ , if θ is an angle in standard position, given that the terminal side of θ passes through the point (5,2).

Answers: $\sin \theta = \frac{2\sqrt{29}}{29}$; $\cos \theta = \frac{5\sqrt{29}}{29}$; $\tan \theta = \frac{2}{5}$; $\csc \theta = \frac{\sqrt{29}}{2}$; $\sec \theta = \frac{\sqrt{29}}{5}$; $\cot \theta = \frac{5}{2}$

Lesson 108

Finding Other trigonometric functional values
(Given one functional value and the quadrant of the angle)

Example   If $\sin \theta = -\dfrac{4}{5}$ and the terminal side of θ is in the fourth quadrant (θ being in standard position)

(a) Find $\cos \theta$, (b) $\tan \theta$, (c) $\csc \theta$

We will determine r, x and y and apply the definitions of the trigonometric functions

Step 1: Since we know that the terminal side is in the 4th quadrant, we will draw θ accordingly.

fig.1

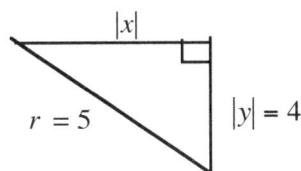

fig.2

Step 2: In this quadrant , the y-coordinate is negative and the x-coordinate is positive

$$\sin \theta = \frac{y}{r} = -\frac{4}{5}$$

$y = -4$ since r is always positive (We give the minus sign to the "4") .
$r = 5$

Step 3:   To find x, we use the Pythagorean theorem.

$$5^2 = (-4)^2 + x^2$$
$$25 = 16 + x^2$$
$$9 = x^2$$
$$3 = x$$
$$\therefore \quad x = 3, \ y = -4, \ r = 5$$

Step 4:

(a) $\cos \theta = \dfrac{x}{r}$

  $\cos \theta = \dfrac{3}{5}$ $(x = 3, r = 5)$

(b) $\tan \theta = \dfrac{y}{x}$

    $= \dfrac{-4}{3}$ $(y = -4, x = 3)$

    $= -\dfrac{4}{3}$

(c) $\csc \theta = \dfrac{1}{\sin \theta}$

    $= \dfrac{r}{y}$

    $= \dfrac{5}{-4}$

    $= -\dfrac{5}{4}$

Note: We could have obtained $\csc \theta$ from $\sin \theta = -\dfrac{4}{5}$ by inversion.

Lesson 108 Exercises

If $\sin \theta = -\dfrac{3}{7}$ and the terminal side of θ is in the third quadrant (θ being in standard position)
(a) Find $\cos \theta$, (b) $\tan \theta$, (c) $\csc \theta$

Solutions: $\cos \theta = \dfrac{-2\sqrt{10}}{7}$; (b) $\tan \theta = \dfrac{3\sqrt{10}}{20}$;   (c) $\csc \theta = -\dfrac{7}{3}$

CHAPTER 46

Trigonometry of Oblique Triangles

Lesson 109: **The Laws of Cosines and Sines**

Lesson 110: **Applications of the Laws of Cosines and Sines**

An **oblique triangle** is a triangle in which no angle is a right angle. We may consider the right triangle as a special case of the oblique triangle and that the laws which are applicable to oblique triangles are also applicable to the right triangle.

In solving oblique triangles, we are usually given three parts of the oblique triangle and we are required to find the remaining three parts. We will usually seek a unique solution (that is only one solution). There may be a unique solution, two distinct solutions or no solutions at all depending on the relative lengths of the sides and the measures of the angles

Lesson 109

The Laws of Cosines and Sines

Use of the Law of Cosines.

Direct use of the Law of Cosines applies to the following cases:
(1) Given two sides and an included angle (abbreviated SAS) we are therefore to find the emaining side and two angles.

(2) Given all three sides (abbreviated SSS) we are therefore to find the remaining three angles. The above cases yield unique solutions. A mnemonic device for the applicability of this law of cosines is **Co-SAS -SSS** - pronounced Kosaszz.

Statement of the Law of Cosines

Using standard notation for corresponding angles and sides, the law of cosines states that :

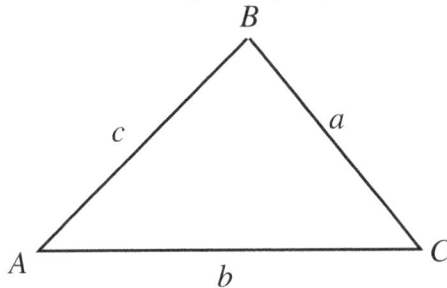

$$a^2 = b^2 + c^2 - 2bc \cos A$$

or

$$b^2 = a^2 + c^2 - 2ac \cos B$$

or

$$c^2 = a^2 + b^2 - 2ab \cos C$$

In words, the square of the length of one side is equal to the sum of the squares of the other two sides minus twice the product (of the lengths) of the other two sides multiplied by the cosine of the angle between these two sides.

Law of Cosine for the Right Triangle (a special case) 504

For a right triangle, let $m\angle A = 90°$. Then the law of cosines, $a^2 = b^2 + c^2 - 2bc \cos A$ becomes

$$a^2 = b^2 + c^2 - 2bc \cos 90°$$

$$a^2 = b^2 + c^2 - 2bc\,(0) \qquad\qquad (\cos 90° = 0)$$

$$a^2 = b^2 + c^2 \; \leftarrow ---\text{The Pythagorean theorem}$$

We may conclude that the law of cosines is a generalization of the Pythagorean theorem.

The Law of Sines

The **direct use** of the Law of Sines applies in the following cases:

(1) Given **two angles** and an **included side** (ASA) and therefore, we are to find the remaining sides and angle. This case yields a unique solution.

(2) Given **two sides** and an **angle** (ASS or SSA) there may be only one solution (one triangle), two solutions (two triangles) or there may be no solution at all (no triangle) depending upon the relative lengths of the sides and the measures of the angles. This case is commonly known as the ambiguous case of the law of sines.

A mnemonic device for its application is:
(1) sin ASA - pronounced "sinasa" **unique case**
(2) ASS - t**he ambiguous case**

Statement of the Law of Sines

The law of sines states that in any triangle, the sides are proportional to the sines of the angles opposite (facing) these sides. That is, the ratio of the length of a side to the sine of the angle opposite this side is equal to the ratio of the length of any other side to the sine of the angle opposite that side.

Using standard notation for angles and sides,

        a is to **sin A** as *b* is to **sin B** as *c* is to **sin C**

Symbolically, $\dfrac{a}{\sin A} = \dfrac{b}{\sin B} = \dfrac{c}{\sin C}$ (equivalent to $\dfrac{\sin A}{a} = \dfrac{\sin B}{b} = \dfrac{\sin C}{c}$)

Lesson 109 Exercises

1. State the Law of Sines from memory.

2. State the Law of Cosines from memory

Lesson 110

Applications of the Law of sines, Law of cosines

Example 1: Solve $\triangle ABC$ given that, $m\angle C = 150°$, $b = 3$, $c = 8$

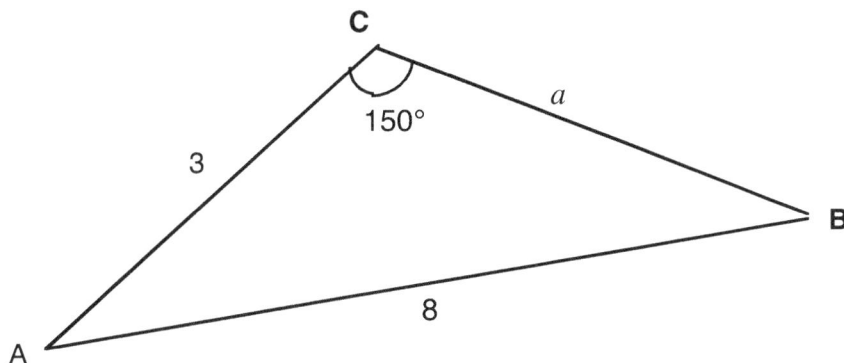

To **solve the triangle**, we will find the remaining three parts, namely the length a, $m\angle A$, and $m\angle B$.

 Solution: We **cannot** apply the Pythagorean theorem, since $\triangle ABC$ is **not** a right triangle.

 We will try the law of sines, or the law of cosines.

We apply the law of sines: $\dfrac{a}{\sin A} = \dfrac{b}{\sin B} = \dfrac{c}{\sin C}$

Step 1: We apply $\dfrac{c}{\sin C} = \dfrac{b}{\sin B}$

$$\frac{8}{\sin 150°} = \frac{3}{\sin B}$$
$$\frac{8}{\sin 30°} = \frac{3}{\sin B} \qquad (\sin 150° = \sin 30° = \tfrac{1}{2})$$

$$\frac{8}{\frac{1}{2}} = \frac{3}{\sin B} \qquad (\, b = 3;\ c = 8;\ \sin 150° = \sin 30° = \tfrac{1}{2})$$

$8 \sin B = \dfrac{3}{2}$ \qquad (by cross-multiplication or otherwise)

$$\sin B = \frac{3}{2} \times \frac{1}{8} = \frac{3}{16}$$
$$= .1875$$

Step 2: Find the angle whose sine is .1875 (From tables or calculator)
 then, $m\angle B = 10.81°$
 $m\angle A = 180° - (150° + 10.8°) = 19.19°$·

Finding *a*

Method 1 Given: $c = 8$, m $\angle$ C= $150°$, $b = 3$

Using the law of cosines:

$$c^2 = a^2 + b^2 - 2ab \cos C \quad (\angle \text{ C is an included angle})$$

$$8^2 = a^2 + (3)^2 - 2a(3) \cos 150°$$

$$64 = a^2 + 9 - 2a(3) \left(-\frac{\sqrt{3}}{2}\right) \qquad (\cos 150° = -\cos 30° = -\frac{\sqrt{3}}{2})$$

$$64 = a^2 + 9 + 2a(3) \frac{\sqrt{3}}{2}$$

$$64 = a^2 + 9 + 3\sqrt{3}\, a$$

$$0 = a^2 + 3\sqrt{3}\, a - 55 \,\text{<-------------quadratic equation}$$

Solve by the quadratic formula.
$$a^2 + 3\sqrt{3}a - 55 = 0$$

$$a = \frac{-3\sqrt{3} \pm \sqrt{(3\sqrt{3})^2 - 4(1)(-55)}}{2} \qquad (a = 1, b = 3\sqrt{3}, c = -55)$$

$$= \frac{-3\sqrt{3} \pm \sqrt{9(3) + 220}}{2}$$

$$= \frac{-3\sqrt{3} \pm \sqrt{247}}{2}$$

$$= \frac{-3\sqrt{3} \pm 15.71}{2} = \frac{-3\sqrt{3} + 15.71}{2} \text{ or } \frac{-3\sqrt{3} - 15.71}{2}$$
$$a = 5.26 \text{ or } -10.45$$

we reject the negative root since the length of a side of a triangle is always positive.
$\therefore a = 5.26$.

Finding a

Method 2   We could also have found a, using the law of sines after having found.

m $\angle$ A = 19.19°

$$\frac{a}{\sin A} = \frac{c}{\sin C}$$

$$\frac{a}{\sin 19.19°} = \frac{8}{\sin 150°} \qquad (\sin 19.19 = .3287)$$

$$(\sin 150° = \sin 30° = .5)$$

$$a = \frac{8 \sin 19.19}{\sin 150}$$

$$= \frac{8(.32870)}{.5}$$

$$a = 5.26$$

Again, we obtain the same value for a. Method 2 is less involved than Method 1. However, the disadvantage of Method 2, in this particular problem, is that we used a value, m $\angle$ A = 19.19°, which we calculated; and if there were error in m $\angle$ A, there would be error in the value of a. The advantage of Method 1 is that we used only given values and no calculated values ($c = 8$, m $\angle$ C = 150, $b = 3$); but then, we had to solve a quadratic equation.

Example 2   Solve $\triangle$ ABC given that   m $\angle$ C = 130° , $a = 6, b = 9.$

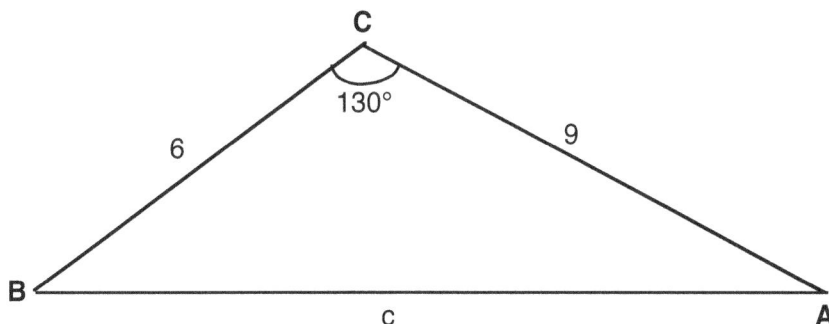

We will find m $\angle$ B, M $\angle$A and length c.

Finding c

$$c^2 = a^2 + b^2 - 2ab \cos C \quad \text{<-------Law of cosines}$$

$$c^2 = 6^2 + 9^2 - 2\,(6)(9) \cos 130°$$

$$= 36 + 81 - 108(-.642) \qquad (\cos 130° = -.642)$$

$$= 36 + 81 + 69.42$$

$$c^2 = 186.42$$

$$c = \sqrt{186.42}$$

$$c = 13.65$$

(Note above that if we had applied the law of sines to find c, we would have had one equation with two unknowns; and we would not be able to obtain a numerical value for c without an additional equation with the same unknowns.)

Finding M ∠ A:

$$\frac{c}{\sin 130^\circ} = \frac{a}{\sin A} \qquad (\text{law of sines}).$$

$$\frac{13.65}{\sin 130} = \frac{6}{\sin A} \quad (c = 13{,}65, a = 6)$$

We solve for sine A.

$$\sin A = \frac{6 \sin 130}{13.65}$$

$$= \frac{6(.766)}{13.65}$$

$$\sin A = .3367$$

Find inverse sine of .3367 (from tables or a calculator)
(That is, find the angle whose sine is .3367)

$$m \angle A = 19.68$$

$$m \angle B = 180 - (130 + 19.68)$$
$$m \angle B = 30.32$$

(We could also use the law of sines to find m ∠ B, and then check that the sum of the measures of the angles equals 180°)

∴ $c = 13.65$; $m \angle A = 19.68^\circ$; $m \angle B = 30.32^\circ$

When do we apply the Pythagorean theorem, the law of cosines, or the law of sines?

We have learned how to solve triangles using the Pythagorean theorem, the law of sines and the law of cosines. Given a triangle to solve, how do we know which theorem or law to apply? Which theorem or law to apply depends on what we are given and what we want to find.

1. We may apply the Pythagorean theorem only if the triangle is a right triangle. However, if by construction, we are able to obtain right triangles, then we can apply the theorem to any such triangles obtained.

2. For the law of cosines:
  (a) Given **two sides and an included angle** (SAS), use of the law of cosines yields a unique solution (i.e. one triangle). In this case, we will find the remaining side and the measures of two angles.

  (b) Given all **three sides** (SSS), the application of the law of cosines yields a unique solution. In this case, we will find the measures of all the angles.

3. For the law of sines:
  (a) Given **two angles** and **any side** (ASA, AAS, SAA), the use of the law of sines will yield a unique solution (one triangle).

  Note that given any two angles, we can easily find the third angle, since the sum of the measures of the angles of a triangle equals 180°.

  (b) **Ambiguous Case:** Given **two sides** and **an angle opposite one of the sides** (ASS or SSA), there may be only one solution (one triangle), two solutions (two Δ's), or there may be no solution at all (no triangle) depending on relative lengths of the sides and the measures of the angles

Note also that if the measure of any angle given is 180° or more there is no solution, since the sum of the measures of the three angles of a triangle is 180°.

Also, if the sum of the measures of any two angles given is 180° or more there is no solution.

Note also that if the triangle is a **right triangle**, we can consider applying any of the six trigonometric functions (ratios): sine, cosine, tangent, cosecant, secant, and cotangent.

Question: How do we remember the above cases?
Answer: By practice, repetition and finding mnemonic devices.

Lesson 110 Exercises

A Solve the triangle:

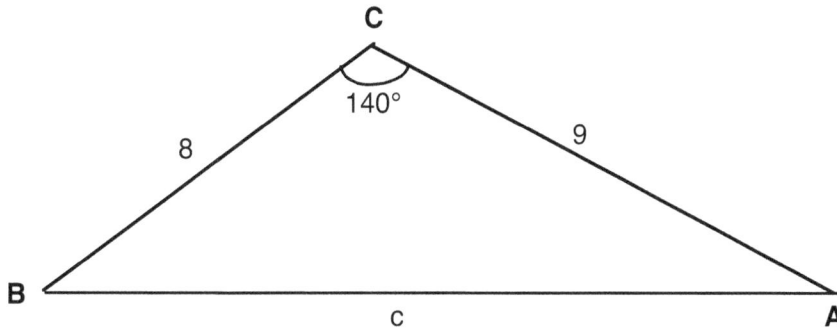

Solutions: c = 15.99; m ∠ B = 21.23° ; m ∠ A = 18.77°

B Solve the following triangles:

1.

2.

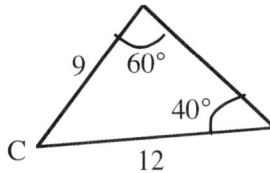

Solve the following triangles using the given dimensions:

3. $a = 15$; $m\angle A = 50°$; $m\angle B = 110°$; **4.** $b = 8$; $m\angle B = 75°$; $m\angle C = 55°$

Answers: 1. $c = 17$, $m\angle A = 20°$; $m\angle B = 25°$; **2.** $c = 13.7$, $m\angle C = 80°$
3. $m\angle C = 20°$; $b = 18$; $c = 6.7$; **4.** $m\angle A = 50°$; $a = 6.3$; $c = 6.8$

C Determine the measure of the largest angle to the nearest degree.

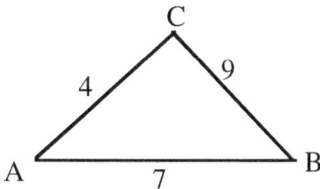

Answer: $m\angle A = 107°$ (Hint: Largest angle opposite longest side)

CHAPTER 47

Lesson 111: **Trigonometric Identities**
Lesson 112: **Proving Trigonometric Identities**

Lesson 111

Trigonometric Identities

An identity is an equation that is true for all permissible replacements of the independent variable.

Trigonometric identities are based on eight fundamental identities which are grouped as reciprocal identities, ratio identities and the Pythagorean identities.

Reciprocal identities

1. $\sec \theta = \dfrac{1}{\cos \theta}$ $\qquad (\cos \theta \neq 0)$

2. $\csc \theta = \dfrac{1}{\sin \theta}$ $\qquad (\sin \theta \neq 0)$

3. $\cot \theta = \dfrac{1}{\tan \theta}$ $\qquad (\tan \theta \neq 0)$

Ratio identities

1. $\tan \theta = \dfrac{\sin \theta}{\cos \theta}$ $\qquad (\cos \theta \neq 0)$

2. $\cot \theta = \dfrac{\cos \theta}{\sin \theta}$ $\qquad (\sin \theta \neq 0)$

Pythagorean identities

1. $\sin^2 \theta + \cos^2 \theta = 1$
2. $1 + \tan^2 \theta = \sec^2 \theta$
3. $1 + \cot^2 \theta = \csc^2 \theta$

For the Pythagorean identities, identities **2** and **3** can be derived from identity **1** by dividing identity **1** by $\cos^2 \theta$ and by $\sin^2 \theta$ respectively. Keep this in mind when trying to memorize these identities.

Addition formulas

1. $\cos(A - B) = \cos A \cos B + \sin A \sin B$

2. $\cos(A + B) = \cos A \cos B - \sin A \sin B$

3. $\sin(A + B) = \sin A \cos B + \cos A \sin B$

4. $\sin(A - B) = \sin A \cos B - \cos A \sin B$

5. $\tan(A - B) = \dfrac{\tan A - \tan B}{1 + \tan A \tan B}$

6. $\tan(A + B) = \dfrac{\tan A + \tan B}{1 - \tan A \tan B}$

Double angle identities

1. $\sin 2\theta = 2 \sin \theta \cos \theta;$ $\qquad$ **3.** $\cos 2\theta = 2\cos^2 \theta - 1$

2. $\cos 2\theta = \cos^2 \theta - \sin^2 \theta ;$ $\qquad$ **4.** $\cos 2\theta = 1 - 2\sin^2 \theta;$ $\qquad$ **5.** $\tan 2\theta = \dfrac{2\tan \theta}{1 - \tan^2 \theta}$

Lesson 111 Exercises

From memory, write down the reciprocal identities, ratio identities, and the Pythagorean identities

Lesson 112
Proving Trigonometric Identities

In proving trigonometric identities, the following guidelines will be helpful:

1. Express all functions in terms of sines and cosines and simplify.
   (However, sometimes, this may not be necessary.)

2. Work **separately** on the left-hand-side and **separately** on the right-hand side.

   Do **not** work on the equation as you would operate on both sides of a conditional equation
     This is not an "if-then statement" proof.

3. Apply the appropriate trigonometric identities or formulas (from above). For example, if there are double angles, use the double angle identities to change to single angles.
   (However, sometimes, this may not be necessary.)

4 . If there are indicated operations such as addition or subtraction of fractions, combine these
     fractions and simplify. (You may have to go back and review how to add rational fractions)
     You may sometimes do the opposite of **4** as in **5** below.

5. If the numerator of a fraction consists of two or more terms and the denominator consists of only
     one term (especially, one function only), divide every term in the numerator by the denominator.

6. If there is a binomial in either the numerator or the denominator, multiply the binomial by its
     conjugate

7. Only practice will make the above guidelines meaningful. Do not be afraid to try and err.
     You will learn from your mistakes. After some mastery, you may even begin to enjoy proving
     trigonometric identities.

Example 1   Prove the following identity: $\dfrac{\sin x}{\tan x} + \dfrac{\cos x}{\cot x} = \cos x + \sin x$.

(**Note** that the right-hand side (RHS) is in terms sines and cosines. Therefore, we do not have to do any work
on the RHS. We therefore work on the LHS and express the LHS in terms of sines and cosines and simplify.)

$$\frac{\sin x}{\tan x} + \frac{\cos x}{\cot x} \overset{?}{=} \cos x + \sin x$$

(We will remove the question mark" ? " when we have

exactly the same expression on both sides of the equation)

$$\frac{\sin x}{\dfrac{\sin x}{\cos x}} + \frac{\cos x}{\dfrac{\cos x}{\sin x}} \overset{?}{=} \cos x + \sin x$$

$$\left(\tan x = \frac{\sin x}{\cos x} \, ; \, \cot x = \frac{\cos x}{\sin x} \right)$$

$$\frac{\sin x}{1} \cdot \frac{\cos x}{\sin x} + \frac{\cos x}{1} \cdot \frac{\sin x}{\cos x} \overset{?}{=} \cos x + \sin x$$

(Inverting the divisor and multiplying)

$$\frac{\cancel{\sin x}}{1} \cdot \frac{\cos x}{\cancel{\sin x}} + \frac{\cancel{\cos x}}{1} \cdot \frac{\sin x}{\cancel{\cos x}} \overset{?}{=} \cos x + \sin x$$

$$\cos x + \sin x = \cos x + \sin x.$$

QED

Example 2   Prove that $\dfrac{\sin 2x}{\sin x} = \sec x + \dfrac{\cos 2x}{\cos x}$

(In this problem, we will work on both sides.)

$$\frac{\sin 2x}{\sin x} \overset{?}{=} \sec x + \frac{\cos 2x}{\cos x}$$

$$\frac{2\sin x \cos x}{\sin x} \overset{?}{=} \frac{1}{\cos x} + \frac{\cos^2\theta - \sin^2\theta}{\cos x} \qquad (\sin 2x = 2\sin x \cos x \text{ for the LHS})$$

$$\frac{2\cancel{\sin x}\cos x}{\cancel{\sin x}} \overset{?}{=} \frac{1}{\cos x} + \frac{\cos^2 x - \sin^2 x}{\cos x} \qquad (\cos 2\theta = \cos^2\theta - \sin^2\theta \text{ for the RHS})$$

$$2\cos x \overset{?}{=} \frac{1 + \cos^2 x - \sin^2 x}{\cos x} \qquad\qquad \text{(Adding the fractions)}$$

$$2\cos x \overset{?}{=} \frac{1 - \sin^2 x + \cos^2 x}{\cos x} \qquad\qquad \text{(Rewriting)}$$

$$2\cos x \overset{?}{=} \frac{\cos^2 x + \cos^2 x}{\cos x} \qquad\qquad (1 - \sin^2\theta = \cos^2\theta)$$

$$2\cos x \overset{?}{=} \frac{2\cos^2 x}{\cos x}$$

$$2\cos x \overset{?}{=} \frac{2\cos x \cancel{\cos x}}{\cancel{\cos x}}$$

$$2\cos x = 2\cos x \qquad\qquad \text{(Note the LHS = the RHS)}$$

Lesson 112 Exercises

Prove the following identities:

1. $\cot x + \tan x = \sec x \csc x$

2. $\tan x + \csc x = \dfrac{\sin x + \cot x}{\cos x}$

3. $\dfrac{\cos 2x}{\cos x} + \sec x = \dfrac{\sin 2x}{\sin x}$;

4. $\dfrac{\cos x}{\sec x} - \dfrac{\sin x}{\cot x} = \dfrac{\cos x \cot x - \tan x}{\csc x}$

CHAPTER 48

Lesson 113
Solutions of Trigonometric Equations

A trigonometric equation is an equation containing a trigonometric function of an unknown angle. A trigonometric equation is a conditional equation. By conditional we mean the equation is true for some but not all permissible replacements of the variable. We can solve a trigonometric equation either graphically or algebraically. We will cover only algebraic methods although on occasion we may complement the algebraic solutions by graphical methods. In solving trigonometric equations, we will make use of trigonometric identities as well as inverse trigonometric relations.
We will cover two types of trigonometric equations, namely,

1. The trigonometric equation which is linear in one trigonometric function of one variable.

2. The trigonometric equation which is quadratic in one variable of the function.

We will restrict our solutions for an unknown angle x such that $0 \le x \le 360^\circ$ ($0 \le x \le 2\pi$).

General Procedure for solving trigonometric equations:

Step 1: Collect all terms on one side of the equation, preferably on the left-hand side of the equation.

Step 2: Solve the equation for the trig function as exemplified below:
Examples in which the functions have been solved for: **1**. $\sin x = .5$
2. $\tan x = 1$
Examples in which the functions have **not** been solved for: **1**. $4\sin^2 x = 1$
2. $\cos^2 x = .5$

In trying to solve for the functions we may have to apply trigonometric identities if the equation contains multiple angles and / or more than one type of a trig function. If there are multiple angles such as double angles, we may apply the double angle identities or we may substitute a new variable for the multiple angle, solve for the function in terms of the new variable, and reconvert to the original variable after having found the measure of the angle.

Step 3: Solve for the angle: Use inverse trigonometric functions (from tables, memory or calculator) to find the measures of the angles involved.

Step 4: Check the solutions if the equation was squared or raised to a power, since we might have introduced extraneous solutions.

Example 1 Solve for x: $\sin x = \frac{1}{2}$ or .5 $0 \le x < 360^\circ$

Solution

Step 1: Find the angle whose sine is $\frac{1}{2}$ (from tables or calculator). The reference angle is 30°

Step 2: Determine the quadrants in which the sine is positive.
The sine is positive in quadrants I and II. ($\frac{1}{2}$ is positive)
Step 3 : Specify the measures of the angles (in standard position).
$x = 30^\circ$ or 150° (The angles are measured from the positive x-axis counterclockwise)
The solution set is $\{30^\circ, 150^\circ\}$

Example 2   Solve for x: $2\sin x - 1 = 0$ $0 \le x < 360^\circ$

Step 1: Solve for $\sin x$

$2\sin x = 1$

$\sin x = \frac{1}{2}$ ....(1)

Step 2: Equation (1) of Step 1 is the same as Example 1 above. We therefore repeat the solution:

The solution set is $\{30^\circ, 150^\circ\}$

Example 3   Solve for x: $\cos 2x = \frac{1}{2}$.

Method 1

Step 1: Let $2x = \theta$

Then, $\cos 2x = \frac{1}{2}$ becomes

$\cos \theta = \frac{1}{2}$

Step 2: Determine the reference angle for θ.

The reference angle is 60° (From tables, memory or calculator)

Step 3: Determine the quadrant in which $\cos \theta$ is positive.

The cosine is positive in quadrant I and IV.   ($\frac{1}{2}$ is positive)

Step 4: Specify the measures of the angles   (in standard position).

$\theta = 60^\circ, \theta = 300^\circ$

Step 5:   The general solutions are $\theta = 60^\circ + 360n$ or $\theta = 300^\circ + 360n$ ($n = 0, 1$ for this problem)

Step 6: Change back to x. That is, replace θ by $2x$

$2x = 60^\circ + n \cdot 360^\circ$ or $2x = 300^\circ + n \cdot 360^\circ$

$x = 30^\circ + n \cdot 180^\circ$ or $x = 150^\circ + n \cdot 180^\circ$ (Dividing by 2 and solving for x)

For $n = 0$, $x = \mathbf{30^\circ}$ or $x = \mathbf{150^\circ}$ (When $n = 0, n \cdot 180 = 0$)

For $n = 1$, $x = 30^\circ + 180^\circ$ or $x = 150^\circ + 180^\circ$

$x = \mathbf{210^\circ}$ or $x = \mathbf{330^\circ}$

The solution set is $\{30^\circ, 150^\circ, 210^\circ, 330^\circ\}$

Note that in Step 6, we considered two values of n ($n = 0, 1$) because of the "2" in $\cos 2x$. If we were given $\cos 3x$, we would have considered three values of n ($n = 0, 1, 2$) because of the "3" in $\cos 3x$.

Method 2: Using the double angle identity, $\cos 2x = 2\cos^2 x - 1$

The given equation $\cos 2x = \frac{1}{2}$ becomes

$2\cos^2 x - 1 = \frac{1}{2}$

$2\cos^2 x = \frac{1}{2} + 1$

$2\cos^2 x = \frac{3}{2}$

$\cos^2 x = \frac{3}{4}$

$\cos x = \pm \sqrt{\frac{3}{4}}$

$\cos x = \pm \frac{\sqrt{3}}{2}$

$\cos x = + \frac{\sqrt{3}}{2}$ or $\cos x = - \frac{\sqrt{3}}{2}$

For $\cos x = +\dfrac{\sqrt{3}}{2}$, the reference angle is 30°.

$\cos x$ is positive in quadrants I and IV.

Therefore, $x = \textbf{30° or 330°}$ (angles in standard position).

Similarly, for $\cos x = -\dfrac{\sqrt{3}}{2}$, the reference angle is 30°.

$\cos x$ is negative in quadrants II and III.

Therefore, $x = \textbf{150° or 210°}$ (angles in standard position).

The solution set is $\{30°, 150°, 210°, 330°\}$.

Again, we obtain the same solution set as by Method 1. You decide which method you prefer.

Example 4 Solve for x: $\quad 2\cos^2 x + 3\sin x = 3 \quad 0 \le x < 360°$

Solution $\quad 2\cos^2 x + 3\sin x - 3 = 0$

Step 1: $\quad 2(1 - \sin^2 x) + 3\sin x - 3 = 0 \qquad\qquad (\cos^2 x = 1 - \sin^2 x)$

$\qquad 2 - 2\sin^2 x + 3\sin x - 3 = 0$

$\qquad -2\sin^2 x + 3\sin x - 1 = 0$

$\qquad 2\sin^2 x - 3\sin x + 1 = 0 \qquad\qquad (1)$

*Step 2: Let $\sin x = t$, Then equation (1) becomes

$\qquad 2t^2 - 3t + 1 = 0$ <--------quadratic equation.

We solve by factoring since the factors are easily recognizable.

$\qquad (2t - 1)(t - 1) = 0$

$\qquad 2t - 1 = 0$ or $t - 1 = 0$

$\qquad t = \dfrac{1}{2}$ or $t = 1$

Now, replace t by $\sin x$.

Then $\sin x = \dfrac{1}{2}$ or $\sin x = 1$

Step 3: Solve for the angles:

$\qquad$ When $\sin x = \dfrac{1}{2}$, the reference angle is 30°,

$\qquad$ and $x = \textbf{30° or 150°}$

$\qquad$ When $\sin x = 1, x = \textbf{90°}$

The solution set is $\{30°, 90°, 150°\}$.

* Note that in Step 2, we could avoid substitution by doing the following: $(2\sin x - 1)(\sin x - 1) = 0$.

Then either $2\sin x - 1 = 0$ or $\sin x - 1 = 0$.

Solving, $\sin x = \dfrac{1}{2}$ or $\sin x = 1$

Note:

To check the solutions to the above equations, substitute in the original equations

Lesson 113 Exercises

Solve for x: 1. $4\cos^2 x - 8\cos x + 3 = 0$; **2.** $2\cos^2 x - \cos x = 1$

Answers: **1** .$\{60°, 300°\}$ **2**. $\{0°, 120°, 240°\}$

CHAPTER 49

Lesson 114: **Distance Formula** (Review)
Lesson 115: **Equation and Graph of a Circle**

Lesson 114
Distance Formula (Review)

The distance, d, between the points $P_1(x_1, y_1)$ and $P_2(x_2, y_2)$ on a line in a plane (Fig.1) is given by

$d = \sqrt{(x_2 - x_1)^2 + (y_2 - y_1)^2}$ (By applying the Pythagorean theorem to the right triangle, and solving for d)

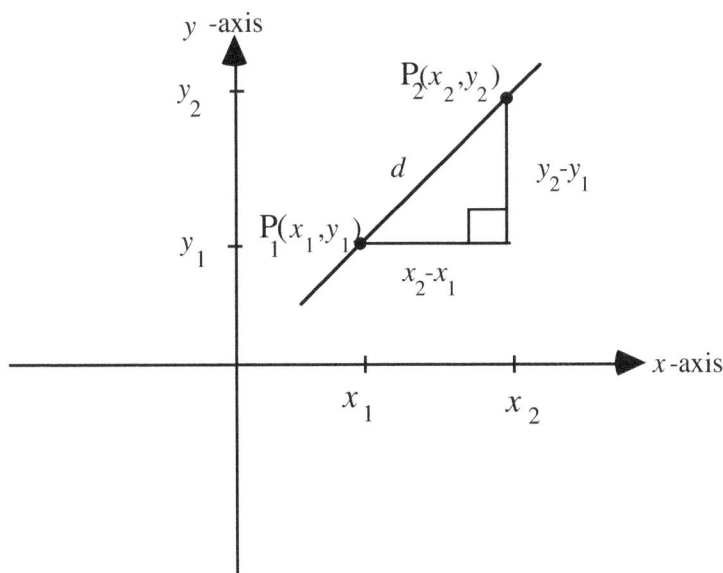

Figure 1

Example   Find the distance between the points $(-2, 3)$ and $(4, -5)$.

Solution: Apply the distance formula:

$$d = \sqrt{(x_2 - x_1)^2 + (y_2 - y_1)^2} \qquad (1)$$

(where d is the distance between the points $P_1(x_1, y_1)$ and $P_2(x_2, y_2)$).

Substituting $x_1 = -2, y_1 = 3, x_2 = 4, y_2 = -5$ in equation (1) above,

$$d = \sqrt{(4 - (-2))^2 + (-5 - 3)^2}$$
$$d = \sqrt{(4 + 2)^2 + (-8)^2}$$
$$= \sqrt{(6)^2 + (-8)^2}$$
$$= \sqrt{36 + 64}$$
$$= \sqrt{100}$$
$$= 10$$

∴ the distance between the given points is 10 units.

Lesson 114 Exercises

1. Find the distance between the points (3,4) and (5, -1) .

2. Find the distance between the points (-4, 2) and (-6, -3)

Answers: **1.** $\sqrt{29}$; **2.** $\sqrt{29}$

Lesson 115

Equation and Graph of a Circle

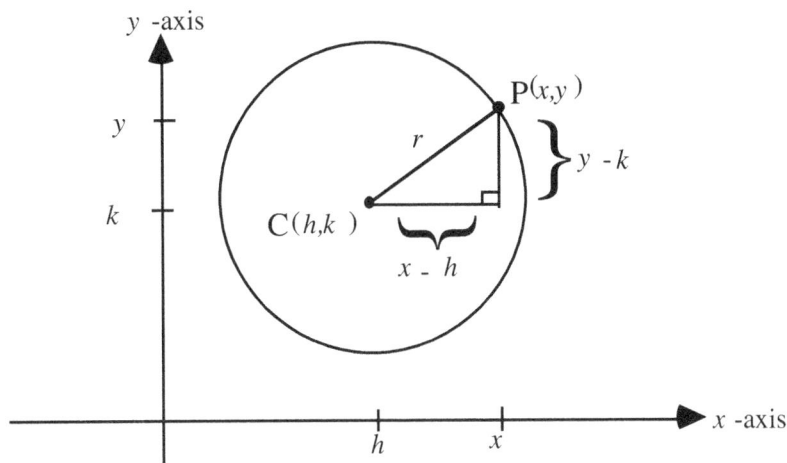

Figure 2

We will derive an equation of a circle with its center at the point $C(h, k)$ (Figure 2, above) and having radius, r, by applying the Pythagorean theorem.

Let the distance between the center $C(h, k)$ of the circle and a point $P(x, y)$ on the circle be r.
Applying the Pythagorean theorem (since we have a right triangle),

$$r^2 = (x - h)^2 + (y - k)^2 \text{ or rewriting,}$$
$$(x - h)^2 + (y - k)^2 = r^2 \text{ <--------- center-radius form of the equation of a circle}$$

This equation could also be obtained from the distance formula as follows:
Replace d by r; x_2 by x; x_1 by h; y_2 by y, and y_1 by k in

$$d = \sqrt{(x_2 - x_1)^2 + (y_2 - y_1)^2} \quad \text{<--------- distance formula.}$$

Then we obtain $r = \sqrt{(x - h)^2 + (y - k)^2}$

Squaring both sides of the equation, $r^2 = (x - h)^2 + (y - k)^2$
Rewriting, $(x - h)^2 + (y - k)^2 = r^2$ <--------- center-radius form

Example 1   (a) Write an equation for the circle with radius 5 and center at $(2, -3)$.
         (b) Sketch the graph of the circle.
The center-radius form of the equation of a circle is given by

$(x - h)^2 + (y - k) = r^2$, where r is the radius of the circle, h = the x- coordinate of the center, and k = the y-coordinate of the center of the circle.

Step 1: Substituting $r = 5$, $h = 2, k = -3$ in
$\qquad (x - h)^2 + (y - k)^2 = r^2$, we obtain
$\qquad (x - 2)^2 + (y - (-3))^2 = 5^2$
$\qquad (x - 2)^2 + (y + 3)^2 = 5^2$ <-------center-radius form.
$\quad$ or $(x - 2)^2 + (y + 3)^2 = 25$.

(b) **Graph**   Step 1: Plot the coordinates (2,-3) of the center of the circle.

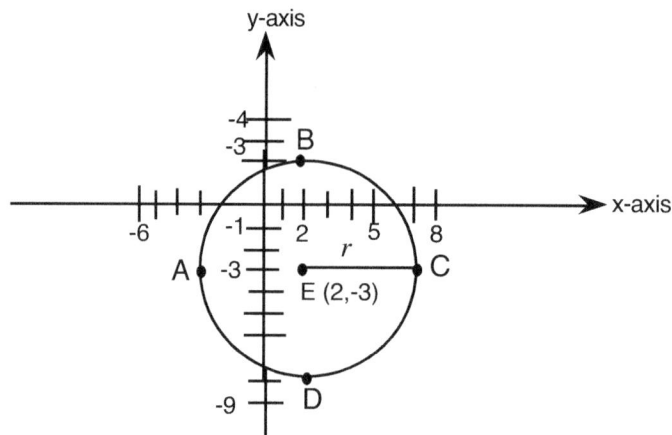

Step 2: Since the radius is 5, count 5 units each , horizontally, from the center (2,-3) to the points A, and C; and also count 5 units each, vertically, from the center (2, -3) to the points B and D. Connect these four points by a smooth "circular " curve to obtain the circle.

Note: Use the same scale for the *x*- and *y*-axes.

Note also that for a circle, the distance from the center of the circle to any point on the circumference is the same as the distance from the center to any other point on the circumference. Keep this in mind when drawing a circle.

Example 2   Find an equation of the circle with center at the point (4, -2) and passing
         through the point (-3, 6).

If we know the radius, r, and the coordinates h, k of the center of a circle we can immediately write down an equation for the circle by applying

$$(x - h)^2 + (y - k)^2 = r^2. \qquad (1)$$

We are given $h = 4, k = -2$, but we do not know r yet; however, we can find it.

Finding *r*:

   Since the point (-3, 6) is given to be on the circle, we may substitute this ordered

   pair in equation (1).

Step 1: Substitute $h = 4, k = -2, x = -3, y = 6$

   in equation (1) and solve for r.

$$(-3 - 4)^2 + (6 - (-2))^2 = r^2$$
$$(-7)^2 + (6 + 2)^2 = r^2$$
$$(-7)^2 + 8^2 = r^2$$
$$49 + 64 = r^2$$
$$113 = r^2$$
$$\sqrt{113} = r$$
$$\text{and } r = \sqrt{113}$$

Step 2: Substitute $h = 4, k = -2, r = \sqrt{113}$ (or $r^2 = 113$) in equation (1) above.

   ∴ an equation for the given circle is $(x - 4)^2 + (y + 2)^2 = 113$ <-------center-radius form.

Finding the center and radius of a circle (Given the center-radius form of the equation of a circle)

Example 1 Find the center and radius of the circle whose equation is given by

$$(x + 4)^2 + (y - 3)^2 = 36$$

Procedure: The center-radius form of the equation of the circle with center at the point (h, k) and with radius r is given by:

$$(x - h)^2 + (y - k)^2 = r^2 \qquad\qquad (1)$$

Compare $(x + 4)^2 + (y - 3)^2 = 36$ with equation (1) above.

Then $(x - h)^2 = (x + 4)^2$; $(y - k)^2 = (y - 3)^2$; and $r^2 = 36$

$x - h = x + 4$; $\qquad y - k = y - 3 \qquad\qquad r = \sqrt{36} = 6$
$-h = 4$; $\qquad\qquad -k = -3$
$h = -4 \qquad\qquad k = 3$

Therefore $h = -4$; $k = 3$; and $r = 6$.

The center of the circle is at (-4,3); and its radius is 6 units.

Shortcut method: Step 1: Set $x + 4 = 0$ and solve to obtain $x = -4$ (Thus $h = -4$
       Step 2: Set $y - 3 = 0$ and solve to obtain $y = 3$ (Thus $k = 3$
       The center of the circle is at (-4,3);

With some practice, we shall be able to read, by comparison with equation (1) above, the coordinates of the center of the circle without performing any calculations.

Finding the center and radius of a circle given the general form of the equation of a circle.

Example 2   Find the center and radius of a circle whose equation is given by
       $x^2 + y^2 + 8x - 6y - 11 = 0$

Procedure: We will group the x-terms, group the y-terms, and complete the square on the x-terms, and on the y-terms. (see page 361 and review how to complete the square.)

Step 1: $x^2 + y^2 + 8x - 6y - 11 = 0$
       $x^2 + 8x + y^2 - 6y = 11$.............................. (1)

Step 2: Complete the squares:

* $x^2 + 8x + 16 + y^2 - 6y + 9 = 11 + 16 + 9$ ....(2)

       $\qquad\qquad\qquad\qquad\qquad\qquad$ (For the x-terms: $x^2 + 8x + (4)^2$)
       $x^2 + 8x + 16 + y^2 - 6y + 9 = 36$ $\qquad$ (For the y-terms : $y^2 - 6y + (-3)^2$)

$(x + 4)^2 + (y - 3)^2 = 36$ <---From this center-radius form, follow the steps exactly as in Example 1 above.

The center of the circle is at (-4, 3); and its radius is 6 units.

* Whenever we add any quantity to the left-hand side of equation (1), we must add the same quantity to the right-hand side of the equation: First, we added 16 (i.e., 4^2) for the x-terms, to both sides of the equation; then we added 9 (i.e.,$(-3)^2$) for the y-terms to both sides of the equation.

Lesson 115 Exercises

A (a) Write an equation for the circle with radius 4 and center at (-2, 3).

(b) Sketch the graph of the circle.

Solution: $(x + 2)^2 + (y - 3)^2 = 16$<------- center-radius form.

B Find an equation of the circle with center at the point (2, 3) and passing through the point (-4, 1).

Answer: $(x - 2)^2 + (y - 3)^2 = 40$

C Find the center and radius of the circle whose equation is given by

$$(x - 5)^2 + (y + 2)^2 = 16$$

Answer: Center at (5, -2); and radius, 4 units

D Find an equation for each circle with the given properties.

1. Radius 4 units with center at the origin.

2. Radius 3 units and center at (-3, 2).

3. Radius 5 units and center at (-2, -4)

Problems 4-6: Find the center and radius of each of the given circle

4. $x^2 - 2x + y^2 - 4y - 11 = 0$;   **5.** $x^2 + y^2 + 4x - 6y = 12$;   **6.** $x^2 + y^2 - x - \dfrac{y}{2} = \dfrac{63}{16}$.

Answers: **1.** $x^2 + y^2 = 16$;   **2.** $(x + 3)^2 + (y - 2)^2 = 9$;   **3.** $(x + 2)^2 + (y + 4)^2 = 25$

     4. Center at (1, 2), radius 4 units;   **5.** Center at (-2, 3), radius 5 units;

     6. Center at $(\frac{1}{2}, \frac{1}{4})$, radius $\dfrac{\sqrt{17}}{2}$ units.

APPENDIX A

The following theorems were proposed by the author in the Summer of 1993, and may be helpful when dealing with problems in which the lengths of the sides and the measures of the angles of the triangles are given.

Similar Triangles

Theorems on similar triangles in terms of the **lengths** of the sides and the **measures** of the **angles**

Theorem 1 (Another form of the **SSS** theorem for similar triangles)
If the lengths of the sides one triangle are x_1, x_2, x_3, where $x_1 \leq x_2 \leq x_3$, and the lengths of the sides of another triangle are y_1, y_2, y_3, where $y_1 \leq y_2 \leq y_3$, and if

$$\frac{x_1}{y_1} = \frac{x_2}{y_2} = \frac{x_3}{y_3}, \text{ then the two triangles are similar.}$$

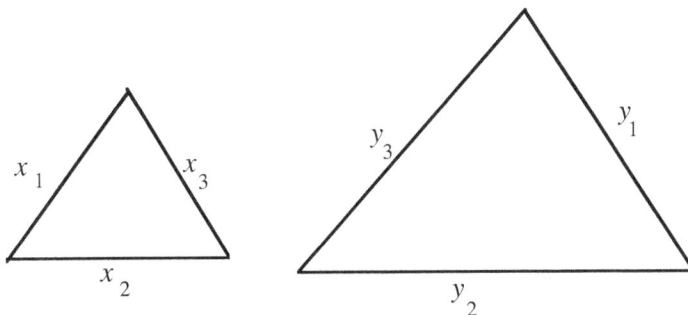

* Note: The "=" in "≤" is for the cases of isosceles or equilateral triangles.

Converse
If two triangles are similar and the lengths of the sides of one triangle are x_1, x_2, x_3, where $x_1 \leq x_2 \leq x_3$, and the lengths of the sides of the other triangle are y_1, y_2, y_3, where $y_1 \leq y_2 \leq y_3$,

then $\dfrac{x_1}{y_1} = \dfrac{x_2}{y_2} = \dfrac{x_3}{y_3}$.

Note above that x_1 and y_1 are corresponding lengths; x_2 and y_2 are corresponding lengths, and x_3 and y_3 are corresponding lengths. (i.e., the smallest length of Δ #1 corresponds to the smallest length of Δ #2, and the largest length of Δ #1 corresponds to the largest length of Δ #2 , and the remaining lengths correspond.)

* **1.** If two triangles are similar and one triangle is isosceles, the other triangle is also isosceles.

* **2.** If two triangles are similar and one triangle is equilateral, the other triangle is also equilateral.
* **3.** Also, all equilateral triangles are similar.

In the figures below:
(a) Find the corresponding lengths .
(b) Form ratios using the corresponding lengths.
(c) Are the two triangles similar?

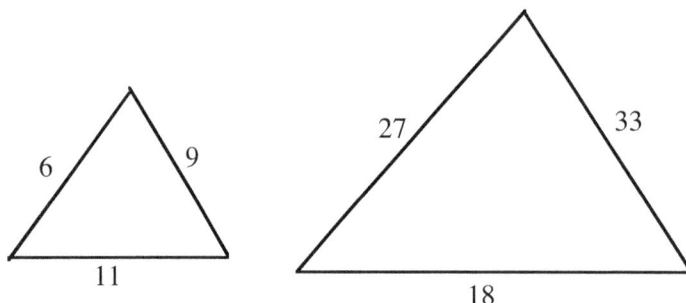

Solutions:
(a) $6 < 9 < 11$ and $18 < 27 < 33$
  6 and 18 correspond; 9 and 27 correspond; and 11 and 33 correspond.
(b) $6{:}18 = \dfrac{6}{18} = \dfrac{1}{3}$; $9{:}27 = \dfrac{9}{27} = \dfrac{1}{3}$; $11{:}33 = \dfrac{11}{33} = \dfrac{1}{3}$.
(c) Since the ratios are the same for each pair of corresponding sides, the corresponding sides are in proportion and the two triangles are similar by SSS.

Theorem 2 (Another form of the **SAS** theorem for similar triangles)

> If the lengths of two sides one triangle are x_1, x_2, where $x_1 \leq x_2$, with an included angle whose measure is α, and the lengths of the sides of another triangle are y_1, y_2, where
>
> $y_1 \leq y_2$, with an included angle whose measure is β; and if $\dfrac{x_1}{y_1} = \dfrac{x_2}{y_2}$ and $\alpha = \beta$, then the
>
> two triangles are similar.

Theorem 3 (Another form of **AA** theorem for similar triangles)
> If the measures of two angles of a triangle are α_1, α_2, where $\alpha_1 \leq \alpha_2$ and the measures of
> two angles of another triangle are β_1, β_2, where $\beta_1 \leq \beta_2$; and if $\alpha_1 = \beta_1$ and $\alpha_2 = \beta_2$,then
> the two triangles are similar.

Congruent Triangles

Theorems on congruent triangles in terms of the lengths of the sides and the measures of the angles

Theorem 1 (Another form of **SSS** for congruent triangles)

If the lengths of the sides of one triangle are x_1, x_2, x_3, where $x_1 \leq x_2 \leq x_3$, and the lengths of the sides of another triangle are y_1, y_2, y_3, where $y_1 \leq y_2 \leq y_3$; and if $x_1 = y_1, x_2 = y_2, x_3 = y_3$, then the two triangles are congruent.

Theorems 2 (Another form of the **SAS** theorem for congruent triangles)

If the lengths of two sides one triangle are x_1, x_2, where $x_1 \leq x_2$, with an included angle whose measure is α, and the lengths of the sides of another triangle are y_1, y_2, where $y_1 \leq y_2$, with an included angle whose measure is β, and if $x_1 = y_1, x_2 = y_2$ and $\alpha = \beta$, then the two triangles are congruent.

Theorems 3 (Another form of the **ASA** theorem for congruent triangles)

If the measures of two angles of one triangle are $\alpha_1 \ \alpha_2$, where $\alpha_1 \leq \alpha_2$, with an included side of length x, and the measures of two angles of another triangle are β_1, β_2, where $\beta_1 \leq \beta_2$, with an included side of length y; and if $\alpha_1 = \beta_1, \alpha_2 = \beta_2$ and $x = y$, then the two triangles are congruent.

Note:

1. If two triangles are congruent and one triangle is isosceles, the other triangle is also isosceles.

2. If two triangles are congruent and one triangle is equilateral, the other triangle is also equilateral

Absolute Value Equations

Example 4 (From p.416)   Solve for x: $|x - 2| = |x + 6|$. We consider four cases.

Case 1: Both expressions within the absolute value bars are positive. Then $x - 2 = x + 6$ (A) **Case 2:** Both expressions within the absolute value bars are negative. Then $-(x - 2) = -(x + 6)$ (B) **Case 3:** The expression within the absolute value bars on the left is positive and that on the right is negative. Then $x - 2 = -(x + 6)$ (C) **Case 4:** The expression within the absolute bars on the left is negative and that on the right is positive. Then $-(x - 2) = x + 6$ (D)	On simplifying, (B) reduces (A). Similarly, (C) and (D) are equivalent, ignoring the domains of the definitions. We therefore solve only Case 1, $x - 2 = x + 6$ and Case 3, $x - 2 = -(x + 6)$ Solving for Case 1, $x - 2 = x + 6$ $-2 = 6$, which is a contradiction. There is no solution for Cases 1 and 2. For Cases 3 and 4, $x - 2 = -(x + 6)$ $x - 2 = -x - 6$, and from which $x = -2$ We check for $x = -2$. Then $|-2 - 2| \overset{?}{=} |-2 + 6|$; $|-4| \overset{?}{=} |4|$; $4 \overset{?}{=} 4$. Yes. $\boxed{\text{The solution is } -2.}$

APPENDIX B

Lesson 116: One-to-One Functions, Composite Functions
Lesson 117: Inverse Functions and Inverse Relations

Lesson 116

One-to-One Functions, Composite Functions

One-to-One Functions

(In Lesson 50, we will learn that if a function is one-to-one, then it has an inverse function.)

We consider three main cases according to how the function is specified.

Case 1: **Given a set of ordered pairs** (or a table of x- and y-values)

A one-to-one (1-1) function is a set of ordered pairs in which for any two ordered pairs, the first elements are different from each other and the second elements are also different from each other.

Example The set, $A = \{(3,2),(4,7),(1,5),(2,3)\}$ is a one-to-one function. However.

The set, $B = \{(3,2),(4,7),(1,5),(5,2)\}$ is **not** a one-to-one function because the first and the last ordered pairs have the same second elements, namely, 2.

Case 2: Given the graph of the function

By the so called **horizontal line test**, the graph of a function is that of a one-to-one function if every possible horizontal line drawn to intersect the graph cuts (intersects) the graph only once (at one point only). **Figures** 2 and 3, p. 223, are graphs of one-to-one functions but **Figure 1** is not one-to-one.

Case 3: Given the equation of the function

A function $f(x)$ is one-to-one if whenever $f(x_1) = f(x_2)$, $x_1 = x_2$. If we let $y_1 = f(x_1)$, and $y_2 = f(x_2)$, then $y_1 = y_2$ **must** imply that $x_1 = x_2$, otherwise, $f(x)$ is not one-to-one.

Example 2: Determine if $f(x) = 2x + 3$ is one-to-one.
Solution

Step 1: $f(x_1) = 2x_1 + 3$
$\qquad f(x_2) = 2x_2 + 3$
$\quad$ Equate RHS of $f(x_1)$ to RHS of $f(x_2)$
$\quad$ (That is, let $f(x_1) = f(x_2)$):
$\qquad 2x_1 + 3 = 2x_2 + 3 \qquad (1)$

Step 2: Solve for x_1. (You may also solve for x_2)
$\qquad 2x_1 + 3 = 2x_2 + 3$
$\qquad\quad 2x_1 = 2x_2$
$\qquad\quad\; x_1 = x_2 \qquad\qquad (2)$

Since from above, whenever $f(x_1) = f(x_2)$ (from equation (1)) $x_1 = x_2$ (from equation (2))

$f(x) = 2x + 3$ is one-to-one. (You may check by sketching its graph and using the horizontal line test)

Example 3 Determine if $f(x) = \sqrt{25 - x^2}$ is one-to-one.

Solution

Step 1: $f(x_1) = \sqrt{25 - x_1^2}$

$f(x_2) = \sqrt{25 - x_2^2}$

Equate RHS of $f(x_1)$ to RHS of $f(x_2)$
(That is, let $f(x_1) = f(x_2)$.)

$$\sqrt{25 - x_1^2} = \sqrt{25 - x_2^2} \qquad (1)$$

Step 2: Solve for x_1.

$$\sqrt{25 - x_1^2} = \sqrt{25 - x_2^2}$$
$$25 - x_1^2 = 25 - x_2^2$$
$$-x_1^2 = -x_2^2$$
$$x_1^2 = x_2^2$$
$$x_1 = \pm \sqrt{x_2^2}$$
$$x_1 = +x_2 \text{ or } -x_2 \qquad (2)$$

Since from above, whenever $f(x_1) = f(x_2)$, $x_1 = -x_2$, (That is for the same y-value, x is not unique: $x_1 = +x_2$, and **also** $x_1 = -x_2$ (from equation (2))

$f(x) = \sqrt{25 - x^2}$ is **not** one-to-one. (You may check by sketching its graph and using the horizontal line test)

Example 4: Determine if $f(x) = |x - 2|$ is one-to-one.

Solution

Step 1: $f(x_1) = |x_1 - 2|$

$f(x_2) = |x_2 - 2|$

Equate RHS's of $f(x_1)$ to $f(x_2)$
(That is, let $f(x_1) = f(x_2)$):

$$|x_1 - 2| = |x_2 - 2| \qquad (1)$$

Step 2: Solve for x_1:

If both $x_1 - 2$ and $x_2 - 2$ are positive,

$\quad x_1 - 2 = x_2 - 2$ and from which $x_1 = x_2$

(same result as if both $x_1 - 2$ and $x_2 - 2$ are negative)

However, if $x_1 - 2$ is positive and $x_2 - 2$ is negative,

$$x_1 - 2 = -(x_2 - 2)$$
$$x_1 - 2 = -x_2 + 2$$
$$x_1 = -x_2 + 4 . \text{ That is, } x_1 \neq x_2$$

Since from above, whenever $f(x_1) = f(x_2)$, $x_1 \neq x_2$

$f(x) = |x - 2|$ is **not** one-to-one. (You may check by sketching its graph and using the horizontal line test)

Another method for determining if a function is one-to-one

The author proposes the following definition for a one-to-one function.

Definition : A function is one-to-one if its inverse relation is a function.
This definition provides another method for determining if a given function is one-to-one.

Example 5

Determine if $f(x) = 2x + 6$ is one-to-one
Solution
Given: $f(x) = 2x + 6$
Required: To determine if $f(x) = 2x + 6$ is one-to-one.
Plan: If it can be shown that $f(x) = 2x + 6$ has an inverse relation which is function, then
$$f(x) = 2x + 6 \text{ is one-to-one.}$$
Determination:
Step 1: Let $f(x) = y$ to obtain $y = 2x + 6$.
Step 2: Interchange x and y to obtain the inverse relation $x = 2y + 6$
Step 3: Solve for y.
$$2y = x - 6$$
$$y = \tfrac{1}{2}x - 3$$
Since clearly, for a given value of x there is exactly one corresponding value of y.

the inverse relation, $y = \tfrac{1}{2}x - 3$, is a function

and therefore, $f(x) = 2x + 6$ is one-to-one.

(You may also check that $y = \tfrac{1}{2}x - 3$ is a function by sketching its graph and applying the vertical line test)

Example 6

Determine if $f(x) = x^2$ is one-to-one.
Solution
Given: $f(x) = x^2$

Required: To determine if $f(x) = x^2$ is one-to-one.

Plan: If it can be shown that $f(x) = x^2$ has an inverse relation which is function, then
$$f(x) = x^2 \text{ is one-to-one.}$$
Determination

Step 1: Let $f(x) = y$ to obtain $y = x^2$
Step 2: Interchange x and y to obtain the

inverse relation $x = y^2$
Step 3: Solve for y.
$$\text{If } y^2 = x$$
$$y = \pm\sqrt{x} \quad \text{(that is, } y = +\sqrt{x} \text{ or } y = -\sqrt{x}\text{)}$$
Clearly, for a given value of x there are two different corresponding y-values. Therefore, y is

not a function of x, and the inverse relation $y^2 = x$ is **not** a function and therefore , the given function

$f(x) = x^2$ is **not** one-to-one.

(You may also check that $y^2 = x$ or $y = \pm\sqrt{x}$ is **not** a function by sketching its graph and applying the vertical line test)

Composite Functions

Some authors refer to composite functions as "product functions". This alternative terminology may be misleading because a reader might be inclined to multiply the given functions. Perhaps, a better alternative terminology for a composite function is " a function within another function".

Use of Composition of Functions:
The principle of composition of functions can be used to determine algebraically if two given functions are inverses of each other.

Definition: If two functions f and g are such that the range of g is in the domain of f, then the composite function of f with g, symbolized $f \circ g$, is specified by $(f \circ g)(x) = f[g(x)]$ which is read "f of g of x". That is, the output of g becomes the input for f.

Similarly, if the range of f is in the domain of g, then the composite function of g with f, symbolized $g \circ f$ is specified by $g \circ f = g[f(x)]$ which is read "g of f of x." That is, the output of f becomes the input for g.

We must **note** that generally, $f \circ g \neq g \circ f$ (i.e., generally, $f \circ g$ is not equal to $g \circ f$)

Example 1

If $f(x) = x^2$, and $g(x) = x + 1$, find (a) $f[g(x)]$; (b) $g[f(x)]$.

Solution

Recall that if, for example,

$$f(x) = x^2$$

then $f(3) = (3)^2$.

Similarly, (a) $f[g(x)] = f[x + 1]$

$$= [x + 1]^2$$

$$f[g(x)] = x^2 + 2x + 1$$

Thus, (in the above problem) wherever there is x in the equation for $f(x) = x^2$, write (substitute) $x + 1$.

(b) $g[f(x)]$

$$= g[x^2]$$

$$= (x^2) + 1 \quad \text{(substitute } x^2 \text{ for } x \text{ in the equation for } g(x) = x + 1)$$

$$= x^2 + 1$$

Thus, wherever there is x in the equation for $g(x) = x + 1$, we substitute x^2.

Example 2: If $f(x) = 2(x + 10)$, and $g(x) = x - 2$, find (a) $f[g(x)]$; (b) $g[f(x)]$.

Solution

(a) $f[g(x)] = 2[(x - 2) + 10]$

$$= 2[x - 2 + 10]$$

$$= 2[x + 8]$$

$$f[g(x)] = 2x + 16$$

(Substituting $x - 2$ in the equation for $f(x)$.

(b) $g[f(x)] = 2(x + 10) - 2$

$$= 2x + 20 - 2$$

$$g[f(x)] = 2x + 18$$

(Substituting $2(x + 10)$ in the equation for $g(x)$.

Domains of Composite Functions

The approach here is similar to that for domains of algebra of functions. However, here, in the first step, we check only for the domain of the "inside" function. In the second step, we check for the domain of the resulting composite function. We add any excluded values to those from Step 1 and if there is an overlap in the domains we use the intersection of the domains from Step 1 and Step 2. Pay attention to radical functions such as $f(x) = \sqrt{x-2}$,,

Given $f(x) = \frac{3}{x}$ and $g(x) = \frac{3x-4}{x+2}$, find the domains of the following:

find (a) $f[g(x)]$; (b) $g[f(x)]$.

(a) Step 1: We check the domain of the "inside function", $g(x) = \frac{3x-4}{x+2}$

Domain of $g : \{x \mid x$ is a real number and $x \neq -2\}$.

Step 2: Form the composite function and check its domain.

$$f[g(x)] = \frac{3}{\frac{3x-4}{x+2}} \qquad (f(x) = \frac{3}{x})$$

$$= \frac{3}{1} \cdot \frac{x+2}{3x-4}$$

$$f[g(x)] = \frac{3(x+2)}{3x-4}$$

Set $3x - 4 = 0$ to obtain $x = \frac{4}{3}$. Therefore $x \neq \frac{4}{3}$.

Combining the domains from Step 1 and Step 2, the excluded values are -2, and $\frac{4}{3}$.; and

the domain of $f[g(x)] = \frac{3(x+2)}{3x-4}$ is

$\{x \mid x$ is a real number and $x \neq -2$ and $x \neq \frac{4}{3}\}$

(a) Step 1: We check the domain of the "inside function", $f(x) = \frac{3}{x}$

Domain of $f : \{x \mid x \neq 0\}$.

Step 2: Form the composite function and check its domain.

$$g[f(x)] = \frac{3(\frac{3}{x})-4}{\frac{3}{x}+2} \qquad (g(x) = \frac{3x-4}{x+2})$$

$$g[f(x)] = \frac{-4x+9}{2x+3}$$

Set $2x + 3 = 0$ to obtain $x = -\frac{3}{2}$.

Therefore $x \neq -\frac{3}{2}$.

Combining the domains from Step 1 and Step 2, the excluded values are 0, and $-\frac{3}{2}$.; and

the domain of $g[f(x)] = \frac{-4x+9}{2x+3}$ is

$\{x \mid x$ is a real number and $x \neq -0$ and $x \neq -\frac{3}{2}\}$

Application of Composition of Functions

If two functions f_1 and f_2 are inverses of each other, then, the following two conditions must be satisfied simultaneously. **1.** $f_1[f_2(x)] = x$ and **2.** $f_2[f_1(x)] = x$. For examples, see page 539.

Lesson 116 Exercises

A Show and determine which of the following functions are one-to-one

1. The set $\{(4,5), (3,4), (1,2)\}$; 2. The set $\{(2,2), (4,3), (5,7)\}$; 3. $f(x) = 3x - 2$;

4. $f(x) = \sqrt{x+3}$; 5. $f(x) = \sqrt{16 - x^2}$; 6. $f(x) = |x|$; 7. $f(x) = |x-2|$; 8. $f(x) = x^3 + 2$.

Answers: 1. Yes; 2. Yes; 3. Yes; 4. Yes; 5. No; 6. No; 7. No.; 8. Yes

B 1. Given that $f(x) = x + 2$, $g(x) = x - 1$, find (a) $f[g(x)]$; (b) $g[f(x)]$

2. Given that $f(x) = 3(x + 2)$, $g(x) = \frac{1}{x}$ find (a) $f[g(x)]$; (b) $g[f(x)]$

3. Given that $f(x) = (x+1)^2 + 3$, $g(x) = -x$, find (a) $f[g(x)]$; (b) $g[f(x)]$

Answers: **1.** (a) $x + 1$; (b) $x + 1$; **2.** (a) $\frac{3}{x} + 6$; (b) $\frac{1}{3x+6}$; **3.** (a) $x^2 - 2x + 4$; (b) $-x^2 - 2x - 4$

Lesson 117
Inverse Functions and Inverse Relations

Recalling the definitions of a relation and a function, a relation is a set of ordered pairs and function is a relation in which no two distinct ordered pairs have the same first elements (components).

Inverse Relation

Given a relation which is specified by the set of ordered pairs (x, y), the inverse relation of this given relation is the set of ordered pairs (y, x). This inverse relation is obtained by interchanging the first and second elements of each ordered pair.

Example 1: Find the inverse relation of the function specified by the set A:
$$A = \{(1, 3), (3, 2), (5, 7) \}$$

Solution   The inverse relation is obtained by interchanging the first and second elements of each ordered pair
The inverse relation of A is the set
$$\{(3, 1), (2, 3), (7, 5)\}$$

From the definition of the inverse, we can conclude that when we form the inverse of a relation, the domain and the range of the original relation are interchanged. Thus, the domain of the original relation becomes the range of the inverse, and the range of the original relation becomes the domain of the inverse. The inverse may be a relation or a function. (See page 369)

Inverse function

Let a function be specified by the set of ordered pairs $\{(x, y)\}$. Then the inverse relation of this function is the set of ordered pairs $\{(y, x)\}$ If this inverse relation also a function, then for the inverse relation ,we symbolize $f^{-1}(x)$, which is read "f inverse of x" and we say that $f(x)$ and $f^{-1}(x)$ are inverse functions of each other

Example 2   Find the inverse function of the function specified by the set B,:
$$B = \{(a, b), (c, d), (e, f)\}$$

Solution: The inverse function of B is the set, denoted by B^{-1}, is given by
$$B^{-1} = \{(b, a), (d, c), (f, e)\}$$

Sometimes, authors ask the question :" Does this function have an inverse?" Such a question may sometimes be misunderstood, since the inverse is found by interchanging the roles of x and y, which is always possible. What is implied in such a question is whether or not the inverse relation obtained is **also** a function.   Perhaps, an u nambiguous form of the question should be "Does this function have an **inverse function**?" If the inverse relation of a given function is also a function, then we also say that the given function is invertible

Note that it is possible that a given relation which is not a function may have an inverse relation which is a function.

Let us elucidate how the terms, " relation, function and inverse", are connected by using the terms "inverse relation" and "inverse function" in the following statements:

Every relation has an inverse relation. This inverse relation may or may not be a function.
Every function has an inverse relation. Some functions have inverse relations which are (inverse) functions.
A function is also a relation , but a relation is not necessarily a function.

Determining if a function has an inverse function

Necessary and sufficient condition for a function to have an inverse function: A function has an inverse function if and only if it is a one-to-one function.

We will consider the different forms in which the rules for specifying a function may be given and determine if the function has an inverse function. We will consider the set form, the tabular form, the graphical form, and the equation form.

Case 1: Given the set form of the function

A function has an inverse function if for any two different ordered pairs, the second elements are different. If a function has an inverse function it is said to be invertible.

The necessary and sufficient condition is a consequence of the definitions of a function (page 369) and of an inverse function. We may note here that the condition for invertibility refers to the differences in the second components while the condition for being a function refers to the differences in the first components.

Example 3 Given the function specified by the set

$A = \{(1,2),\ (2,3),(4,5),(6,7)\}$, determine if the inverse relation of this set is a function.

Solution Since the second components are all different from one another, the inverse relation is a function.

In fact, the inverse of the set A is given by $A^{-1} = \{(2,1),(3,2),(5,4),(7,6)\}$, which is clearly a function, since the first components are all different from one another.

An example of a function whose inverse relation is **not** a function is the set $\{(3,4),(5,4),(6,5)\}$, because the first and the second ordered pairs have the same second component, 4.

Case 2: Given the tabular form of the function

If any two y-values are the same, then the inverse relation is not a function (but only a relation) otherwise, it is a function. This case corresponds to the set form of a function.

Example 4 **Table 3** represents a function whose inverse relation is a function, while **Table 4** represents a function whose inverse relation is **not** a function.

Tables 3

x	y
1	2
2	3
3	4
4	5
5	6
6	7

Table 4

x	y
1	2
2	**3**
3	**3**
4	5
5	7
6	8

} same two y-values

Case 3: Given the graphical form of the function

The horizontal line test

If any **horizontal line** drawn to intersect the graph intersects the graph at only one point, then the inverse relation of the function (graph) is a function. However, if a horizontal line meets the graph in more than one point, then the inverse of the graph is not a function but only a relation. See Figs 1, 2 and 3 below.

Example 3 Figure **1** is **not** the graph of a one-to-one function, Figures **2** and **3** are the graphs of one-to-one functions.

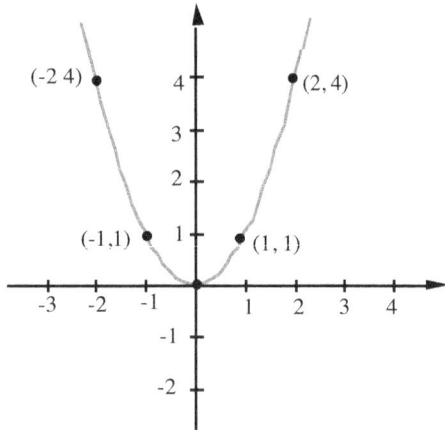

Figure 1: The graph of $y = x^2$,
The inverse relation of this graph is **not** a function

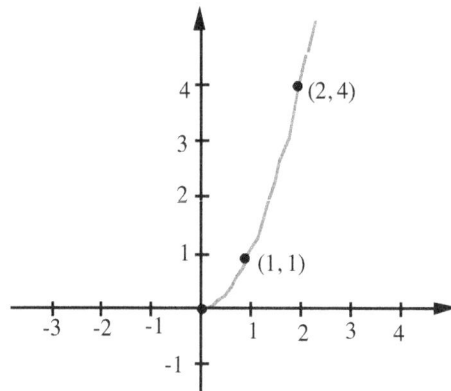

Figure 2: The graph of $y = x^2$, $x \geq 0$.
The inverse relation of this graph is a function
(This is the graph of a one-to-one function)

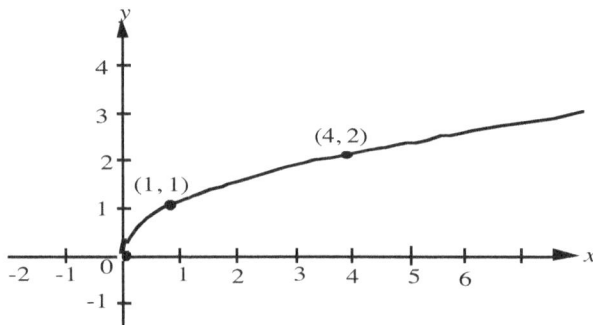

Figure 3: The inverse relation of this graph is a function. (This is the graph of a one-to-one function)

Case 4: Given the equation of the function

By definition, y is a function of x if there is a rule which gives only one corresponding value of y for a given value of x.

Example Determine if the inverse relation of $y = x^2$ is a function.
Method 1: By interchanging the roles of x and y in the given equation, we obtain the inverse relation

$$x = y^2 \text{ or } y^2 = x. \text{ Solving } y^2 = x \text{ for } y, \text{ we obtain, } y = +\sqrt{x} \text{ or } y = -\sqrt{x}$$

Since for the same x-value (say, 4) we have two different y-values (+2 and -2),. the inverse relation of $y = x^2$ is not a function. However, we could make the inverse relation become a function by restricting the domain of this function (see Figures 1 and 2, above)
For some functions, we will graph the function and use the graphical method.

Method 2: Let $y = f(x)$

Step 1: $f(x_1) = x_1^2$

$\qquad f(x_2) = x_2^2$

$\quad$ Equate RHS of $f(x_1)$ to RHS of $f(x_2)$ $\qquad$ (That is, let $f(x_1) = f(x_2)$):

$\qquad x_1^2 = x_2^2 \qquad$ (1)

Step 2: Solve for x_1. $\qquad\qquad$ (You may also solve for x_2)

$\qquad x_1^2 = x_2^2$

$\qquad x_1 = \pm\sqrt{x_2^2}$

$\qquad\qquad x_1 = +x_2 \text{ or } -x_2 \qquad\qquad$ (2)

Since from above, $f(x_1) = f(x_2)$, does **not** imply $x_1 = x_2$ (from equation (2)), $f(x)$ is **not** one-to-one and therefore does **not** have an inverse function.

Finding the inverse of a function specified by an equation

Example 3. Find the inverse function of $f(x) = 3x + 2$. $\qquad$ (1)

Solution

$\quad$ Step 1: Let $f(x) = y$. Then $y = 3x + 2$.

$\qquad$ Interchange x and y in the given equation.

$\qquad$ Then, we obtain $x = 3y + 2$. $\qquad\qquad$ (2)

$\qquad$ By tradition, we want to keep the x-axis horizontal and express y as a function of x.

$\quad$ Step 2: Solve equation (2) for y.

$\qquad$ Then from $x = 3y + 2$.

$$\frac{x-2}{3} = y$$

$$\text{or } y = \frac{x-2}{3}$$

The inverse $f^{-1}(x) = \frac{x-2}{3}$

$\quad$ Alternatively,

$\quad$ Step 1: Solve $y = 3x + 2$. for x.

$\qquad$ Then $x = \frac{y-2}{3}$

$\quad$ Step 2: Interchange x and y $\qquad$ (by definition of the inverse)

$$y = \frac{x-2}{3}$$

The inverse $f^{-1}(x) = \frac{x-2}{3}$

We can observe from above that Steps 1 and 2 are interchangeable.

Since the inverse $y = \frac{x-2}{3}$ is also a function, we can say that $y = 3x + 2$. and $y = \frac{x-2}{3}$ are inverse functions (of each other). We may also add that $f(x) = 3x + 2$. is invertible.

Example 4 Find the inverse function of $f(x) = x^2 + 6$.

Solution

Step 1: Let $f(x) = y$

Step 2: Solve for x.

$$y = x^2 + 6.$$
$$x = \pm\sqrt{y - 6} \qquad\qquad (2)$$

Step 3: Interchange x and y in equation (2).

Then $y = \pm\sqrt{x - 6}$ (same as $y = +\sqrt{x - 6}$ or $y = -\sqrt{x - 6}$) $\qquad$ (3)

Clearly, equation (3) does not represent a function, since for a given value of x, there are two y-values. We can test this by graphing and using the so called **vertical line tes**t (see page 370).

We can say that the given function has an inverse relation specified by $y = \pm\sqrt{x - 6}$. However, the given function does not have an inverse function.

In this example, if we were asked to find $f^{-1}(x)$, we would say that there is no $f^{-1}(x)$ (no inverse function) for the given function.

Example 5 Find the inverse function of $f(x) = x^2 + 6$ $\qquad\qquad$ $x \geq 0$

Solution

Step 1: Let $f(x) = y$

$\qquad$ Then $y = x^2 + 6$ $\qquad\qquad$ $x \geq 0$

Step 2: Solve for x.

$\qquad$ Then $x = \pm\sqrt{y - 6}$ $\qquad$ (That is, $x = +\sqrt{y - 6}$ or $x = -\sqrt{y - 6}$) $\qquad$ (1)

$\qquad$ Since we are only interested in the domain $x \geq 0$, we reject the negative part of equation (1).

$\qquad$ Then, we obtain $x = +\sqrt{y - 6}$ $\qquad\qquad$ (2)

Step 3: Interchange (to obtain the inverse) x and y in equation (2)

$\qquad$ Then $y = +\sqrt{x - 6}$ $\qquad\qquad$ (3)

$\qquad$ Equation (3) is the inverse of $y = x^2 + 6$ $\qquad\qquad$ $x \geq 0$

$\qquad$ Therefore, $f^{-1}(x) = +\sqrt{x - 6}$ $\qquad\qquad$ $x \geq 6$

The graph of this inverse (Figure **2**) indicates that the inverse is a function.

Similarly, if the given function were $f(x) = x^2 + 6$ $\qquad\qquad$ $x \leq 0$,

the inverse relation would be $f(x) = -\sqrt{x - 6}$ and this also is a function.

We may note from Examples 2 and 3 that, sometimes, by redefining (or restricting) the domain of a given function, a function which does not have an inverse function can be made to have an inverse function. See Figures 1, 2 and 3.

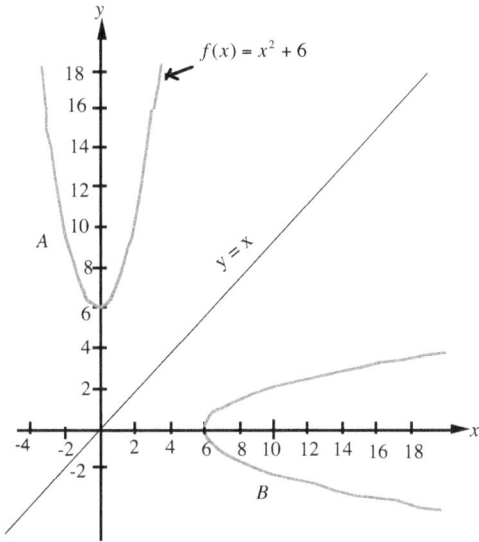

Figure 1: B, the inverse relation of A, is **not** a function .

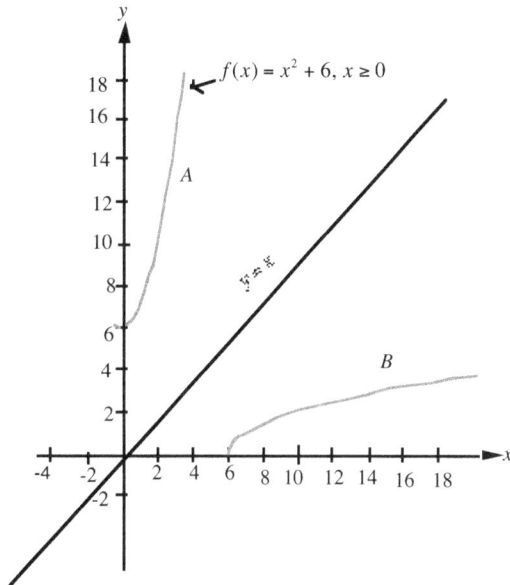

Figure 2: B, the inverse relation of A, is a function (by the vertical line test.)

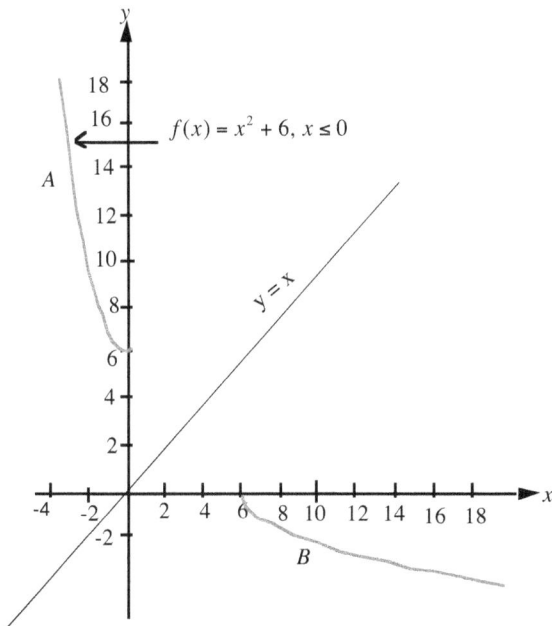

Figure 3: B, the inverse relation of A, is a function (by the vertical line test)

Given the graph of a function, how to sketch the inverse of the graph

Example 5 Given the graph of $y = 3x + 2$, sketch the graph of the inverse of this function.

Geometrically, a function (or relation) and its inverse are symmetric with respect to the line $y = x$. To obtain the inverse graph, we will reflect the given graph about the line $y = x$. The given function is a straight line and so, we will reflect two points about the line $y = x$, and then connect these points by a straight line.

Step 1: Interchange the x- and y-coordinates of any two points on the given line, say, $(1, 5)$ becomes $(5, 1)$; and $(-2, -4)$ becomes $(-4, -2)$.

Step 2: Plot the points $(5, 1)$ and $(-4,- 2)$ on the same coordinate system of axes (as the given graph

Step 3: Connect the points from Step 2 by a straight line to obtain the inverse graph (Fig.)

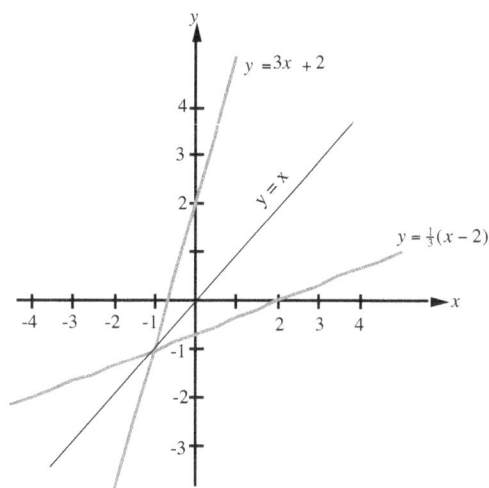

Figure 4: The graphs of $y = 3x + 2$ and $y = \frac{1}{3}(x - 2)$

Note above that, we could also find the equation of the graph, find its inverse equation and then use this equation to sketch the inverse graph.

Determining if two functions are inverses of each other
Case 1: Given the equations of the functions

Example Are the functions $y = 4x + 2$ and $y = \dfrac{x-2}{4}$ inverses of each other?

Solution

Method 1 We will use the principle of composition of functions which states that:

Two functions f_1 and f_2 are inverses of each other if

1. $f_1[f_2(x)] = x$ and **2.** $f_2[f_1(x)] = x$ (Note that inverse functions reverse the action of each other.)

Let $f_1 = 4x + 2$ and $f_2 = \dfrac{x-2}{4}$

$$f_1[f_2(x)] = 4[\frac{x-2}{4}] + 2$$

$$= 4[\frac{x-2}{4}] + 2$$

$$= x - 2 + 2$$

$$= x$$

$$f_2[f_1(x)] = [\frac{(4x+2)-2}{4}]$$

$$= \frac{4x+2-2}{4}$$

$$= \frac{4x}{4}$$

$$= x$$

Since $f_1[f_2(x)] = f_2[f_1(x)] = x$,

$y = 4x + 2$ and $y = \dfrac{x-2}{4}$ are inverses of each other.

Method 2 Find the inverse of one of the functions and compare this inverse with the other function.

Solution

Step 1: We find the inverse of $y = 4x + 2$ (by interchanging x and y and solving for y)

$$x = 4y + 2$$

$$y = \frac{x-2}{4} \quad \text{(solving for } y)$$

Step 2: Clearly, this inverse is identical with the other function.

Therefore, $y = 4x + 2$ and $y = \dfrac{x-2}{4}$ are inverses of each other.

Case 2: Given the graphs of the functions

Procedure: Fold the page along the line $y = x$ and if the two graphs coincide, then the functions are inverses of each other. See Figures 2 and 3 (page 537, 538)

Lesson 117 Exercises

A Find the inverse relation in each of the following and state if the inverse relation is a function:

1. $A = \{(2,3),\ (4,5),\ (6,7)\}$ **2.** $B = \{(0,1),\ (2,5),\ (4,10)\}$ **3.** $C = \{(b,a),\ (c,d),\ (a,b)\}$

4. $\{(1,3),(2,3),(4,5)\}$

5. Given the x- and y-values of a function, determine if the inverse relation in the following table represents an inverse function.

x	2	4	8	9
y	5	6	10	11

Determine if the inverse relation in each of the following graphs represents an inverse function:

6.

7.

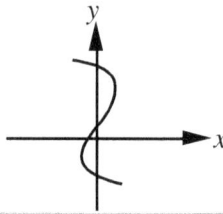

Answers: 1. $\{(3,2),(5,4),(7,6)\}$. It is a function;   **2.** $\{(1,0),(5,2),(10,4)\}$ .It is a function.
3. $\{(a,b),(d,c),(b,a)\}$ .It is a function;   **4.** $\{(3,\ 1),(3,\ 2),(5,\ 4)\}$ .It is not a function.
5. The inverse relation is a function;   **6.** The inverse relation is **not** a function;
7. The inverse relation is a function; even though the given relation is not a function.

B Find the inverse relation of the following and indicate which of the inverse relations are functions.

1. $y = -4x + 1;$ **2.** $y = x^2 - 3$ **3.** $y - 5 = x^2;$ **4.** $y = x^3;$ **5.** $x - 2y = 6$

Determine which of the following pairs are inverses of each other.

6. $y = \dfrac{1}{x+2} - 3$ and $y = \dfrac{1}{x+1} - 4;$ **7.** $y = 5x + 2$ and $y = \dfrac{x-2}{5};$ **3.** $y = x^3$ and $y = x^{\frac{1}{3}}$

Answers: **1.** $y = -\dfrac{x}{4} + \dfrac{1}{4}$ (a function); **2.** $y = \pm\sqrt{x+3}$ (**not** a function);

3. $y = \pm\sqrt{x-5}$ (**not** a function); **4.** $y = x^{\frac{1}{3}}$ (a function); **5.** $y = 2x + 6$ (a function);
6. No;   **7.** Yes;   **8.** Yes.

C

In each of the following graphs, sketch its inverse by reflecting the graph in the line $y = x$:

1.

2.

Answers:

1.

2.

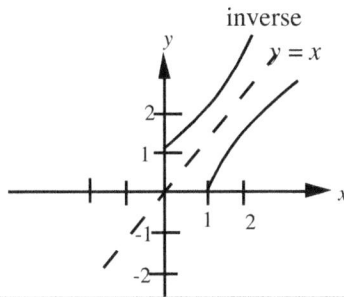

APPENDIX C

Lesson 118: **Graphs of Exponential Functions and Applications of Exponential Functions**
Lesson 119: **Graphs of Logarithmic Functions and Applications of Logarithmic Functions**

Introduction to Exponential and Logarithmic Functions

We use exponential functions in numerical calculations.

1. In physical chemistry, the exponential function is an important function. For example, the exponential equation $C = C_0 e^{-kt}$ relates the decrease in concentration, C, of a reacting material, per unit time, t, where C_0 is the concentration at time $t = 0$. There are similar equations for the decay of a radioactive material.

2. In optics, $I = I_0 e^{-kt}$, where I is the decrease in the intensity of light, l is the depth of the absorbing material.

3. In barometry, $P = P_0 e^{-kh}$, where P_0 is the pressure when h is zero (say at sea level).

4. In physics, the equation of the vibration of a material could be given by $y = A e^{-kt} \cos 2\pi bt$

5. In chemistry, $\log_e k_p = -\dfrac{\Delta G^0}{RT}$ where k_p is the equilibrium constant, ΔG^0 is the free energy, R is the gas constant, and T is the temperature in degrees Kelvin.

5. In business, we use logarithms, to evaluate the compound interest formula $A = P(1 + \frac{r}{n})^{nt}$ where A is the amount accumulated after t years: P is the principal invested at r % annual interest. rate; and n is the number of times a year the interest is compounded.

Lesson 118
Graphs of Exponential Functions and Applications of Exponential Functions

Let b be a positive constant ($b > 0$), such that b is not equal to 1 ($b \neq 1$). Then the exponential function with a base b is defined by $f(x) = b^x$ \qquad (1)
The domain of the exponential function consists of all real numbers and the range consists of all **positive** real numbers.

If we let $f(x) = y$, then equation (1) above becomes
$$y = b^x \qquad (2)$$
We shall cover two cases of the values of b.

Case 1: $b > 1$ and $y = b^x$. Example: $y = 2^x$

Case 2: $0 < b < 1$ and $y = b^x$. Example: $y = \left(\frac{1}{2}\right)^x$

Case 1: Properties of $y = b^x$, $b > 1$

l. The curve crosses the y-axis at $(0, 1)$. That is, the y-intercept $= 1$. (Note: $y = b^0 = 1$).

2. For all values of x, the function is positive (the curve is above the x-axis or all y-values are positive). The function is increasing

3. As x decreases without bound ($x \rightarrow -\infty$), y approaches 0 ($y \rightarrow 0$.). Thus the negative x-axis is a horizontal asymptote to the curve. Graphically, the curve is concave up to the left.

4. As x increases without bound ($x \rightarrow \infty$), the function increases without bound ($y \rightarrow \infty$)

5. When $x = 1$, $y = b$.

How to Sketch the Graph of an Exponential Function

Example 1 Sketch the graph of $f(x) = 2^x$.

> Note: Do not be frightened by the type of equation involved in this example. Sketching the graph of an exponential equation is **not** more difficult than sketching the graphs of, say, the line $y = 3x - 5$ or the parabola $y = x^2$. What we need are some points (ordered pairs) on both sides of the y-axis to plot and connect. In this graph, always include $x = 0$ and $x = 1$.

Step 1: Let $f(x) = y$, Then $y = 2^x$.

Calculations   Ordered pairs (points)

⇓   ⇓

Step 2: Find the y-intercept by letting $x = 0$.

$$\left.\begin{array}{l} x = 0, y = 2^0 \\ \qquad = 1 \end{array}\right\} \Rightarrow (0, 1)$$

Step 3: Choose other convenient x-values (some positive and some negative values) and calculate the corresponding y-values.

$$\left.\begin{array}{l} x = 1, y = 2^1 \\ \qquad = 2 \end{array}\right\} \Rightarrow (1, 2)$$

$$\left.\begin{array}{l} x = 2, y = 2^2 \\ \qquad = 4 \end{array}\right\} \Rightarrow (2, 4)$$

$$\left.\begin{array}{l} x = 3, y = 2^3 \\ \qquad = 8 \end{array}\right\} \Rightarrow (3, 8)$$

$$\left.\begin{array}{l} x = -1, y = 2^{-1} \\ \qquad = \dfrac{1}{2} \end{array}\right\} \Rightarrow (-1, \tfrac{1}{2})$$

$$\left.\begin{array}{l} x = -2, y = 2^{-2} \\ \qquad = \dfrac{1}{2^2} \\ \qquad = \dfrac{1}{4} \end{array}\right\} \Rightarrow (-2, \tfrac{1}{4})$$

Step 4: Plot the points from Steps 2, and 3 (Figure...): That is plot $(0,1)$, $(1,2)$, $(2,4)$, $(3,8)$, $(-1,\frac{1}{2})$ and $(-2,\frac{1}{4})$, noting that as x decreases without bound ($x \to -\infty$), y approaches 0 ($y \to 0$). Thus the negative x-axis is a horizontal asymptote to the curve. Graphically, the curve is concave up to the left. Also, note that as x increases without bound, the function increases without bound.

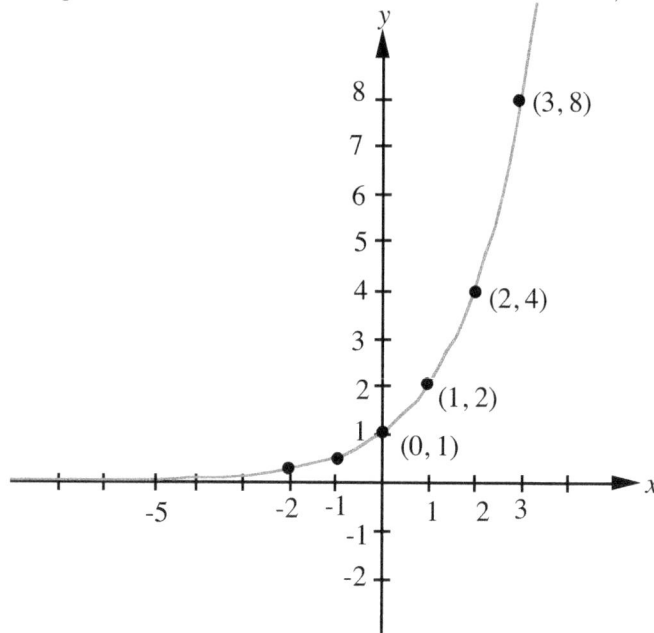

Figure 1: The graph of $y = 2^x$

Case 2: $y = b^x$, $0 < b < 1$

Examples:

1. $y = \left(\frac{1}{2}\right)^x$ or $y = 2^{-x}$ Note: $\left(\frac{1}{2}\right)^x = \left(2^{-1}\right)^x = 2^{-x}$

2. $y = \left(\frac{1}{3}\right)^x$ or $y = 3^{-x}$

Properties of $y = b^x$, $0 < b < 1$

1. The y-intercept is 1. That is, the curve crosses the y-axis at $(0,1)$.

2. For all values of x the function is positive (All the y-values are positive).

3. The function is decreasing.

4. As x increases without bound ($x \to \infty$), the function approaches zero ($y \to 0$). Thus the positive x-axis is a horizontal asymptote to the curve.

5. As x decreases without bound, ($x \to -\infty$), the function increases without bound ($y \to \infty$).

6. Graphically, the graph is concave up to the right (Figure)

Example 2 Sketch the graph of $f(x) = \left(\frac{1}{2}\right)^x$ or $f(x) = 2^{-x}$

Step 1: Let $f(x) = y$, Then $y = \left(\frac{1}{2}\right)^x$ or $y = 2^{-x}$

Step 2: Find the y-intercept by letting $x = 0$. $\left. x = 0, \; y = \left(\frac{1}{2}\right)^0 = 1 \right\} \Rightarrow (0, 1)$

Step 3: Choose some other convenient x-values (positive and negative values) ; calculate the corresponding y-values

$x = 1, y = \left(\frac{1}{2}\right)^1 = \frac{1}{2} \Big\} \Rightarrow (1, \frac{1}{2})$

$x = 2, y = \left(\frac{1}{2}\right)^2 = \frac{1}{4} \Big\} \Rightarrow (2, \frac{1}{4})$

$x = 3, y = \left(\frac{1}{2}\right)^3 = \frac{1}{8} \Big\} \Rightarrow (3, \frac{1}{8})$

$x = -2, y = \left(\frac{1}{2}\right)^{-2}$
$= \frac{1}{2^{-2}} \Big\} \Rightarrow (-2, 4)$
$= 2^2$
$= 4$

$x = -3, y = \left(\frac{1}{2}\right)^{-3}$
$= \frac{1}{2^{-3}} \Big\} \Rightarrow (-3, 8)$
$= 2^3$
$= 8$

$x = -1, y = \left(\frac{1}{2}\right)^{-1} = \frac{1}{2^{-1}}$
$= 2^1 \Big\} \Rightarrow (-1, 2)$

Step 4: Plot the points from Steps 2, and 3 (Fig.2). That is plot $(0, 1)$ $(1, \frac{1}{2})$ $(2, \frac{1}{4})$ $(3, \frac{1}{8})$, $(-1, 2)$, $(-2, 4))$ and $(-3, 8))$, noting that as x increases without bound ($x \to \infty$), y approaches 0 ($y \to 0$). Thus the positive x-axis is a horizontal asymptote to the curve. Graphically, the curve is concave up to the right. Also, note that as x decreases without bound ($x \to -\infty$), the function increases without bound ($y \to \infty$).

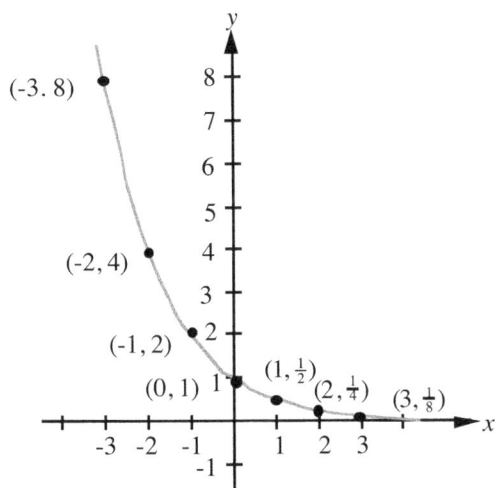

Fig. 2: The graph of $y = \left(\frac{1}{2}\right)^x$ or $y = 2^{-x}$

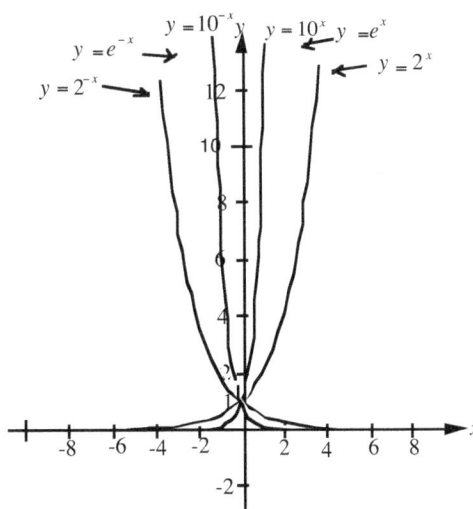

Fig. 3: The graphs of $y = e^x$, $y = 10^x$, $y = 2^x$, $y = e^{-x}$, $y = 10^{-x}$ and $y = 2^{-x}$

Other Exponential Graphs; **The graphs of** $y = e^x$, $y = 10^x$ $y = 2^x$, $y = 2^{-x}$,

The graph of $y = e^x$ is a special case of $y = a^x$ in which $a = e$ ($e = 2.72$, and e is the base of the natural logarithms. (see page 442). Similarly, when $a = 10$, we obtain $y = 10^x$. The graphs of $y = 10^x$, $y = 2^x$, $y = e^{-x}$, $y = 10^{-x}$ and $y = 2^{-x}$ on the same set of coordinate axes, are presented in Fig 3, above.

Applications of Exponential Functions

Exponential Growth and Decay

The exponential growth or decay model is given by the equation $y(t) = y_0 e^{kt}$, where y_0 is the initial value, $y(t)$ is the value at time t. If $k > 0$, y is said to grow exponentially, and k is then called the growth constant. If $k < 0$, y is said to decay exponentially, and k is then called the decay constant. The applications of this model include population growth, bacterial growth, radioactive decay and compound interest.

Compound Interest

The formula relating the principal P, the time, t, taken by P to grow to the amount A is given by

$A(t) = P(1 + i)^t$

Example If $30,00 is invested at 5% interest and compounded annually, in t years, it will grow to an amount, A, given by $A(t) = 30,00(1.05)^t$

(a) How long will it take to accumulate $65,000 in the account?
(b) Find the time required for $30,000 to double itself.

Solution

a) $A = 30,000$

$65,000 = 30,000(1.05)^t$

$\dfrac{65,000}{30,000} = (1.05)^t$

$2.17 = (1.05)^t$

$\log 2.17 = \log 1.05^t$

$\log 2.17 = t \log 1.05$

$\dfrac{\log 2.17}{\log 1.05} = t$

$15.9 = t$

Scrapwork

$\log 2.17 = 0.3365$

$\log 1.05 = 0.0212$

$\dfrac{\log 2.17}{\log 1.05} = \dfrac{0.3365}{0.0212} = 15.9$

The time taken by $30,000 to grow to $65,000, at the interest rate of 5%, is 15.9 years.

(b) $30,000 doubles to $60,000 at the interest rate of 5% in t years.

Therefore, $P = 30,000$, $A = 60,000$; and substituting in $A(t) = P(1 + i)^t$, we obtain

$60,000 = 30,000(1.05)^t$, and next solve this exponential equation for t.

$\dfrac{60,000}{30,000} = (1.05)^t$

$2 = (1.05)^t$

$\log 2 = \log(1.05)^t$

$\log 2 = t \log(1.05)$

$\dfrac{\log 2}{\log 1.05} = t$

$14.2 = t$

Scrapwork

$\log 2 = 0.3010$

$\log 1.05 = 0.0212$

$\dfrac{\log 2}{\log 1.05} = \dfrac{0.3010}{0.0212} = 14.2$

Therefore, at the interest rate of 5%, $30,000 doubles in 14.2 years.

Lesson 118 Exercises

Sketch the graphs of the following pairs on the same set of rectangular axes:

1. $y = 4^x$ and $y = 4^{-x}$;　**2.** $y = e^x$ and $y = e^{-x}$;　**3.** $y = 10^x$ and $y = 10^{-x}$

4. What can be said with respect to transformations about the relationship between each pair in Exercises 1-3?

5. If \$30,000 is invested at 5% interest and compounded annually, in t years, it will grow to an amount, A, given by $A(t) = 30,000(1.05)^t$

(a) How long will it take to accumulate \$65,000 in the account?

(b) Find the time required for \$30,000 to double itself.

Answers 1-3

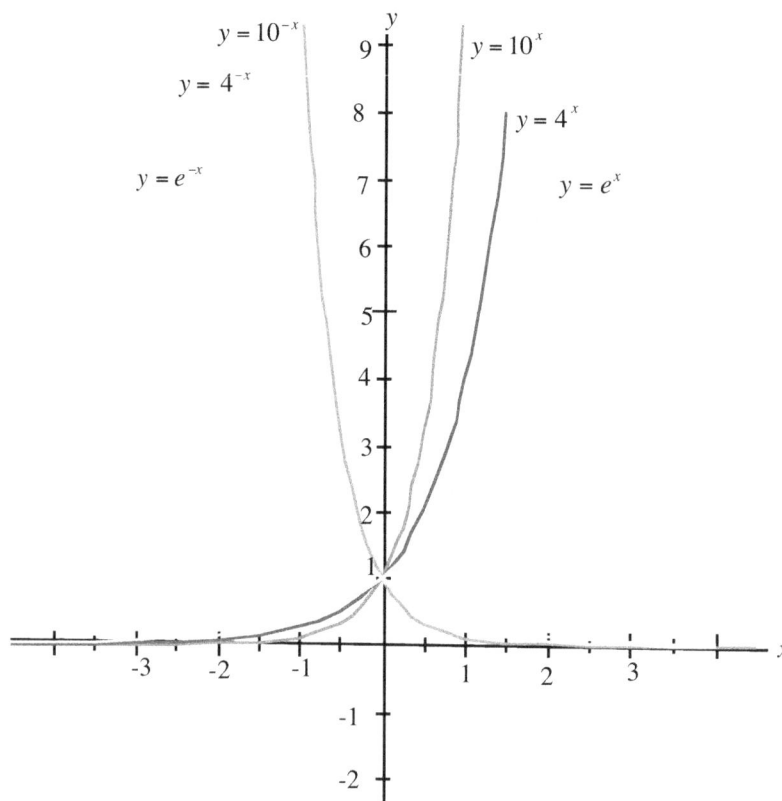

5. (a) The time taken by \$30,000 to grow to \$65,000, at the interest rate of 5%, is 15.9 years.

(b At the interest rate of 5%, \$30,000 doubles in 14.2 years.

Lesson 119
Graphs of Logarithmic Functions and Applications of Logarithmic Functions

The logarithmic function is the inverse of the exponential function.

The exponential function is given by $y = b^x$, $b > 0$, $b \neq 1$.. The inverse of this function is obtained by interchanging the x- and y-coordinates of the exponential function. If we interchange the coordinates, we obtain the logarithmic function given by

$$x = b^y \qquad (A)$$

Now, we solve equation (A) for y using the basic properties of logarithms,

$$\log_b x = \log_b b^y \qquad \text{(Taking logs of both sides of equation (A))}$$
$$\log_b x = y \log_b b$$
$$\log_b x = y \qquad (\log_b b = 1)$$

We call the function $\log_b x$ the logarithm function of x to the base b. We must note that

$$x = b^y \text{ is equivalent to } \log_b x = y \qquad \text{(Memorize this equivalent relationship)}$$

Note that the logarithmic function is one-to-one.

Example 1 The graph of $y = \log_2 x$ or $x = 2^y$

The graph of $y = \log_2 x$ can be obtained from the graph of $y = 2^x$ by reflecting $y = 2^x$ about the line $y = x$.

Properties of $y = \log_2 x$ or $x = 2^y$

1. The x-intercept is at $(1, 0)$.

2. As $x \to 0$ (x approaches zero) $\log_2 x \to -\infty$. The curve decreases without bound. The negative y-axis is an asymptote to the curve (in the 4th quadrant).

3. The domain consists of the set of all positive numbers.

4. The range consists of the set of all real numbers.

5. The graph has the negative y-axis as a vertical asymptote (in the 4th quadrant).

6. The graph is concave down to the right (Figure 4)

Figure 4: $y = \log_2 x$ or $x = 2^y$

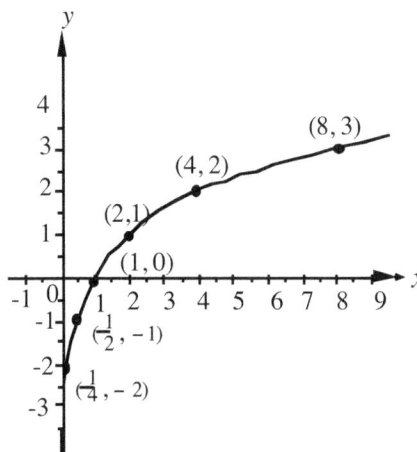

Example Sketch the graph of $f(x) = \log_2 x$.

Step 1: Let $y = f(x)$.

Then $y = \log_2 x$.   (A)

Step 2: Write equation (A) in the equivalent exponential form.

Then $x = 2^y$ <--------- Exponential form.

Calculations   Ordered pairs (points)

$\Downarrow$ $\Downarrow$

Step 3: Choose y and calculate the corresponding values of x. It is more convenient to choose y and calculate x..(This may be the first time you are choosing y and calculating x.)
Find the x-intercept by letting $y = 0$. Note that there is no y-intercept for this function.

$$\left.\begin{array}{l} y = 0, x = 2^0 \\ \qquad = 1 \end{array}\right\} \Rightarrow (1,0)$$

Becareful of how you write the ordered pairs, since you are choosing y and calculating x. The first element of each ordered pair is always the x–value irrespective of the order in which the values are obtained.

Step 4: Choose other convenient y-values (positive and negative values) and calculate corresponding the x-values

$$\left.\begin{array}{l} y = 1, x = 2^1 \\ \qquad = 2 \end{array}\right\} \Rightarrow (2,1)$$

$$\left.\begin{array}{l} y = 2, x = 2^2 \\ \qquad = 4 \end{array}\right\} \Rightarrow (4,2)$$

$$\left.\begin{array}{l} y = 3, x = 2^3 \\ \qquad = 8 \end{array}\right\} \Rightarrow (8,3)$$

$$\left.\begin{array}{l} y = -1, x = 2^{-1} \\ \qquad = \dfrac{1}{2} \end{array}\right\} \Rightarrow (\tfrac{1}{2},-1)$$

$$\left.\begin{array}{l} y = -2, x = 2^{-2} \\ \qquad = \dfrac{1}{2^2} \\ \qquad = \dfrac{1}{4} \end{array}\right\} \Rightarrow (\tfrac{1}{4},-2)$$

Step 5: Plot the points from Steps 3, and 4 (Figure 4). (i.e., is plot $(1,0)$, $(2,1)$, $(4,2)$, $(8,3)$, $(\tfrac{1}{2},-1))$, $(\tfrac{1}{4},-2))$ and connect them, noting that as x approaches zero the function decreases without bound, and thus the negative y-axis is a vertical asymptote to this curve. Graphically, the curve is concave down to the right. Also, note that as x increases without bound, the function increases without bound.

Other Logarithmic Graphs

The graphs of $y = \log_a x$ and in particular the graphs of $y = \log_{10} x$ and $y = \log_e x$

The graph of $y = \log_{10} x$ or $x = 10^y$. can be obtained from the graph of $y = 10^x$ by reflecting $y = 10^x$ about the line $y = x$. Similarly, the graph of $y = \log_e x$ or $x = e^y$ can be obtained from the graph of $y = e^x$ by reflecting $y = e^x$ about the line $y = x$.

General Properties of $y = \log_a x$, $y = \log_{10} x$ and $y = \log_e x$

1. The x-intercept is at $(1, 0)$.

2. As $x \to 0$, $\log_{10} x \to -\infty$ and $y = \log_e x \to -\infty$

3. The domain consists of the set of all positive numbers.

4. The range consists of the set of all real numbers.

5. The graph has the negative y-axis as a vertical asymptote (in the 4th quadrant).

6. The graph is concave down to the right (Figure)

The graphs of $y = \log_{10} x$ and $y = \log_e x$ are presented below.

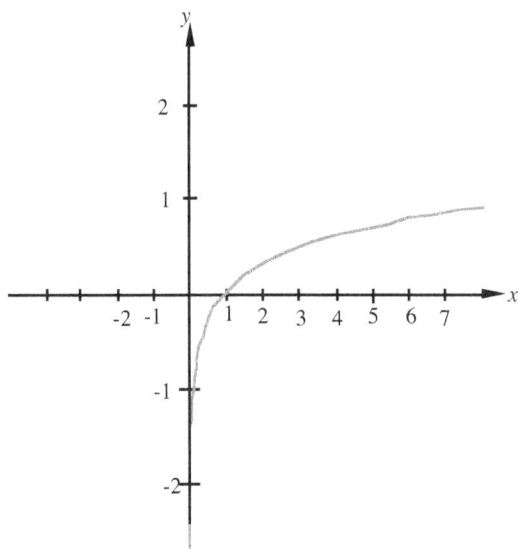

Figure 5: The graph of $y = \log_{10} x$ or $x = 10^y$.

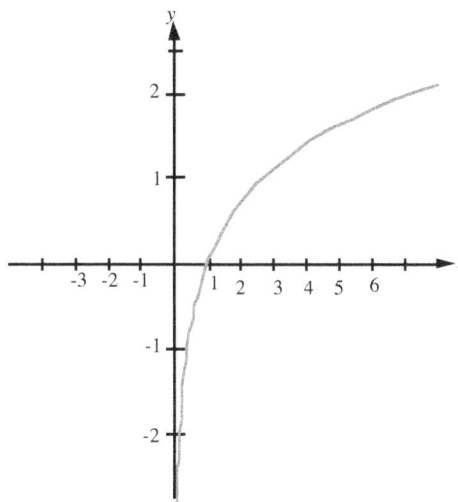

Figure 6: The graph of $y = \log_e x$ or $x = e^y$

We must note the following:

1. For an **exponential equation**, we must be careful not to confuse its equivalent equation in the logarithmic form with its inverse.

   Thus $y = e^x$ is equivalent to $x = \log_e y$ (but **not** equivalent to $y = \log_e x$, its inverse).

2. For a **logarithmic equation**, we must be careful not to confuse its equivalent equation in the exponential form with its inverse.

Thus $y = \log_e x$ is equivalent to $x = e^y$, (but **not** equivalent to $y = e^x$, its inverse).
See Figures below.

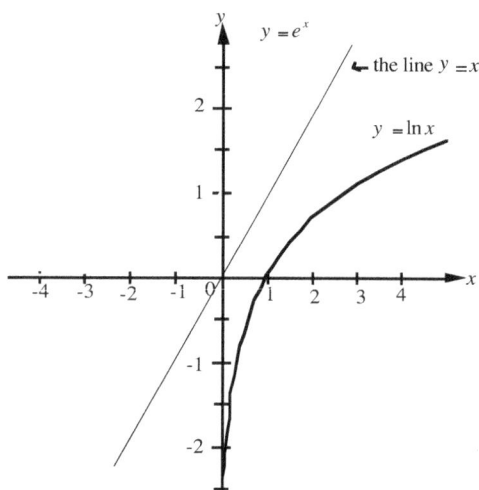

Figure 7: The graph of $y = e^x$ and $\ln x$

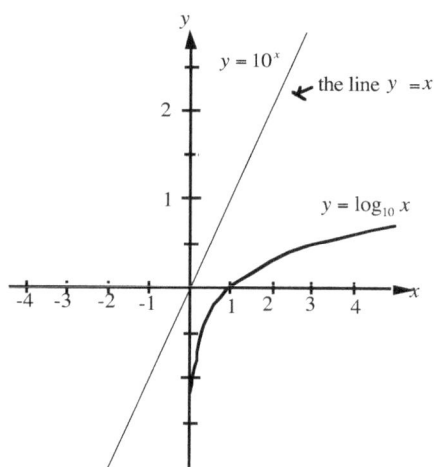

Figure 8 : Graph of $y = 10^x$ and $\log x$

Applications of Logarithmic Functions

Sound Level

Formula for sound level $L = 10\log\frac{I}{I_0}$, where L (in decibels, dB) is the loudness of sound, I is the sound intensity in watts per square meter, and $I_0 = 10^{-12}\,W/m^2$

(a) If $I = 10^{-3}\,W/m^2$, find L. (b) If $L = 75$, find I.

Solution

(a) $L = 10\log_{10}\frac{10^{-3}}{1}\frac{w}{m^2}\cdot\frac{m^2}{10^{-12}\,w}$

$\quad = 10\log_{10}\frac{10^{-3}}{10^{-12}}$

$\quad = 10\log_{10}10^{9}$

$\quad = 9(10\log_{10}10)$

$\quad = 10(9)$ (Note: $\log_{10}10 = 1$)

$\quad = 90$ decibels (dB)

(b) $L = 75$

Step 1; Substituting in

$\quad L = 10\log\frac{I}{I_0}$

$\quad 75 = 10\log\frac{I}{10^{-12}}$

$\quad 75 = 10\log 10^{12}\,I$

$\quad \frac{75}{10} = \log_{10}(10^{12}\,I)$

Step 2: $7.5 = \log_{10}10^{12} + \log_{10}I$

$\quad 7.5 = 12 + \log_{10}I$

$\quad (\log_{10}10^{12} = 12\log_{10}10 = 12)$

$\quad 7.5 - 12 = \log_{10}I$

$\quad -4.5 = \log_{10}I$

$\quad 10^{-4.5} = I$

$\quad I = 10^{-4.5}\frac{w}{m^2}$

Hydrogen Ion Concentration

The *pH* of a solution is a measure of the acidity or alkalinity of the liquid. The *pH* of a solution is defined as follows: $pH = -\log_{10}\left[H^+\right]$, where $\left[H^+\right]$ is the hydrogen ion concentration (concentration per liter).

Example 1 If $\left[H^+\right]$ for a solution is 2.45×10^{-8} moles/liter, find the *pH* of this solution.

Solution $\quad pH = -\log_{10}\left[2.45\times10^{-6}\right]$

$= -\left(\log_{10}2.45 + \log_{10}10^{-6}\right)$

$= -\left(\log_{10}2.45 + (-6)\log_{10}10\right)$

$= -\left(\log_{10}2.45 + (-6)(1)\right)$ Note: $\log_{10}10 = 1$

$= -(0.389 - 6)$ Note: From a calculator, $\log_{10}2.45 = 0.389$

$= -(-5.61)$

$= 5.61$

The *pH* of the 2.45×10^{-8} moles/liter solution is 5.61

Example 2 If the *pH* of a solution is 7.5, find the $\left[H^+\right]$ of this solution.

Solution

Step 1: $pH = -\log_{10}\left[H^+\right]$

$7.5 = -\log_{10}\left[H^+\right]$ (substituting for $pH = 7.5$)

Step 2: $-7.5 = \log_{10}\left[H^+\right]$

$\quad 10^{-7.5} = \left[H^+\right]$

The $\left[H^+\right]$ of the solution is $10^{-7.5}$.

Lesson 119 Exercises

Sketch the graphs of the following:

1. $y = \log x$.;   **2.** $y = \log(x-1)$;   **3.** $y = \log(x+2)$;   **4.** $y = \log_3 x$;   **5.** $y = \log_2(x+3)$.

6. If the hydrogen ion concentration, $\left[H^+\right]$ for a solution is 2.45×10^{-8} moles/liter, find the pH of this solution.

7. If the pH of a solution is 7.5, find the $\left[H^+\right]$ of this solution.

8. Formula for sound level $L = 10\log\dfrac{I}{I_0}$,where L (in decibels, dB) is the loudness of sound, I is the sound intensity in watts per square meter, and $I_0 = 10^{-12} W/m^2$

   (a) If $I = 10^{-3} W/m^2$, find L. (b) If $L = 75$, find I .

Answers:

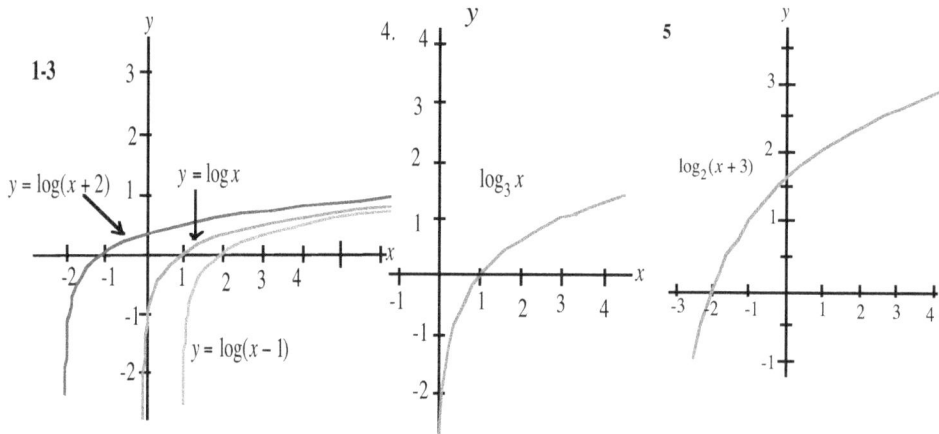

6.: The pH of the 2.45×10^{-8} moles/liter solution is 5.61

7. The hydrogen ion concentration, $\left[H^+\right]$ of the solution is $10^{-7.5}$.

8. (a) $= 90$ decibels (dB);   (b) $I = 10^{-4.5} \dfrac{w}{m^2}$

APPENDIX D
(see also p.115-117)
About Measurements
Standard Unit, Error, Rounding-off Numbers, Significant Digits, Scientific Notation

To determine the size of a physical quantity, we compare its size with a standard quantity called a unit. Example: To determine the length of the cover of a book in inches or in meters, we can use a ruler with its scale in inches or in meters.

A measurement is the ratio of the magnitude of a physical quantity to that of a standard unit

Standard unit

A standard unit is a measure with which other quantities are compared. A standard unit of measure is defined by a legal authority (such as the US Bureau of Weights and Measures) or by a conference of scientists.

Some universally accepted standards:

1. For mass, the standard (primary standard) is the kilogram (kg).
2. For length, the standard is the meter (m).
3. For time, the standard is the second (s)

Some devices for taking measurements

Examples: Rulers (for length), chemical balances (for mass), stop watches (for time), ammeters, (for electric current) voltmeters (for electric voltage) , thermometers (for temperature) and barometers (for pressure).

Experimental Errors (or Uncertainties)

There are two main types of errors, namely, systematic errors, and random errors.

Systematic Errors (constant errors):

These errors are due to faulty measuring devices. Systematic errors make the measurements either too small or too large:

Examples of faulty devices:

1. Instruments with needles off the zero mark;   2. Faulty clocks (stop clock)
3, Corroded weights;   4. Faulty thermometers. Heat leaking equipment

Random errors (accidental errors or indeterminate errors)

Random errors may be due to chance variations of the physical quantity being measured, or chance variations in the measuring device. Random errors may also be due to failure to take into account variables such as temperature fluctuations; and environmental effects. We can reduce random errors by making a large number of measurements and taking the average of the measurements.

Absolute error (or absolute uncertainty)

Absolute error Experimental value − accepted (true value)

Example: In an experiment to determine the acceleration due to gravity, g, the experimental value of g, was 986 cm/s^2. The accepted value of g is 980 cm/s^2. Find the absolute error.

Solution   Experimental value = 986 cm/s^2

Accepted value = 980 cm/s^2
Absolute error = experimental value - accepted value

= (986 - 980) cm/s^2

= 6 cm/s^2

The positive value indicates that the experimental value is greater than the accepted value.

Note: If the experimental value = 964 cm/s^2

The absolute error = (964 - 980) cm/s^2

$$= -16 \text{ cm/} s^2$$

The negative value means that the experimental value is less than the accepted value.

Relative error (or relative uncertainty)

Relative error $= \dfrac{\text{absolute error}}{\text{accepted value}}$

Example 1: If the absolute error = 6 cm/s^2, and

the accepted value = 980 cm/s^2,

Relative error $= \dfrac{\text{absolute error}}{\text{accepted value}}$

Then relative error $= \dfrac{6 \text{ cm / } s^2}{980 \text{ cm / } s^2}$

$= 0.00612$

Example 2: If the absolute error = -16 cm//s^2 and

the accepted value = 980 cm/s^2

Then relative error $= \dfrac{-16 \text{ cm / } s^2}{980 \text{ cm / } s^2}$

$= -0.00612$

$= -0.0163$

Note: If an accepted value is not known, and we have two or more experimental values, then the average of the experimental values would be used as the "accepted value" in calculations.

Rounding-Off Numbers

The rules for rounding-off a number may differ slightly depending upon the field. For instance, in accounting the rule may be slightly different from the rule in chemistry.

1. If the digit or group of digits to be dropped is more than 500...,hen drop that digit or group and add 1 to the last digit retained.

2. If the digit or group of digits to be dropped is less than 500...,then drop that digit or group and leave the last digit retained unchanged.

3. If the digit or group of digits to be dropped is exactly 500..., then drop that portion and add 1 to the last digit retained if this digit is odd but if this digit is even, then this digit remains unchanged.

Rounding off Whole numbers

Procedure:

Step 1: Locate the digit in the round-off place.
(The round-off place is the place to which we want to round-off the number)

Step 2: Drop all digits to the right of the round-off place, and if the digit immediately to the right of the round-off place is more than 5 or is 5 followed by non-zero digits, , add 1 to the round-off place digit (i.e. we round-up); but if the digit immediately to the right of the round-off place is less than 5, the round-off place digit remains unchanged. However, if the digit immediately to the right of the round-off place digit is 5 or 5 followed by zeros, we add 1 to the round-off place digit if it is odd, but if it is even, it remains unchanged.(i.e. we round-down). Also, replace each digit dropped by a zero.

Rounding-off Decimals

The procedure is the same as that for rounding-off whole numbers, except that after the decimal point, we do not replace any digits dropped by zeros.

Procedure:

Step 1: Locate the digit in the round-off place.
(The round-off place is the place to which we want to round-off the number)

Step 2: Drop all digits to the right of the round-off place, and if the digit immediately to the right of the round-off place is more than 5 or is 5 followed by non-zero digits, add 1 to the round-off place digit (i.e., we round-up); but if the digit immediately to the right of the round-off place is less than 5, the round-off place digit remains unchanged. However, if the digit immediately to the right of the round-off place digit is 5 or 5 followed by zeros, we add 1 to the round-off place digit if it is odd, but if it is even, it remains unchanged.(i.e., we round-down).

Rounding-off (Alternatively)

When we round-off a number, we drop some of the digits explicitly or implicitly specified. We must distinguish between rounding-off to a specified number of decimal places (or significant digits) and the implicit rounding-off which we must determine from the numbers involved in the calculation.

The rules for rounding-off a number may differ slightly depending upon the field. For instance, in accounting the rule may be slightly different from the rule in chemistry.

1. If the digit or group of digits to be dropped is more than 500...then drop that digit or group and add 1 to the last digit retained.

2. If the digit or group of digits to be dropped is less than 500...then drop that digit or group and leave the last digit retained unchanged.

3. If the digit or group of digits to be dropped is exactly 500..., then drop that portion and add 1 to the last digit retained if this digit is odd but if this digit is even, then this digit remains unchanged.

Example: The following have been rounded-off to three decimal places.

(1) $.4398. \approx$ **.440**

(2) $.43652 \approx$ **.437**

(3) $.43637 \approx$ **.436**

(4) $.43750 \approx$ **.438**

(5) $.43650 \approx$ **.436**

(6) $.43946 \approx$ **.439**

(7) $.43650001 \approx$ **.437**

(8) $.4365001 \approx$ **.437**

Example We round off **85376.7463** to the following places, using the simple "5 or greater or less than 5 rule"

1. 85376.7463 to the nearest **thousandth** becomes **85376.746** (We do **not** replace the 3 dropped by a zero)

2. 85376.7463 to nearest **hundredth** becomes **85376.75** (We added 1 to the digit in the round-off place)

3. 85376.7463 to the nearest **tenth** becomes **85376.7** (The 7 is unchanged since the 4 dropped is less than 5)

4. 85376.7463 the nearest **unit** becomes **85377.** (Adding 1 to the 6)

5. 85376.7463 to the nearest **ten** becomes **85380.** (Replacing the 6 dropped by a zero)

6. 85376.7463 to the nearest **hundred** becomes **85400.** (Replacing the digits (6 and 7) dropped by zeros)

7. 85376.7463 to the nearest **whole number** becomes **85377.** (same as to the nearest unit)

Estimation

In estimation, we round-off the numbers before carrying out the operations of addition, subtraction, multiplication , division etc. For convenience, we will round-off each number to the first non-zero digit, unless specified.

Approximate Numbers, Significant Digits, Scientific Notation,

A measurement consists of a numerical value and a unit of that measurement. Example: 4 kilograms, where the 4 is the numerical value and the kilograms is the unit of the measurement.

Numbers obtained from a measurement are never exact (i.e., are approximate) due to the limitations of the measuring instrument as well as the skill of the person making the measurement. As such, when one records a measurement, one should indicate the reliability of the measurement. All measurements may be assumed to have an uncertainty in at least one unit in the last digit of the measurement, since in making a measurement, we usually estimate the last digit.

Results obtained from calculations using measurements are also as uncertain as the measurements themselves.

In summary, the numbers that we deal with in calculations are obtained from observations. Some of the numbers are exact and some are approximate, The approximate numbers are those numbers obtained from making measurements.

Exact Numbers: An exact number is a number that contains no uncertainties. It is assumed to be infinitely accurate. We can obtain exact numbers from definitions and from direct count. For example, the number of students in a math class by count is 25. In this case, there is no uncertainty, since we know that there are exactly 25 students. Similarly, when one counts 200 dollars, one knows that one has exactly 200 dollars, and there is therefore, no uncertainty. Also by definition, 60 minutes = 1 hour; 2.54 centimeters = 1 inch. Since these numbers are defined, this 60 and 2,54 are exact and contain no uncertainties. We can also add that this 60 has an infinite number of significant digits, (we can write 60 as 60.000...) and therefore the zero in the 60 is significant. However if you make your own measurement and by coincidence obtain 2.54 centimeters, then this 2.54 would not be exact. We can generalize that all the conversion factors (from tables) are exact.

Significant Digits (Significant Figures), Digits obtained in a measurement

A significant digit (or figure) is one which is known to be reasonably reliable (or correct).

When we make measurements, the digits we read and estimate on a scale are also called significant digits (or significant figures). These digits include digits that we are certain of, and one additional digit that we are uncertain of. This uncertain digit is obtained by the estimation of the fractional part of the smallest subdivision on the scale being used. As such, the rightmost digit is assumed to be uncertain.

Significant figure notation is an approximate method of indicating the uncertainty of a measurement. when recording a measurement.

We agree to the following:

1. The digits $1, 2, 3, 4, 5, 6, 7, 8, 9$ are always significant.

2. The digit zero, $0.$ may or may not be significant according to its position in the number as follows:

   (a) Zeros before the first non-zero digits are **not** significant.

For example: (i) .0450 has **three** significant digits: The first zero is not significant; but the last zero is significant since if it were not we would not write it.

(ii) .0012 has **two** significant digits. The first two zeros are not significant

. (b) Zeros between non-zero digits **are** significant.

Example: 3.045 has **four** significant digits. The zero between 3 and 4 is significant.
40.240 has **five** significant digits. The last zero is significant because if it were not we would not write it.

More examples: The numbers referred to below are assumed to have been obtained from measurements.
23.00 has four significant digits: the zeros in this case are significant since we do not have to write the zeros if they were not significant. If the zeros were not significant we would have written 23.
2300. has two significant digits, and 600 has one significant digit; however, in each of these two examples, the number of significant digits is sometimes ambiguous.
It is suggested that when the number of significant digits is in doubt , the maximum number of significant digits is to be assumed. Also, in recording data, if we know the number of significant digits, we will use the scientific notation.
Note above that if 600 and 2300 had been obtained by counting, or by definition, all the zeros would have been significant. As a reminder, significant notation generally pertains to numbers obtained from measurements.
Using **scientific notation** avoids all ambiguities with respect to the number of significant digits..
For instance, if we know that the above number, 600 were measured to two significant digits, then we would write 6.0×10^2. If the measurement were to three significant digits, we would write 6.00×10^2
When we deal with very large or very small numbers we prefer to write the numbers in scientific notation form. In this form, the significant digits (digits) including the zeros can be unambiguously indicated.

In scientific notation:

(1) 125000 would be written 1.25×10^5

(2) 1467 would be written 1.467×10^3

(3) .032500 would be written 3.2500×10^{-2}

(4) .0325 would be written 3.25×10^{-2}

(5) If 125000 were known to four significant digits it would be written 1.250×10^5.

Note: Some authors indicate which zeros are significant by underlining the last significant zero.
For example, in 23000 the first two zeros are significant but the last zero is not significant.
Note also that in some books, a decimal point placed after the last zero makes all the zeros significant. For example, 23000. has five significant digits, but 23000 has two significant digits. However, but there may still be ambiguity if the number is at the end of a sentence.

Accuracy and Precision in Measurements

Two contributions to uncertainty in measurement are limitations of precision, and limitations of accuracy. **Accuracy** indicates how close a measured value is to the true value but **precision** indicates how close two measurements of the same quantity are close to each other. Generally, more precision implies more accuracy. However, there are instances in which numbers may be more precise, but may not be more accurate.. For example, if a measuring device is incorrectly calibrated (having incorrect scale).

Accuracy and Precision in Calculations

With respect to significant digits, **accuracy** refers to the number of significant digits but **precision** refers to the number of decimal places. The larger the number of significant digits in a number, the more accurate the number. The larger the number of decimal places, the more precise the number.

Note: In calculations, the approximate numbers determine how the rounding-off is done

Rounding-off to Significant Digits or Figures in Arithmetic Operations
(Implicit Specification)

1. In **multiplication and division** involving significant digits, the product or quotient (answer) should be rounded-off so that the number of significant digits in the answer is equal to the number of significant digits of the number with the least number of significant digits. (In other words, the answer should not be more accurate than the number with the least accuracy)

Example 1: Multiply 2.34 cm by 5.6 cm

Step 1: $2.34 \times 5.6 = 13.104$

Step 2: The number with the fewest number of significant digits is 5.6 and it has two significant digits. Therefore, the product (answer) should contain only two significant digits.
Thus, 13.104 becomes 13.
Answer: 13 cm^2

In the above case the "4" in 2.34 is considered to be reasonably reliable. The " 6" in 5.6 is considered to be reasonably reliable.

Example 2: Maria determined the length of a piece of wood to be 6.47 yards. What is the length of this wood in feet?
Solution
By definition, 3 feet = 1 yard
6.47 yards is used to determine the number of significant digits in the answer,
(The 3 feet is exact and has an infinite number of significant digits)

$$\frac{6.47 \text{ yards}}{1} \times \frac{3 \text{ feet}}{1 \text{ yard}}$$

= 19.41 feet

= 19.4 feet (6.47 has three significant figures)

2. **In addition and subtraction**, we shall round-off so that the number of decimal place in the answer equals the number of decimal places in the number with the least number of decimal places.(In other words, the answer should not be more precise than the number with the least precision)

Example 1: Add: 143.54, 172.3, and 64.62
Solution
143.54 < - - - - has two decimal places
172.3 < - - - - -has one decimal place (determines the number of decimal places in answer)
  64.62 < - - - - - has two decimal places
380.46
Answer: 380.5 (has one decimal place)

Example 2 Subtract 12.4 from 143.63
Solution
143.63 < - - - - has two decimal places

  12.4 < - - - - -has one decimal place (determines the number of decimal places in answer)

131.23
Answer: 131.2 (has one decimal place)

3. **In finding powers and roots,** the root or power should be rounded-off so that the number of significant digits in the answer is equal to the number of significant digits in the number.

Example 1: Find $\sqrt{26.9}$
Radicand has three significant digits , and therefore the root should have three significant digits.
From a calculator, $\sqrt{26.9} = 5.1865$

$\sqrt{26.9} = 5.19$ (has three significant digits. Same as in the radicand)

Note: When two or more different operations are involved, the final operation determines how the final result is rounded-off.
Example Simplify: $38.3 + 12.9(3.58)$
Solution According to order of operations, we multiply 12.9 by 3.58 first and then add 38.3
   $38.3 + 12.9(3.58)$

$= 38.3 + 46.182$

$= 84.485$

$= 84.5$ (has one decimal place as in 38.3)

Since addition was the last step, we use precision (the number of decimal places) to round-off.

Addition and Subtraction Involving Scientific Notation

Before adding or subtracting the numbers must have the **same powers** of 10. We will rewrite the expression so that the power of 10 is that of the highest power in the expression

Example: **1.** $4 \times 10^2 + 3 \times 10^2 = (4 + 3) \times 10^2$
$$= 7 \times 10^2$$

Example: **2.** $4 \times 10^2 + 2 \times 10^3 = 0.4 \times 10^3 + 2 \times 10^3$
$$= (0.4 + 2) \times 10^3$$
$$= 2.4 \times 10^3$$

or

$4 \times 10^2 + 2 \times 10^3 = 4 \times 10^2 + 20 \times 10^2$
$$= (4 + 20) \times 10^2$$
$$= 24 \times 10^2$$
$$= 2.4 \times 10^3 \quad \text{(Again, we obtain the same result)}$$

Order of Magnitude (for comparing relative sizes using powers of 10).

The order of magnitude is the power of 10 closest to the given number.

(It is an approximation to the number. Note the sequence, $..., 10^{-2}, 10^{-1}, 10^0, 10^1, 10^2, ...$)

If a given quantity is 1000 times another quantity, the given quantity is larger by three orders of magnitude.

Examples:

1. The order of magnitude of 123 is 10^2, since 123 is closer to 100 than to 1000.

2. Find the order of magnitude of 0.00352.

Solution

Step 1: Write the number in scientific notation.

$0.00352 = 3.52 \times 10^{-3}$.

Step 2: Since the integer before the decimal point, 3, is less than 5, we replace 3.52 by 1 (since this is closer to 1 or 10^0, than it is to 10 or 10^1)

Step 3: 3.52×10^{-3}
$$= 10^0 \times 10^{-3}$$
$$= 1 \times 10^{-3}$$
$$= 10^{-3}$$

The order of magnitude of 0.00352 is 10^{-3}. (since by definition, the order of magnitude is the power of 10 closest to the given number.).

We can use the order of magnitude in estimation by rounding-off to the orders of magnitude.

More examples: Round-off to the nearest order of magnitude.

1. 1.32×10^2
2. 8.02×10^4
3. 0.0009
4. 0.0302

Solution:

1. Since 1.32 is closer to 1 than to 10,

1.32×10^2
$= 10^0 \times 10^2$
$= 1 \times 10^2$
$= 10^2$

The order of magnitude of 1.32×10^2 is 10^2

2. 8.02 is closer to 10 than to 1

8.02×10^4
$= 10^1 \times 10^4$
$= 10^5$

The order of magnitude is 10^5

3. $\quad 0.0009 = 9 \times 10^{-4}$
$= 10^1 \times 10^{-4}$
$= 10^{-3}$

The order of magnitude is 10^{-3}.

4. $0.0201 = 2.01 \times 10^{-2}$
$= 10^0 \times 10^{-2}$
$= 1 \times 10^{-2}$
$= 10^{-2}$

The order of magnitude is 10^{-2}

Summary for rounding-off a number to the order of magnitude.

Step !: Write the number in scientific notation

Step 2: Ignoring the power of 10, if the integer before the decimal point is 5 or greater, replace the non-power of 10 part by 10 ((i.e. 10^1); but if the integer is less than 5, replace the non-power of 10 part by 1 (10^0)

Step 3: Simplify

International System of Units

The International System of Units (SI) has adopted a set of seven base (or primary) units.

Quantity	Unit	Symbol
Length	meter	m
Mass	kilogram	kg
Time	seconds	s
Electric current	ampere	A
Temperature	Kelvin	K
Amount of substance	mole	mol
Luminous Intensity	candela	cd

Derived units

In addition to the seven base units, there are derived units which are combinations of the base units

Example:

From the SI base unit, m (meter). for length, the unit for area is $m \times m = m^2$, Since area = length $\times$ width, and the unit of length is m and the unit of with is m.

For more practical or convenient units, we use prefixes and multiplication factors (in powers of 10) to express other units.

Example: 1 kilometer = $10^3 m$, where kilo = 10^3

$\qquad$ 1 km = 10^3 m or I km = 1000m.

To show that if $\frac{a}{b} = \frac{c}{d} = \frac{e}{f}$, then $\frac{a}{b} = \frac{c}{d} = \frac{e}{f} = \frac{a+c+e}{b+d+f}$

We proceed in two steps. In Step 1, we consider $\frac{a}{b} = \frac{c}{d}$. In Step 2, We consider the result

from Step 1 and $\frac{e}{f}$.

Step 1: $\frac{a}{b} = \frac{c}{d}$ (Given)

$\frac{a}{b} + \boxed{\frac{c}{b}} = \frac{c}{d} + \boxed{\frac{c}{b}}$ (Adding $\frac{c}{b}$ to both sides of the equation)

$\frac{a+c}{b} = \frac{bc+cd}{bd}$ (Adding on the LHS and on the RHS of the equation)

$\frac{a+c}{b} = \frac{c(b+d)}{bd}$ (Factoring the RHS)

$\frac{a+c}{b+d} = \frac{bc}{bd}$ (Dividing both sides by $b+d$ and multiplying both sides by b)

$\frac{a+c}{b+d} = \frac{c}{d}$ (Dividing out the b on the RHS)

Step 2: $\frac{a+c}{b+d} = \frac{e}{f}$ (Since $\frac{a}{b} = \frac{c}{d} = \frac{e}{f}$, replace $\frac{c}{d}$ on the RHS by $\frac{e}{f}$)

$\frac{a+c}{b+d} + \boxed{\frac{e}{b+d}} = \frac{e}{f} + \boxed{\frac{e}{b+d}}$ (Adding $\frac{e}{b+d}$ to both sides of the equation)

$\frac{a+c+e}{b+d} = \frac{e(b+d)+ef}{f(b+d)}$ (Adding on the LHS and on the RHS of the equation)

$\frac{a+c+e}{b+d} = \frac{be+de+ef}{f(b+d)}$

$\frac{a+c+e}{b+d} = \frac{e(b+d+f)}{f(b+d)}$ (Factoring out the e on the RHS)

$\frac{a+c+e}{b+d+f} = \frac{e(b+d)}{f(b+d)}$ (Dividing both sides by $b+d+f$ and multiplying by $b+d$)

$\frac{a+c+e}{b+d+f} = \frac{e}{f}$ (Dividing out the $b+d$ on the RHS)

Since $\frac{a}{b} = \frac{c}{d} = \frac{e}{f}$, (given)

$\frac{a}{b} = \frac{c}{d} = \frac{e}{f} = \frac{a+c+e}{b+d+f}$

Application of the above to similar triangles.
Consider two similar triangles ABC and DEF.

If the lengths of the sides of ΔABC are a, b, c and the corresponding lengths of the sides of

Δ DEF are d, e, f, then for the ratios of corresponding sides, $\frac{a}{d} = \frac{b}{e} = \frac{c}{f}$ and

$\frac{a}{d} = \frac{b}{e} = \frac{c}{f} = \frac{a+b+c}{d+e+f}$. Thus $\frac{a}{d} = \frac{b}{e} = \frac{c}{f} = \frac{\text{perimeter of }\Delta \text{ ABC}}{\text{perimeter of }\Delta \text{ DEF}}$

Example Let the lengths of sides of Δ ABC be 2, 3, 4; and let the corresponding lengths of the

sides Δ DEF be 10, 15. 20,. Then $\frac{2}{10} = \frac{3}{15} = \frac{4}{20} = \frac{2+3+4}{10+15+20} = \frac{9}{45} = \frac{1}{5}$

INDEX

A

D

G

N

O

P

S

T

INDEX

Trinomial 159
Types of Numbers 1

5 8 0

U

Union of two sets 346
Unit fraction 5, 26
Unitary Method 69
Units
    of Measuring angles 451
Units Label Method 65
Unlike Fractions 337

V

Variable 158
Variables 150
Variation
    compound variation 85
    direct variation 59
Variation Method 66
Vertex
    parabola 392
Vertical angles 454
Vertical line
    equation 264

W

whole numbers 3, 4, 9
    dividing by a power of 10 45
    multiplying by a power of 10 44
    powers 13
    rounding-off 19
    writing using numerals 9
    writing using words 10
Word Problems 190
    age problems 211
    aircraft flight problems 292
    average problems 216
    averages 219
    base finding 211, 214
    coin problems 286
    coin-like problems 288
    consecutive even integers 209
    consecutive integer problems 209
    consecutive odd integer problems 210
    inequalities 352
    integer problems 209
    mixture problems 216
    motion problems 197
    number problems 192
    percent finding 213
    perimeter problems 194

X

Y

Z

Some useful conversion factors

Units of length

1 cm = .3937 in = .0328ft = .01094 yd
1 m = 100 cm = 39.3701 in = 3.2808 ft.= 1.0936 yd
1 in. = 2.54 cm = .0833 ft = .0254 m
1 ft = 12 in. = 30.48 cm = .3048 m = .3333 yd
1 mile = 1760 yd = 5280 ft = 1.6093 km
1 yd = 3 ft = 36 in.= .9144 m = 91.44 cm
1 km = .62137 mile = 1000 m = 100,000 cm

Units of mass

1 kg = 1000 g = 2.2046 lb =35.274 oz
1 lb (avdp) = 453.592 g = 16 oz
1 metric ton = 1000 kg = 10^6 g=1.1023 ton
1 ton = 907.1847 kg = 2000 lb = .9072 metric ton
1 gm = .03527 oz = 1000 mg =.0022046 lb
1 oz = 28.3495 g = .0625 lb = 16 drams
1 long ton (British) = 2240 lb

Units of volume

1 liter = 1000 cm^3= 61.0237 $in.^3$= .26417 gal = 1.0567 qt = .03531 ft^3= 2.113 pt.
1 gal (U.S.) = 4 qt = 3.7854 liter = 8 pt. = 231 $in.^3$ = .13368 ft^3
1 qt = 2 pt.= .946353 liter = 946.353 cm^3 = 57.75 $in.^3$ = .25 gal = .034201 ft^3
1 cord = 128 ft^3
1 pt = .473 liter
1 ft^3 = 1728 $in.^3$
1 yd^3= 27 ft^3 = 46656 $in.^3$

Units of area

1 ft^2 (sq. ft.) = 144 $in.^2$(sq. in.)
1 yd^2 (sq. yd.) = 9 ft^2 (sq. ft.) = 1296 sq. in.
1 $mile^2$(sq. mile) = 640 acres
1 m^2 = 10^4 cm^2
1 acre = 4840 yd^2

Symbols for units

cm = centimeter	gal = gallon	g = gram
m = meter	qt = quart	kg = kilogram
in. = inch	pt = pint	lb = pound
ft = foot	oz = ounce	mg = milligram
yd = yard		
km = kilometer		
mi = mile		

Some prefixes (International System)

Prefix	Power
tera	10^{12}
giga	10^9
mega	10^6
kilo	10^3
hecto	10^2
deka	10^1
deci	10^{-1}
centi	10^{-2}
milli	10^{-3}
micro	10^{-6}
nano	10^{-9}
pico	10^{-12}

Some useful conversion factors

Conversion Factors for Measurements

American System (British System)   Interconversion (Factors)   Metric System

↓   ↓   ↓

Some " **bridge**s" for converting from one system to the other

↓

Length

| 12 inches (in) = 1 foot (ft.) |
| 3 feet (ft.)   = 1 yard (yd) |
| 5280 feet   = 1 mile (mi) |
| 1760 yards   = 1 mile |

| 1 in. = 2.54 cm |
| 1 yd = 0.9144 m |
| 1 mi = 1.61 km |
| 1 km = 0.62 mi |

1 kilometer (km) $= 10^3$ m = 1000 m
1 hectometer (hm) $= 10^2$ m = 100 m
1 dekameter (dam) $= 10^1$ m = 10 m
1 meter (m) $= 10^0$ m = 1m
1 decimeter (dm) $= 10^{-1}$ m $= 0.1$m
1 centimeter (cm) $= 10^{-2}$ m $= 0.01$m
1 millimeter (mm) $= 10^{-3}$ m $= 0.001$m

Some " **bridge**s" for converting from one system to the other

↓

Mass

| 1 lb = 16 oz |
| 1 ton = 2000 lb |
| 1 long ton = 2240 lb |

| 1 kg = 2.2 lb |
| 1 lb = 454 g |
| 1 oz = 28.4 g = 16 drams |
| 1 ton = 0.9072 metric ton |

1 kilogram (kg) $= 10^3$ g = 1000 g
1 hectogram (hg) $= 10^2$ g = 100 g
1 dekagram (dag) = 10^1 g = 10 g
1 gram (g) $= 10^0$ g = 1 g
1 decigram (dg) $= 10^{-1}$ g $= 0.1$g
1 centigram (cg)) $= 10^{-2}$ g $= 0.01$g
1 milligram (mg) $= 10^{-3}$ g $= 0.001$g

Some " **bridge**s" for converting from one system to the other

↓

Volume

| 16 fluid oz (fl-oz)= 1 pint (pt) |
| 2 pints (pt) = 1 quart (qt) |
| 4 quarts   = 1 gallon (gal) |

| 1 liter (l) = 1.057 qt |
| 1 gal = 3.785 l |
| 1 liter = 2.1 pt |
| 1 pt = .473 l |

1 kiloliter (kl) $= 10^3$ l = 1000 l
1 hectoliter (hl) $= 10^2$ l = 100 l
1 dekaliter (da$l$) = 10^1 l = 10 l
1 liter (l) $= 10^0 l$ = 1 l
1 deciliter (dl) $= 10^{-1} l$ $= 0.1l$
1 centiliter (cl) = $10^{-2} l$ $= 0.01 l$
1 milliliter (ml) $= 10^{-3}l$ $= 0.001$ l

Must remember the following (metric system:)

| 100 cm   = 1 m |
| 1000 m = 1 km |

| 1000 mg = 1 g |
| 1000 g   = 1 kg |

| 1000 ml = 1 l |
| 1 ml = 1 cc = 1cm^3 |
| 1000 cc = 1 l |

Mnemonic device (metric system)

k – ilo – 10^3
h – ecto – 10^2
d – eka – 10^1
d – eci – 10^{-1}
c – enti – 10^{-2}
m – illi – 10^{-3}

Say the following aloud:
Step 1: First go down vertically as kei-eitch-dii-dii-see-em, then Step 2
Step 2: Kilo-hecto-deka-deci-centi-milli, and then note how the powers
     decrease vertically downwards.
   Examples: 1 Kilometer = 10^3meter; 1 milligram = 10^{-3}gram;
   1 centimeter = 10^{-2} meter $= \frac{1}{100}$ meter --->100 centimeters = 1 meter.

Trigonometric Identities

Reciprocal Identities

1. $\sec\theta = \dfrac{1}{\cos\theta}$

2. $\csc\theta = \dfrac{1}{\sin\theta}$

3. $\cot\theta = \dfrac{1}{\tan\theta}$

Ratio Identities

1. $\tan\theta = \dfrac{\sin\theta}{\cos\theta}$

2. $\cot\theta = \dfrac{\cos\theta}{\sin\theta}$

Pythagorean Identities

1. $\sin^2\theta + \cos^2\theta = 1$

2. $1 + \tan^2\theta = \sec^2\theta$

3. $1 + \cot^2\theta = \csc^2\theta$

Sum and Difference Identities (Addition and Subtraction Formulas)

1. $\cos(A - B) = \cos A \cos B + \sin A \sin B$; **4.** $\sin(A - B) = \sin A \cos B - \sin B \cos A$

2. $\cos(A + B) = \cos A \cos B - \sin A \sin B$; **5.** $\tan(A + B) = \dfrac{\tan A + \tan B}{1 - \tan A \tan B}$

3. $\sin(A + B) = \sin A \cos B + \sin B \cos A$, **6.** $\tan(A - B) = \dfrac{\tan A - \tan B}{1 + \tan A \tan B}$

Double Angle Identities

1. $\sin 2\theta = 2\sin\theta\cos\theta$; **2.** $\cos 2\theta = \cos^2\theta - \sin^2\theta$; **3.** $\cos 2\theta = 2\cos^2\theta - 1$;

4. $\cos 2\theta = 1 - 2\sin^2\theta$; **5.** $\tan 2\theta = \dfrac{2\tan\theta}{1 - \tan^2\theta}$

Half Angle Identities

1. $\cos\frac{1}{2}\theta = \pm\sqrt{\dfrac{1 + \cos\theta}{2}}$; **2.** $\sin\frac{1}{2}\theta = \pm\sqrt{\dfrac{1 - \cos\theta}{2}}$;

3. $\tan\frac{1}{2}\theta = \dfrac{1 - \cos\theta}{\sin\theta} = \dfrac{\sin\theta}{1 + \cos\theta}$

The $\pm$ sign indicates the sign to use depending on the quadrant location of $\frac{\theta}{2}$

Product or Product as a Sum Formulas

1. $\cos(A - B) + \cos(A + B) = 2\cos A \cos B$⎤

2. $\cos(A - B) - \cos(A + B) = 2\sin A \sin B$⎦ Derived from sum and difference identities}

3. $\sin(A + B) + \sin(A - B) = 2\sin A \cos B$⎤

4. $\sin(A + B) - \sin(A - B) = 2\cos A \sin B$⎦ Derived from sum and difference identities}

Sum Identities

1. $\cos A + \cos B = 2\cos\dfrac{A + B}{2}\cos\dfrac{A - B}{2}$⎤

2. $\cos A - \cos B = -2\sin\dfrac{A + B}{2}\sin\dfrac{A - B}{2}$

3. $\sin A + \sin B = 2\sin\dfrac{A + B}{2}\cos\dfrac{A - B}{2}$

4. $\sin A - \sin B = 2\sin\dfrac{A - B}{2}\cos\dfrac{A + B}{2}$⎦

Cofunction Relationships

$\sin A = \cos(90° - A)$⎤
$\cos A = \sin(90° - A)$⎦ cofunctions.

$\tan A = \cot(90° - A)$⎤
$\cot A = \tan(90° - A)$⎦ cofuntions.

$\sec A = \csc(90° - A)$⎤
$\csc A = \sec(90° - A)$⎦ cofuntions.

Other identities (Useful when finding functional values of negative angles)

For any θ, ⎧ **1.** $\cos(-\theta) = \cos\theta$ **4.** $\sec(-\theta) = \sec\theta$

⎨ **2.** $\sin(-\theta) = -\sin\theta$ **5.** $\csc(-\theta) = -\csc\theta$

⎩ **3.** $\tan(-\theta) = -\tan\theta$ **6.** $\cot(-\theta) = -\cot\theta$

Mathematical Modeling
Some Reciprocal Relationships

1. Arithmetic If A working alone can do a piece of work in time t_A; B working alone can do the same work in time t_B; C working alone can do the same work in time t_C, and if A, B, and C working together, can do the same work in time t_{ABC}, then

$$\frac{1}{t_{ABC}} = \frac{1}{t_A} + \frac{1}{t_B} + \frac{1}{t_C}$$

That is, the reciprocal of the working-together time equals the sum of the reciprocals of working-alone times (individual times).

2. Geometry: For any triangle, the reciprocal of the inradius (R) equals the sum of the   reciprocals of the exradii $(r_1, r_2,$ and $r_3)$.

Thus
$$\frac{1}{R} = \frac{1}{r_1} + \frac{1}{r_2} + \frac{1}{r_3}$$

3. Physics (Electricity) For electrical resistances in parallel (in an electric circuit), the reciprocal of the combined resistance, R, equals the sum of the reciprocals of the separate resistances, r_1, r_2, and r_3.

Thus
$$\frac{1}{R} = \frac{1}{r_1} + \frac{1}{r_2} + \frac{1}{r_3}$$

4. Physics (Optics)

For two thin lenses in contact, the reciprocal of the combined focal length, F, equals the sum of the reciprocals of the separate focal lengths, f_1 and f_2, .

Thus
$$\frac{1}{F} = \frac{1}{f_1} + \frac{1}{f_2}$$

5. Physics (Optics) For spherical mirrors and thin lenses, the reciprocal of the focal length F equals the sum of the reciprocals of the object distance, d_o and the image distance d_i.

Thus
$$\frac{1}{F} = \frac{1}{d_o} + \frac{1}{d_i}$$

6. Physics (Mechanics). If two bubbles of radii r_1, r_2, coalesce into a double bubble, the radius, R, of the partition is given by

$$\frac{1}{R} = \frac{1}{r_1} - \frac{1}{r_2}$$

www.ingramcontent.com/pod-product-compliance
Lightning Source LLC
Chambersburg PA
CBHW081756200326
41597CB00023B/4044